100 YEAR STARSHIP®

canopus 310 LY

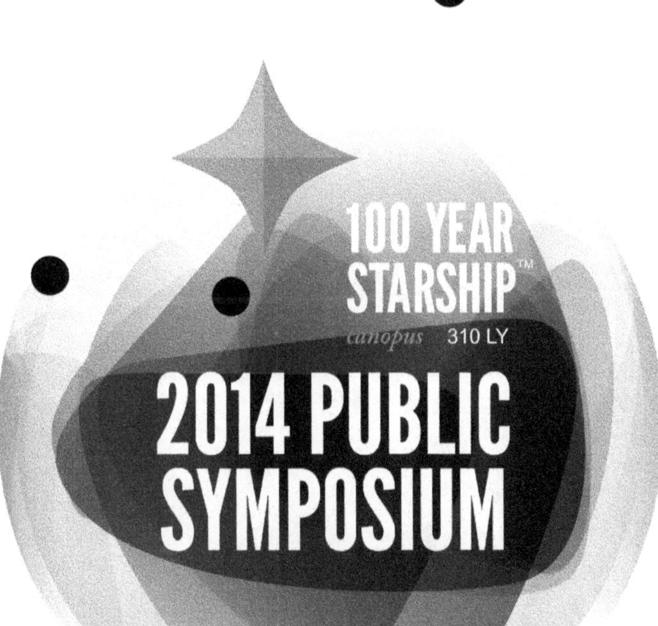

Pathway *to the* **Stars,**
Footprints *on* **Earth**

100 YEAR STARSHIP®

2014 Conference Proceedings

Pathway to the Stars, ***Footprints on Earth***

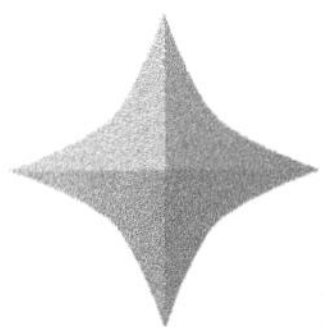

Published by 100 Year Starship®
www.100yss.org

ISBN-13: 978-0-9903840-1-4
ISBN-10: 0990384012

The 100 Year Starship® exists to make the capability of human travel beyond our solar system a reality within the next 100 years. We unreservedly dedicate ourselves to identifying and pushing the radical leaps in knowledge and technology needed to achieve interstellar flight, while pioneering and transforming breakthrough applications that enhance the quality of life for all on Earth. We actively seek to include the broadest swath of people and human experience in understanding, shaping and implementing this global aspiration.

For more information, visit www.100yss.org

Edited by Mae Jemison, M.D., Jason D. Batt, and Alires J. Almon
Design and Layout by Jason D. Batt

Cover Logo Credit:
the barbarian group

PRINTED IN THE UNITED STATES OF AMERICA

100 YEAR STARSHIP®

Table of Contents

TECHNICAL TRACKS

Data, Communications, and Information Technology

Interstellar Innovations Enhancing Life on Earth

Propulsion and Energy

Poster Session

100 YEAR STARSHIP®

2014 Public Symposium

Pathway to the Stars, ***Footprints on Earth***

Across the globe, calls are being made for bolder human expansion into space beyond earth's orbit. It is against this backdrop that the 100 Year Starship 2014 Public Symposium is held.

Achieving the interstellar human journey in many ways, must necessarily build upon, promote and establish fundamental research, technology development, societal systems and capacities that facilitate ready access to our inner solar system. To truly have the best opportunity for the aspiration to be long-lived enough to actually be accomplished, organizations and individuals involved must work to ensure it leaves a positive, indelible mark upon life right here on Earth.

100 Year Starship (100YSS) is working to create new avenues that foster innovative, robust collaborative, transdisciplinary research, project design and technological development. The 100YSS 2014 Public Symposium—*Pathway to the Stars, Footprints on Earth*—seeks to highlight both the small incremental steps and radical leaps required to make significant progress on the way to interstellar space.

Thank you for coming. The 100YSS Symposium Team has worked diligently to create a singular experience. I for one, certainly look forward to your participation and unique contributions. Welcome to the third 100YSS Public Symposium—two and half years into the journey!

"Meeting the challenge of 100YSS® stands to be even more transformative to our world than Sputnik..."

- Dr. Mae Jemison

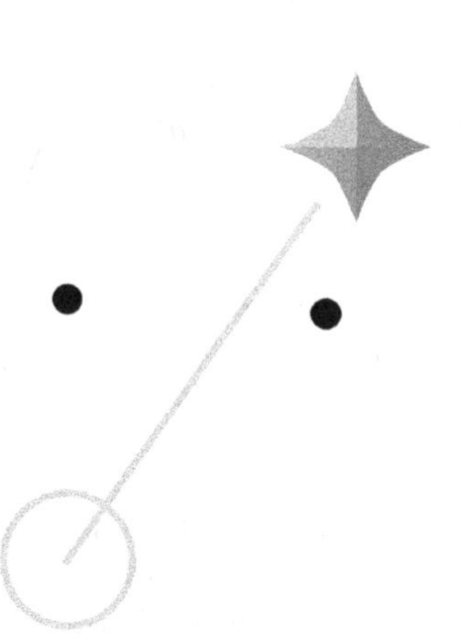

canopus 310 LY

Symposium Overview

CURVING. CROOKED. 90 DEGREES LEFT TURN. SPLITS. UPHILL. DOWNHILL. RETROGRADE. DEAD ENDS. FRESHLY HEWN. BUMPY. SMOOTH. CROWDED. ISOLATED.

The road to achieve the capabilities for human interstellar travel will not be straight, flat, unidirectional, one dimensional or even predictable from year to year.

(continued on next page)

Yet, I believe the requirements to discover a pathway, and to travel upon it until the ability to establish human life outside our solar system is assured, are quite clear. Requirements include, first and foremost, flexibility and openness. As we purposefully commence the journey, it is especially essential to reconsider how knowledge and ideas are categorized and presented. It is vital that we are willing to re-evaluate the "accepted methods" to achieve objectives. We must question whether technology and society rubrics, as currently "evolved," are the optimal concepts and techniques for an interstellar undertaking or *even living here on Earth. Such were major elements of the 100YSS 2014 Public Symposium.*

First, the Technical Tracks included four new sessions. The former *"Time and Distance"* track was split into *"Propulsion and Energy"* and *"Communications, Data and Information Technology."* A session focused on *"Education"* was added. A separate *"Poster Session"*—a forum for nascent ideas, business concepts, stories and student presentations— also debuted at the 2014 Symposium.

The Symposium introduced the *"Radical Leaps"* plenary session to examine potential technologies and ideas that could leapfrog forward capabilities and be achievable with current knowledge, societal structure and engineering concepts, but are discontinuous from accepted directions or infrastructure. The very exciting "Space Elevator" concept presented by Peter Swan, Ph.D. is one such undertaking. Concurrently held *"Deep Dive"* breakout sessions allowed Symposium attendees to explore in more detail specific constructs, projects and discussion with speakers.

Karl Aspelund, Ph.D., led the *"Perils of Futuring"* plenary session that considered the success and shortcomings of the early predictions of the course of space exploration compared to the actual history and against the framework of other technological achievements during and since the 1960s. Aspelund was joined by noted space historian Jim Oberg, former NASA Chief Technology Officer Mason Peck, Ph.D., and Marianne Caponnetto, technology branding, marketing and start-up expert. Similarly, discussion during *Trending Now* and *State of the Universe* included compelling topics. Amalie Sinclair, Ph.D. around the need to re-examine and update international space treaties and laws provoked thought around governance issues. Steampunk as a backdrop for interstellar technological development was explored with the help of world expert Mike Perschon, Ph.D. Steampunk arises from a "what if" of science fiction and fantasy writing that builds complex, advanced technologies and systems—airships, computers such as Babbage's analytical engine, optics, protective clothing, etc.—based on 19th century steam powered, mechanical engineering. 100YSS asked "Is there a role for non-silicon dependent technologies in interstellar? The Firaxis Games franchise development team previewed their new game, Civilization: Beyond Earth, that ponders "How will humanity engage other intelligent life?" Conceptualization exercises considering space exploration in the next 50 years by teams at NASA Marshall Spaceflight Center were discussed.

Science Fiction Stories Night engaged award-winning science fiction authors Les Johnson, Yoon Ha Lee, Edward M. Lerner, and Nisi Shawl, and editor Jaym Gates on the importance of getting the science as right as possible alongside the need for social and cultural narratives "getting it right". Arguably, one of the best and most influential science fiction movies, the 1957 "The Day the Earth Stood Still", was shown. **Accelerating Creativity** treated participants to a pre-release showing of PBS' acclaimed documentary series "MAKERS: Women in Space" and a discussion with astronauts Bonnie Dunbar, Ph. D., (Professor Engineering University of Houston) Peggy Whitson, Ph.D. (Head of Astronaut Office and Commander of Space Station) and Mae Jemison, MD (Principal 100YSS). We were honored to have special commentary by the legendary actress and activist Nichelle Nichols.

EXPO Inspire! brought to life the creativity, energy and wonder that is needed to fuel the ambition and achievement of major human advancements and exploration. The EXPO focused on K thru 14 students and included Alka Rocket contest sponsored by Bayer Corporation, for 11-14 grade students, hands-on experimentation in Kids Zones, and emphasized not just on science and technology literacy, but engagement in reading and art as well.

The purpose of the 100YSS Public Symposium is to bring together individuals and organizations with varying types and levels of expertise, disciplines, experiences, skills and access to focus on one singular goal—*An Inclusive Audacious Journey to Transform Life Here on Earth and Beyond*—human interstellar travel.

I sincerely thank and applaud all the individuals and organizations who gave so generously of their time, knowledge and resources to organize, staff, create and participate in the 2014 Symposium. From the Technical Track Chairs, to the moderators, speakers, designers, sponsors, volunteers, presenters and admin support—the work highlighted here is the result of their work.

The pathway to the stars is evolving and taking shape right now. And though we do not have a roadmap—we do know the ever-changing landscape will be fantastic, and the footprints we leave on Earth, transformational.

I know that these Proceedings from the 100YSS 2014 Public Symposium are a good step among the many we will take. Enjoy!

Mae Jemison, M.D.
Principal, 100YSS

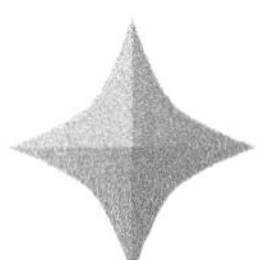

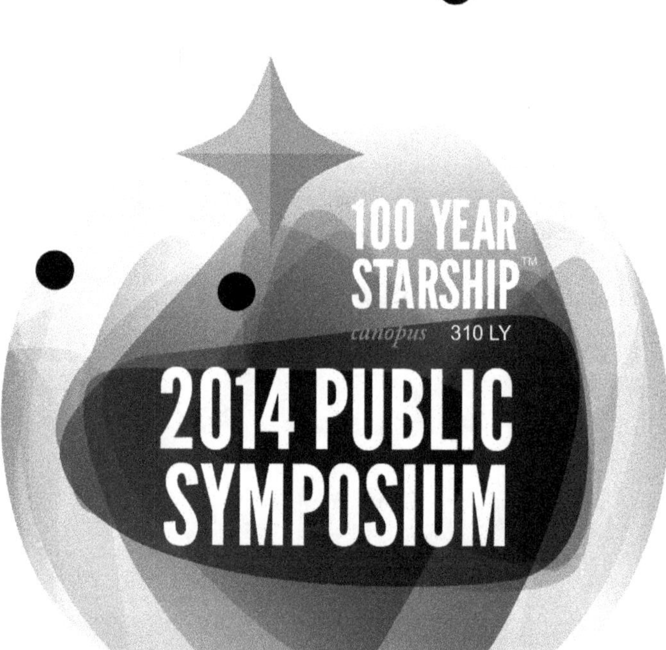

Pathway *to the* Stars,
Footprints *on* Earth

SCHEDULE

THURSDAY

8AM **REGISTRATION OPENS** **340**

3:30PM **SPECIAL INTEREST GROUP (SIG) MEETINGS** **332F**

100YSS Special interest groups meet and review progress and plan.

4PM **100YSS CLASSES** **332A-E**

Classes taught by 100YSS team members and focus on basic/fundamental principles of specific disciplines that underlie the capabilities of interstellar flight. The classes help attendees build a greater understanding and appreciation of the 100YSS mission and challenges. Classes are also a part of the 100YSS volunteer program. Classes are in the areas of Engineering, Astronomy & Space Sciences, Social Sciences, Policy, Economics and Governance, Space Life Sciences, and Education.

6PM **100YSS MEMBERS EVENT** **332AB**

7PM **OPENING RECEPTION FOR 100YSS 2014 PUBLIC SYMPOSIUM** **3RD FLOOR PREFUNCTION**

8:30PM **HALO LOUNGE** **330AB**

A meeting and networking place for symposium attendees.

FRIDAY

8AM **OPENING PLENARY SESSION** **PLENARY ROOM**

Dr. Mae Jemison, Principal 100YSS

9AM **TECHNICAL TRACK SHOW & TELL** **PLENARY ROOM**

Preview of Technical Tracks through featured papers picked by the Chairs of each track. Moderated by Pamela R. Contag, PhD, 2014 Overall Technical Track Chair.

9:45AM **TECHNICAL TRACKS • SESSION A** **332A-332E**

Presentations of submitted papers in the eight Technical Tracks: *BECOMING AN INTERSTELLAR CIVILIZATION • PROPULSION & ENERGY • UNCHARTED SPACE & DESTINATIONS • DESIGNING FOR INTERSTELLAR • DATA COMMUNICATIONS & INFORMATION TECHNOLOGY*

12PM **LUNCHEON—THE PERILS OF 'FUTURING'** **342A-D**

Experts from disparate fields discuss the problem of creating and maintaining a future vision. Identifying current and potential areas of inadequate knowledge, errors, blind spots, fixed beliefs and shifting ground from a historical perspective sheds light on the path ahead for 100YSS. Space exploration and science/technology projects will be considered through the lens of Robert K. Merton's "Unanticipated Consequences of Purposive Social Action" Moderated by Karl Aspelund, Ph.D., Assistant Professor, Department of Textiles, Fashion Merchandising and Design, The University of Rhode Island; Yvonne Cagle, M.D., Aerospace Researcher, NASA Ames Research Center; Peter Swan, Ph.D., President, International Space Elevator Consortium; Marianne Caponnetto, Digital Age Entrepreneur.

1:30PM **TECHNICAL TRACKS • SESSION B** **332A-332E**

Presentations by authors who have submitted research in the one of eight Technical Tracks: *INTERSTELLAR INNOVATIONS ENHANCING LIFE ON EARTH • PROPULSION AND ENERGY • BECOMING AN INTERSTELLAR CIVILIZATION • DESIGNING FOR INTERSTELLAR*

3:30PM **TECHNICAL TRACKS • SESSION C** **332A-332C**

Presentations by authors who have submitted research in the one of eight Technical Tracks: *INTERSTELLAR EDUCATION • PROPULSION & ENERGY • BECOMING AN INTERSTELLAR CIVILIZATION*

6PM **POSTER SESSION RECEPTION** **340AB**

Hosted session for Technical Track poster submissions.

7PM **SCIENCE FICTION STORIES NIGHT** **PLENARY ROOM**

Screening of movie classic, "The Day the Earth Stood Still" (Original 1957). Science Fiction Authors Panel with four award winning authors, discuss context, story and technology motivation; moderated by Jason Batt. Panel includes hard science writers like NASA exotic propulsion expert Les Johnson and Edward M. Lerner, to cultural science fiction author Yoon Ha Lee, and science fiction and fantasy short stories author, Nisi Shawl. Sit back and enjoy the authors movie critique with Ronke Olabisi, PhD., Assistant Professor, Department of Biomedical Engineering, Rutgers University, School of Engineering. Popcorn and candy included at movie screening.

SATURDAY

10AM **EXPO INSPIRE!** **GRB HALL B3**

All day event for children of all ages. This is an exciting, hands-on learning experience for children, students and adults to explore the world of space and science. Try your hand at launching an Alka Rocket, be the bee that transfers the pollen, visit the Bayer "Making Science Make Sense" Kid Zone, learn the Science of Magic, sit under the stars in the planetarium, check out the H.E.B. Starship Quest, walk over to The Edge exhibit, catch the Scholastic Book Fair, sing with astronomers, meet real astronauts, with talks, live entertainment and much, much more.

8:15AM **100YSS CLASSES** **332A-E**

Classes taught by 100YSS team members that focus on basic/fundamental principles in specific disciplines that underlie the capabilities of interstellar flight.

9:30AM **STATE OF THE UNIVERSE** **PLENARY ROOM**

Facilitated by Jill Tarter, Ph.D., Bernard M. Oliver Chair for SETI Research, Director, Center for SETI Research, hear of the latest finds in astronomy and space technology with Mason Peck, Ph.D., Associate Professor, Cornell University, Department of Mechanical and Aerospace Engineering; Director, Cornell Space Systems Design Studio; and former NASA Chief Technology Officer.

10:30AM **TRENDING NOW** **PLENARY ROOM**

Facilitated by Hakeem Oluseyi, Ph.D., Astrophysicists and Associate Professor, Florida Institute of Technology, Department of Physics and Space Sciences, be part of a discussion of current issues or trends that may be instrumental in facilitating or stymying interstellar exploration with Kurt Zatloukas, M.D., Professor Medicine and Pathology, Graz University, Austria; Hear the latest on Outer Space Treaties and Policies in need of revamping with Amalie Sinclair, PhD, Director of the Leeward Space Foundation, and author; and think Steampunk - the new retro futurism in design with Mike Perschon, Ph.D., Professor of Literature, Grant MacEwan University, Alberta, Canada and Steampunk Scholar.

12PM **KEYNOTE LUNCHEON—CIVILIZATION: BEYOND EARTH** **342A-D**

Moderated by Alires Almon, Guest: Peter Murray, Firaxis Games.

1:30PM **PLAN B—HUMANITY'S HEALTH** **PLENARY ROOM**

Our panelists answer the question: Can humanity survive if we leave the planet? Humans have a symbiotic relationships with earth, going into space breaks that relationship. Our bodies will have to adapt to life in a environment free of our earthy relatives. Top medical doctors, researchers, and bio-medical entrepreneurs talk about health and wellness in space. Moderated by Yvonne Cagle, MD, NASA Astronaut and Flight Surgeon. Panelists: Pamela Contag, PhD, Microbiologist, CEO of Cygnet and ConcentRx; Barry Holtz, Phd, President G-CON, LLC; Kurt Zatloukal, PhD, Medical Center of Graz.

2:30PM **RADICAL LEAPS PLENARY SESSION** **PLENARY ROOM**

A look at cutting edge research applied to interstellar research. Facilitated by Les Johnson, PhD (Author, Tennessee Valley Interstellar Workshop Founder). Topics: Space Elevators, Gravitational Lens, Space Entrepreneurship

3:30PM **DEEP DIVES** **332A-C**

Independent two-hour workshops making deep dives into subject matter from the symposium—Space Elevator with Peter Swan, Ph.D.; Updating Space Treaties and Policy with Amalie Sinclair; and Designing Interstellar Capabilities through Steampunk-Tech, i.e., minimal silicon with Mike Pershon and Mae Jemison, M.D. *// SPACE POLICY AND TREATIES (332A) • SPACE ELEVATORS (332B) • STEAMPUNK STARSHIP (332C)*

6PM **ACCELERATING CREATIVITY SPECIAL EVENT—MAKERS: WOMEN IN SPACE** **PLENARY ROOM**

6PM ACCELERATOR RECEPTION • 7PM DINNER DOORS OPEN // A spectacular evening to celebrate the intersection of art and science centered around a premiere screening of the documentary MAKERS: Women In Space narrated by Jodie Foster and panel discussion documentary Director Michael Epstein with women pioneers in space and featuring Nichelle Nichols and Mae Jemison, M.D. It includes dinner, classical guitarist and dancing. Hosted by KUHT Executive Director and General Manager Lisa Shumate and moderated by Linda Lorelle, CEO Lorelle Media, Inc.

8AM **100YSS CLASSES** **332A-E**

Classes taught by 100YSS team members that focus on basic/fundamental principles in specific disciplines that underlie the capabilities of interstellar flight.

9:15AM **EVOLUTION TO AN INTERSTELLAR CIVILIZATION** **PLENARY ROOM**

Discussions around the impact of society on space exploration and how society changes as a result of scientific, technological and perspective change and advances that result directly and indirectly from humanity's push beyond the Earth. Topics include ethics, morality, culture, politics and economics as well as envisioning exercises of the challenges and opportunities for space exploration in the future.

11:15AM **100YSS: THE YEAR AHEAD** **PLENARY ROOM**

Discussion on the next year and beyond.

12:15PM **2014 SYMPOSIUM CLOSE** **PLENARY ROOM**

CLASSES

Engineering // PROPULSION: PROVEN, PROBABLE, OR PREPOSTEROUS?

Since the dawn of spaceflight, we have propelled all of our craft with the reliable power that rockets can provide, but even the fastest rocket-powered craft ever made would take millennia to reach another star system. What else might we use to power a ship, what else could we build, that would safely get us there sooner than rockets can? This lesson will cover the various technologies currently available that have the potential to power an interstellar spaceflight. Along the way, we will acquaint ourselves the various known fundamental particles, available materials, and physical forces that might be harnessed to send us to the stars.

Engineering // SHIPWRIGHT'S WORKSHOP

How can we design a Star Ship built to face challenges we have not yet foreseen? What can we learn from more than half a century of flying through space? Will the interstellar environment pose threats not known in the near-Earth orbits that humans have already experienced? In this lesson, we will attempt to cover all of the design questions we must answer in order to build a safe, multi-generational Star Ship. How will we sleep, eat, and bathe? How will we repair damage to the ship without any hope of being "resupplied?" Can we simulate or substitute the force of gravity? To what degree will we rely on automation and robots? How many people should our ship carry? This lesson, as an introduction, will not attempt to answer every question, but instead to identify the design questions that must be answered before we are cleared for lift-off!

Life Sciences // LEAVING THE NEST

All life on Earth, from Bacteria to Blue Whales, has evolved on the same planet, with similar conditions and materials that have supported that life. Our species, being the first to leave this planet intentionally, will face dangers and challenges unknown to all other earthling creatures. Our survival has always depended on countless species of microscopic life living within us: the Human Micro-biome. We also cultivate and consume numerous species of plants and animals that provide our food. If we are to spend generations in space, who needs to come with us? Can we exclude the rogues, viruses and harmful bacteria, without endangering the beneficial companions in the human habitat? How do diseases behave in space? Will our current medical techniques work aboard a Star Ship? This lesson will survey the fields of Biology and Medicine to address what happens to our bodies when we leave the Earth behind.

Space Sciences // THE VERY, VERY BIG PICTURE

In this new age of discovery, esoteric terms such as hadron, boson, exoplanets, dark matter, and dark energy have now become everyday topics of conversation! This lesson will focus on the recent revelations and scientific endeavors that have shaped our understanding of the structure of the universe, with a particular emphasis on the potential dangers to the human body and the prospects for finding habitable places outside of the Solar System. With this new knowledge, might we better refine our search for intelligences like ourselves? How does the current state of cosmology inform our ideas for where other life might be found?

Space Sciences // GETTING TO KNOW THE NEIGHBORHOOD

Next year, the New Horizons mission will give us our first close-up pictures of Pluto and its entourage. Rosetta is now orbiting a comet, and its Philae probe will be the first robot to ride a dirty snowball as it dives around the sun! With Curiosity on Mars, Cassini around Saturn, and Juno on her way to Jupiter, there has never been a better time to go sightseeing in the Solar System! This lesson will tour the planets, asteroids, and comets as well as the star that serves as their central hearth: The Sun. How does our Solar System compare to the others: the extra-solar planetary star systems we've been finding by the thousands with the Kepler Space Telescope and other planet-hunting missions? What should we be looking for if we are ever to call another star system our home?

Social Sciences // CAN WE ALL JUST GET ALONG?

If our pioneers are to survive a multi-generational spaceflight in a ship of limited size, an effective method for the peaceful resolution of conflicts will be as essential as oxygen or water. Obviously, we could use this on a larger ship with seven billion passengers too! What have we learned from studies in Psychology, Anthropology, Sociology, Mediation, and even Pathology that will help forge new, reliable methods in conflict resolution? What have we learned in Neuroscience and even Psychopharmacology that might help temper humanity's propensity for violence, or help us predict and prevent it? This lesson will explore the frontiers of multiple disciplines to shed light on the path to engineering social harmony.

Culture // THE SCIENCE FICTION DREAM FACTORY

Jules Verne's Nautilus scoured the seafloor long before real submarines came along, and the inventor of the notorious Taser named it for Tom Swift's Electric Rifle! It is clear that many of the greatest inventions and accomplishments of our time are inspired by works of science fiction. Perhaps a graphic novel, motion picture, or video game being made now will inspire and guide the builders of the 100YSS? This lesson will review several historical examples of seemingly fanciful fictions that have become essential realities, ending with a review of current trends in popular culture that might inform our goals and plans for 100YSS.

SPEAKERS

Science Fiction Stories Night Authors

YOON HA LEE

Science Fiction & Fantasy Author

Yoon Ha Lee's short story collection Conservation of Shadows was published by Prime Books in 2013. Lee's short fiction has appeared in Tor.com, Clarkesworld, Lightspeed, The Magazine of Fantasy and Science Fiction, and other venues. Her stories "Ghostweight" and "Flower, Mercy, Needle, Chain" were Theodore Sturgeon Award finalists. Lee was born in Houston, Texas and has also lived in South Korea, Missouri, upstate New York, Boston, Washington state, and Southern California. Currently she lives in Louisiana with her family and has not yet been eaten by gators. Her website is http://www.yoonhalee.com.

EDWARD M. LERNER

Science Fiction Author

Author EDWARD M. LERNER has degrees in physics, computer science, and business. Before turning to full-time writing, he spent thirty years in high tech at every level from engineer to senior VP. He's worked at such techie havens as Bell Labs, Hughes Aircraft, and Northrop Grumman. Lerner's novels range from technothrillers, like Energized (asteroid deflection and solar power satellites) to traditional SF, like the InterstellarNet series, to (with NY Times best-selling author Larry Niven) the grand space epic Fleet of Worlds series. He writes nonfiction, too, on topics as varied as nanotechnology, privacy in the Internet era, defending Earth from asteroids—and technologies that might enable interstellar travel. Lerner is a member of the Science-fiction and Fantasy Writers of America (SFWA), the Institute of Electrical and Electronics Engineers (IEEE), and SIGMA ("The Science Fiction Think Tank"). His website isedwardmlerner.com.

NISI SHAWL

Science Fiction Author

Nisi Shawl's story collection Filter House was a James Tiptree, Jr., Award winner. She co-authored Writing the Other: A Practical Approach, and co-edited the 2014 Locus Award finalistStrange Matings: Science Fiction, Feminism, African American Voices, and Octavia E. Butler. Shawl was the 2011 Guest of Honor for the feminist SF convention WisCon and a 2014 Guest of Honor for the Science Fiction Research Association. She has spoken at Stanford and Duke Universities and Smith College, and is a co-founder of the Carl Brandon Society, a nonprofit supporting the presence of minorities in the fantastic genres. Her website iswww.nisishawl.com.

Plenary Speakers

MICHAEL EPSTEIN

MAKERS: Women in Space Panelist,

Director MAKERS' Documentary

Michael Epstein is an Academy Award-nominated documentary producer, director and writer whose work has been awarded two George Foster Peabody Awards, two Primetime Emmy Awards, a Writers Guild Award, a Clio, as well as numerous other distinctions. His films have screened in dozens of international film festivals, and been broadcast throughout the world.

LES JOHNSON

Radical Leaps Facilitator, Science Fiction Author, Scientist,

NASA Technologist

Les Johnson is a scientist, an author, and a NASA technologist. He is an author of several popular science books about space exploration, including "Living Off the Land in Space," "Sky Alert: When Satellites Fail," and "Harvesting Space for a Greener Earth;" and three science fiction books "Back to the Moon," "Going Interstellar," and "Rescue Mode." You might have seen him on NatGeo or The Science Channel where he has appeared in numerous programs about space and science. In his day job, he serves as the Senior Technical Advisor for NASA's Advanced Concepts Office at the Marshall Space Flight Center in Huntsville, Alabama. Les is the Principal Investigator (PI) of NEA Scout, an asteroid reconnaissance mission being considered for launch in 2017 and the NASA Co-Investigator (Co-I) for the European Union's Deploytech Solar Sail demonstration mission planned for launch in 2015. He thrice received NASA's Exceptional Achievement Medal and has 3 patents. Les is a frequent contributor to the Journal of the British Interplanetary Society and a member of the National Space Society, the World Future Society, and MENSA. He serves on the Editorial Advisory Board for the British Interplanetary Society and is Chairman of the Tennessee Valley Interstellar Workshop. Les was the featured "interstellar explorer" in the January 2013 issue of National Geographic magazine and a technical consultant for the movie, "Europa Report."

NICHELLE NICHOLS

MAKERS: Women in Space Panelist, Actress, Activist

Nichelle Nichols is an American actress, singer, and voice artist. Nichols' most famous role is that of communications officer Lieutenant Uhura aboard the USS Enterprise in the popular Star Trek television series, as well as the succeeding motion pictures, where her character was promoted in Starfleet to the rank of commander.

MASON PECK, PHD

Perils of Futuring Panelists, and State of the Universe Panelist

Dr. Peck is an Associate Professor in Mechanical and Aerospace Engineering at Cornell University and the Director of Cornell's Space Systems Design Studio. His research interests include space-systems architecture and satellite dynamics and control. From late 2011 through 2013 he served as NASA's Chief Technologist in Washington, DC. In that role, he served as the agency's chief strategist for technology investment and prioritization and chief advocate for innovation in aeronautics and space technology. Dr. Peck received an undergraduate degree in Aerospace Engineering at the University of Texas at Austin, a Master's degree in English at the University of Chicago, and a Ph.D. in Aerospace Engineering at UCLA. Since 2004 he has been on the Faculty at Cornell University, where he is jointly appointed to the Systems Engineering program and the School of Mechanical and Aerospace Engineering. His lab currently leads three spaceflight technology demonstrations, the most recent of which is Kicksat, a crowd-funded technology demonstrator for satellites-on-a-chip.

MIKE PERSCHON, PHD

Trending Now Panelist

Deep Dive Workshop Leader Steampunk Scholar

Dr. Mike Perschon began studying steampunk for his Ph.D. dissertation in the fall of 2008. That October, he attended Steam Powered, the Northern California Steampunk convention. In addition to sharing panels with Hugo-award winning fanzine editor Christopher J. Garcia, authors J. Daniel Sawyer, and the yet-unknown Gail Carriger, presented content that would become his first published article on steampunk: "Finding Nemo: Verne's Antihero as Original Steampunk." Perschon has published numerous articles on steampunk in fanzines, websites, and magazines, including Exhibition Hall, Tor.com, On-Spec, and Locus. He has been featured in many interviews, from small personal blogs to USA Today. In the fall of 2012, Perschon defended his Ph.D. dissertation titled The Steampunk Aesthetic: Technofantasies in a Neo-Victorian Retrofuture. Perschon currently resides in Edmonton, Alberta, in the Great White North of Canada, where he works as a full time instructor of English at Grant MacEwan University.

AMALIE SINCLAIR

Trending Now Panelist

Deep Dive Workshop Leader Space Treaty Working Group

Amalie Sinclair was born in London on April 12th 1951. On her 10th birthday, Yuri Gagarin became the first man in space. With background in the arts and humanities her childhood years were formative ones, developing an insight for world culture, at an early age she took up studies for the extant philosophies of Asia, with many significant teachers including the late Karmapa XVI. In the 70's she was introduced to the deep sematic structures of Wittgenstein through the work of the Cambridge Language Research Unit. In 1984 she located to California with her family where she wrote, directed and produced a program for the Millennial celebrations which included the participation of Lakota Sioux and peoples of Taos Pueblo, together with artists from Mongolia, Japan, South India, Tibet, Cambodia and Indonesia. In 2004 she was invited to address the formulation of an advanced US Space Policy with an opening program in Washington DC. Since then she has worked to promote international collaboration in space, providing various talks and presentations including for the planetary decadal survey and the Lunar Science Institute. Amalie is a Director of the Leeward Space Foundation, a member of the Star Voyager group, Lifeboat advisory board, and the UNISPACE program for advanced concepts. She is currently proposing for the placement of a specialized unit at the US Library of Congress dedicated towards forthcoming expansions of the 1967 Treaty on the Peaceful Uses of Outer Space

PETER SWAN, PHD

Radical Leaps Panelist

Deep Dive Workshop Leader Space Elevators

President of the International Space Elevator Consortium. As such, he leads a team who further the concept with incremental studies and yearly conferences. Over the last ten years he has published five books on the topic as co-author and/or co-editor. They are: Space Elevators: An Assessment of the Technological Feasibility and the Way Forward [2013], Design Considerations for Space Elevator Tether Climbers [2014], Space Elevator Concept of Operations [2013], Space Elevator Survivability – Space Debris Mitigation [2012], and Space Elevators Systems Architecture [2005]. He graduated from the US Military Academy in 1968 with a Bachelor of Science degree and served 20 years in the Air Force with a variety of research and development positions in the space arena. He taught at the Air Force Academy and retired as a Lieutenant Colonel. Upon retirement in 1988, he joined Motorola on the Iridium satellite program. He lead the team responsible for the development of the Iridium spacecraft bus. Pete received his Ph.D. from the University of California at Los Angeles in Mechanical Engineering with a specialty in space systems. He has published many papers and a few books; two of which are on preparing for SCUBA trips.

JILL TARTER, PHD

State of the Universe Moderator, SETI, Institute

Jill Tarter holds the Bernard M. Oliver Chair for SETI Research at the SETI Institute in Mountain View, California. Tarter received her Bachelor of Engineering Physics Degree with Distinction from Cornell University and her Master's Degree and a Ph.D. in Astronomy from the University of California, Berkeley. She served as Project Scientist for NASA's SETI program, the High Resolution Microwave Survey, and has conducted numerous observational programs at radio observatories worldwide. She is a Fellow of the AAAS and the California Academy of Sciences, she was named one of the Time 100 in 2004, and one of the Time 25 in Space in 2012, received a TED prize in 2009, public service awards from NASA, multiple awards for communicating science to the public, and has been honored as a woman in technology. Since the termination of funding for NASA's SETI program in 1993, she has served in a leadership role to design and build the Allen Telescope Array and to secure private funding to continue the exploratory science of SETI. Dr. Tarter was recently awarded the prestigious Jansky Lectureship, which honors outstanding contributions to the field of Radio Astronomy. Many people are now familiar with her work as portrayed by Jodie Foster in the movie Contact.

KURT ZATLOUKAL, PHD

Medical University of Graz, Austria; Plan B - Humanity's Health

Kurt Zatloukal's research work focuses on molecular pathology of metabolic liver diseases and cancer. He coordinated the preparatory phase of a European biobanking and biomolecular research infrastructure (BBMRI) within the 7th EU framework programme. BBMRI should provide access to high quality human biological samples to enable future needs of large genetic epidemiology and sequencing studies. In this context it is crucial to establish Europe-wide harmonized processes and quality criteria that are compliant with the requirements of latest -omics technologies as well as with ethical and legal regulations. Furthermore, he leads in the FP7-funded large integrated project SPIDIA the development of new European standards and norms for tissue-based biomarkers, and leads the medical platform of the FET Flagship project IT Future of Medicine. Kurt Zatloukal was member of the OECD task force on biological resource centres and the Roadmap Working Group of the European Strategy Forum on Research Infrastructures. Moreover, he contributed to the OECD best practice guidelines for biological resource centres, the regulations for genetic testing of the Austrian Gene Technology Law, and the opinion on Biobanks for research of the Bioethics Commission at the Austrian Federal Chancellery.

BARRY HOTLTZ, PHD

President and Co-Founder, G-CON, LLC

Plan B - Humanity's Health

Dr. Holtz is the President and Co-Founder of G-CON,LLC and also serves as the Chief Science and Technology Officer for Caliber Biotherapeutics. Dr. Holtz was the Senior Vice President of Biopharmaceutical Development for Large Scale Biology Corporation (LSBC) for 15 years and was an integral member of the management team that successfully took the company from a start-up to a publicly-traded company. Dr. Holtz was responsible for the product development, long-term regulatory strategy, and clinical development and manufacturing compliance of the company's proprietary therapeutics portfolio. He has extensive expertise in providing consulting services to the biopharmaceutical development and manufacturing industry including the design, validation, and quality systems development for personalized cancer therapies and other biologics. Dr. Holtz has held research management positions at Foremost-McKesson and was on the faculty of Ohio State University. He received his Ph.D. at Pennsylvania State University and was an NSF Postdoctoral Fellow at Scripps Institution of Oceanography. Dr. Holtz has been awarded 23 U.S. patents and has published more than 50 scientific papers. He was awarded the Pennsylvania State University, Outstanding Alumni Award in 2003.

MARC GARVIN

Interlude Presentation—Music, Analytics and the Creative Process.

Marc Garvin is a guitar soloist, chamber music performer and well-known instructor. He's performed regularly with Houston's classical music triad of the Houston Symphony, Houston Grand Opera and Houston Ballet. Whether performing with the Pittsburgh Symphony Chamber Music Ensemble, at the Kennedy Center in Washington, D.C., or as a recitalist through the Texas Commission on the Arts, his enthusiasm and musicality has charmed audiences throughout the United States and Canada. As a graduate with a degree in music from Carnegie-Mellon University, Marc Garvin has placed a high emphasis on music education. He currently is the principle classical guitar instructor at Lamar University and Houston Community College (Central Campus). He has also worked as an artist on the roster of Young Audiences of America, and was an active teaching artist for the Texas Institute for Arts in Education. As an instructor, Mr. Garvin has been on the faculties of Carnegie-Mellon University, Houston Baptist University and San Jacinto Jr. College. Many of his students have been accepted to prestigious music schools including The New England Conservatory of Music, The University of Southern California's Thornton School of Music, The Butler School of Music at The University of Texas, The University of Miami's Frost School of Music and The Beinen School of Music at Northwestern University. Marc will also be performing for 100YSS at our Accelerating Creativity reception and dinner.

CYNTHIA FERGESON

Cynthia "Tia" Kaiser Ferguson was born and raised in Natchez, MS. She graduated from Tulane University in 1989 with a Bachelor of Science in Mechanical Engineering, and from the University of Alabama in Huntsville in 2003 with a Master of Science in Electrical Engineering. She has been a Professional Engineer since 1998 and holds a patent for a Micro-Electro-Mechanical-Systems (MEMS) translation stage. Ms. Ferguson has worked at NASA for almost 25 years. She started at Kennedy Space Center, employed from 1990 – 1995 as an integration and test engineer for various space shuttle payloads before moving to Huntsville, AL to work at NASA Marshall Space Flight Center (MSFC). At MSFC, her current position is Project Manager for SERVIR, which is a joint NASA-USAID international initiative to help developing countries use information provided by Earth observing satellites and geospatial technologies for managing climate risks and land use. SERVIR has an imaging instrument on the International Space Station (ISERV), and hub organizations in Mesoamerica, Eastern and Southern Africa, and Hindu Kush Himalaya regions, with plans to expand to South East Asia and West Africa regions. Prior to her SERVIR management role, Ms. Ferguson served as Assistant Manager for the MSFC Science and Technology Office, Branch Chief of the Structural and Mechanical Design branch for the Space Systems Department, Project Manager for Cargo Element Integration of the Multi-Purpose Logistics Module (MPLM) for ISS during the development phase, and Structural and Mechanical Design engineer for space and ground hardware. At MSFC she has operated a science Earth observing payload on a Spacelab mission, been a scuba diver to support Hubble Space Telescope astronauts training for a repair mission, and recovered a cosmic radiation experiment 200 miles from the South Pole. She currently lives in Huntsville with her husband and two teenage children.

Technical Track Chairs

PAMELA CONTAG, PHD

Overall Technical Track Chair

Pamela R. Contag, PhD, has founded four early stage technology companies, most recently ConcentRx, a cell-based immune therapy company and Cygnet Inc. commercializing a platform technology to discover beneficial microbes for applications in food, renewable fuels, low cost therapeutics, industrial enzymes, and carbon dioxide capture. With more than 25 years of microbiology research experience, Dr. Contag is widely published in the field of Microbiology and Optical imaging and has over 35 patents in Biotechnology. Dr. Contag received her Ph.D. in Microbiology.

JOHN C. MCKNIGHT, JD, PHD.

Becoming an Interstellar Civilization Track Chair

John Carter McKnight is a former corporate finance lawyer with a PhD in Human & Social Dimensions of Science and Technology . His research focuses on emergent community governance, ethics and social norm enforcement in technologically-mediated spaces from internet gaming communities to spacecraft. He is currently a post-doctoral Research Associate in the Department of Sociologyat Lancaster University in the U.K. Dr. McKnight is working on the Third Party Dematerialization and Rematerialization of Capitalproject on peer-to-peer finance communities.

YVONNE D. CAGLE, MD

Life Sciences in Space Exploration Track Chair

Dr. Yvonne Cagle is an astronaut for the National Aeronautics and Space Administration (NASA), a Colonel, U.S. Air Force (retired), a Family Physician, and consulting professor for Stanford University's department of cardiovascular medicine and its department of electrical engineering. Dr. Cagle is currently the Chief scientist for the Level II Program Office of NASA's Commercial Reuseable Sub-orbital Research program. Her groundbreaking work is preserving NASA legacy data while galvanizing NASA's lead in global mapping, sustainable energies, green initiatives and disaster preparedness.She was assigned as Stanford's lead astronaut science liaison and strategic relationships manager for Google and other Silicon Valley programmatic partnerships.

DAVID ALEXANDER, PHD

Uncharted Space and Destinations Track Chair

David Alexander , PhD is a professor in the Department of Physics and Astronomy, where his primary area of research is solar astrophysics. Professor Alexander is author of "The Sun" part of the Greenwood Press "Guide to the Universe" Series. Professor Alexander has served on many national and professional committees including the NASA Advisory Council's Heliophysics Subcommittee, the NASA Solar Heliospheric Management and Operations Working Group (SH-MOWG), ESA/NASA Solar Orbiter Payload Committee and the Science Advisory Board of the High Altitude Observatory Coronal Solar Magnetism Observatory.

HAKEEM OLUSEYI, PHD

Propulsion & Energy Track Co-Chair

Hakeem M. Oluseyi, PhD is an internationally recognized astrophysicist, science TV personality, and global science education activist. His research interests span the fields of astrophysics, cosmology, and technology development. He currently has 7 U.S. patents, 4 EU patents and over 60 scholarly publications in the areas of astrophysics, optics and detector technologies development; nanotechnology manufacturing; observational cosmology; and the history of astronomy. Dr. Oluseyi leads a group studying processes by which electromagnetic fields and plasmas interact in order to understand solar atmospheric heating and acceleration, which has resulted in a new in-space propulsion technology.

ERIC DAVIS, PHD

Propulsion & Energy Track Co-Chair

Eric Davis, PhD is a Senior Research Physicist at the Institute for Advanced Studies at Austin and is the co-editor/author of the first-ever academic research monograph on breakthrough propulsion physics for interstellar flight: Frontiers of Propulsion Science (AIAA Press). His research specializations include breakthrough propulsion physics, beamed energy and nuclear propulsion, general relativity theory, and quantum field theory. Since 1984, Dr. Davis has been a contractor/consultant to the USAF, Air Force Research Laboratory, Department of Defense, Department of Energy, and NASA. He has been featured in American and UK television and film productions, as well as in numerous news media articles, on interstellar flight.

MARIANNE CAPONNETTO

Interstellar Innovations Enhances Life on Earth Track Co-Chair

Ms. Caponnetto's career has included the roles of global strategist, marketer and sales leader. As Founder of MCW Group, she acts as Strategic Advisor and Consultant with a focus on technology start-ups, Venture Capital and Fortune 500 organizations with significant transformational and growth objectives in both B2B and B2C businesses. She is also a Founding Partner in HyperBlue Lab, a technology IP consulting firm. From 2005-2008, as Chief Sales and Marketing Officer of DoubleClick, she was a key member of the management team that recaptured digital market leadership, resulting in the sale of the company to Google.

DAN HANSON

Interstellar Innovations Enhance Life on Earth Track Co-Chair

Dan is a principal with Technology Innovation Group, Inc. (TIG). Dan has a keen interest in leveraging art, science, and education infrastructure to promote economic development in regional economies across the globe. TIG pursues it mission through two primary service offerings: advising governments, foundations, and communities wanting to build technology-based economies; and serving as translational consultants with institutions and private companies to commercialize specific technologies, primarily those with public health or economic development benefits. TIG works with clients that are developing products and services based on complex technologies, and universities and research institutions that desire to move discoveries from the laboratory to businesses.

RON COLE

Data, Communications, and Information Technology Track Chair

Ron finished a 50-year career with the US Intelligence Community in May 2013. Ron received the Civilian Defense Meritorious Award upon leaving the NSA to begin working with Scitor Corporation as a systems engineer technical advisor to the NSA and the National Reconnaissance Office (NRO) on system development and data processing. Ron left Scitor to go to work for Riverside Research as a senior advisor for five years to the National Geospatial-Intelligence Agency on technology developments for mission execution and then moved to support the NRO on policy and management issues.

KARL ASPELUND, PHD

Designing for Interstellar Chair

Karl Aspelund, PhD, is an anthropologist with a design background. He is assistant professor at the Department of Textiles, Fashion Merchandising and Design at the University of Rhode Island and visiting professor of ethnography at the University of Iceland. His interests lie in examining the role of textiles and design in identity-creation, the environmental impact of the textile life-cycle, and how designers may contribute to environmental sustainability. He is currently investigating the design and cultural needs and constraints of apparel in long-term space exploration. Dr. Aspelund was recently a speaker at TEDx Reykjavik in Iceland.

KATHLEEN TOERPE, PHD

Interstellar Education Track Chair

Kathleen D. Toerpe, PhD, is a social and cultural historian who researches the human dimension of outer space through an emerging field called "astrosociology." She is the Deputy CEO for Programs and Special Projects with the Astrosociology Research Institute, and volunteers as a NASA/JPL Solar System Ambassador. She has served as a Historian-in-Residence and a museum Educational Curator and has provided local outreach programming, oral history program management and exhibit curation. As a research and applied astrosociologist, she investigates how individuals and societies react and respond to space exploration and astrobiological discoveries, and how those responses can reflect, predict, inform or mitigate social and cultural conflict here on earth.

TIMOTHY MEEHAN, PHD

Poster Sessions Chair

Timothy Meehan, PhD has extensive expertise in biomolecular analytical science and diagnostics. He has applied nanotechnology approaches to bioengineering through a collaborative effort with the Australian Stem Cell Centre in order to achieve large scale human synthetic whole blood production. Tim is a seasoned small business leader with experience in genetic diagnostics, microbial parasite detection and commercial New Space startup ventures. At Saber Astronautics he is working with NASA performing reduced gravity hardware flight tests and developing the next generation of autonomous fault detection and recovery solutions for spacecraft.

BOBBY FARLICE-RUBIO

Fairbanks Museum and Planetarium

100YSS Class Organizer

Bobby Farlice-Rubio has been a Science Educator at the Fairbanks Museum & Planetarium in St. Johnsbury, Vermont since 2003. There he teaches classes, to visiting students and the public at large, on a wide variety of subjects ranging from Astronomy and Natural Sciences to History and Culture. Mr. Farlice-Rubio may also be seen in his monthly "Star Struck" segments on WCAX-TV's news show "The :30," on which he presents the latest happenings in the field of Astronomy. Raised in Hialeah, Florida from Cuban and African-American roots, Bobby is also an avid musician who plays in a local band called Tritium Well, as well as his solo musical endeavor, Bobby & The Isotopes. He currently resides in Barnet, Vermont with his partner and their four children.

MAKERS : Women In Science Panelists

MAE JEMISON, MD *Principal, 100YSS, Former NASA Astronaut*

NICHELLE NICHOLS *Actress, and former NASA Astronaut Recruiter*

MICHAEL EPSTEIN *Director, MAKERS: Women In Space*

BONNIE DUNBAR, PHD *Lead, University of Houston STEM Center, Retired NASA Astronaut*

ANNA FISHER, PHD *NASA Astronaut*

Event Moderators

ALIRES ALMON *Sid Meier's Civilization: Beyond Earth, Moderator*

KARL ASPELUND, PHD *Perils of Futuring, Moderator*

JASON D. BATT *Science Fiction Stories Night, Moderator*

LINDA LORELLE *Accelerating Creativity, Moderator*

RONKE OLABISI, PHD *Science Fiction Night Movie Reviewer*

SPONSORS

Sponsors
BAYER CORPORATION

SCHOLASTIC, INC.

HOUSTON FIRST CORPORATION

DOROTHY JEMISON FOUNDATION FOR EXCELLENCE • TEWS SPACE RACE

H•E•B, INC.

SCHLUMBERGER, INC.

CREDIT SUISSE

THE JEMISON GROUP, INC.

AIRBORNE VISUALS

JASON D. BATT

Student Sponsors
THE DREAM MAKER FOUNDATION, LOS ANGELES, CA

WEALTH DEVELOPMENT STRATEGIES
MS. CHERYL CREUZOT

MAE JEMISON, M.D.

Underwriters
OPTIMUM CARE, SACRED HOPE FUNERAL HOME

In-Kind
MACY'S

JUS JAZZ, SHELDON NUNN

TYCOON ADVERTISING GROUP

RIVER OAKS BOOKSTORE OF HOUSTON

HOUSTON COMICPALOZZA

GRAPHICSDEPARTMENT.COM

REGINA JOHNSON MUSIC

Special Thanks
PAUL DAVID VAN ATTA SPECIAL EVENTS

HILTON OF THE AMERICAS, HOUSTON

SOUTHERN SOUND

BELAY MEDIA, LLC

TECHNICAL TRACKS

DESIGNING FOR INTERSTELLAR
CHAIR: KARL ASPELUND, PHD

SESSION A· 9:45AM · 332C

VICTORIA ADAMS, CENK TUNASAR, MARK GERNER, AND WALTER JANSEN • *In Space No One Can Hear You Deal*

ANTOINE G. FADDOUL • *Space Architecture: Design Aspects for Extended Space Travel*

KARL ASPELUND • *Research Imperatives for Apparel in Long Duration Space Flight*

SESSION B · 1:30PM · 332C

KERAVOLOS • *Organic Matter and Molecular Sensor*

KENNETH WISIAN • *Military Planning for Interstellar Flight*

UNCHARTED SPACE & DESTINATIONS
CHAIR: DAVID ALEXANDER, PHD

SESSION A · 9:45AM · 332D

T. MARSHAL EUBANKS • *Nomadic Planets near the Solar System: Detecting the Natural Early Targets for an Interstellar Voyage*

DAVID ALEXANDER • *Planetary Magnetic Fields and Exoplanet Habitability*

PAULI ERIK LAINE • *Exoplanets and Survival Kits. Where to go and what to pack for an Interstellar Voyage?*

EMMERSON EDWARDS • *Deep Space Life_Bot*

DATA COMMUNICATIONS & INFORMATION TECHNOLOGY
CHAIR: RON COLE

SESSION A · 9:45AM · 332C

CLAUDIO MACCONE • *Interstellar Links Created by Two Focal Space crafts*

JAYM GATES • *Loading, Please Wait: Long-Distance Crisis Communications*

CLAUDIO MACCONE • *KLT Filtering, Even from Relativistic Sources*

DAVID BURKE • *Cutting the Umbilical Cord: Building Trustworthy Software for a Starship*

ANTOINE G. FADDOUL • *Contemporary Classification of Astronomy and Space Encoding*

LIFE SCIENCES IN INTERSTELLAR
CHAIR: YVONNE CAGLE, MD

SESSION B · 1:30PM · 332D

RALPH ACOSTA • *How to Create the Perfect Crew. Or Someone else.*

TERRY MULLIGAN • *Global Emergency Care and the Global Emergency Medicine Initiative*

MARCEL DIRKES AND MICHAL HEGER • *Artificial Induction of Suspended Animation*

EDWARD V. WRIGHT • *Searching for Extraterrestrial Life at the Edge of Space*

DAVID ALMANDSMITH AND CARMEN NEVAREZ • *When Biology Meets Exobiology*

PAMELA R. CONTAG • *Exoplanet Environmental remediation: Lessons from Earth*

INTERSTELLAR INNOVATIONS ENHANCING LIFE ON EARTH

CHAIR: DAN HANSON

SESSION B · 1:30PM · 332E

JO ANN ORAVEC • *Celebrtiy and Social Media in 100YSS Missions: Another Kind of "Dancing with the Stars"?*

RANDALL CHUNG • *Removing Atmospheric Carbon Dioxide Using Terraforming Technology*

KURT ZATLOUKAL • *Minimal Requirements for a Fully Autonomous Health Care in a Closed System*

KIRK FRENGER • *Molten Salt Reactors: A Case for Enhancing Life on Earth . . . and in Space*

BUCK FIELD • *Paradigm Hack: A 2015 Workshop Toward FTL Physics*

INTERSTELLAR EDUCATION

CHAIR: KATHLEEN TOERPE, PHD

SESSION C · 3:30PM · 332C

MARSHALL BARNES • *STEM and the Oppenheimer Strain: Why Children are Ready for Space Education*

ADRIENNE PROVENZANO • *Telling HERstory: A Vital Aspect of Interstellar Space Education*

BARBRA SCHUESSLER MAHER AND JOHN SCHUESSLER • *Broadening the Perspective of Undergraduate Students to Prepare them for Interstellar Travel: First Steps*

BUCK FIELD • *Enterprise in Space*

MARK DARIUS JUSZCZAK • *Fact-Fiction Convergence in Adult Education in Interstellar Travel*

JANET ALEXANDRA DE VIGNE • *Education in Space: A Chance for Utopia?*

BECOMING AN INTERSTELLAR CIVILIZATION

CHAIR: JOHN CARTER MCKNIGHT, PHD

SESSION A· 9:45AM · 332B

DONNA A. DULO • *Interstellar Treaties, Intellectual Property, and Prime Directives: Creating the Unique Legal Foundations for an Interstellar Space Mission*

JEREMY CHAO • *Space Colonization: Exploring Lessons from History*

JOHN CARTER MCKNIGHT • *Beyond Heinlein: Facing the Space-Libertarian Shortage*

PAUL ZIOLO • *Characteristics of an Asimovian "'econd Foundation"*

SESSION B· 1:30PM

ALIRES J. ALMON • *Breaking Bad: Creating predictive models for psychological breaking points during long-term space travel.*

ILYA I. ALEKSEYEV • *The Bioethics of Space Travel*

JASON D. BATT • *The Future of Religion in Space as Glimpsed by Interstellar Fiction*

BOB HAWKINS • *The Enduring Challenge*

SESSION C · 3:30PM

ALBERT A. JACKSON • *Extreme SETI*

PAUL ZIOLO • *The 100YSS as a Potential Chrysalis within a "Universal State"*

JASON D. BATT • *Interstellar Travel in the New Science Fiction*

PROPULSION & ENERGY

CO-CHAIRS: ERIC W. DAVIS, PHD AND HAKEEM OLUSEYI, PHD

SESSION A · 9:45AM · 332A // ANTIMATTER ANNIHILATION PROPULSION FOR INTER-STELLAR FLIGHT

HAKEEM OLUSEYI • *Energy and Propulsion: The Nature of Advancement*

SUMONTRO LAL SINHA • *Industrial Production of Antimatter in the Van Allen Belts*

MARC H. WEBER • *Novel Method To Store Charged Antimatter Particles for Interstellar Propulsion*

T. MARSHALL EUBANKS • *Antimatter From Condensed Quark Matter in the Solar System*

SESSION B · 1:30PM · 332A // INNOVATIVE ELECTROMAGNETIC PROPULSION FOR INTERSTELLAR FLIGHT

GEORGE REITER • *Propulsion Using Microwave Beams From The Moon?*

FRIEDWARDT WINTERBERG • *Thermonuclear Operation Space Lift*

DAVID L. CHESNY • *The Magnetic Reconnection Rocket: Advanced Ion Propulsion Inspired by Solar Particle Acceleration*

CLYDE ALBERT JR. BARRY STOUTE • *Enhanced Magnetoplasmadynamic Thruster*

SESSION C · 3:30PM · 332A // INNOVATIVE IDEAS FOR INTERSTELLAR PROPULSION AND POWER

JEFF S. LEE • *Fermion Re-Inflation of a Schwarzschild Kugelblitz*

LES JOHNSON • *Solar Sail Propulsion for Interplanetary and Interstellar Travel*

ALEXANDER I. KAZYKIN • *Forecast of Breakthrough Trends in Technique and Technology of Interstellar Flights*

POSTER SESSION

CHAIR: TIMOTHY MEEHAN, PHD

6PM · 340AB

ROBERT MANNING • *Interstellar Travel*

MARSHALL BARNES • *STDTS™ and the Door to Opening Space*

BUCK FIELD • *Historical Guidance for Spacetime Breakthroughs Today*

ERIKA GASPER AND CHANDA CUMMINGS • *The Adventures of Miss Q, Spacecat*

KATHERINE HERLEMAN • *Developing Pre-Flight Culture: 100 Year Starship University-Level Student Organization Proposal*

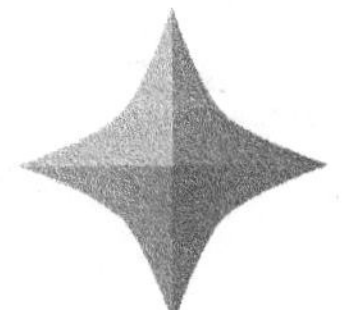

100 YEAR STARSHIP™

Experts from disparate fields discuss the problem of creating and maintaining a future vision. Identifying current and potential areas of inadequate knowledge, errors, blind spots, fixed beliefs and shifting ground from a historical perspective sheds light on the path ahead for 100YSS. Space exploration and science/technology projects will be considered through the lens of Robert K. Merton's "Unanticipated Consequences of Purposive Social Action" Moderated by Karl Aspelund, Ph.D., Assistant Professor, Department of Textiles, Fashion Merchandising and Design, The University of Rhode Island; Yvonne Cagle, M.D., Aerospace Researcher, NASA Ames Research Center; Peter Swan, Ph.D., President, International Space Elevator Consortium; Marianne Caponnetto, Digital Age Entrepreneur.

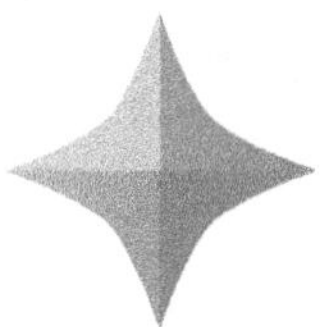

Seeing the Space Elevator in the Future

Peter A. Swan, Ph. D.

President and Member Board of Directors, International Space Elevator Consortium

Summary

The discussions started with a basic overview of the space elevator concept and how it will change everything for cheap access to space. However, it has far more to offer than just inexpensive access to GEO and beyond. The discussion focused on the topic of what are the perils of trying to foresee the future. Indeed, the projection of the concept of a space elevator, and how it would change the human condition, is a definite "look into the future," without the knowledge that it will occur. As such, parallels were used during the presentation, discussion. The obvious one that brings forth some very real concerns about projections is the one where the US built a transcontinental railroad during the war years of 1863-65. The dream of almost all of the investors was the realization of Eastern [China/Japan/India] market products being transported to the east coast of the United States. In fact, this did not drive the profit of the cross-country infrastructure. The majority [>90 %] of the product moved by the transcontinental railroad was intra-continental, from the middle of the country and all the towns and communities established during the final push of the railroads. This inability to see the future in not unusual, and can be projected for a space elevator. This is just one of the perils of futuring that was discussed during the luncheon.

100 YEAR STARSHIP™

Our panelists answer the question: Can humanity survive if we leave the planet? Humans have a symbiotic relationships with earth, going into space breaks that relationship. Our bodies will have to adapt to life in a environment free of our earthy relatives. Top medical doctors, researchers, and bio-medical entrepreneurs talk about health and wellness in space. Moderated by Yvonne Cagle, MD, NASA Astronaut and Flight Surgeon. Panelists: Pamela Contag, PhD, Microbiologist, CEO of Cygnet and ConcentRx; Barry Holtz, Phd, President G-CON, LLC; Kurt Zatloukal, PhD, Medical Center of Graz.

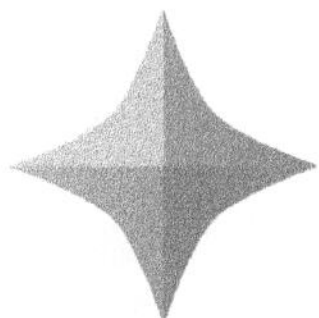

Starship Medicine: Not as "Far, Far Away . . ."

Barry Holtz, Ph.D.

Chief Executive Officer, Intervexion Therapeutics

www.intervexion.com

- Conventional medicines may not be necessary
  - Low threat from Earth Borne disease
  - Bacterial control likely in the spacecraft environmental systems
  - Screening for conventional viral and bacterial disease will be very robust for space travelers
  - Genetic screening may elucidate most likely problems in the population
  - Some drugs might be harmful for long term space travel

A New Paradigm is necessary

- Stressors may cause new disease syndromes
- The Immune system is the new focus
  - Strengthen the immune system
  - Help the immunes system fight disease
  - Understand the patient specific problems
- New threats may occur in the long term environment
- Radiation, chemical, and biological threats will by dynamic
  - Mutation of human genes
  - Mutation of bacterial genes

Adaptive Medicine
Diagnose and Manufacture

- Advanced Diagnosis of disease will be required
  - Mutational Testing
  - Proteomic Testing
  - Understanding of mutations of microorganisms
  - Advanced environmental analysis systems
- Manufacturing of Therapeutics will be mandatory
  - Rapid manufacturing of therapeutics
  - Robust production system
  - Rapid production of "Patient-Specific" therapeutics

These Principles are Guiding Modern Medicine

Three Examples of New Technologies

Breast Cancer Diagnosis and

Patient Specific Medicine

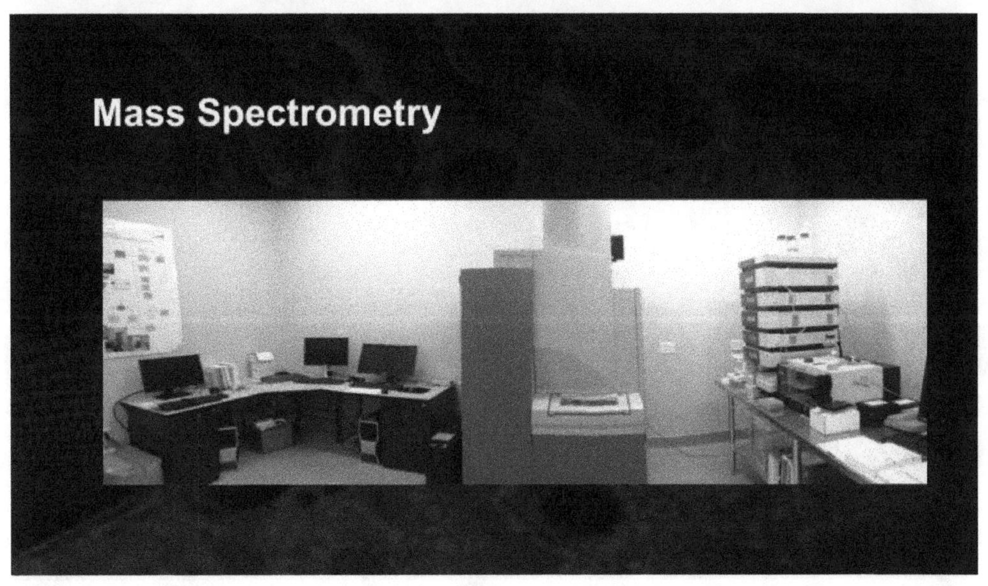

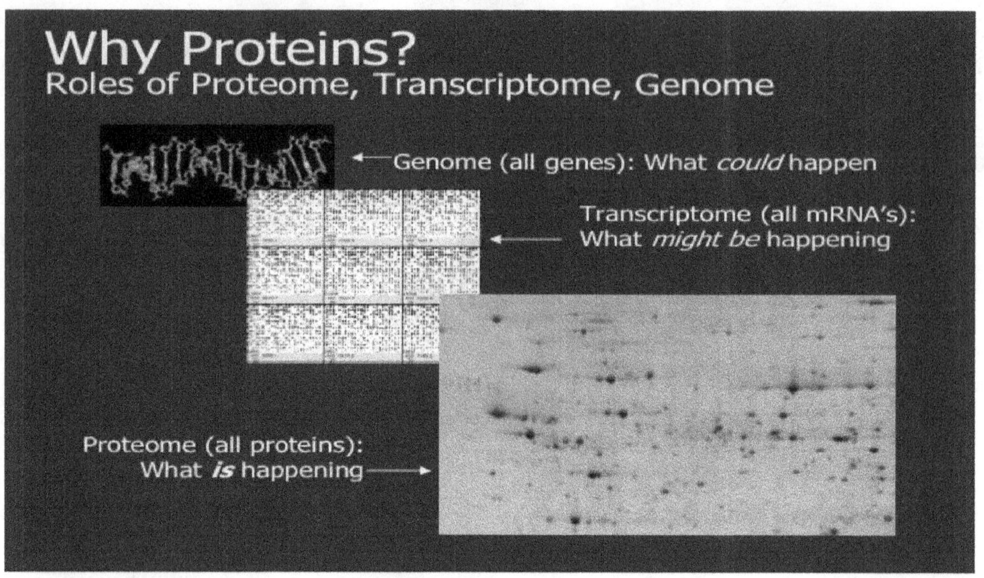
Why Proteins?
Roles of Proteome, Transcriptome, Genome
Genome (all genes): What *could* happen
Transcriptome (all mRNA's):
What *might be* happening
Proteome (all proteins):
What *is* happening

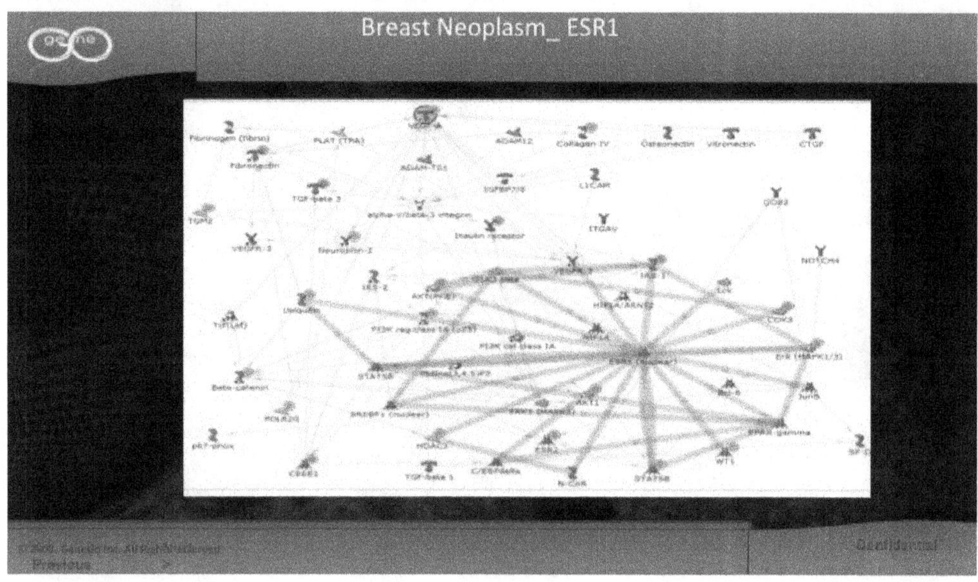
Breast Neoplasm_ ESR1

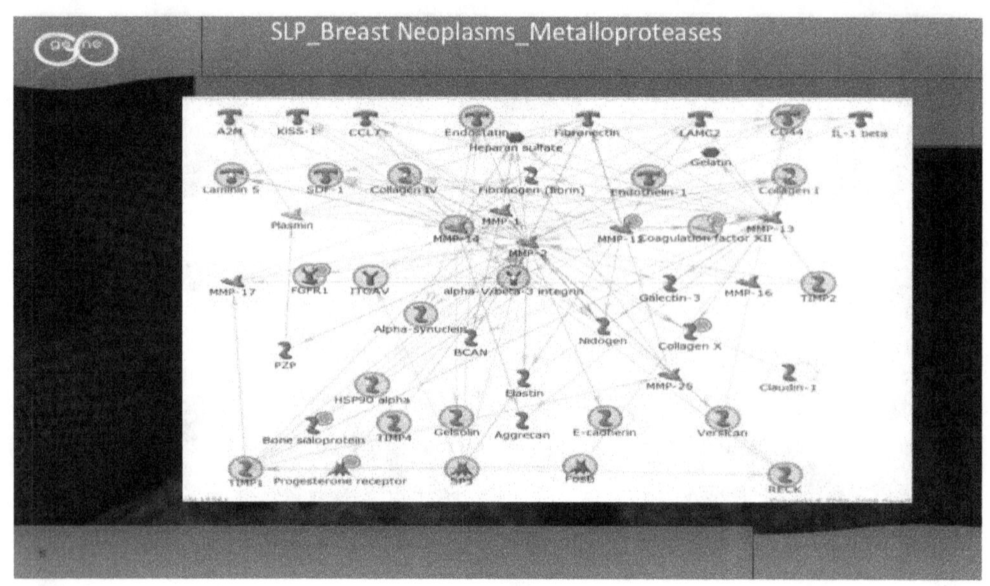
SLP_Breast Neoplasms_Metalloproteases

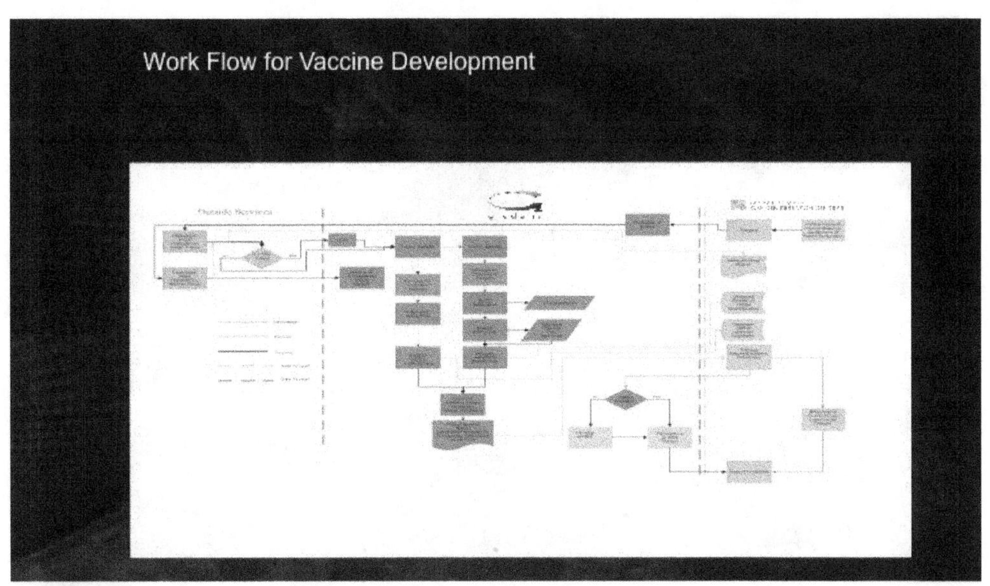
Work Flow for Vaccine Development

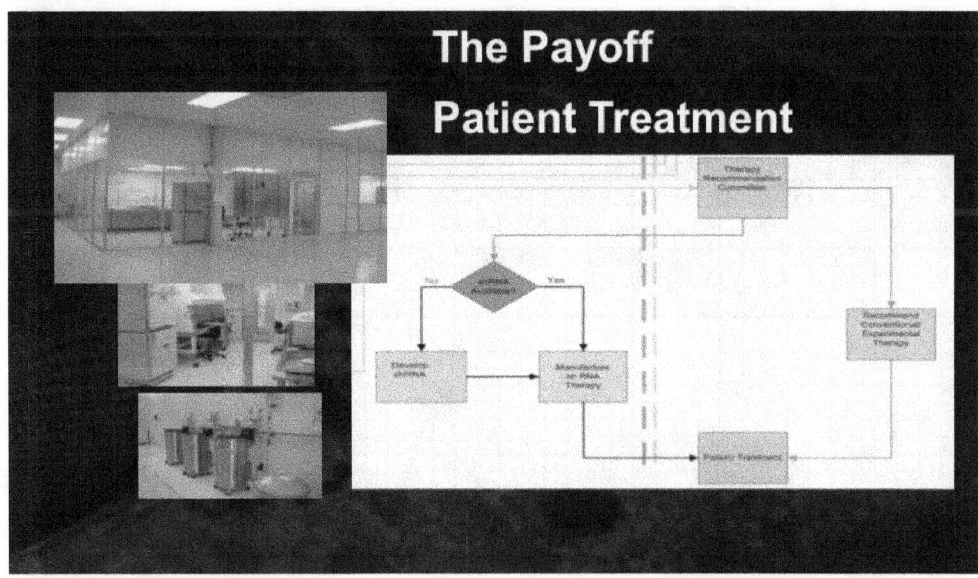
The Payoff
Patient Treatment

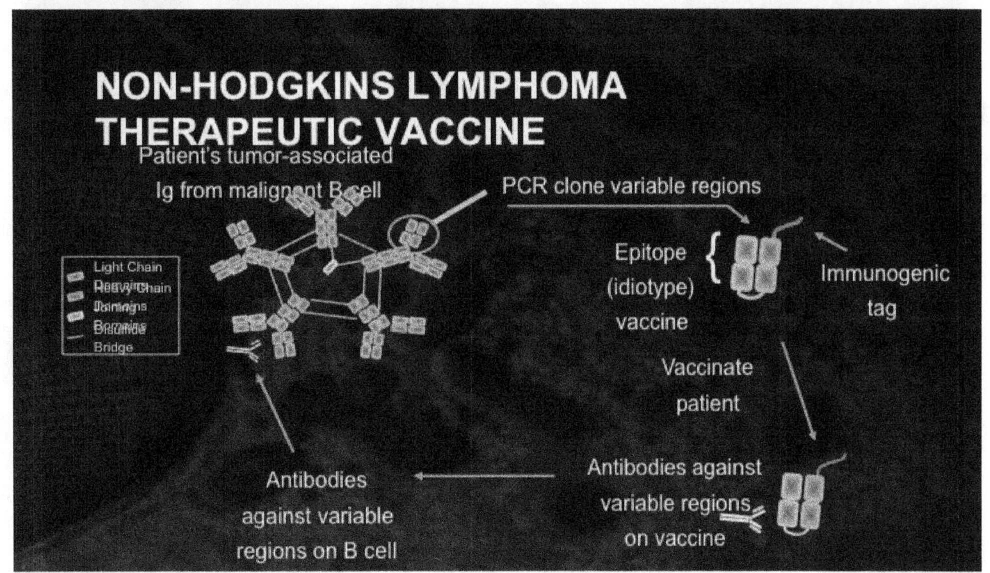
NON-HODGKINS LYMPHOMA
THERAPEUTIC VACCINE
Patient's tumor-associated
Ig from malignant B cell
PCR clone variable regions
Epitope
(idiotype)
vaccine
Immunogenic
tag
Vaccinate
patient
Antibodies against
variable regions
on vaccine
Antibodies
against variable
regions on B cell

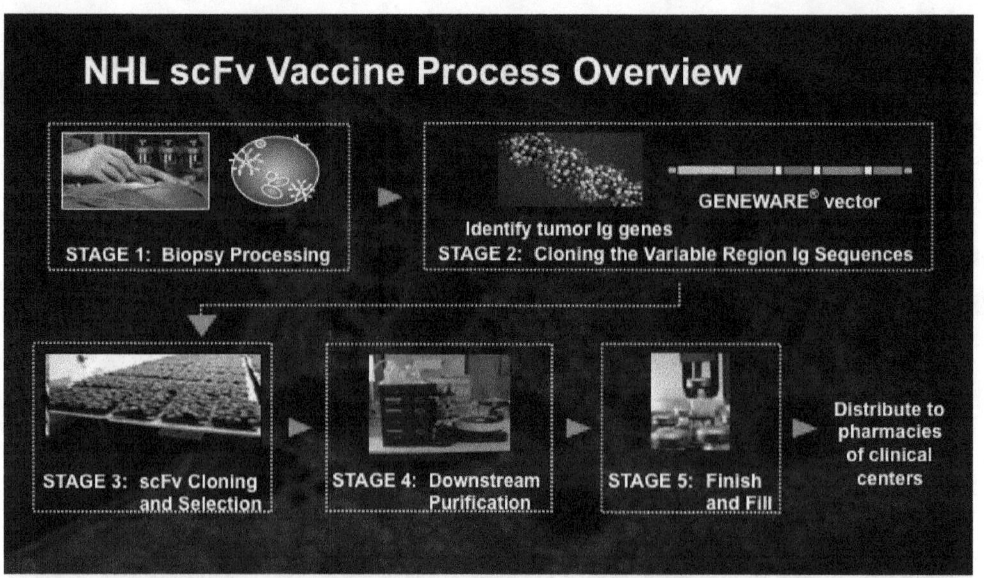

Making an Ally from an Enemy

Using a Virus to make vaccines

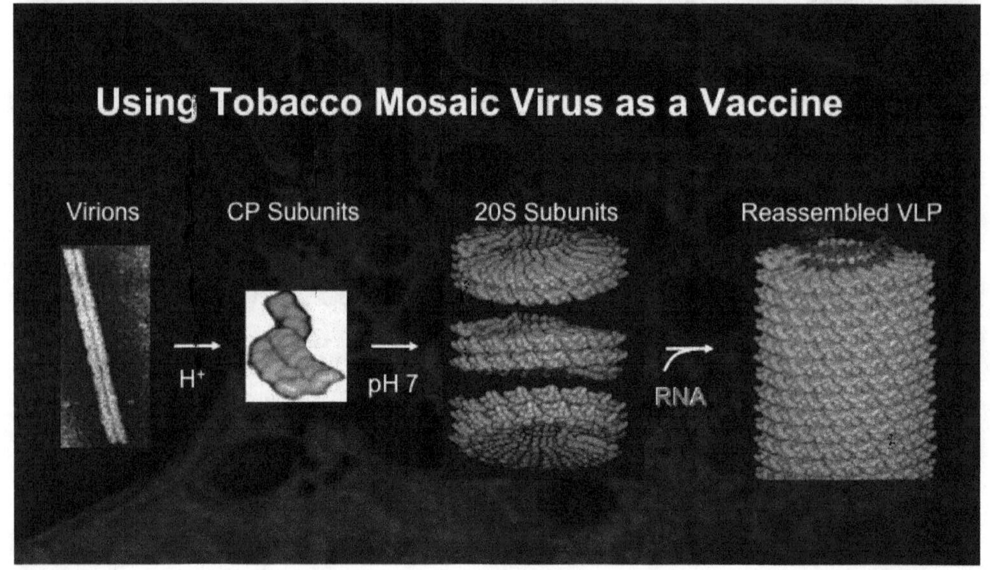

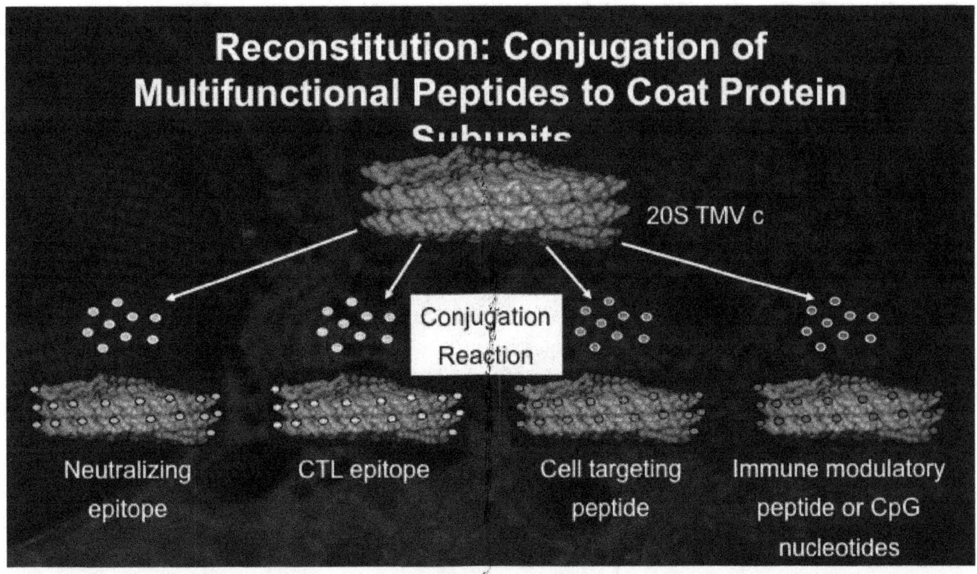

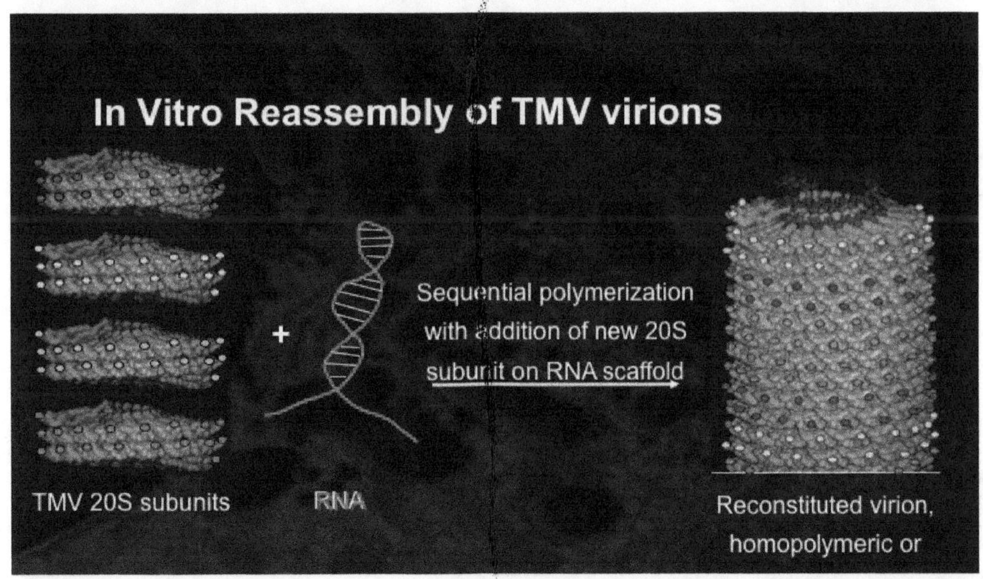

Current Medical Need : Addiction to Small Molecules

- Drug addiction is a long-term, recurring illness
- No approved medicines for treatment
- Patients in METH treatment programs increased 340% from 1994-2004
- ...and still growing
- >60% of patients cannot quit

The Need in Space Travel:

Sequester Toxic small molecules not found on Earth

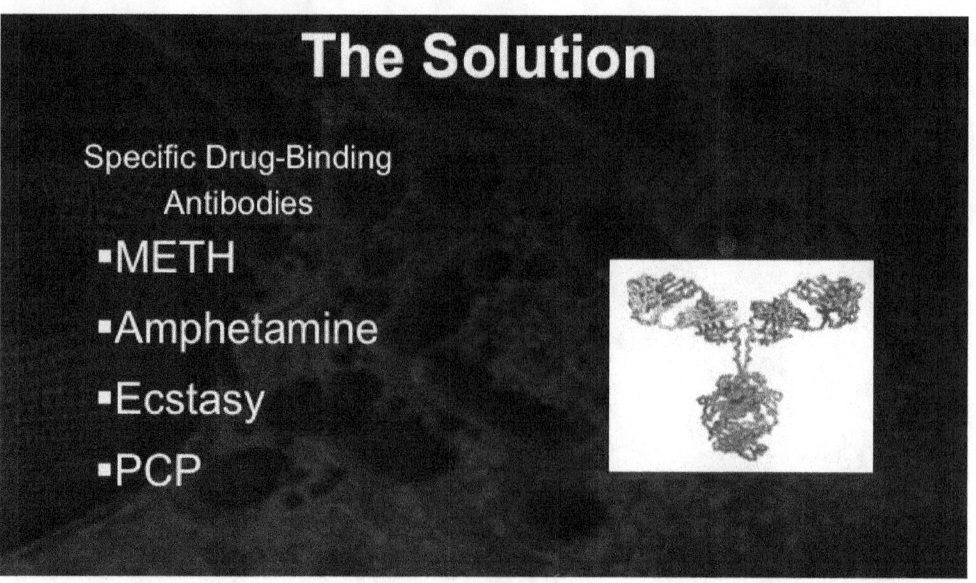
The Solution
Specific Drug-Binding Antibodies
▪METH
▪Amphetamine
▪Ecstasy
▪PCP

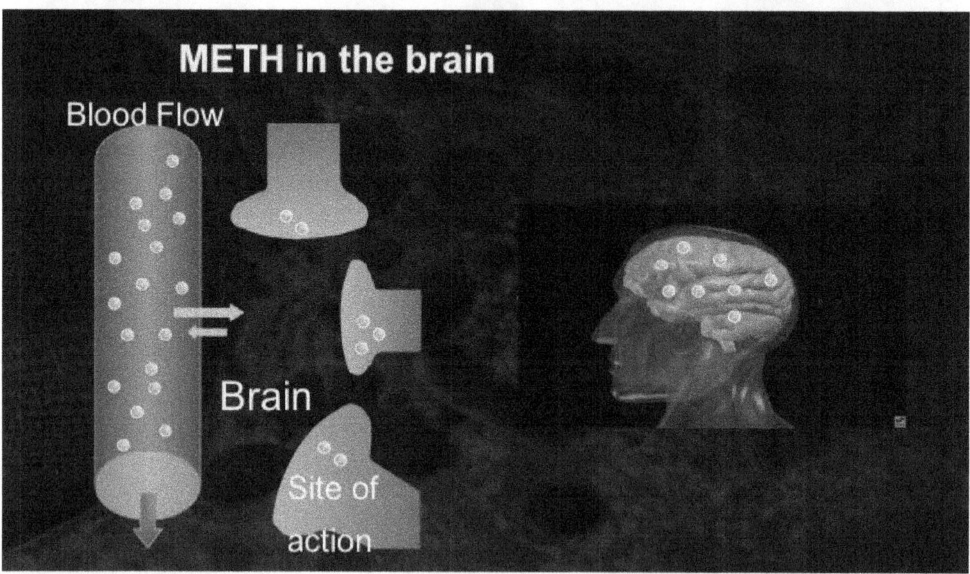
METH in the brain
Blood Flow
Brain
Site of action

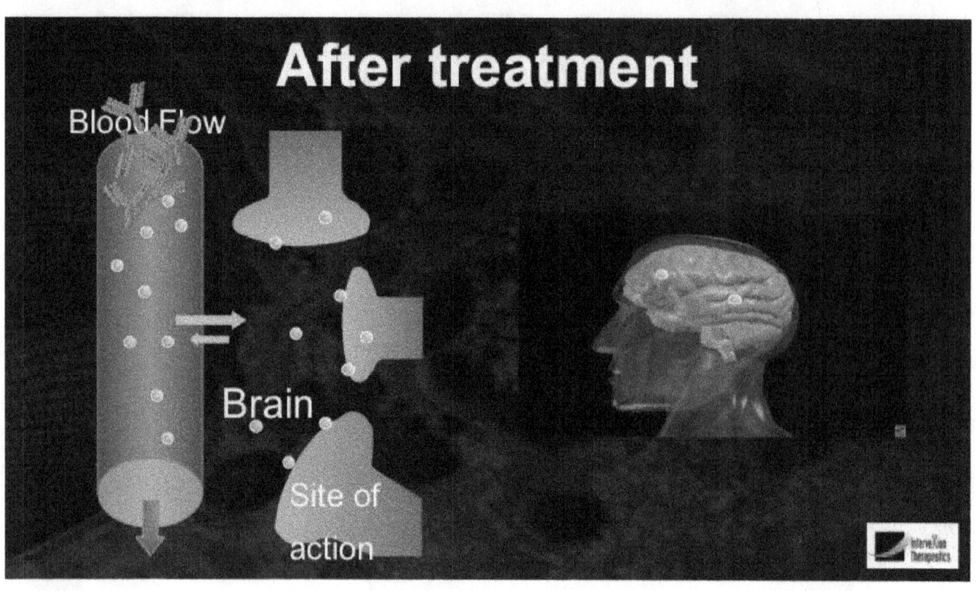
After treatment
Blood Flow
Brain
Site of action

Bioreactors

Completed March 22, 2011

CALIBER
biotherapeutics

Germination Room Laminar Flow Walls

Minimal Requirements for Fully Autonomous Health Care in a Closed Envionrment

Kurt Zatloukal, M.D.

Institute of Pathology, Medical University of Graz, Austria

The European Strategy Forum for Research Infrastructures Roadmap

- One infrastructure for biobanks and biomolecular resources
- Scientific excellence
- Pan-European
- Access to resources and services
- Long-term sustainability
- International integration

Global Health Challenges

- Diseases associated with ageing societies
- (re)emerging infectious diseases and pandemics
- Increased health care costs

There is a Need for New Solutions

„Rising spending on health care will jeopardise the creditworthiness of leading industrialized countries by the middle of this decade unless reforms are enacted"

Standard & Poor's, FT Feb 1st, 2012

"Cancer therapy costs will raise from $125 billion last year to at least $158 billion in 2020. If drugs become pricier, as seems likely, that bill could rise to $207"

The Economist, 2011

Roadblocks for Health Care Innovation

- Regulatory issues (FDA, EMA)
- Financing (insurances, social security systems)
- Acceptance by health care professionals

A Case for 100YSS

- Solutions developed might also revolutionize health care on earth
- The 100YSS might provide an experimental environment for disruptive innovation by avoiding existing health innovation roadblocks

Challenges

- From a remote to fully autonomous health care
- Suited to address all currently known diseases
- New challenges because of a closed system
  - Microbial flora
  - Adaption of pathogens
  - Drug resistance
- New environmental exposures

Need for Paradigm Shifts

- From a reductionistic view to understanding the complexity of biological systems and diseases

- From therapy to prevention

Investigation of Complex Biological Systems will Require Computational Models: The virtual patient

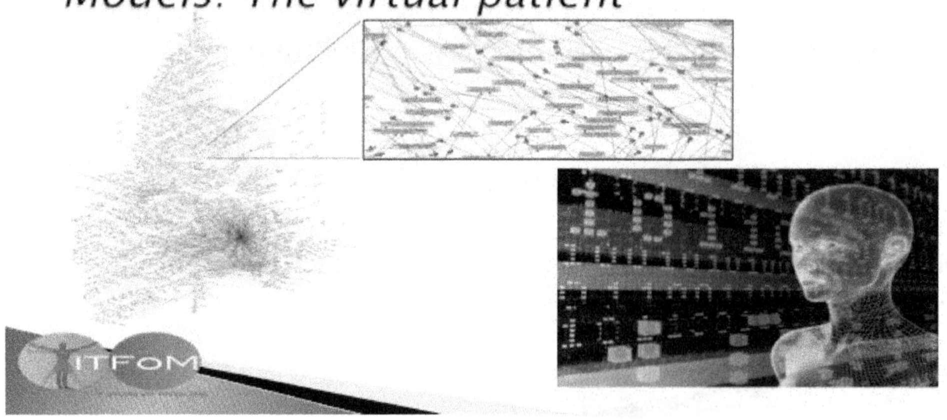

Understanding Genetic and Environmental Factors in Diseases:

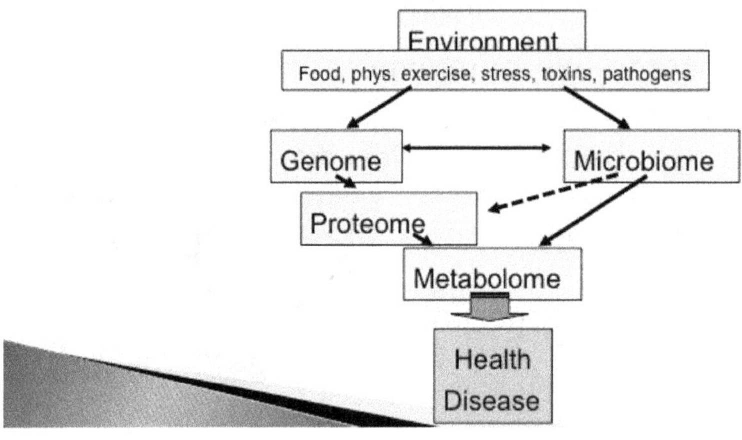

NMR-Based Metabolomics

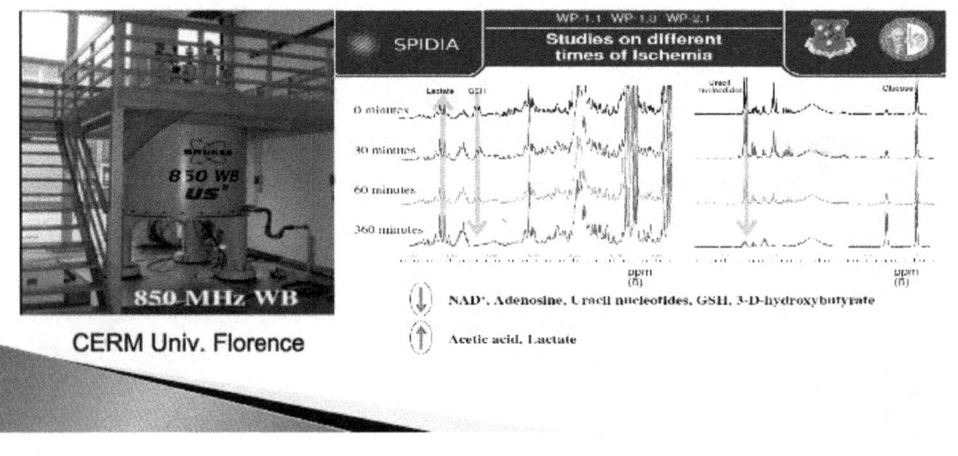

Metabolomic Fingerprints

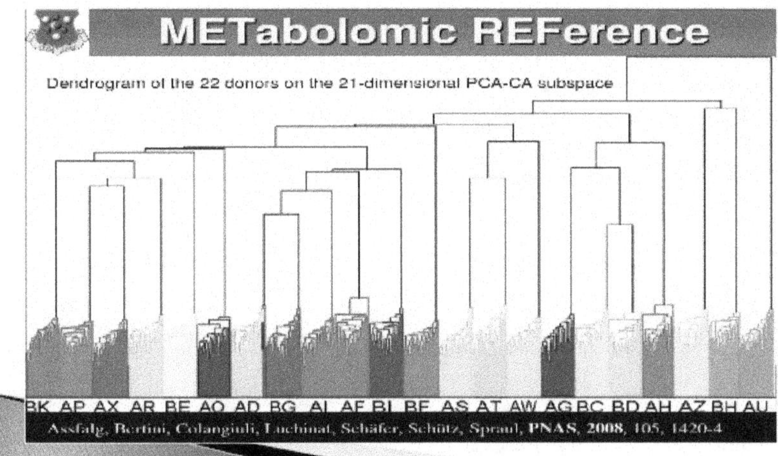

Health Monitoring and Disease Prevention by Metabolomics

Health assessment from metabolic trajectories

BRUKER

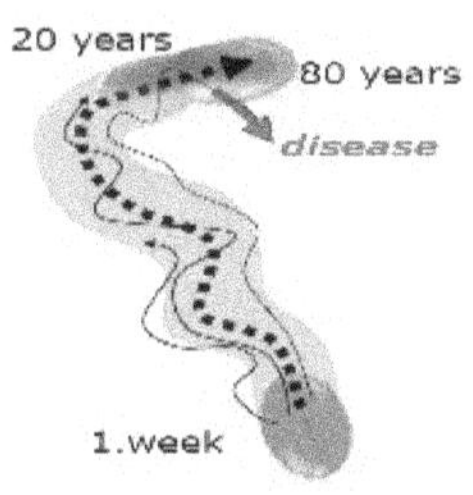

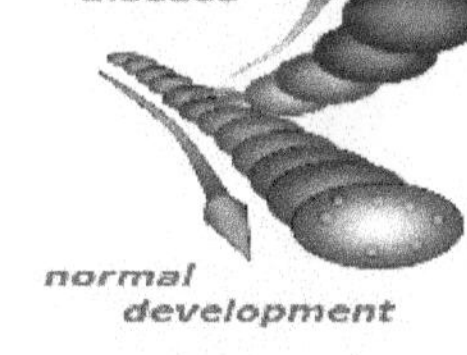

Personalized health assessment via comparison to metabolic trajectory of reference cohort

Self referencing personalized Health assessment

Thank you

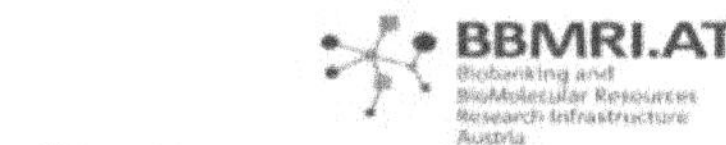

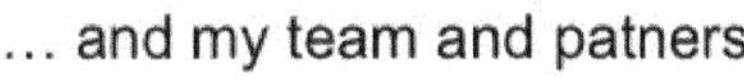

Plan B: Traveling to Exotopia

Pamela Reilly Contag, Ph.D.

Molecular Sciences Institute

prcontag@molsci.org

Here is what we do know---Human Interstellar Travel Will Require Innovation, New Technology, New Knowledge And New Capabilities.

Here is what we do not know:
What is out there living or not in the Cosmos.

How do we measure it if we don't know it. (Will we, by default, measure everything "earth-like".

Can we develop and utilize our technology and knowledge to "do no harm"?

Can we live harmoniously with whatever environment we encounter. What does harmonious mean?

OUR KNOWLEDGE OF:
CLOSED LOOP SYSTEMS
SPACE HABITATS
EXOPLANETS

TO KNOW WHAT TO "CARRY" TOWARDS EXOPLANETS, WE MUST "KNOW WHAT WE DO NOT KNOW" ABOUT EXOPLANETS

EARTH 1.0

WE CAN USE EARTH'S EVOLUTIONARY TIMELINE TO LEARN ABOUT IMPORTANT FEATURES FOR HUMAN SURVIVAL

How Can Our Knowledge Of The Earth And Human Evolution Help Us?

TIME FROM EVIDENCE OF PHOTOSYNTHETIC LIFE TO OXYGEN ATMOSPHERE IS 2.2 B YEARS

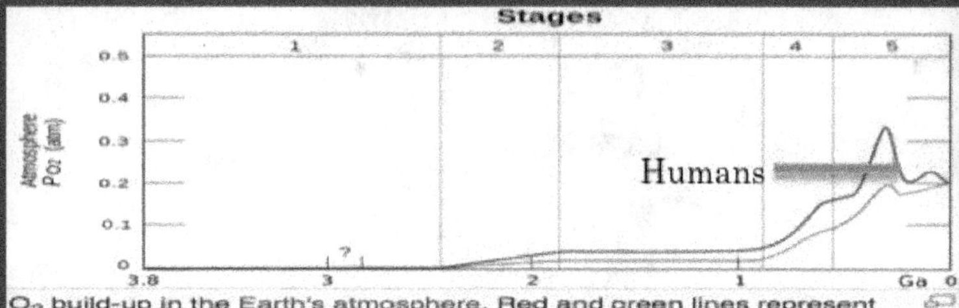

O_2 build-up in the Earth's atmosphere. Red and green lines represent the range of the estimates while time is measured in billions of years ago (Ga).
Stage 1 (3.85–2.45 Ga): Practically no O_2 in the atmosphere.
Stage 2 (2.45–1.85 Ga): O_2 produced, but absorbed in oceans & seabed rock.
Stage 3 (1.85–0.85 Ga): O_2 starts to gas out of the oceans, but is absorbed by land surfaces.
Stages 4 & 5 (0.85–present): O_2 sinks filled and the gas accumulates.[1]

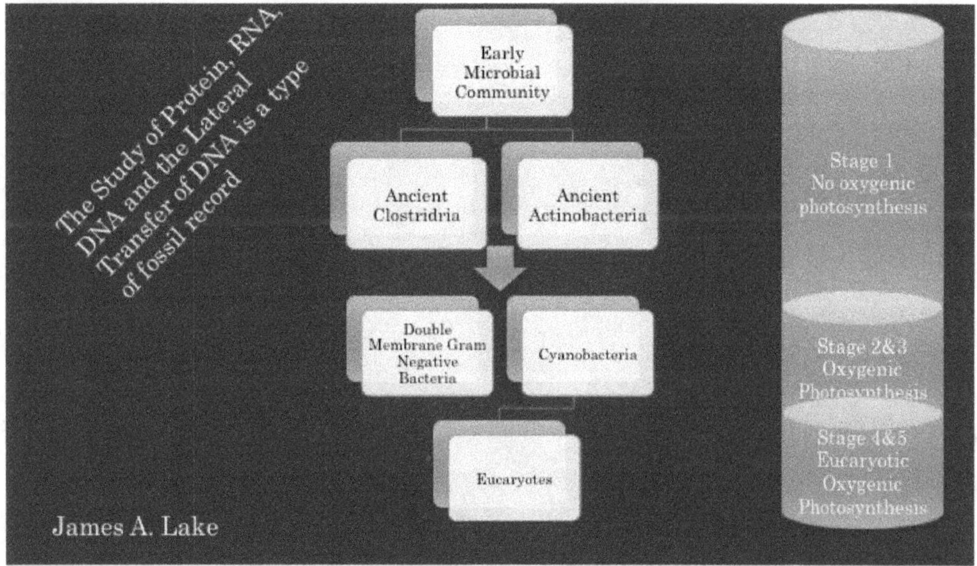

ONLY IN THE BACTERIA - AND AMONG THE BACTERIA WITHIN A SINGLE WINOGRADSKY COLUMN - DO WE FIND ALL FOUR BASIC LIFE STRATEGIES.

- All life on earth can be categorized in terms of the organism's carbon and energy source:
  - energy can be obtained from light reactions (phototrophs)
  - or from chemical oxidations (of organic or inorganic substances) (chemotrophs);
  - the carbon for cellular synthesis can be obtained from CO2 (autotrophs)
  - or from preformed organic compounds (heterotrophs).
- Combining these categories, we get the four basic life strategies: photoautotrophs (e.g. plants), chemoheterotrophs (e.g. animals, fungi), photoheterotrophs and chemoautotrophs.

2 BILLION YEAR OLD STROMATOLITES "FOSSILIZED MICROBIAL MATS" ARE THE ANCIENT MICROBIAL COMMUNITIES AND THESE COMMUNITIES STILL EXIST TODAY:

THUS WE HAVE BEEN ABLE TO CONNECT THEORY WITH CURRENT COMMUNITIES

CONCLUSION I: MICROBIAL ECOSYSTEMS ARE IMPORTANT FOR STABILITY OF THE ENVIRONMENT AND THUS FOR HUMAN SURVIVAL

- Microbes can fight other microbes.
- Microbes can share nutrients with plants and animals, including humans.
- Shared signals between host organisms and microbes can induce shared metabolism and relationship such as parasitism or symbiosis
- Adaptation by gene expression and epigenetic mechanisms has created emerging infectious diseases
- There is huge diversity in microbes, over 30,000 taxonomically differentiated microbes live on or near plants alone.

We have begun to develop platforms to measure the interaction between hosts and their microbiome, however, we need better approaches to determine functionality.

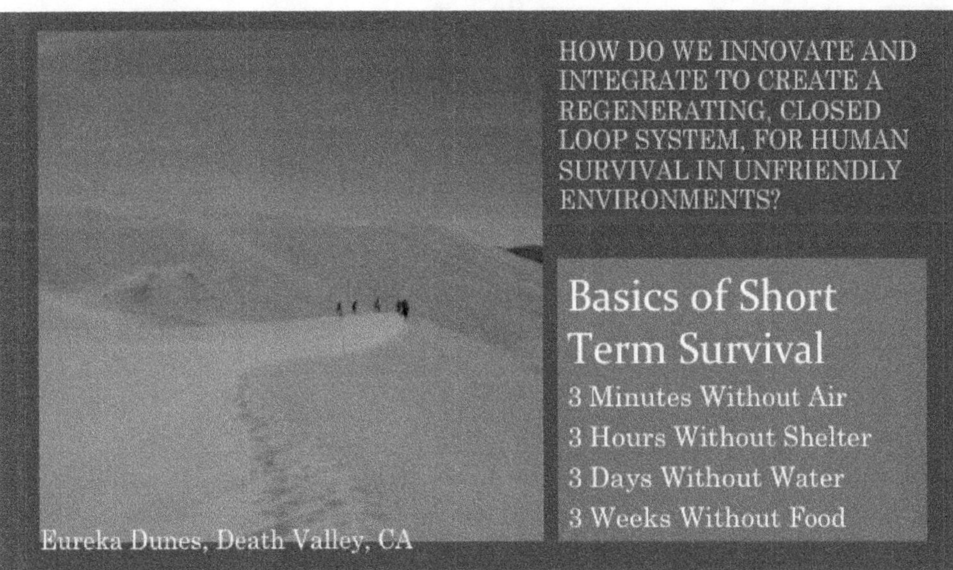

EARLY DATA FROM HUMANS LIVING IN SPACE AND IN DOMED COMMUNITIES SUGGEST THAT WE DO NOT YET KNOW THE INDIVIDUALS ABILITY TO ADAPT TO ANYTHING ELSE.

- We need:
  - Narrow temperature range
  - Water
  - Oxygen
  - Amino Acids, Fats, Carbohydrate
  - Electron Acceptors (Nitrogen, Sulfates, Carbohydrates)
  - Vitamins
  - Minerals
  - Gravity
  - Magnetic Field
  - Other? What don't we know that we need?

FEASIBILITY FOR SURVIVAL ON EXOPLANET DESTINATIONS

Need:

- A method to get there. (P&E).
- Survival in route. (Life Sciences)
- Protection, basic needs, planning and development once you arrive and determine the profile of the actual environment (we might not know until we get there).
- **Methodology to remediate any environmental deficiencies in order to grow to a critical mass of humans.**

TWO QUESTIONS

- What technology is needed to understand the chemistry and biochemistry of an exoplanet before we arrive?
- What features are needed for long term support of human life in order to gain critical mass for colonization?

CONCLUSIONS

- Take information and knowledge of our past and present: As context for future knowledge
- The ability to build "stuff" like Simple Machines
- Take Microbes
- Inspiration: Because inspiration comforts us
- Flexibility : We will need to live with ambiguity

"All the pests that out of earth arise, the earth itself the antidote supplies." Lithica c. 400 B.C.

100 YEAR STARSHIP™

A look at cutting edge research applied to interstellar research. Facilitated by Les Johnson, PhD (Author, Tennessee Valley Interstellar Workshop Founder). Topics: Space Elevators, Gravitational Lens, Space Entrepreneurship

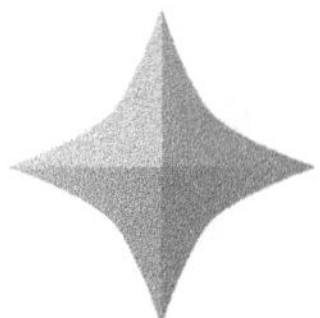

Radical Leaps Panel Summary

Peter A. Swan, Ph. D.

President and Member Board of Directors, International Space Elevator Consortium

Summary

The discussions started at the end point, a Starship departing by 2114, and illustrated some of the necessary steps to get there. The charts worked from backwards, from future to present with the following basic steps:

- Initial charts: A starship arriving at a new solar system.
- Middle charts: Assembly of starship with asteroid mining occurring next to a manufacturing facility at the Earth Moon L-1 spaceport.
- Next charts: Step-by-step process to arrive at assembly at EM L-1 spaceport, including the assumption that asteroid mining would be a robust commercial enterprise by then.
- Final set of Charts: Space Elevator development, a key element in low cost access to space. The image portrayed during these final charts was one of a robust architecture established around the Earth with 16 space elevators [8 pairs], each with a capacity of 100 metric tons of payload per day to "beyond GEO" for release towards the EM L-1 spaceport.

Earth-Moon Infrastructure: Building the 100 Year Starship

Peter A. Swan, Ph. D.

President and Member Board of Directors, International Space Elevator Consortium

Topics to be discussed

- Flight to the Stars
- Test Flights
- Manufacturing
- Material Gathering
- Depots, Villages, Transportation
- Leaving the Earth Cheaply
- Space Elevators

George Whiteside 2004 stated:

"Until you build an infrastructure, you are not serious."

9/20/2014

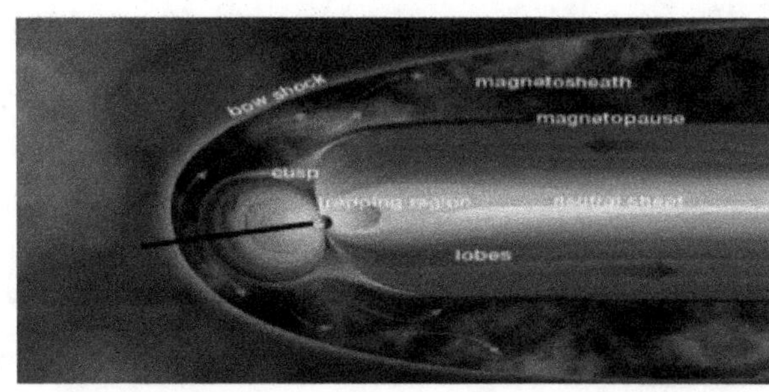

How do we get to a Nearby Star?

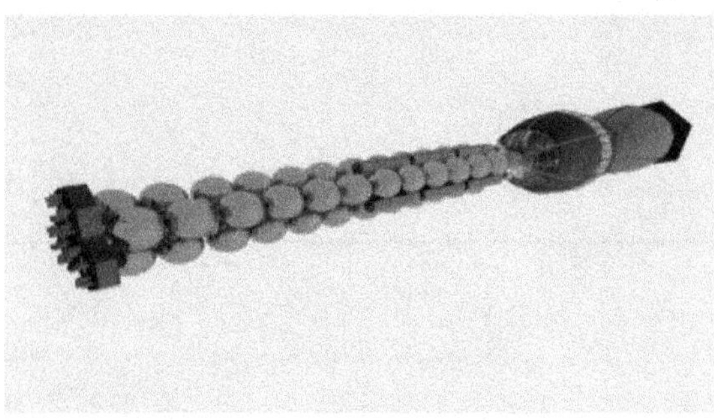

Images by chasedesignstudios.com

Lets Think Radical

We must believe and conceive of an infrastructure that will enable flights to the stars. To be feasible, we must Initiate:

- Assembly in zero-G [at the EM L-1 location]
- Minerals & supplies from asteroids or lunar mining
- Oxygen, water & fuel from asteroids or lunar processing
- Live the dream at EM L-1 or on a long duration Colony

Remember – Think Big or Go Home

9/20/2014 156

Inside Viewing

What will it look like?

- Similar to Earth
- Artificial gravity [0.6, 0.75 or 1.0?]
- Radiation shielding
- Recycled Ox, H2O, etc.
- Growing Food
- Lots of animals, insects, microbes
- Energy efficiency
- Shuttle craft for movement
- Diverse Culture
- Oceans, Lakes?

A World Ship should reflect the World we are leaving and the World we want to create.

157

Cruise Ships

Name	*Passengers*	*Crew*	*Total*	*Mass (tons)*	*Cost*
Disney Wonder	2700	900	3600		
Carnival Legend	2124	930	3054	88,500	
Carnival Elation	2052	920	2972	70,367	
RC Oasis of the Seas	5400	2100	7500	225282	$ 1.2 billion US
RC Allure of the Seas	6296	2100	8396	225282	

7,000 people with 225,282 Metric Tons For seven days

Allure of the Seas from Wikipedia

9/20/2014

Test Flights Where to?

- Flight to the Stars
- Test Flights
- Manufacturing
- Material Gathering
- Depots, Villages, Transportation
- Leaving the Earth Cheaply
- Space Elevators

Image from 100 YSS

9/20/2014 159

Test Flights

Images by chasedesignstudios.com

- Preliminary Flights to Mars and Back
- Cyclers Earth-Mars [3 + years]
- Cycler for Earth Asteroid Belt
- In-Parallel, Colonies at EM L1 and in Orbit around Mars

9/20/2014 160

Manufacturing

- Flight to the Stars
- Test Flights
- Manufacturing
- Material Gathering
- Depots, Villages, Transportation
- Leaving the Earth Cheaply
- Space Elevators

9/20/2014 Images by chasedesignstudios.com 161

Final Assembly at Earth Moon L-1

Material Gathering

- Flight to the Stars
- Test Flights
- Manufacturing
- Material Gathering
- Depots, Villages, Transportation
- Leaving the Earth Cheaply
- Space Elevators

Hugh Cook – Aug 2014

"The currency of Space will be Water!"

From IAA study report draft, 2014

9/20/2014

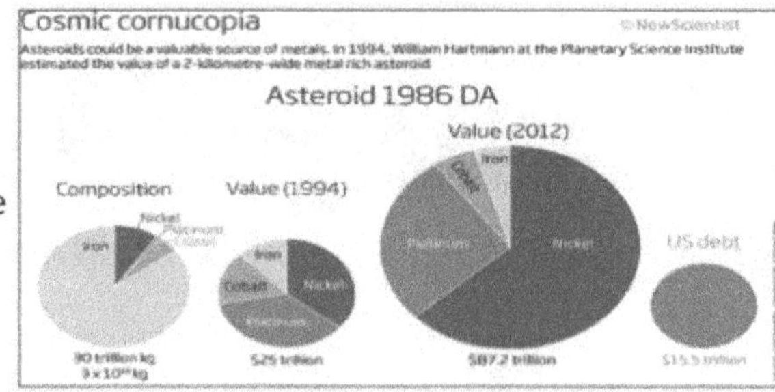

Figure 1-4, Asteroid 1985 DA Value

Materials Mining & Manufacturing

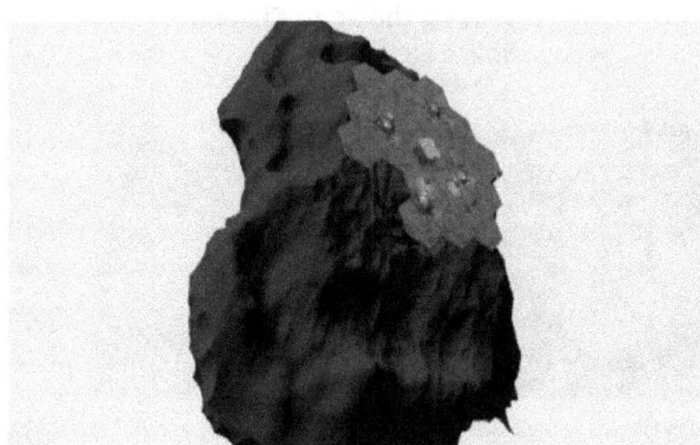

One significant key is that asteroids and the lunar surface have all the resources needed. The content of water allows living while the minerals available will be sufficient to manufacture all sophisticated electronics and structures. Bringing a near-Earth asteroid to EM L-1 manufacturing Facility will ensure required resources.

Images by chasedesignstudios.com 164

Capture and Move Resources

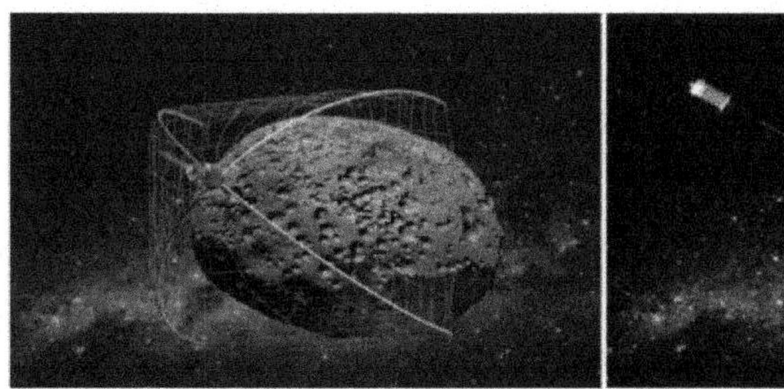

WRANGLER system - Asteroid Capture [Tethers Unlimited]

9/20/2014 165

Lunar Resources

From Project Horizon Study, US Army, 1959]

- Resources are endless on the Moon
- Low gravity enables movement by multiple methods
- One key development is the Lunar Elevator for moving supplies to the EM L-1 Manufacturing arena
- A key resource is Lunar water for living and for fuel.

9/20/2014 166

Apex Anchor Assembly

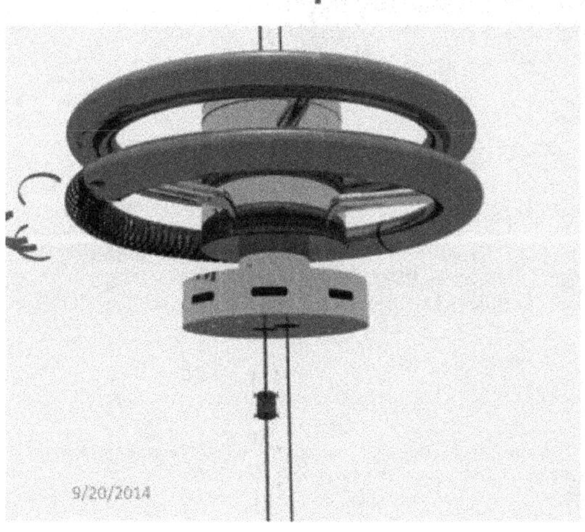

- Apex Anchor at end of logistics tether
- Some artificial gravity to assist in living and manufacturing
- Manufacturing leverages easy access to Earth's people and resources
- Quick release goes to EM L-1
- EM L-1 is assembly area for all Starships.

Images by chasedesignstudios.com

9/20/2014 167

Topics to be discussed

- Flight to the Stars
- Test Flights
- Manufacturing
- Material Gathering
- Depots, Villages, Transportation
- Leaving the Earth Cheaply
- Space Elevators

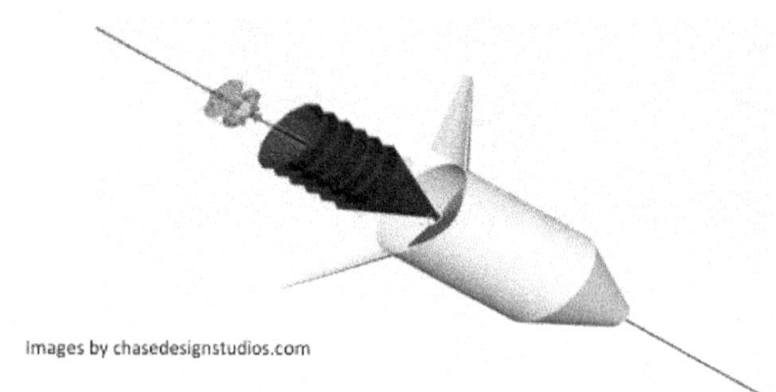

Images by chasedesignstudios.com

9/20/2014

Earth Moon Infrastructure Velocity Requirements:

Delta-Vs in Earth's Neighborhood [Mankins, 2012].

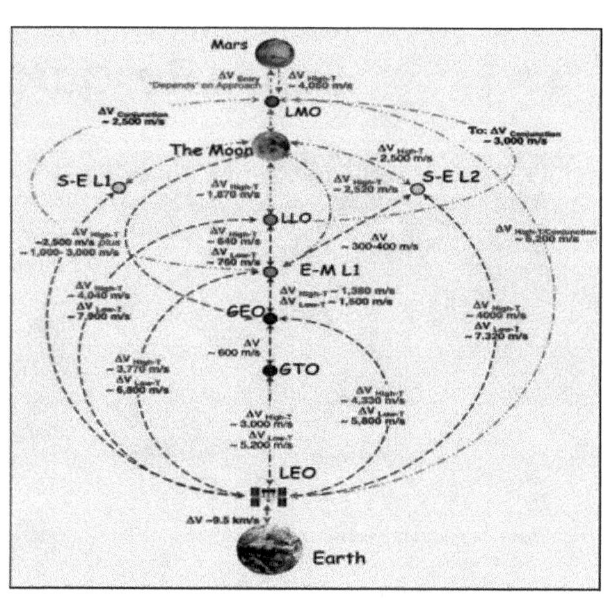

9/20/2014

Flow of Resources Across Space Elevators

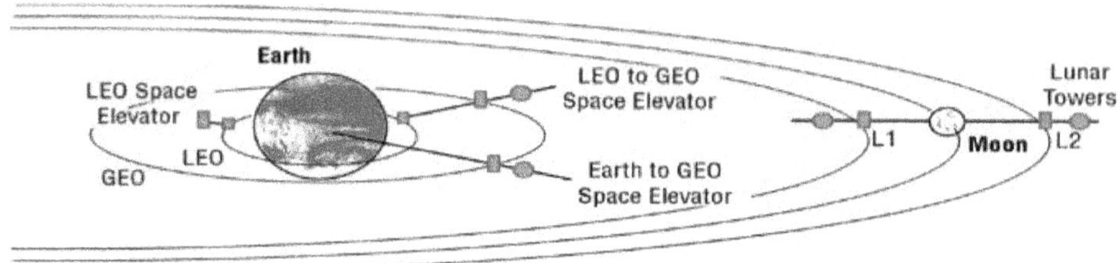

Image from: David Smitherman Space Elevators Concept Overview, Marshall Space Flight Center Flight Projects Directorate Advanced Projects Office 256-961-7585 David.Smitherman@nasa.gov http://flightprojects.msfc.nasa.gov/fd02.html

9/20/2014 170

Leaving the Earth Cheaply

- Flight to the Stars
- Test Flights
- Manufacturing
- Material Gathering

- Depots, Villages, Transportation
- Leaving the Earth Cheaply
- Space Elevators

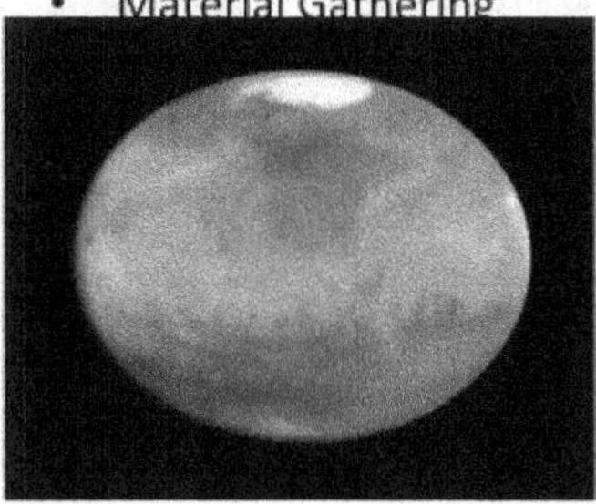

Destinations:

- Moon
- Earth-Moon L-1
- Mars

9/20/2014 171

Elevator Delivery for the Starship Fabrication

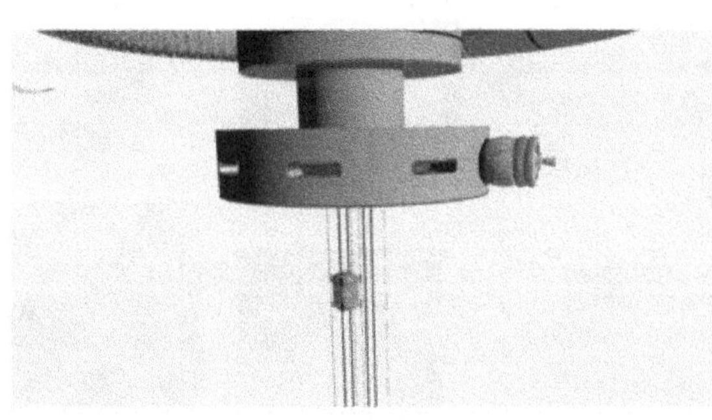

Images by chasedesignstudios.com

By 2060:

- 16 Elevators
- Each capable of 100 Metric Tons per day
- No size restrictions
- Human and robotic delivery
- GEO stations for Assembly
- Apex Anchor Nodes for Assembly

9/20/2014 172

Massive Space Elevator Car

- Human Rated
- Three days to GEO, two more to Apex Anchor
- 100 Metric Tons of Payload
- Less than $10 per Kg
- Eight Pairs of Space Elevators operating around the world
- Probably solar powered
- Safe, routine [1 lift per day per elevator]
- No Shake Rattle and Roll
- Elevator Music to Stimulate

Images by chasedesignstudios.com

9/20/2014 173

- Flight to the Stars
- Test Flights
- Manufacturing
- Material Gathering
- Depots, Villages, Transportation
- Leaving the Earth Cheaply
- Space Elevators – today and the near future

Remember –

Think Big or Go Home

9/20/2014 Images by chasedesignstudios.com

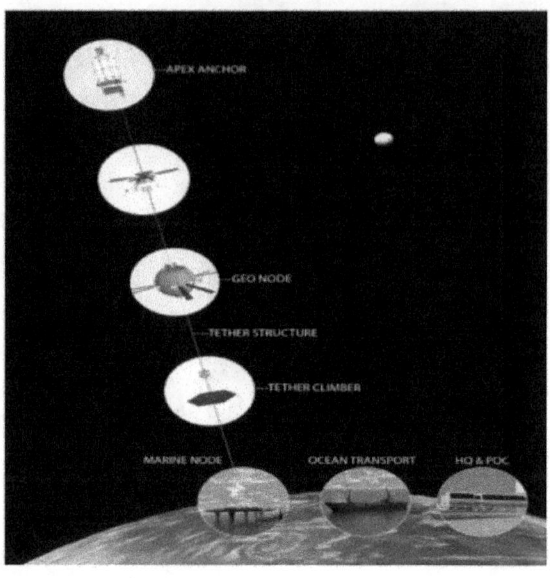

SE in Literature & Art

- Arthur C. Clarke introduced concept of space elevator in his 1978 novel *Fountains of Paradise*. Engineers construct rigid connection between GEO and Earth – for his cable Clarke envisioned hyperfilament made of diamond crystal – though he later expressed belief that another form of carbon would be used
- *Web Between the Worlds* by Charles Sheffield written at same time as Clarke's book describes construction of space elevator
- Kim Stanley Robinson's *Mars Trilogy* published in 1990s features space elevators on Earth and Mars with jointed segments each of which is diamond crystal

From: The Space Elevator: It's Place in History, Literature and Art ' A Presentation by Dr. David Raitt European Space Agency IAC-05-D4.3.02

9/20/2014 175

Dr. Edwards' Space Elevator [2000 -2003]

- Length: 100,000 km, anchored on the Earth with a large mass floating in the ocean and a large counterweight
- Width: One meter, curved
- Design: Woven with multiple strands to enable localized damage; and curved to ensure that edge on, small size, hits do not sever the ribbon.
- Cargo: The first few years will enable 25 ton payloads without humans with five concurrent payloads on the ribbon
- Production: The space elevator can, and will, be produced in the near future because the human condition demands it and the materials are almost ready to enable its construction today.
- Construction Strategy: The first space elevator will be built the tough, and only, way – from GEO; then, once the gravity well has been overcome it will be replicated from the ground up leading to multiple elevators appearing around the globe. This redundancy will reduce the magnitude of the impact if one is lost.

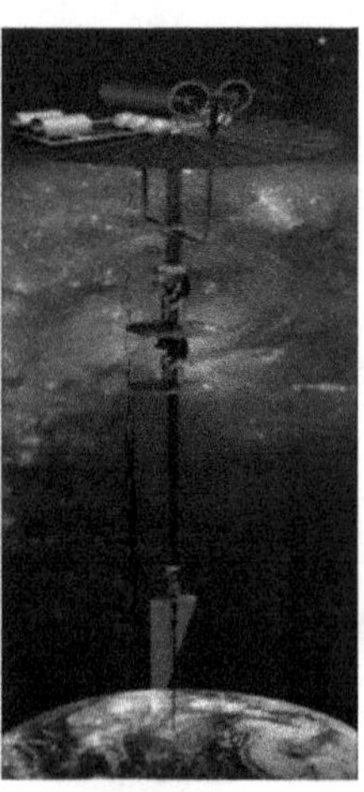

Dr. Edwards in Space Elevators [*Edwards, Bradley C. and Eric A. Westling, 2003*]
Space Elevator Systems Architecture, [*Swan, Peter A. and Cathy W. Swan, 2007*]

9/20/2014

Cosmic Study Assessment

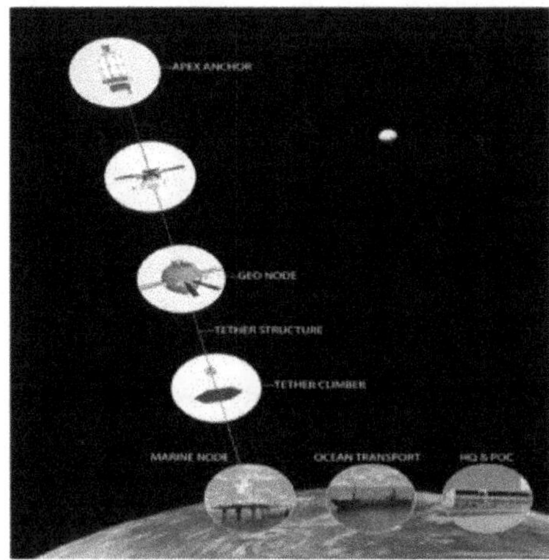

The authors have come to believe that the operation of a space elevator infrastructure will lead to a "game changing" experience in the space world. Each of the authors considers that the space elevator can be developed when the material is mature enough for the demands of the space elevator. Our final assessment is:

A Space Elevator
Seems Feasible.

Images by chasedesignstudios.com

177

Cosmic Study Outline

Executive Summary
1 Introduction
2 Systems Infrastructures
Part I – Major Elements
3 Tether Material
4 Tether Climber and Spacecraft
5 End Station Infrastructure (Base & Apex Anchor)
Part II – Systems Approach
6 Dynamics of Operation Tether
7 Systems Design for Environment
8 Systems Design for Space Debris
9 Operations Concept
10 Technology Assessment
Part III – Future Considerations
11 Developmental Roadmaps
12 Legal Perspective
13 Market Projection
14 Financial Perspective
Part IV – Future Considerations
15 Study Findings and Conclusions
16 Study Recommendations and Next Steps
Appendix (Glossary, Participants, Study Form, Space Elevator History, acronyms, tether history, Myuri, safety factor, tether buildup, laser option

Figure 4-1. Climber as Spacecraft [chasedesignstudios.com]

9/20/2014

178

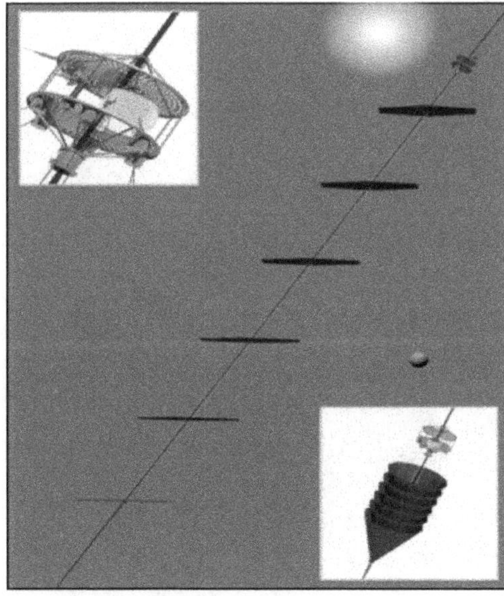

Executive Summary

"Don't undertake a project unless it is manifestly important and nearly impossible."

Edwin Land, quoted in the Coral Reef Alliance letter, March 30, 2011. www.coral.org

Major Questions:

Why a space elevator?
Can it be done?
How would all the elements fit together to create a system of systems?
What are the technical feasibilities of each major space elevator element?

Images by chasedesignstudios.com

179

Major Conclusion: Seems Feasible

The detailed conclusions from this study fall into distinct categories:

Legal: The space elevator can be accomplished within today's arena!

Technology: It can be accomplished with today's projection of where materials science and solar array efficiencies are headed.

Space Elevators will open up human spaceflight and decrease space debris and environmental impact.

Business: This mega-project will be successful with a positive return on investment within 10 years of installation.

Cultural: This project will drive a renaissance on the surface of the Earth with its solutions to key problems, stimulation of travel throughout the solar system, with inexpensive and routine access to GEO and beyond.

9/20/2014 180

Ribbon Design

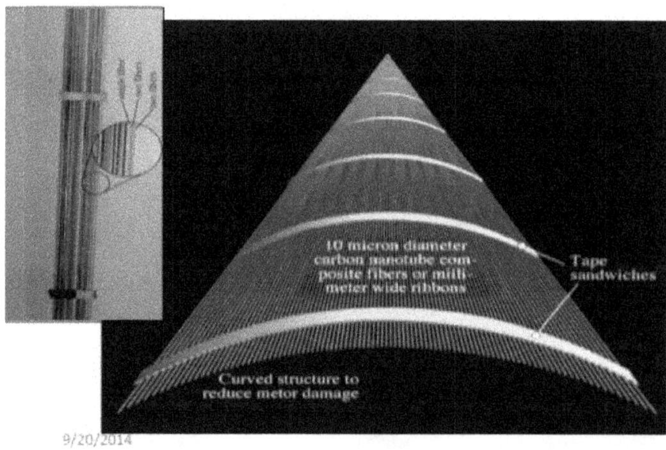

- The final ribbon is one-meter wide and composed of parallel high-strength fibers
- Interconnections maintain structure and allow the ribbon to survive small impacts
- Initial, low-strength ribbon segments have been built and tested

Images from Dr. Edwards

9/20/2014 181

Tether Climber

First 40 kms protected through atmosphere
Above 40 kms, solar arrays for power

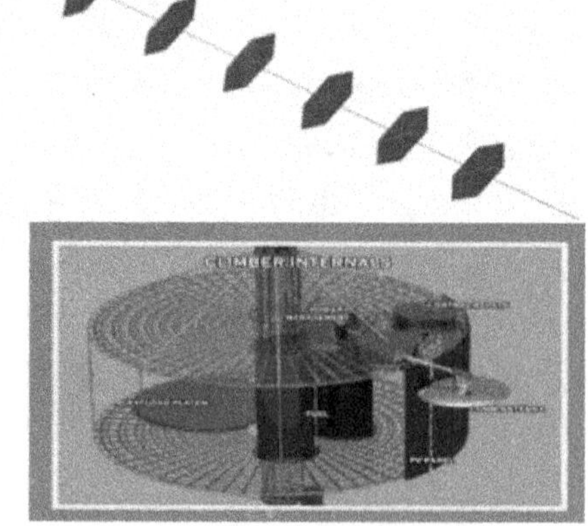

Images by chasedesignstudios.com

Tether Terminus Anchor

- Marine Node is a mobile, ocean-going platform identical to ones used in oil drilling
- Marine Node is located in eastern equatorial pacific, weather and mobility are primary factors

Images by chasedesignstudios.com

Three Options to Reach 40 Km Altitude

- Option One: Marine Stage One
  - 1a – MSO: Box Protection
  - 1b – MSO: Spring Forward
- Option Two: High Stage One

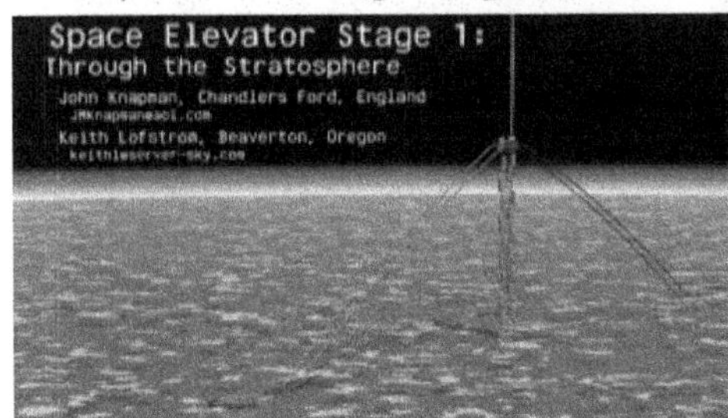

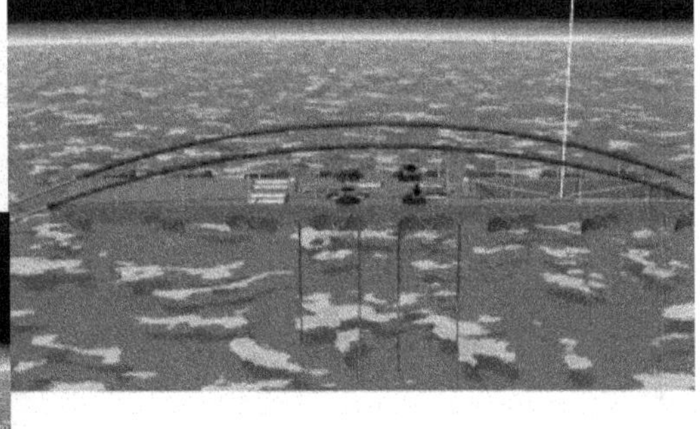

184

Geosynchronous Operations Complex

- Infrastructure at GEO: along the ribbon, where most payloads will be destined. There will be autonomous operations to off-load, adjust, and monitor satellites that are designed for GEO arc mission locations. In the future, this operations center could become a manufacturing / assembly location as well as a location for human habitat.
- GEO Station:. Autonomous operations will ensure that the satellite is healthy and then assist in the release of the system. In addition, this location will be where returning GEO payloads are collected and prepared for the trip back to the Earth's surface. Also, it will be a location for refueling spacecraft.

Images by chasedesignstudios.com

9/20/2014

185

Challenges

- Magnetosphere
- Induced oscillations
- Radiation
- Atomic oxygen in Earth's upper atmosphere
- Environmental Impact: Ionosphere
- Malfunctioning climbers
- Lightning, wind, clouds
- Meteors and space debris
- Satellites
- Health considerations

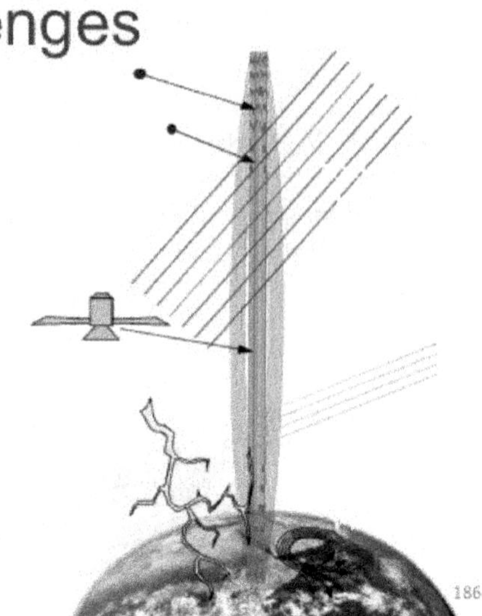

186

Technology Assessment

	Expected year for Space Elevator System	TRL Level	TRL Level by 2030	Remarks
The Tether	2035+ with estimates varying to 2060 (JSTM, 2010)	2	7	Major development funding required Terrestrial version will be available by 2030 in greater than 1,000 km lengths with appropriate strength
Apex Anchor	2025	5	8	Reel-out in vacuum of long material will require design and testing of components in orbit.
Geosynchronous Station	today	6	9	routine
Tether Climber	2025	4	8	Major design effort, however, not out of the knowledge of current satellite designers
Marine Node	2015	8	9	Deep Ocean Drilling Platforms and Sea Launch platform can be a models.
High Stage One	2025-30	3	6	Major design and development effort. Major breakthroughs needed in timely manner for many of its major components.
Ocean Going cargo Vessel	today	9	9	Routine
Helicopter Transport	today	9	9	Routine
Operations Centers	today	9	9	Routine

Table 10-XVI. Integrated System Realizable Time and TRLs

Figure 10-8 Project Risk Position Reporting

9/20/2014

187

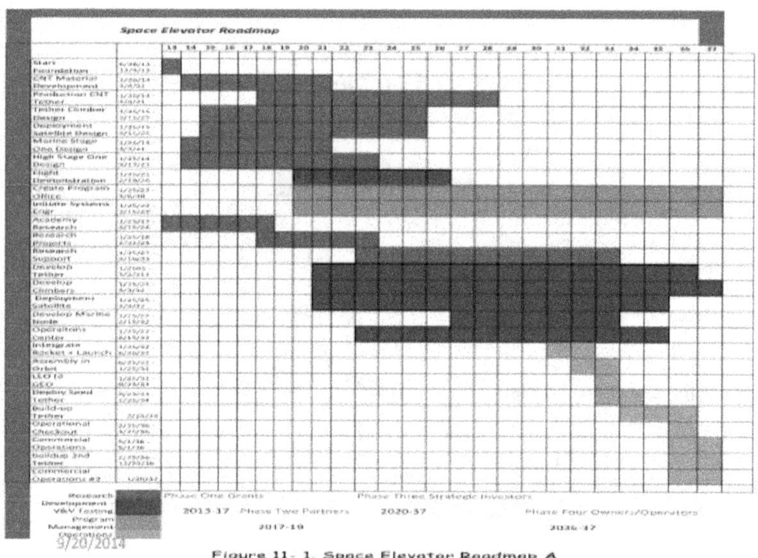

9/20/2014

Figure 11- 1. Space Elevator Roadmap A

188

Roadmap A: Assumption [based upon chapter 3] is that the tether material matures rapidly and supports a 2036 space elevator deployment.

Roadmap B: Assumptions support a space elevator deployment in the 2050s.

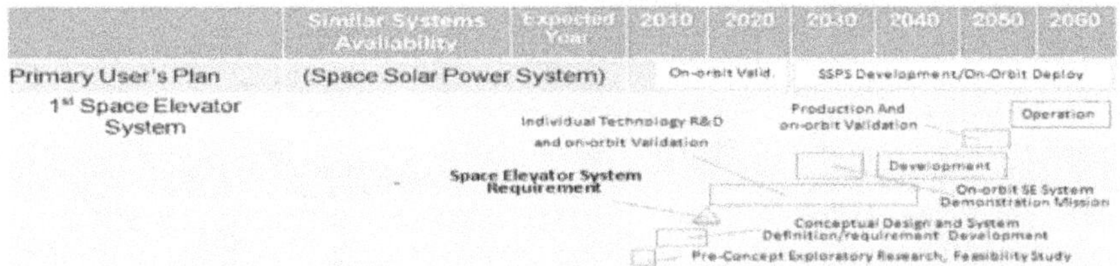

Figure 11.2. Summarized Space Elevator Roadmap B (Tsuchida, 2011)

WHY – The Real Reason

- **The human spirit needs no restrictions:** Once the Apollo 8 picture of the Earthrise from the lunar orbit was broadcast, the world was sensitized to our limitations and the realization that we were on a fragile "Big Blue Marble." We must soar beyond our boundaries and expand into the solar system.
- **The recognition that the "*Space Option*" will enable solutions to Earth's current limitations:** The space option is an alternative that is will open access to space to humanity.
- **The realization that chemical rockets can not get us to and beyond Low Earth Orbit economically:** The rocket equation requires that approximately 80% of the mass on the launch pad is fuel and 14% is structure, control equipment and other essential elements of a launch vehicle. This leaves roughly 6% for payload (mission satellite).

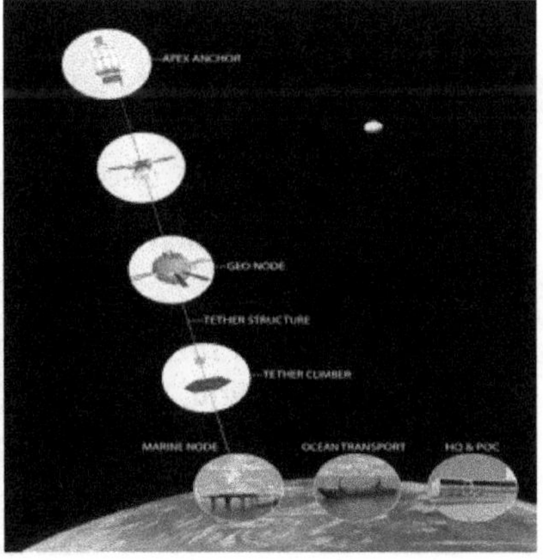

Rationale:

- **Routine:** Space will become boring and routine with lift-offs occurring every day with 20 ton tether climbers.
- **Price:** The price for a pound of payload to be delivered to GEO will be below $ 500/kg. This change from $ 20,000/kg will alter the clientele for space liftoff and open businesses that are not even considered today.
- **Safety:** Elevators have inherent safety vs. the dangerous practice of mounting valuable payload on top of huge explosive tanks.
- **Delivery Dynamics:** Space elevators will have vibrations in the region of cycles per day and shock loads of marshmallows dropping into a pool instead of explosive potential and the rock & roll during liftoff of rockets.

Space Elevators vs. Transcontinental Railroads

- "President Andrew Jackson traveled no faster than Julius Caesar," and that... "Thoughts or information could [not] be transmitted any faster than in Alexander the Great's time."*
- By 1869, human movement had advanced to a heart stopping 90 kilometers an hour while word and ideas had leaped to light speed, telegraphed across a continent.*

NY to San Francisco by Railroad	Pre- Golden Spike	Railroad Operations
	Via Panama 3/14 – 8/30/1849 Via Straights Magellan 7/14 – 1/26/1849 Across the plains approx. 6 months. Multiple deaths along the way Mail costs dollars per ounce Trip cost about $ 1,000.00	Government Awarded Loans to build Right of way over public land Five alternate sections per mile awarded Trip time – approx. 7 days Trip cost - $70.00 Mail costs at pennies per ounce
Space Elevator	**Traditional Launching**	**Space Elevator Operations**
	Launch costs Commercial: $25,000/kg Government $40,000/kg Space Shuttle - $2.4 B per Launch Rate About 80 launches per year Probability of Success – 95% Launch on time rate – approx. 0%	Lift Costs $500 per kilogram for materials Human launch by rockets initially Lift Rate Seven carriers on space elevator Each carrier at 14 tons payload Week trip – estimate 7 days Probability of Success –estimate >99% Launch on time rate – estimate >99%

*Ambrose, S. (2000), "Nothing like it in the World". Simon & Schuster, New York, 2000.

9/20/2014 192

Topics to be discussed

- Flight to the Stars
- Test Flights
- Manufacturing
- Material Gathering
- Depots, Villages, Transportation
- Leaving the Earth Cheaply
- Space Elevators

Questions

George Whiteside 2004 stated:

"Until you build an infrastructure, you are not serious."

9/20/2014

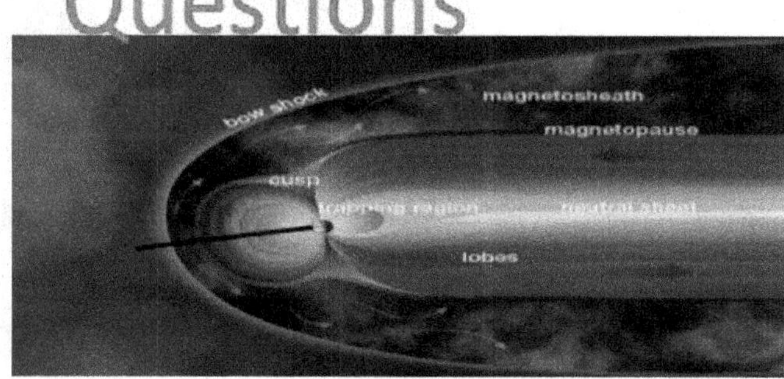

Transcontinental Railroad Analogy

- Planning began in the 1850's
- Built from 1863-1869 in a "wilderness"
- The Union was fighting the Civil War when it began this project
- Huge initial cost to build the line from Omaha, Nebraska to Sacramento, California
- Built the railroad line as well as infrastructure such as coaling stations and water sources for the steam locomotives

Transcontinental Railroad Analogy

- Created towns in the middle of nowhere
- Unified the United States across the continent and opened the west
- America's greatest engineering feat of the 19th century
- New York to San Francisco travel fell from 6 months to 7 days and $1000 to $70
- Owners became the some of the richest men in America

Peter A. Swan

A Magnifying Lens as Big as the Sun: Exploring Exoplanets Without Getting There

Dr. Claudio Maccone

Istituto Nazionale di Astrofisica, Italy

clmaccon@libero.it

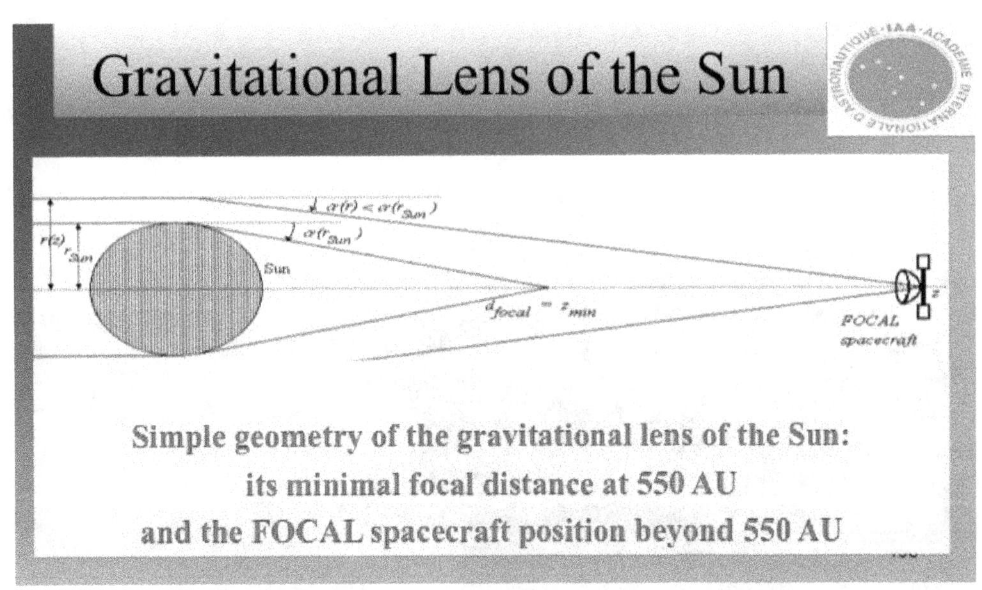

Gravitational Lens of the Sun

- The geometry of the Sun gravitational lens is easily described: incoming electromagnetic waves (arriving, for instance, from the center of the Galaxy) pass *outside* the Sun and pass within a certain distance *r* of its center.
- Then a basic result following from General Relativity shows that the corresponding *deflection angle (r)* at the distance *r* from the Sun center is given by (Albert Einstein, 1915):

$$\alpha(r) = \frac{4GM_{Sun}}{c^2 r}.$$

Gravitational Lens of the Sun

- Let's set the following parameters for the Sun:
  1. Mass of the Sun: $1.9889164628 \cdot 10^{30}$ kg, that is μSun = 132712439900 $kg^3 s^{-2}$
  2. Radius of the Sun: 696000 km
  3. Sun Mean Density: 1408.316 kgm^{-3}
  4. Schwarzschild radius of the Sun: 2.953 km

  One then finds the BASIC RESULT:

MINIMAL FOCAL DISTANCE OF THE SUN:

548 AU ~ 550 AU ~ 3.17 light days

~ 13.86 times the Sun-to-Pluto distance.

2009 BOOK by this author

TWO TETHERED ANTENNAS

Claudio Maccone

TWO TETHERED ANTENNAS

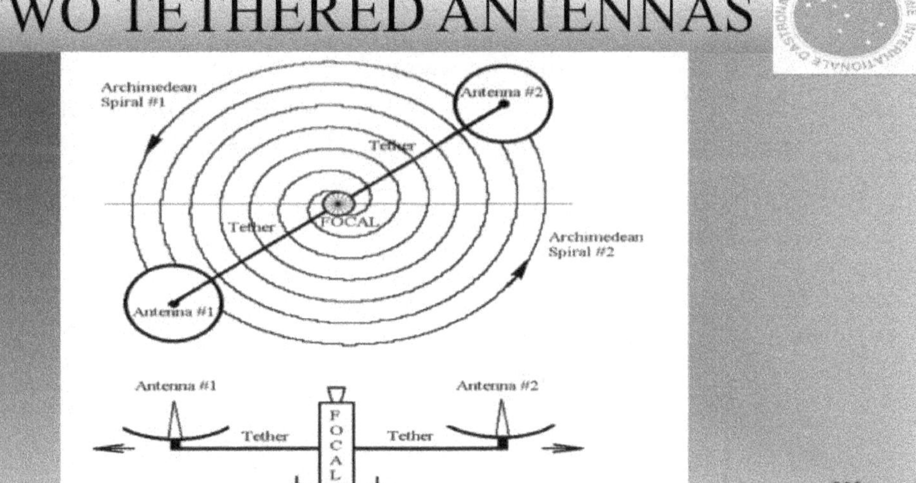

203

Infinity of Focuses > 550 AU

- There is an infinity of focuses from 550 AU outward in any direction. Thus, 550 AU is actually the minimal focal sphere.
- In the practice, we won't have to stop a spacecraft just at 550 AU, but we just let it go.
- The further the spacecraft goes beyond 550 AU, the better it is. In fact, radio waves impinging on the spacecraft at distances higher than 550 AU will have to cross less and less dense layers of the Solar Corona.
- The Solar Corona is difficult to model. Essentially, because of the radially decreasing electron density, the Solar Corona acts as a divergent lens opposing the convergent lens of gravity.
- We shall study the Corona later. Just the Naked Sun for now.

204

Gain of Any Star as a Lens

- The "gravity lens" concept means that the Sun (and any other massive celestial body) is an antenna since it can increase the intensity of the signal, by virtue of its deflection.
- We define the Gain associated to any star, G_{star}, as the ratio between the intensity of the signal in presence of the star compared to the intensity of the signal without the star. It can be proven that, along the focal axis, one has

$$G_{star}(\lambda) = 4\,\pi^2\,\frac{r_g}{\lambda} = \frac{8\,\pi^2\,G\,M_{star}}{c^2}\cdot\frac{1}{\lambda}$$

The gain is constant along the focal axis but is wavelength-dependent. There, r_g is the Schwarzschild radius of the star.

205

Sun Gain at Some Frequencies

- In the following table we remind on-axis GAIN of the gravitational lens of the naked Sun lens for seven important frequencies:

Line	Neutral Hydrogen	OH maser	Water maser	Ka Band	CMB peak	**Visible red**	**Visible Violet**
Frequency **ν (GHz)**	1.420	1.6	22	32	160	**4.3 10^5**	**5.5 10^5**
Wavelength **λ (cm)**	21	18	1.35	0.937	0.106	**700 nano m**	**400 nano m**
Naked Sun Gain (dB)	57.4	57.9	69.3	71.46	80.40	112.22	114.65

Off-Axis Gain for any Star

- Off axis, when the spacecraft is at a distance ρ from the focal axis, and when it is at a distance z from the star, *the gain* can be proved to be:

$$G_{star}(\lambda,\rho,z) = 4\,\pi^2\,\frac{r_g}{\lambda}\cdot J_0^2\left(\frac{2\,\pi\,\rho}{\lambda}\sqrt{\frac{2\,r_g}{z}}\right)$$

where $J_0(x)$ is the Bessel function of order zero and argument x

- The *total gain* for the combined (Star + receiving antenna) system is:

$$G_{Total}(\lambda) = G_{star}(\lambda)\cdot G_{antenna}(\lambda)$$

and so it increases with the CUBE of the frequency (next slide).

Total Gain of Star+Spacecraft

- If a spacecraft (S/C) has an antenna with radius $r_{antenna}$ and efficiency $k_{antenna}$, the spacecraft antenna gain is given by

$$G_{antenna}(\lambda) = 4\pi \frac{A_{physical} \cdot k_{antenna}}{\lambda^2} = \frac{4\pi^2 r_{antenna}^2 \cdot k_{antenna}}{\lambda^2}$$

and is proportional to the inverse of λ^2.

- The *total gain* for the combined (Star+S/C receiving antenna) system is proportional to the inverse of λ^3 and is:

$$G_{Total}(\lambda) = G_{star}(\lambda) \cdot G_{antenna}(\lambda) = \frac{32\pi^4 \, G \, M_{star} \cdot r_{antenna}^2 \cdot k_{antenna}}{\lambda^3}$$

Sun comes BEFORE interstellar!

- ***Sun's Focus Comes FIRST.***
- ***Interstellar Target Comes SECOND.***

1) The Sun's gravity focus is ***MUCH CLOSER*** than the target star, actually hundreds or thousand of times closer according to the target star (for α Cen it is 253 times closer than 1000 AU, where the "true" focus is found by taking the CORONA into account.
2) ***BEFORE*** any interstellar probe is launched towards a nearby star, we need a highly magnified radio-map of whatever lies around that star. And this can be achieved only by sending a probe to the opposite direction to let the Sun magnify!
3) It is much ***CHEAPER*** to reach 550 AU or 1000 AU than hundreds of AU, and it takes so much time less!

209

EXAMPLES: Three Targets

- We now provide THREE EXAMPLES of Targets for three different FOCAL missions:
- 1) The Galactic Black Hole (i.e. a FOCAL mission for Astrophysics and Cosmology).
- 2) The Alpha Centauri system of three stars, Alpha Cen A, B and C (Proxima) at 4.37 ly (basically a FOCAL mission to radio-explore the first target for any really interstellar mission).
- 3) Any Extrasolar Planet, for instance the Earth-size recently discovered Gliese 581 e.

NASA-JPL Study (Sept.8-11,'14)

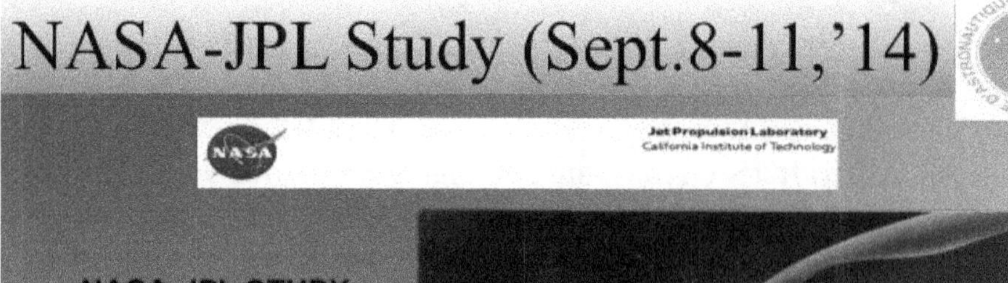

NASA-JPL STUDY of the first Probe to 1000 AU

- *Science and Enabling Technologies to Explore the Interstellar Medium, CALTECH, Pasadena,*
- *September 8-11, 2014*

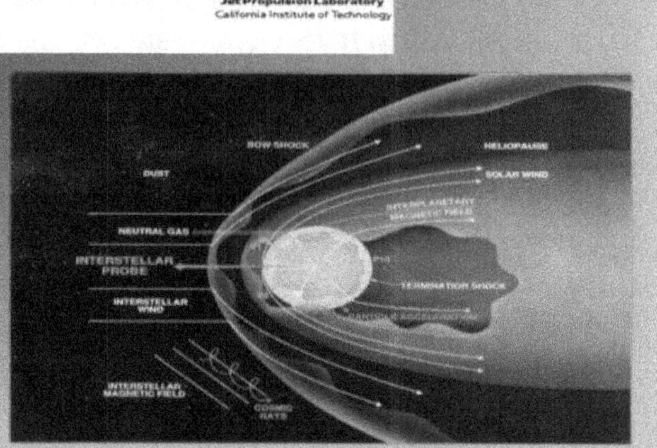

The Radio Link

- This is equation of the radio link between any two antennas, such a Deep Space Network antenna on Earth and the FOCAL s/c antenna:

$$P_r = \frac{P_t\, G_t\, G_r}{(4\pi)^2 \frac{d^2}{\lambda^2}} = \frac{P_t\, G_t\, G_r}{L(d,\lambda)}.$$

- This expresses the RECEIVED power P_r at a distance d as a function of the TRANSMITTED power P_t and of the ANTENNA GAIN G_t and G_r of both the transmitting and receiving antenna.
- The wavelength λ of the radio waves used for the transmission is appearing in the equation also.

The Bit Error Rate (BER)

- The Bit Error Rate (BER) is the number of ERRONEOUS bits received divided by the TOTAL number of bits transmitted, and is:

$$BER(d,\nu,P_t) = \frac{1}{2} \cdot erfc\left(\sqrt{\frac{E_b(d,\nu,P_t)}{N_0}}\right)$$

- Here $erfc(x)$ is the complementary error function used in Statistics.
- $N_0 = k$ *Noise_Temperature_of_Empty_Space.*
- And $E_b(d,\nu,P_t)$ is the received energy per bit, that is the ratio:

$$E_b(d,\nu,P_t) = \frac{P_r(d,\nu,P_t)}{Bit\_rate}$$

213

BER without & with the Sun's Gravitational Lens (1/3)

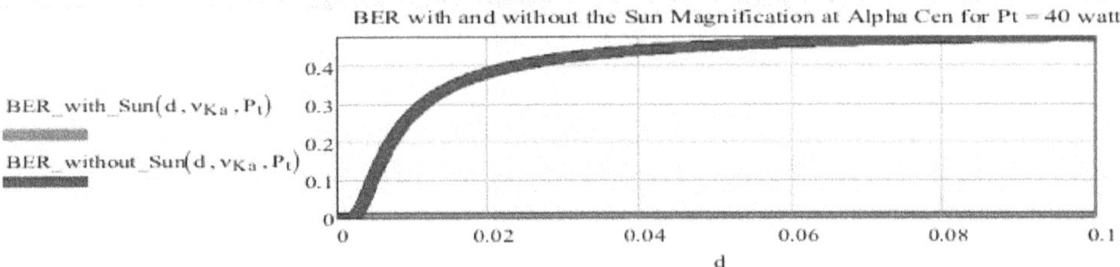

DISTANCE of interstellar probe from the Sun (light years)

- **Figure 4. The Bit Error Rate (BER) (upper, blue curve) tends immediately to the 50% value (BER = 0.5) even at moderate distances from the Sun (0 to 0.1 light years) for a 40 watt transmission from a DSN antenna that is a DIRECT transmission, i.e. without using the Sun's Magnifying Lens. On the contrary (lower red curve) the BER keeps staying at zero value (perfect communications!) if the FOCAL space mission is made, so**

BER without & with the Sun's Gravitational Lens (2/3)

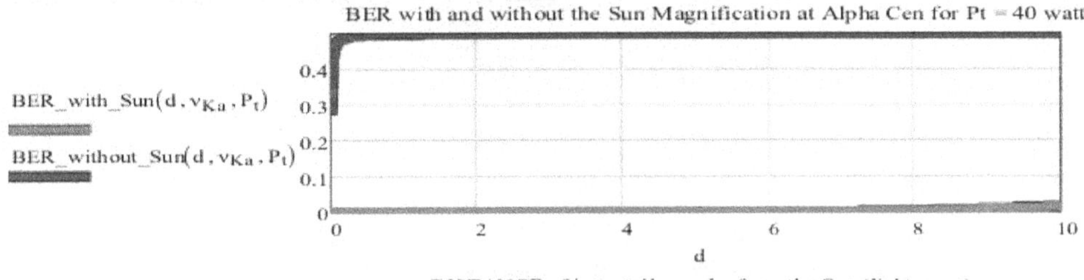

DISTANCE of interstellar probe from the Sun (light years)

Figure 5. Same as in Figure 4, but for probe distances up to 10 light years. We see that at about 9 light years away the BER curve starts being no exactly flat any more, and starts increasing slowly.

215

BER without & with the Sun's Gravitational Lens (3/3)

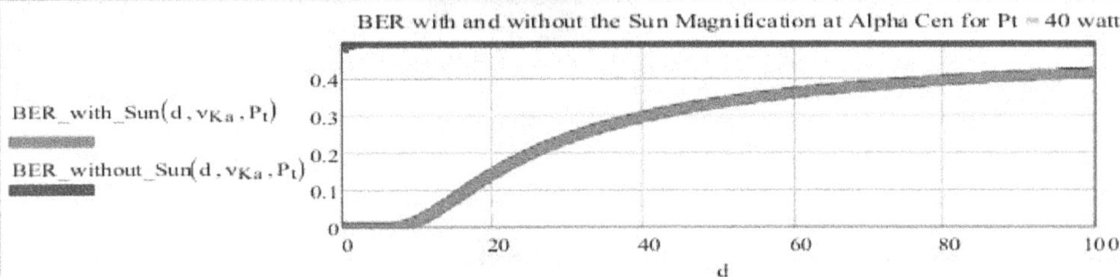

DISTANCE of interstellar probe from the Sun (light years)

Figure 6. Same as in Figure 5, but for probe distances up to 100 light years. We see that, from 9 light years onward, the Sun-BER increases, reaching the dangerous level of 40% (Sun-BER = 0.4) at about 100 light years. Namely, at 100 light years even the Sun's Lens cannot cope with this very low transmitted power of 40 watt.

216

Alpha Centauri A as a Gravitational Lens

- The following data apply to Alpha Centauri A:
  1. Mass of Alpha Centauri A: $M_{Alpha\_Cen\_A} = 1.100\ M_{Sun}$.
  2. Radius of Alpha Centauri A: $R_{Alpha\_Cen\_A} = 1.227\ R_{Sun}$.
  3. Schwarzschild radius of Alpha Centauri A: 3.248 km
  4. Minimal Focal Distance of naked Alpha Centauri A: 749 AU .
  5. What is the Effective Minimal Focal Distance ?
  6. Distance of Alpha Cen A from the Sun: 4.37 light year.
  7. The GAIN of Alpha Centauri A at 32 GHz (Ka band) is:

$$\mathbf{G_{Alpha\_Cen\_A}(\nu_{Ka}) = 13{,}689{,}321 = 71\ dB}$$

Sun - Alpha Centauri A RADIO BRIDGE

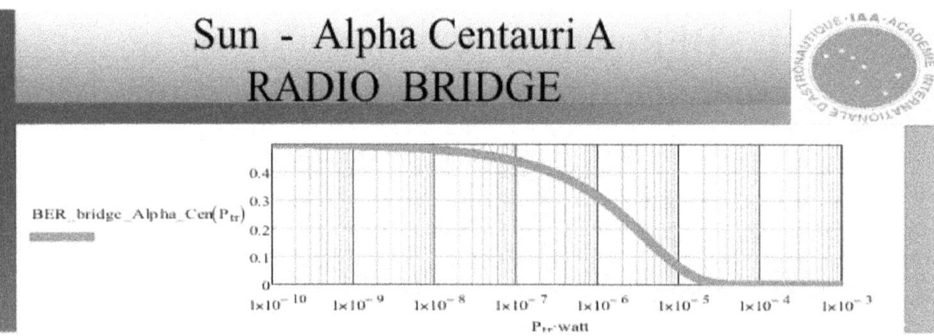

•Figure 7. Bit Error Rate (BER) for the double-gravitational-lens system giving the radio bridge between the Sun and Alpha Cen A. In other words, there are two gravitational lenses in the game here: the Sun one and the Alpha Cen A one, and two 12-meter FOCAL spacecrafts are supposed to have been put along the two-star axis on opposite sides at or beyond the minimal focal distances of 550 AU and 749 AU, respectively. This radio bridge has an OVERALL GAIN SO HIGH that a miserable 10^-4 watt transmitting power is sufficient to let the BER get down to zero, i.e. to have perfect telecommunications! Notice also that the scale of the horizontal axis is logarithmic, and the trace is yellowish since the light of Alpha Cen A is yellowish too.

Sun - Barnard's Star RADIO BRIDGE

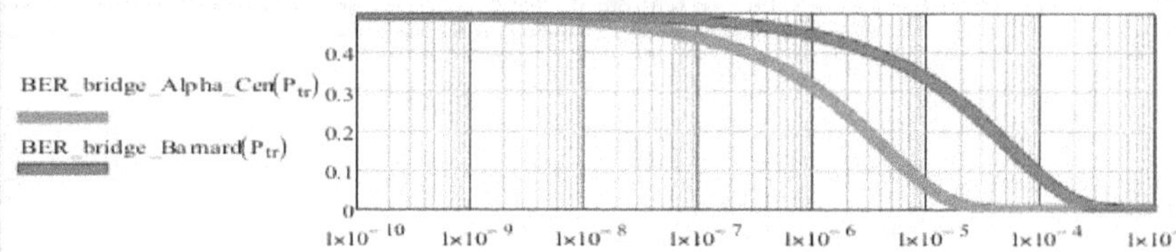

Figure 8. Bit Error Rate (BER) for the double-gravitational-lens of the radio bridge between the Sun and Alpha Cen A (yellowish curve) plus the same curve for the radio bridge between the Sun and Barnard's star (reddish curve, just as Barnard's star is a reddish star): for it, 10-3 watt are needed to keep the BER down to zero, because the gain of Barnard's star is so small when compared to the Alpha Centauri A's.

219

Sun - Sirius A
RADIO BRIDGE

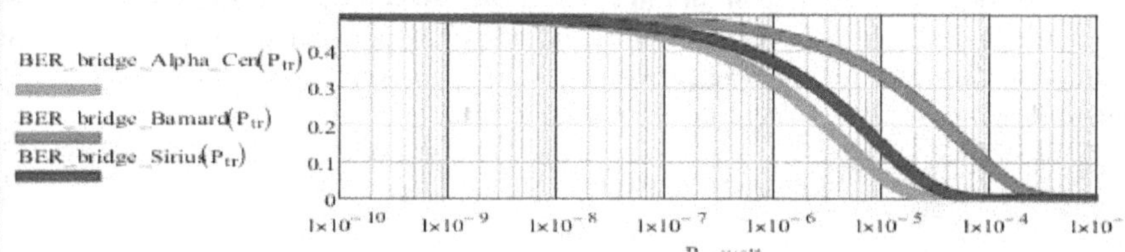

Figure 9. Bit Error Rate (BER) for the double-gravitational-lens of the radio bridge between the Sun and Alpha Cen A (yellowish curve) plus the same curve for the radio bridge between the Sun and Barnard's star (reddish curve, just as Barnard's star is a reddish star) plus the same curve of the radio bridge between the Sun and Sirius A (blue curve, just as Sirius A is a big blue star). From this blue curve we see that only 10-4 watt are needed to keep the BER down to zero, because the gain of Sirius A is so big when compared the gain of the Barnard's star that it "jumps closer to Alpha Cen A's gain" even if Sirius A is so much more further out and the Barnard's star! In other word, the star's gain and the size combined matter even more than its distance!

Sun - Sun-like Star at Bulge
RADIO BRIDGE

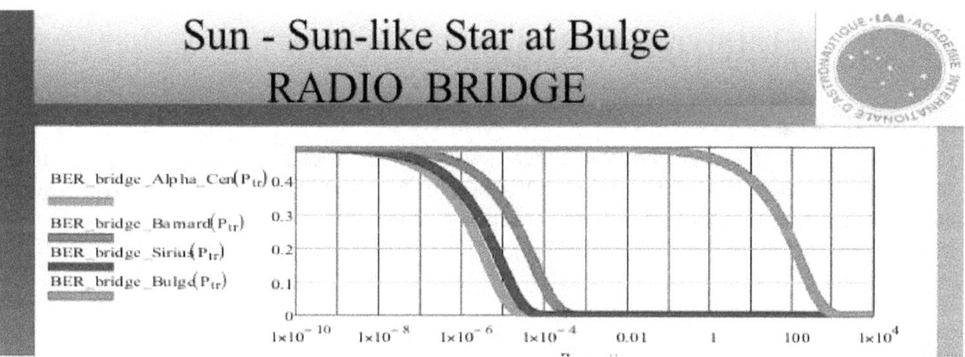

Figure 10. Bit Error Rate (BER) for the double-gravitational-lens of the radio bridge between the Sun and Alpha Cen A (orangish curve) plus the same curve for the radio bridge between the Sun and Barnard's star (reddish curve, just as Barnard's star is a reddish star) plus the same curve of the radio bridge between the Sun and Sirius A (blue curve, just as Sirius A is a big blue star). In addition, to the far right we now have the pink curve showing the BER for a radio bridge between the Sun and another Sun (identical in mass and size) located inside the Galactic Bulge at a distance of 26,000 light years. The radio bridge between these two Suns works and their two gravitational lenses works perfectly (i.e. BER = 0) if the transmitted power higher than about 1000 watt.

Sun - Sun-like in Andromeda
RADIO BRIDGE

Figure 11. The same four Bit Error Rate (BER) curves as shown in Figure 9 plus the new cyan curve appearing here on the far right: this is the BER curve of the radio bridge between the Sun and another Sun just the same but located somewhere in the Andromeda Galaxy M 31. Notice that this radio bridge would work fine (i.e. with BER = 0) if the transmitting power was at least 10^7 watt = 10 Megawatt. This is not as "crazy" at it might seem if one remembers that recently (June 2009) the discovery of the first extrasolar planet in the Andromeda Galaxy was announced, and the method used for the detection was just GRAVITATIONAL LENSING !

Interstellar Radio Bridges as INFORMATION CHANNELS

• **Information Theory by Claude Shannon (1948) enabled to find how much information (*C* in bit/sec) could possibly be sent across an information channel embedded in Additive Gaussian White Noise if the received power is *Pr* , the receiver has a noise power *Nr* and the electromagnetic waves are received over a bandwidth *W* :**

$$C = W \log_2\left(1 + \frac{P_r}{N_r}\right)$$

• **The receiver's noise is thermal and so its power is given the Johnson-Nyquist formula (where *kb* is Boltzmann's constant and *Tr* is the receiver's noise temperature:**

$$N_r = W k_b T_r$$

• **Upon replacing the second equation into the first one, we thus get the channel capacity as a function of the bandwidth (please notice that the bandwidth appears twice):**

$$C(W) = W \log_2\left(1 + \frac{P_r}{W k_b T_r}\right)$$

223

Interstellar Radio Bridges as INFORMATION CHANNELS

• **One might naively think that, for *W* tending to infinity, the channel capacity would tend to infinity also. But this is NOT the case because of the noise in which all transmissions are embedded. Actually, upon letting *W* tend to infinity, one finds:**

$$\lim_{W\to\infty} C = \lim_{W\to\infty}\left[W \log_2\left(1 + \frac{P_r}{W k_b T_r}\right)\right] = \infty \cdot 0 = \lim_{W\to\infty}\left[\frac{\log_2\left(1 + \frac{P_r}{W k_b T_r}\right)}{\frac{1}{W}}\right] = \lim_{W\to\infty}\left[\frac{-\frac{P_r}{W^2 k_b T_r}}{\ln(2)\left(1 + \frac{P_r}{W k_b T_r}\right)} \cdot \frac{1}{-\frac{1}{W^2}}\right] = \frac{1}{\ln(2)} \cdot \frac{P_r}{k_b T_r},$$

• **Thus, the HIGHEST POSSIBLE CHANNEL CAPACITY for any interstellar bridge, even upon using gravitational lensing, is given by**

$$C_{\max} = \frac{1}{\ln(2)} \cdot \frac{P_r}{k_b T_r}$$

224

Interstellar Radio Bridges as INFORMATION CHANNELS

• **Thus, the HIGHEST POSSIBLE CHANNEL CAPACITIES for the previous five interstellar bridge, are given by :**

Radio Bridge	Bandwidth = 1 Hz	Bandwidth = 1 kHz	Infinite Bandwidth
Sun – Alpha Cen A	37.6 bit/sec	27.6 kbit/sec	210 Gbit/sec
Sun – Barnard's Star	33.9 bit/sec	23.9 kbit/sec	16.3 Gbit/sec
Sun – Sirius A	36.5 bit/sec	26.5 kbit/sec	99.7 Gbit/sec
Sun – Sun at Bulge	12.3 bit/sec	2.4 kbit/sec	5.4 kbit/sec

225

CONCLUSION

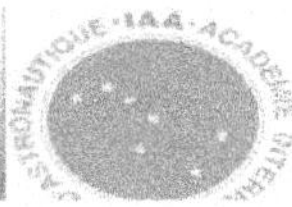

The conclusions reached by this author in this paper are thus:

- **A Galactic Internet constructed by advanced Aliens by exploiting the gravitational lenses of stars may already exist in the Galaxy.**
- **The CHANNEL CAPACITY for each radio bridge between any two stars communicating by virtue of their two gravitational lenses has the UPPER PHYSICAL LIMIT (in bits/sec) :**

$$C_{\max} = \frac{1}{\ln(2)} \cdot \frac{P_r}{k_b T_r}$$

226

Deep Space Industries

Rick N. Tumlinson

DSI: Deep Space Industries

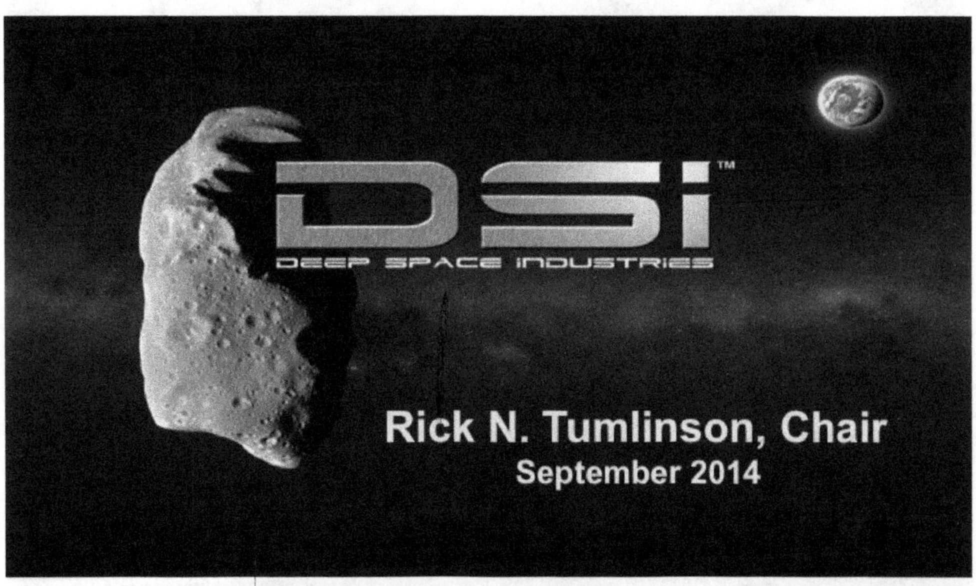

A Renaissance Team to Begin a New Renaissance

Rick Tumlinson
Chairman
MIR space station
Founding X Prize Trustee
Signed first citizen explorer

Dr. John Lewis
Chief Scientist
Global authority on asteroid resources and potential threats

David Gump
CEO
Co-founded t/Space and Astrobotic Technology
First TV ad on Intl. Space Station

Stephen Covey
Director of Research & Development
Inventor of MicroGravity Foundry and chair of asteroid track for 2013 International Space Development Conf.

Kirby Ikin
President
Satellite market analysis and risk management; co-founder of Orbital Recovery LLC, Chairman, National Space Society

Mark Sonter
Director of Mining and Processing
Scientific consultant in Australian mining and metallurgical industries; asteroid research at U. of Arizona

John Mankins
Chief Technical Officer
Directed $850M space technology program at NASA; global leader in Space Solar Power satellites

Dr. Marc Rayman
Senior Advisor
Chief engineer of two NASA asteroid missions, including first to use ion propulsion and artificial intelligence

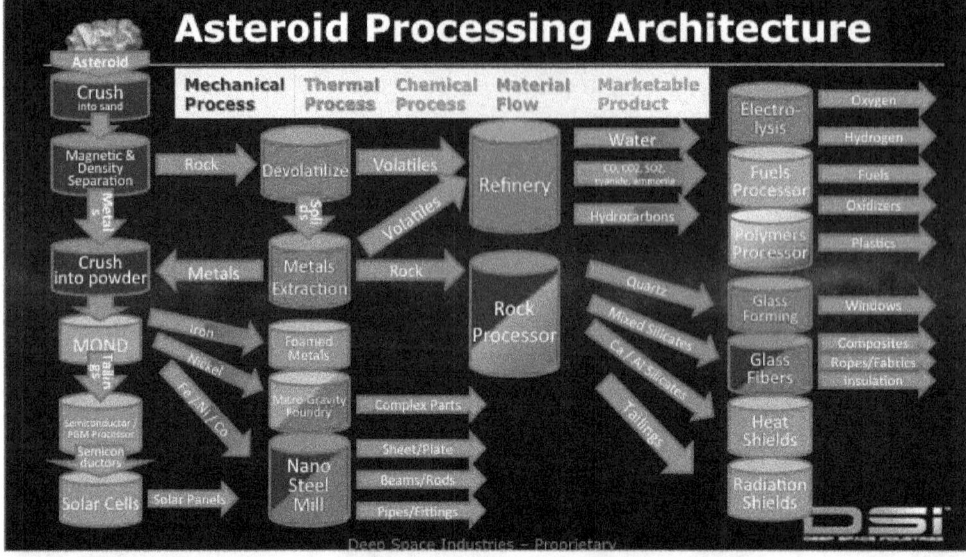

Stuff we aren't going to tell you here.

(X 10 slides)

If you are a potential investor, sponsor or customer let's talk after.

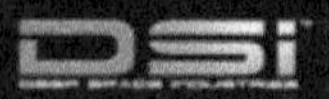

THE HIGH
FRONTIER
GERARD K.
O'NEILL
THE HIGH
FRONTIER
GERARD K.
O'NEILL

Why?

Why?

The Space Settlement *and* Development Act of 2015

A BILL

To require the National Aeronautics and Space Administration to investigate and promote human settlements *and economic development* in space, and for other purposes.

What is Settlement?

Settlement is what occurs when people move into a place that is beyond the borders of civilization and begin to live there and create a community – not just as visitors, but as residents who consider the new place to be their *HOME*.

What is a Settler?

You are Not a settler if you are on a shift, tour of duty, or are going home to another place when you are done doing what you are doing where you are.

You *are* a settler when the place you are is your *HOME.*

Not Settlers

Settlers

Settlers

Not a Settlement

This is a Settlement

AlphaTown - A new partnership
First we build a government facility on the Frontier...

AlphaTown
The job is done

AlphaTown Phase 1
The Government
Fort on the Frontier

Buy stuff.

Buy stuff from more than one "contractor"

Help us make better stuff.

"Let a thousand rockets fly!"
The sayings of Chairman Rick/Vol. 4

Give us a place to try out our stuff.
(Bigelow module on ISS)

Then come and rent some stuff from us!
(with any luck we might be overbooked...)

Alpha Town
(new commercial buildings on the orbital street)

Then let's do it again a bit further out...

And yes,
we build gas stations...

...and as we all get better at it...
the next community will be born....

The tipping point is now....
(GLXP, VG, PRI, DSI - others without acronyms...)

GOLDEN SPIKE
sometimes
we may get
ahead of you...

On the Moon
we get to practice twice

MoonTown One - a port and mixed government/private sector community on the pole of the Moon.

Home to scientists, entrepreneurs and is the control and

For the folks over on FarSide who are practicing for Mars.
Roughing it, removed from anyone, out of sight of Earth, yet like camping in the backyard before hitting the wilderness trail...close enough for mom and dad to come save the day if it gets spooky...

...and again, a new baby is born....

A bridge too far (for governments)

Some might want stop for gas/recon here...

Then go for the glory later...

Repeat the recipe...

Why humans?
(the left brain reasons)

Free Space/LEO
- Politics/Strategic
- Science
- Planetary Defence
- Economics ($ for Settlement)
  - Environment (micro-gravity/Solar energy)
  - Resources
  - Tourism
  - Energy

Moon
- Politics/Strategic
- Science
- Economics ($ for Settlement)
  - Resources
  - Tourism

There is No economic, direct strategic or planetary defence basis to support the settlement of Mars.

(that can't be done someplace else closer and cheaper...)

Everything else is just stuff we are making up because we are afraid we might sound stupid.

It is time to come out of the airlock and admit the Real Reasons.

We Go because it is there.
We Go because we want to.
We Go because it's inspiring.
We Go because we feel called to go.
We Go because it is what we as living creatures do.
We Go because it is why we exist.

This time we go together

This time we take it from no one

This time we give it everyone.

To carry the seeds of life to worlds now dead

To carry the light of life to worlds now dark

Why?

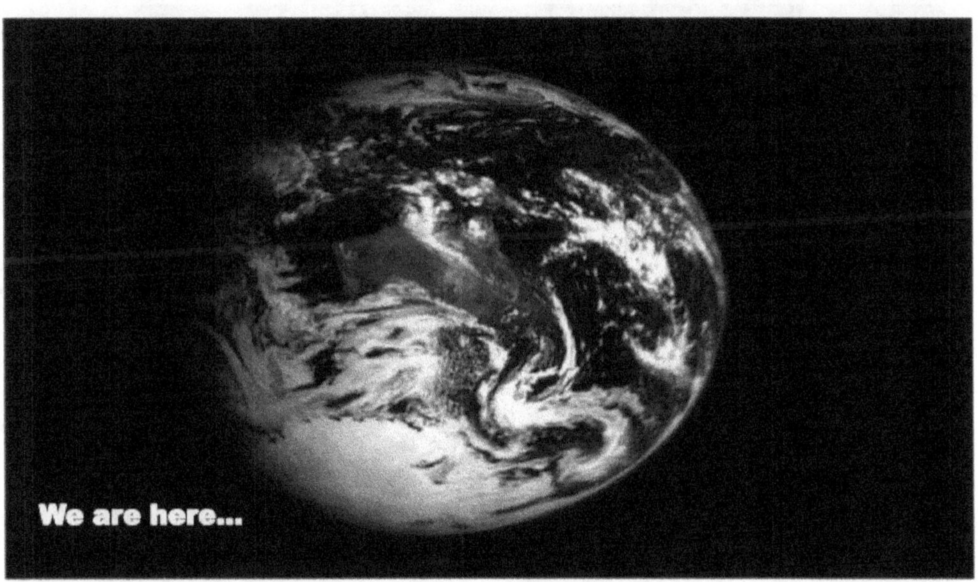
We are here...

To Go There!

SPACE
FRONTIER
FOUNDATION

NEW WORLDS
INSTITUTE

100 YEAR STARSHIP™

Facilitated by Jill Tarter, Ph.D., Bernard M. Oliver Chair for SETI Research, Director, Center for SETI Research, hear of the latest finds in astronomy and space technology with Mason Peck, Ph.D., Associate Professor, Cornell University, Department of Mechanical and Aerospace Engineering; Director, Cornell Space Systems Design Studio; and former NASA Chief Technology Officer.

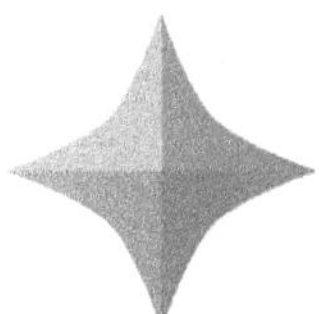

Summary:
State of the Universe

with Mason Peck, Ph. D.
moderated by Jill Tarter, Ph. D.

21st Century:
"The Century of
Biology"
On Earth and Beyond

Life Beyond Earth

- Discover it
  - In situ biomarkers
  - Remote biosignatures
- Look for it
  - Technosignatures
  - Serendipitous observations
- Export it
  - LEO, Moon, Mars, Asteroids,
  - 100 Year Starship Study

What's the State of Technology for:

- Biomarkers?
  - In situ life detectors
  - Sample return
- Biosignatures?
  - Transmission spectroscopy
  - Disequilibrium chemistry
  - Planetary characterization
- Technosignatures?

Is This a Shortcut?

Sample return:
Geysers of
Enceladus
& Europa

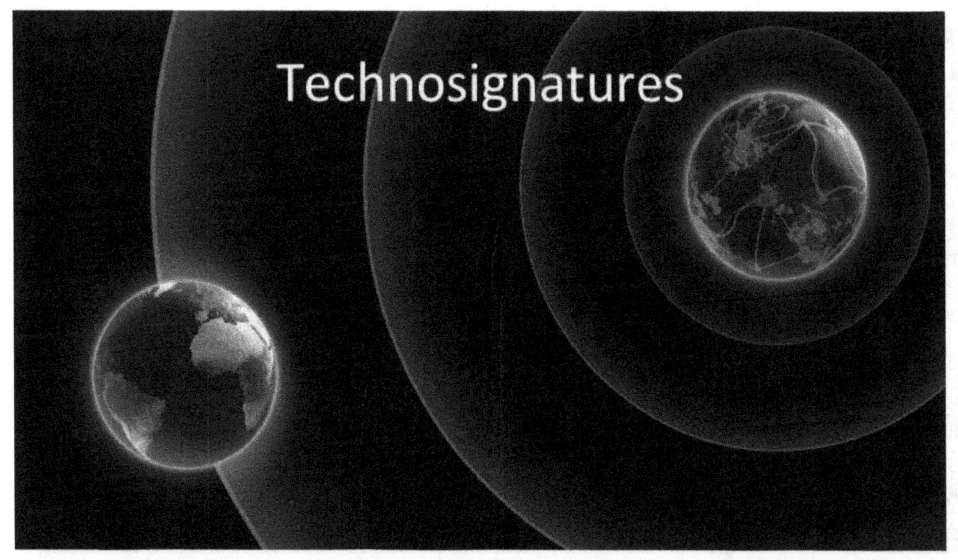
Technosignatures

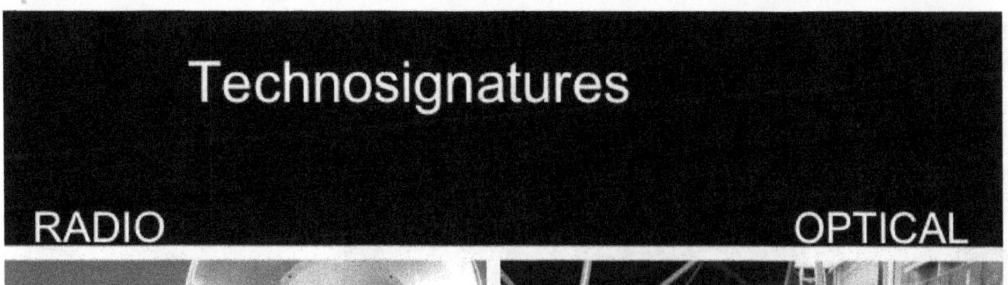
Technosignatures
RADIO
OPTICAL

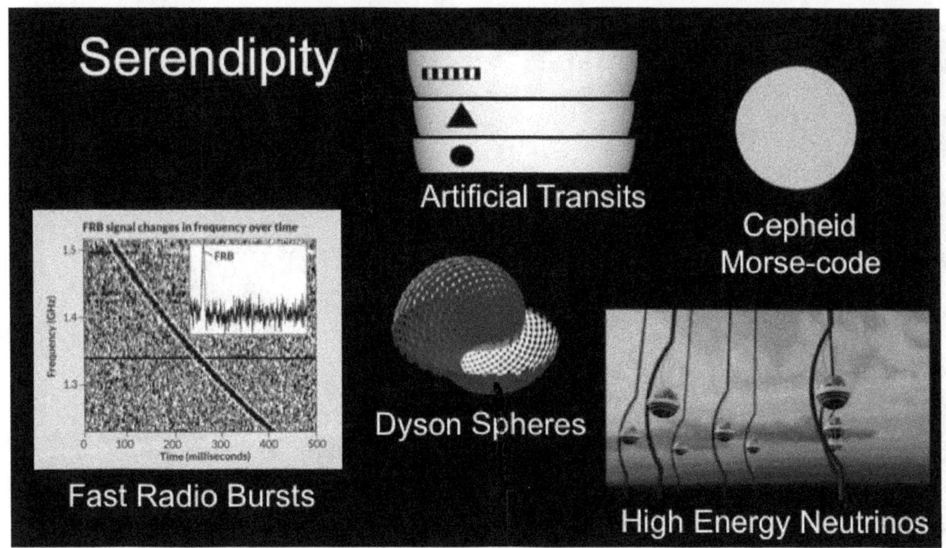
Serendipity
Artificial Transits
Cepheid
Morse-code
FRB signal changes in frequency over time
FRB
Frequency (GHz)
1.5
1.4
1.3
0
100
200
300
400
500
Time (milliseconds)
Fast Radio Bursts
Dyson Spheres
High Energy Neutrinos

What's the State of Technology for:

- Precursor missions?
  - Destinations
  - Mineral extraction
  - Settlements
  - Entrepreneurial engagement – driver or partner?
  - One way only?
- What do you imaging 100 years from now?

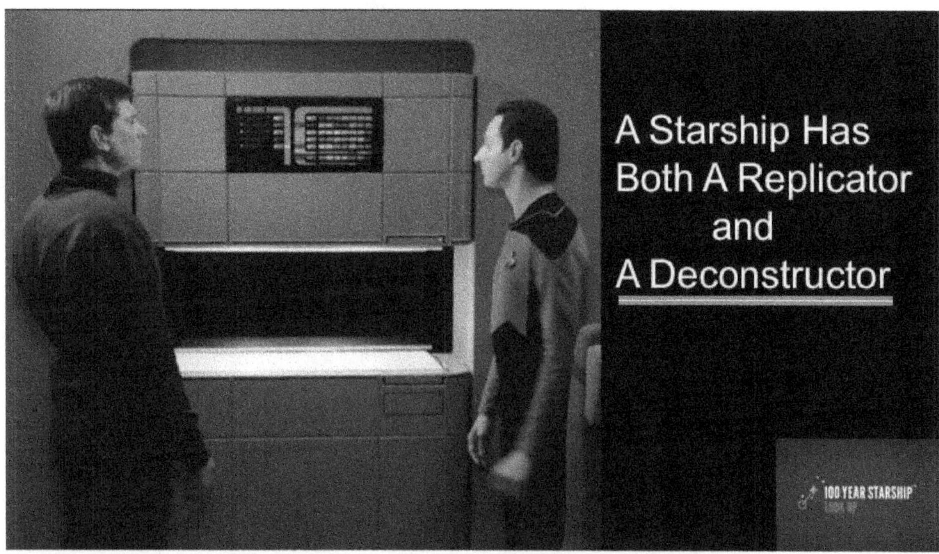
A Starship Has
Both A Replicator
and
A Deconstructor
100 YEAR STARSHIP

Fallacy Of The Replicator-Only
NYC
1894: Drowning in manure
(2.5 million pounds/day)
Today: Drowning in mechanical horses and CO_2

Fallacy Of The Replicator-Only
NYC
Tomorrow: Drowning
Treat Our Megacities Like Starships
Use Biomimicry To Engineer Their Defense

100 YEAR STARSHIP™

Facilitated by Hakeem Oluseyi, Ph.D., Astrophysicists and Associate Professor, Florida Institute of Technology, Department of Physics and Space Sciences, be part of a discussion of current issues or trends that may be instrumental in facilitating or stymying interstellar exploration with Kurt Zatloukas, M.D., Professor Medicine and Pathology, Graz University, Austria; Hear the latest on Outer Space Treaties and Policies in need of revamping with Amalie Sinclair, PhD, Director of the Leeward Space Foundation, and author; and think Steampunk - the new retro futurism in design with Mike Perschon, Ph.D., Professor of Literature, Grant MacEwan University, Alberta, Canada and Steampunk Scholar.

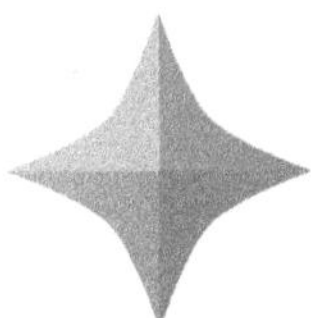

Space: A Healthcare Ecosystem

Dorit Donoviel, Ph.D.

Assistant Professor and Pharmacology, Center for Space Medicine, Baylor College of Medicine

Deputy Chief Scientist Industry Forum Lead, National Space Biomedical Research Institute (NSBRI)

donoviel@bcm.edu

The take-home message:

Learning how to administer healthcare in space will improve healthcare on Earth

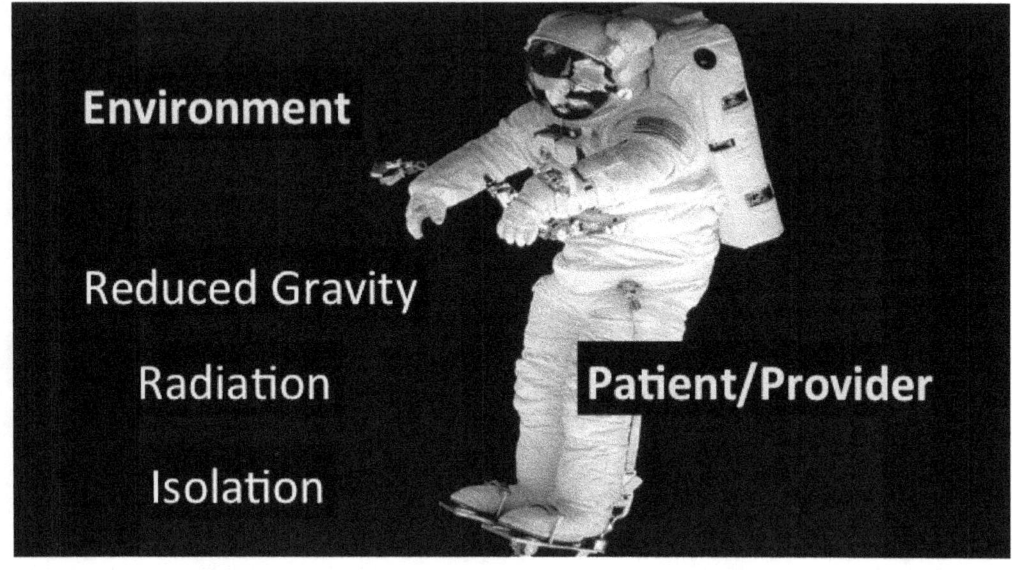

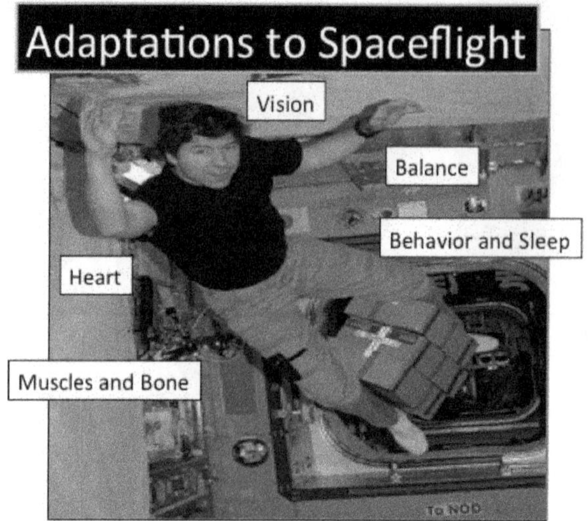

Space as a Healthcare Ecosystem

Routine Health Surveillance is not routine!

- Non-invasive real-time diagnostics
- Autonomous/semi-autonomous care
- Clinical-decision support
- Telementoring

Space as a Healthcare Ecosystem

- On-site clinical lab analysis
  blood, urine, saliva, stool
- Robust dependable technologies
- Simplified procedures

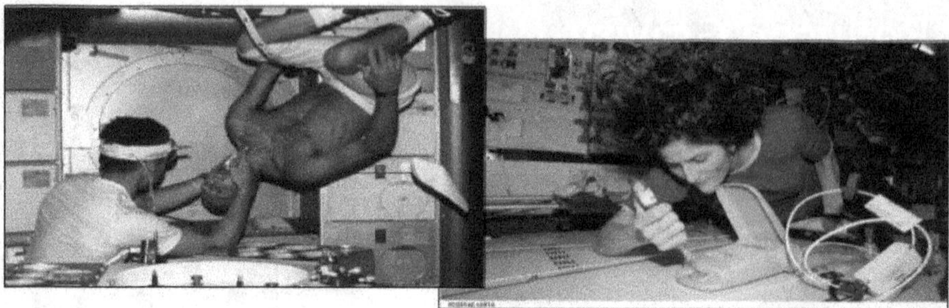

Space as Healthcare Ecosystem

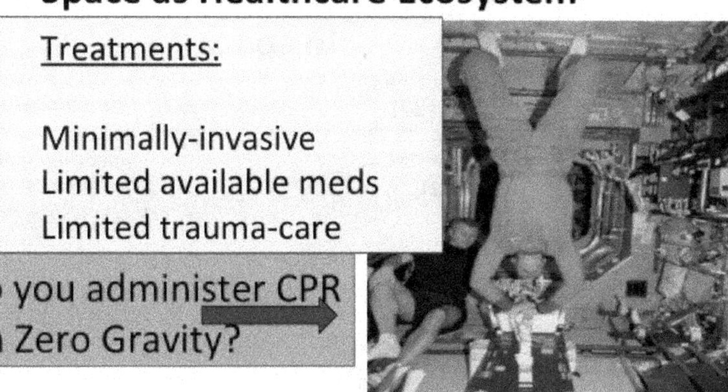

Medical Innovations for Space: Impact on Earth

Feature for Space	Benefit on Earth	↑Access	↓Cost
Small footprint (size, power, consumables)	Can be used in non-traditional settings	✓	✓
Robust and reliable	Require no infrastructure for maintenance	✓	✓
Simple to use	Low-cost providers	✓	✓
Minimally-invasive	Low-cost providers	✓	✓

Reduced Costs **Increased Access** **Improved Outcomes**

Learning how to administer healthcare in space will improve healthcare on Earth

Example 1: Bone Health

Astronauts Lose Bone Density In space

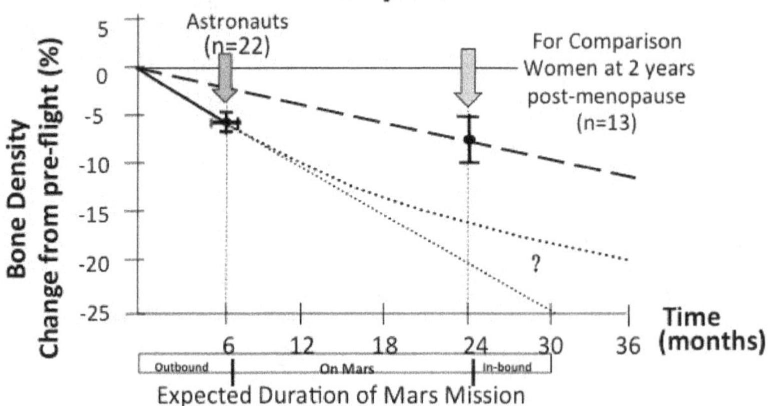

Current Clinical Practices for Bone Surveillance and Fracture Detection Are Inappropriate in Space

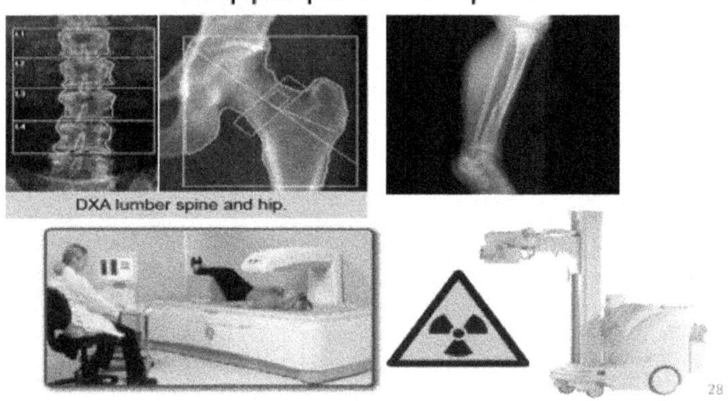

New Ultrasound Technology to Monitor and Heal Bones in Space

Scanning Confocal Acoustic Navigation (SCAN)

Diagnostic mode assesses bone health in real-time

- Determines bone density
- Charts bone structural properties
- Predicts bone strength
- Detects fractures

Therapeutic mode

- Targeted via ultrasound imaging
- Stimulates bone growth
- Improves fracture healing

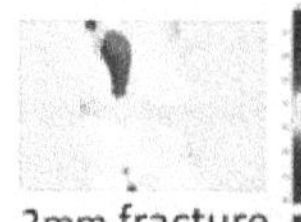

2mm fracture

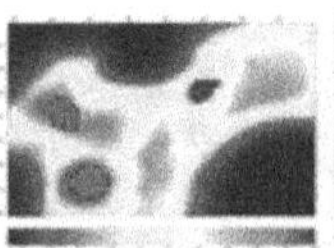

SCAN

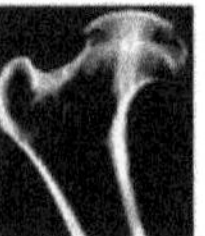

CT

Qin *et. al.*, Acta Astronaut. 2013; 92(1):79-88

SCAN Can Transform Bone Clinical Care on Earth

- Safe for longitudinal and frequent surveillance
- Safe for pregnant and pediatric patients
- Portable with small energy footprint
- Appropriate for resource-limited environments
- Inexpensive to buy and easy to administer
- Can prevent bone-loss & injury; heal fractures

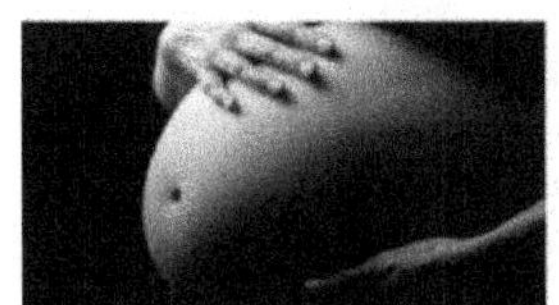

30

Learning how to administer healthcare in space will improve healthcare on Earth

Example 2: Kidney Stones

Astronauts are susceptible to kidney stones

- Increased bone breakdown leads to increased concentration of minerals in urine filtered by kidneys
- 14 astronauts developed kidney stones before or after space flight
- One symptomatic in-flight case in Russian cosmonaut
- Based on Lifetime Surveillance of Astronaut Health data, 15 to 20% of kidney stones are expected to require surgical intervention
- Symptoms are often debilitating: severe pain, dysuria, hematuria, nausea or vomiting
- Stones can lead to urine obstruction, acute renal failure, infection, and sepsis

Current Clinical Practices to Treat Kidney Stones Are Inappropriate for Space

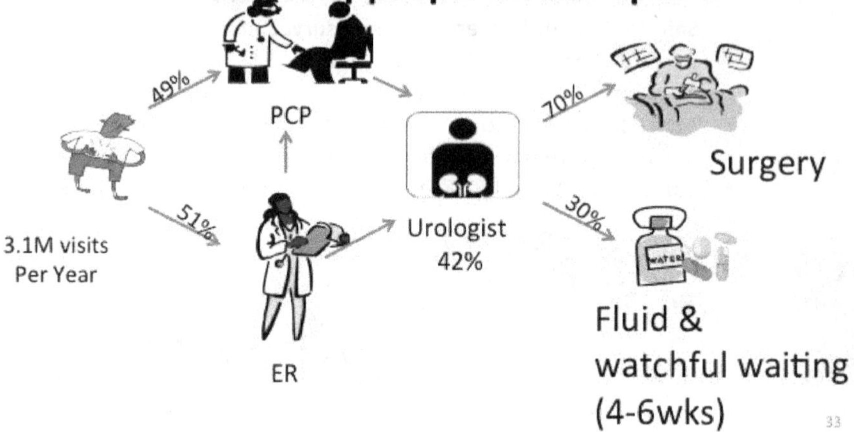

33

New Ultrasound Technology to Detect and Treat Kidney Stones

- Detects and expulses kidney stones using focused ultrasound energy at "diagnostic" levels
- "First-in-man" clinical trial of repositioning kidney stones ongoing

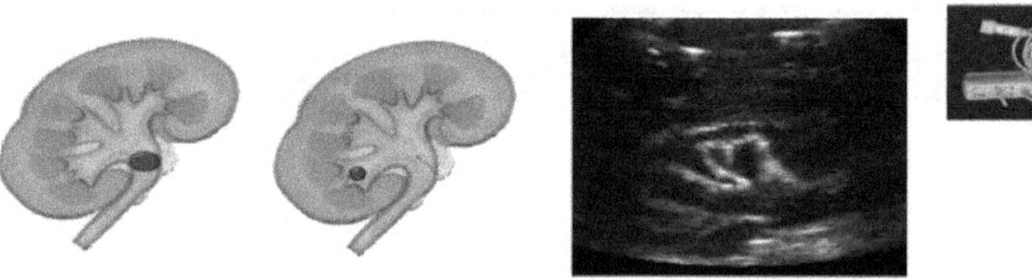

Harper JD *et al.*, J Urol. 2013 Sep;190(3):1090-5

34

Technology for Space will Change Clinical Practice for Kidney Stones on Earth

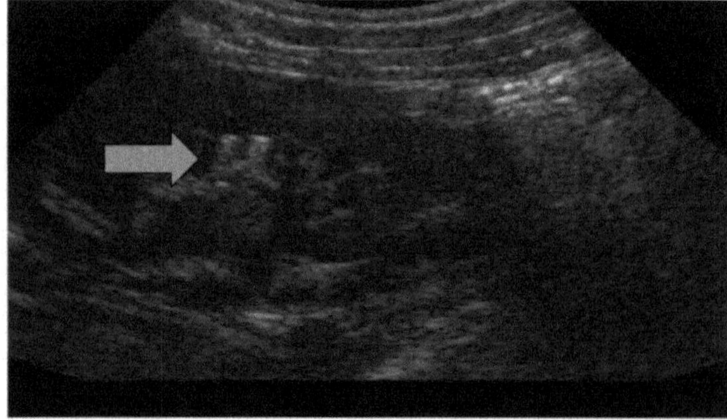

35

Learning how to administer healthcare in space will improve healthcare on Earth

Example 3: Diagnostic Imaging

Expanding And Simplifying Ultrasound Diagnostic Protocols For Space and Earth

- Computer-based just-in-time training modules for astronauts
- Musculoskeletal, CNS, cardiovascular & abdominal scans
- Integrated into Wayne State University School of Medicine curriculum and a course administered by the American College of Surgeons

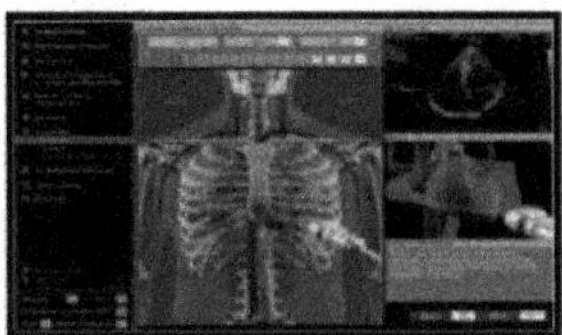

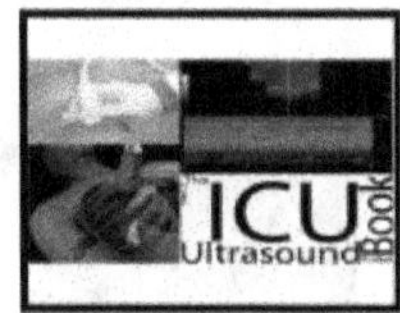

37

Learning how to administer healthcare in space will improve healthcare on Earth

Example 4: Brain Health

New Medical Syndrome in Astronauts is Driving Neurology Innovations

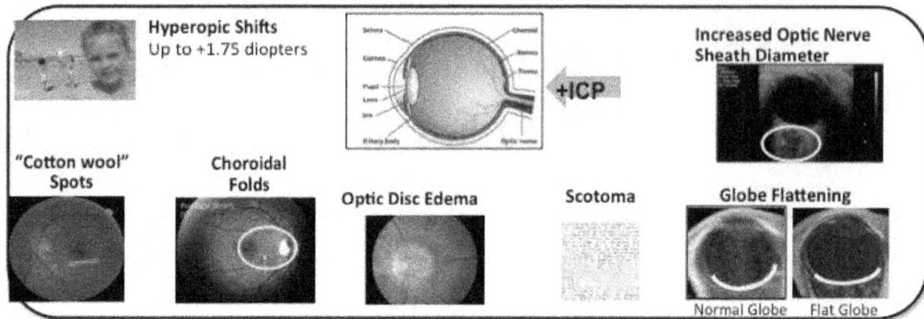

- Novel diagnostic technologies and imaging methods
- New paradigms and models for understanding brain fluid and pressure regulation
- New paradigms and models for understanding visual structures and function

Developing and Testing New Non-Invasive Brain and Eye Monitoring Technologies for Space and Earth

- 3-D ultrasound probe for the eye
- Brain fluid volume analyzer (blood, CSF, tissue)
- Brain oxygenation analyzer
- Intracranial pressure measurement devices
  - Two-depth transcranial Doppler
  - Cerebral cochlear fluid pressure device
  - Distortion product otoacoustic emissions

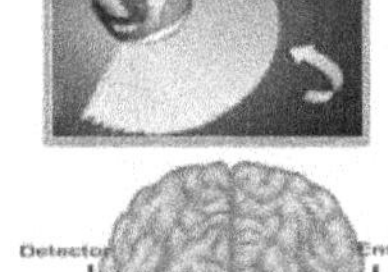

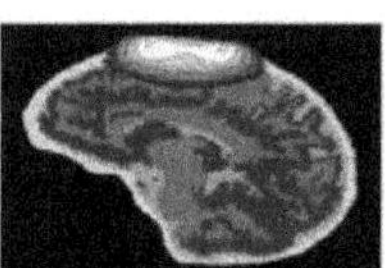

Simplifying Brain Health Monitoring for Earth

- Brain Fluid or pressure levels can increase due to stroke, brain injury (hemorrhage), infections, tumors, or other conditions
- Untreated, elevation in fluid or pressure can cause irreversible brain damage
- Early detection is key to good outcomes; conditions can deteriorate with time
- Current standard of care is MRI, CT scan, and lumbar puncture (LP) for measuring pressure on the brain
- LP has risks; CT and MRI are usually performed once due to costs; hence, many patients are not adequately monitored
- Non-invasive methods of monitoring brain health developed for astronauts will be cheaper and less risky to implement

41

All examples shown:

- Non-invasive
- Safe
- Easy to use
- Can be administered by lower-cost providers
- Low footprint (power, size)
- Robust

Reduced Costs ➡ **Increased Access** ➡ **Improved Outcomes**

We are looking for the next transformative health and performance technology

SMARTCAP: Space Medical And Related Technologies Commercialization Assistance Program (www.smartcap.org)

- Seed funding to small U.S.-based companies
- $100K - $250K awards
- Fulfills a need in space
- Impact on Earth
- Applications (1-2 pages) accepted year-round
- Awards made at least 3 times a year
- Requires a match – can be in-kind
- Investors are looking over our shoulder

Treaty Making for Global Exploration and the Long-term

A. H. Sinclair

Director, Leeward Space Foundation

Abstract

The dawn of the space age has opened up new and original long term vistas. As space development become the leading medium for international collaboration and the collaborative global exploration program quickly expands, significant legislative conditions and features will come to the fore. Many of our most important institutions will be affected by the rapid shift towards space based engagement.

The future development of technologies such as space based solar power, nuclear propulsion and laser based communications, will require additional recommendation through security based international agreements. This international dialog will also include definitions for comprehensive issues such as space debris, security interchange and the development of global informational assets against climate change.

Process towards an accommodating review and expansion of the 1967 Treaty on the Peaceful Uses of Outer Space can be obtained by the placement of a specialized unit at the US Library of Congress with view to the establishment of the 2017 UNISPACE IV forum representing the 50 year anniversary of the treaty since ratification. A working basis through treaty based outlooks will open up the collaborative global opportunities and support the fast track development of the star ship program.

The global exploration paradigm will enable the preparation of important and interrelated resources and assets including an asteroid mitigation platform, a new generations space station and lunar and Mars settlement. As the technologies and infrastructures employed for interstellar travel continue to advance, international legislation will keep pace resulting in further clarification of the 1967 treaty basis and the inclusive coverage of many additional aspects.

1. Discussion and Analysis

This paper will attempt to define some of the arenas which can be directly addressed through the establishment of a revisionary basis within the 1967 Treaty on the Peaceful Uses of Outer Space. Included is a brief discussion on the critical role of US leadership and engagement in such a dialog for both national and international space development enterprise going forward.

An international treaty is often viewed as being primarily a legislative instrument which comprises a focused objective for the enforcement of binding international standards and commitments. Although such a definition is generally appreciated as worthwhile, treaty based structures and resources can also be acknowledged as representing a fluid and readily available invitation for developmental process, interchange and the enhancement of international relations.

Space development over the past 50 years has generated a remarkable ability, while the exploration programs of space faring nations continue to expand, a pervasive global connectivity through the availability of space based communications and informational assets becomes the established and overriding milieu. We find ourselves facing a watershed of singular type, confronting complex problems in critical areas while effective problem solving contingencies are immediately sought and quickly developed. The pace of such a historic global transformation may be said to be exponential, even so it is apparent that despite underlying factors such as climate change and the depletion of natural resources, the will to achievement is well demonstrated within societies around the world.

Such a transitional phase in global affairs will rapidly lead towards a newer world and the burden for a successful passage may be readily assumed, yet we are sure to properly appreciate the impartial nature of scientific and technological inquiry and the continuing need for science to serve the highest ideals of humanity. At this juncture the valid assessment for the way forward will rely upon an expedient rationale towards the designation of optimal benefits. Recommendations within the 1967 treaty basis can provide both leading opportunity and functionality in many key areas, nationally and across the world, and will quickly lead to significant results; the laying out of an equitable and motivated space development formulation will not depend so much upon insularity as upon logic. The fragile nature of the global development paradigm is well appreciated, a fast track momentum towards sustainability and global productivity will depend upon the establishment of collaborative and responsive values and these can be demonstrated and upheld in many ways.

2. Justification

Our expectations are not limited ones, a catalytic program for global space exploration can certainly design and create a fully manned and long term lunar settlement within 50 years, along with the opening segues to Mars development and the placement of many key space based infrastructures, such as an asteroid mitigation shield. Such expectations are surely historic ones and they imply historic transformations at least equivalent to those which have been demonstrated since the onset of the space age. The inclusive dimensions of our expansion into the solar system and an original space based industrialization are both plausible and possible, and indicate the growing partnerships of agency, inter-governmental and commercial providers. Although the push for solar system expansion will in itself result in many newer types of technological asset and formative levels of scientific discovery and advancement, an underlying justification and motivation can be directly assumed in terms of the directly related problem solving effect within the leading concerns for planetary sustainability and global development.

The linked paradigm, for this world and the worlds beyond, clearly provides an adequate impetus for the incremental global space program which lies ahead. An exploration roadmap which properly fulfils the ambitious agendas over the next 50 or 100 years will need to be empowered and enabled through an equivalent global enterprise condition and capable global commitments. Investment and development on a continuous basis will become an essential, as space based infrastructures and incremental objectives increase, investment and participation for space based utilities will need to keep pace with the growing demands. As with any other investment the returns may be estimated according to the benefits obtained. The economics of space development and space industrialization present as a complexity of affairs, although productivities can be calculated within many related issues and aspects, it is also understood that these may not necessarily be similar to what have been more typical and previous evaluations. The drivers for space development comprise as diverse opportunities which are detailed, qualified and substantiated through multiple national and international governmental entities and forums. The primary considerations within topics for international space based alignment and partnership include technological transfer, security interchange, expansion of informational assets and national, international and global developmental utilizations. We can thus conclude that the rapid expansion of the unique space industry unlike other type of identified economic commodity will be formally positioned within consolidated policy driven commitments and objectives and will rely upon the availability and emergence of compatible assets and resources which will bring return of value through more intangible and interrelated attributes.

The returns from space are often seen as being fairly remote, although unilateral space exploration may obtain a measure of national prestige and a related technological and scientific proficiency, governmentally obtained deep space exploration is often seen as being a single ended investment. Money spent will realize results but these are not necessarily viewed within the terms of a specific or directly related immediate capital appreciation. In some sense space exploration is often perceived as a one way street, an affordable and forward looking expenditure for space faring nations but not necessarily comprising for a globally available market place.

US based commercial space development has brought forward a newer type of perception in these areas. Various propositions for the development of in-situ resources and further opportunities such as space based mining have described how commercially oriented space industrialization will provide a leading element within the global exploration continuum. Independent commercial providers are quickly achieving proficiencies in many fields including launch and habitat abilities. While directed enterprise within the commercial space industry moves ahead, associated development issues within commercial high tech communities are focusing for the provision of universal global communications and informational access through the placement of innovative space based utilities. Additional topics within many of these US based communities evolve around space based collation of Big data and ensuing applications through mapping and computational basis.

It is certainly true that innovative commercial propositions for launch, access and communications provide a stimulating and unique outlook. Data based platforms together with comprehensive earth observation techniques have ability to make a vast contribution to the global development outlooks. Such utilities together with computational abilities will support large scale governmental planning and the implementation of new phases of resource and infrastructure development, enabling productivities in critical areas such as for energy, agriculture and water. Social utilizations are also feasible, and can be negotiated for education, health care and other essential needs.

We should not forget the invasive circumstances; the radical demands of our time imply that problem solving potentials must be quickly taken up across the world. Rapid achievement for space development can be obtained through many lines of approach; a cohesive policy based condition will enable commercial enterprise for space, while supporting the structuring of comprehensive internationalized models for the formative ventures which lie ahead. The opening up of the international markets for space based development through commitment to the global space exploration venues will support participation and productivity across the board.

For a paradigm shift towards a global space exploration program, a paradigm shift in global space policy will lead the way. Space based enterprise must be fully empowered through international consensus and subscription, key to this purpose is the essential linkage between space exploration and planetary development, which will address the global communications and security mediums. The recommendation for well related informational assets and resources within extension of the 1967 treaty profiles will bring value based opportunities, and optimize problem solving outlooks at the widest scales. American leadership in this respect can not only ensure a creative outlook for technological dissemination it may also facilitate and enable a worthwhile and highly expedient short track towards equitable and balanced global productivity.

3. Security

Although the usage of space based utilities within the security basis continues to provide the pervasive background, the formal definition and consolidation of space based security attributes within international profiles remains as an indeterminate value. The genuine potential of space based assets for security, crisis and containment is apparent, as space based surveillance and communications structures may readily construe for worthwhile problem solving contingency even across national borders. The availability of an international basis for space based security topics poses for a radical and historic referendum, while formulations for bi-lateral agreements may engage for newer national alignments such initiatives will also take place within the international context.

The linkage of space based security into both national and international objectives and functions will remain as an underlying principle, which can be directly addressed and properly negotiated through the designations of an original treaty level prospectus. Such a type of policy oriented momentum can be readily achieved through the investigation of an inclusive methodology of means which moves beyond the a priori expectations and which will quickly obtain an inclusive and participatory global platform, bringing equitable and original opportunities in many key areas. A productive and fast track initiative towards the future world will rely upon the expedient usage of near earth space and informational resources in particular for essential process such as governmental and inter-governmental planning and developmental parameters. Although resources are limited, the scalable and exponential nature of the space based informational paradigm will support the establishment and dissemination of collaborative international and global security features and will define a durable working basis going forward.

Reform of this type can address many critical areas, including the evaluation of climate change parameters. Space based observation, along with the associated analytics can support comprehensive problem solving attributes, in particular for the placement and dissemination of well related informational assets and the dissemination of expertise. Addressing climate change through an internationalized space based model requires the creation of both a real time platform and a focused commitment. The consolidation and participation of the international community into climate change considerations will be greatly facilitated through the motivated conferral of an advanced UN technological basis within the treaty level perspectives. Not only does climate change continue to pose as dedicated criteria for sustainable economic development, a consolidated and globalized technological purview of this type will obtain an incremental capacity to address and negotiate unforeseen attributes such as any radical modification of deep ocean currents.

In facing towards the immediate technological watersheds, potentials can be described in many ways. Collaborative international models may typically provide and uphold expedient solutions when set against isolate national features. A substantial international commitment towards non-proliferation may also be considered as providing a realistic and appropriate venue. Modernization and the development of newer generations of security equipment imply the expectations for an ongoing redundancy within security features worldwide. Although the upgrading and modification of national security entities will be considered as being the necessary allowance, collaborative features and agreements within treaty basis will help avoid the dangers posed by national isolation and escalation scenarios. The arena of space clearly poses for both defensive and offensive capabilities, of particular significance remains the vulnerability of space based assets and the achievement of a durable cyberspace continuum. The 1967 treaty basis can readily address such critical topics while space based surveillance can provide an accurate and comprehensive space utility assurance. Additional features can extend to worthwhile details such an enhanced global nuclear monitor and the provision of a protected international cyberspace as a requisite global utility.

It is apparent that space based security is of paramount import for the international community, the historic choices between national isolation and international consensus within newer conditions are pressing, and US leadership may readily establish and fulfill the creative perspective for the way forward. The opening up of the 1967 treaty basis towards equitable extensions and further formulations implies the designation and placement of a number of highly productive international venues and resources, thereby providing engagement into international relations and obtaining substantial global development structures. Space based technological and informational resources can be located, contributed and utilized in unique ways, offering a remarkable potential for the terms ahead. Primarily these specialized space based security attributes will propose for climate change evaluation and problem solving, the non-proliferation basis and cyberspace assurance.

The engagement for technological development runs parallel to the purpose for global development. Scalable assets towards global productivity can be quickly obtained and distributed through a treaty basis, offering a comprehensive space based developmental profile. The advantages of such an approach are various, while the establishment of international participation, subscription and oversight recommends an essential responsibility towards the many critical issues of our time. The generations of space based communications and informational technologies which lie ahead will permit the description of an unprecedented medium of interaction and dissemination and a guided and definitive international policy which recognizes the identity and value of such key attributes is clearly required. Formulating as a unique social and cultural perspective and informed by earth observation, Big data, computational analysis and real time meta-mapping , the evolving horizons of the extemporary space age are tangible ones, the task ahead is to ensure that an adequate international resource to meet such occasion obtains the most inclusive and worthwhile expectations.

The rapidly changing profile of space based security may be quickly addressed through invitation towards the global dimensions. The leading issues describe pervasive and mutable prospects that can be formally defined and met through insight for an accommodating policy of means. The utilization of space based assets can provide benefit to populations in all places, supporting the growing objectives of a comprehensive space based development. Although underlying security requirements will encompass many areas of concern, including the protection of critical infrastructures, an adequate international consensus is readily available through the extension of inclusive treaty based guidelines and recommendations. The key features of such a revisionary basis will also include formative aspects such as an international platform for asteroid mitigation and the usage of advanced technologies in space including within the nuclear, laser and fusion basis.

4. Proposal

The modern culture of space represents both an advanced scientific endeavor and a universal and humanistic outlook. We will certainly find that exponential technologies provide catalytic yet impartial perspectives and newer dangers are also apparent. A world in which the informational basis becomes a primary focus is a realistic proposition, yet mass data will reflect not only the detailed physical attributes of the world around us but also our diversified social and cultural features. There is no doubt that a successful policy orientation around forthcoming phases can be achieved and the pathway to optimal benefit clearly obtained but such a purpose is best approached and structured within the most inclusive and compatible framework.

Space development prospects across the world are highly complex and will address a multiplicity of features and topics, as specialized communities and agencies along with educational and commercial entities continue to engage for the way forward. Taking into account both the multiplicity of aspects and the need for a consolidated focus, a democratic and readily available US agenda towards the 1967 treaty basis may be directly established.

This proposal for a revisionary basis suggests the establishment of a central Library of Congress specialized unit dedicated to the 1967 treaty parameters, thereby providing a working basis for on-going academic and agency collations and serving for the essential congressional inquiry and administrative permissions. Owing to the inclusive aspects of the space development paradigm, and the comprehensive nature of treaty language and literature, recommendations for policy topics and deliberations can also be initiated through the description of a number of additional clauses which will detail the various arenas into further elaboration and engagement. Such a working basis within the US is expected to be achieved in three participatory phases, request and formal LOC authorizations, establishment of working group structures and originating presentations, the solicitation and collation of academic and agency contributions and formulation for public education and outreach. Although a time line is variable, the proposal is targeted towards the preparation of a comprehensive US based portfolio with early presentation at UNISPACE IV in 2017, the 50th anniversary of the 1967 treaty ratification.

A treaty construct is the most sensitive and effective of policy instruments, with historic reach, the provisions of such a notable medium may be easily referenced and suitably obtained. The reform of the space development basis into international alignment and an equitable consensus and conferral can be considered as providing leading criteria with significant import for US national and economic objectives. A productive expansion of the 1967 basis will provide opportunity and engagement in particular for the undertakings of the global space exploration platform and the closely related global development agenda. In this respect US leadership will be the determining factor, an ensuing alignment of global space technologies in particular within the communications and security sectors is well recommended and can be obtain through treaty based and incremental designations according to ongoing evaluations and recommendations .

The comprehensive features of treaty based extensions will provide valuable references and structures across the board. These formulas will include a guided basis for international participation into next generation space utilities, such as advanced utility for space debris mitigation and removal together with the utilization of solar and nuclear energies in fields such as space based energy and communications.

Collaborative international engagement into large scale space ventures will find genuine opportunity fulfilling the growing expectations for permanent lunar settlement and the phases ahead. The creation of a protected international zone on the far side of the moon will ensure that pristine fields for radio astronomy remain intact for the uses of posterity. In less than 100 years as the interstellar probes set out on the furthest journey of discovery, we will look back again at the turning point of human history. At the onset of the 21st century our focus towards a global society has already reached the mid-line, set against the past 50 years of man's venture into space, the 50 years ahead offers us an era of unprecedented aspect. American leadership can certainly undertake and carefully ensure a remarkable transition across the age, providing a notable basis for the space faring histories of an ascendant humanity.

> "I am not an advocate for frequent changes in laws and constitutions, but laws and institutions must go hand in hand with the progress of the human mind. As that becomes more developed, more enlightened, as new discoveries are made, new truths discovered and manners and opinions change, with the change of circumstances, institutions must advance also to keep pace with the times. We might as well require a man to wear still the coat which fitted him when a boy as civilized society to remain ever under the regimen of their barbarous ancestors." —Thomas Jefferson

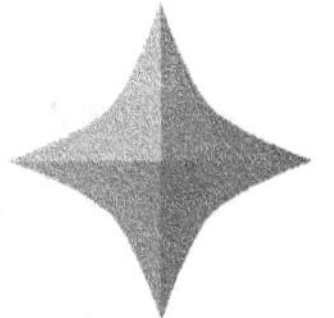

Pamela Reilly Contag, PhD

Overall Technical Track Chair, 2014 100YSS Public Symposium

CEO, Cygnet Biofuels

The 100YSS 2014 Public Symposium had an all star cast of speakers carrying forward the theme of interstellar travel. From energy sources to communications, health care to education and designing for life in space and the civilization we may create. The result was four days of morning plenaries, presentations, evening events that generated deep conversations about where, how and when will we enter space with the intent to colonize. At the same time, participants never forgot the charge to forge technology that will be used beneficially here on earth.

Two things were embedded in every conversation. First, our biggest challenge is the vastness of space. The emptiness that is not really empty but contains a medium that we have little skill in analyzing, dark matter and dark energy. Secondly, our venture into space requires the kind of energy source that is currently theoretical even hypothetical. To cover the vast distances or even hover in Earth's inner space requires a source of energy, whether antimatter, dark matter, quark nuggets, moonlight or starlight, that we are currently not able to harvest.

We have a technology gap in space and here on Earth that will require mass partnership, participation and dedication in order to close. The timeline we have set for ourselves is short and the gap is large and will require us to leap frog over the incremental and strive for the breakthrough science and technology. If we want to survive as a civilization in space, our breakthrough must also be in our civilization here on Earth.

I want to thank all of the speakers, participants and track chairs for making each year, a colossal event.

Overall Track Chair Biography

Pamela Reilly Contag, PhD

Assistant Professor, Biomedical Engineering Department, Rutgers University

Pamela R. Contag, Ph.D., founded and is Managing Partner of the Starting Line Group, LLC a virtual ecosystem for the commercialization of advanced technologies. She is also currently the CEO of Cygnet Inc. Cygnet develops technology platforms for the research and development of advanced materials, biologics and industrial enzymes.

Dr. Contag founded Xenogen Corporation in 1995 and took Xenogen public in 2004. She served as CEO, President and Founder at Xenogen from 1995 to 2006, and concurrently, the CEO of Xenogen Biosciences from 2000-2006 when Xenogen merged with CaliperLS. In 2000, Xenogen Corporation was listed as one of the "Top 25 Young Businesses" by Fortune Small Business and in both 2001 and 2003 received the R&D 100 award for achievements in Physics. In 2004, Xenogen was named in one of the top 100 fastest growing companies by the San Francisco Times and received the Frost and Sullivan Technology Innovations awards. Dr. Contag was named one of the "Top 25 Women in Small Business" by Fortune magazine. She was also awarded the Northstar Award from Springboard Enterprises.

In 2005, Dr. Contag founded Cobalt Technologies, Inc., a venture backed company that produces biobutanol from renewable feedstock. She was the Chairman and CEO of Cobalt Biofuels from 2005-2008. In 2008 Cobalt was named one of the top 20 Cleantech Companies and in 2009 one of the top 100 Cleantech Companies. In 2007, Dr. Contag co-founded ConcentRx, Inc. a biotechnology company developing a unique cancer therapy developed by three Researchers from Stanford University. Dr. Contag founded Cygnet BioFuels in 2009. Cygnet BioFuels, her second biofuels company, is a company focused on the utilization of novel organisms for feedstock and biofuel production. In 2011 Dr. Contag was awarded "Cleantech Innovator of the Year" award for Cygnet technology.

Dr. Contag has held board positions, public, private and not-for-profit sectors. Dr. Contag was a Director of Xenogen Corporation (Nasdaq) (1995-2005) and a Delcath (Nasdaq) Board Member (2008-2011). In the private sector she was CEO and Chairman of Cobalt Technologies (2005-2008), Cygnet Biofuels (2009-present), Director at ConcentRx (2007-present) She also joined in 2009 the DOE Biomass technology Advisory Committee and two nonprofit boards, Springboard Enterprises, an accelerator of women entrepreneurs and the Molecular Sciences Institute as executive chairman and in 2011 merged that entity into MSI/VTT, and remains a Director. Dr. Contag also consults in biotechnology for academics and industry, including consulting Professorship at Stanford School of Medicine in the Department of Pediatrics (1999-present), the Dean's Advisory Board of the Johns Hopkins Bloomberg School of Public Health (1999-2005). In 2010 Dr. Contag joined the Merrick Engineering Consultancy specializing in the energy field and in 2011 Dr. Contag was named to the Start-up America Foundation National Board.

With more than 25 years of microbiology research experience, Dr. Contag is widely published in the field of Microbiology and Optical imaging and has over 35 patents in Biotechnology. Dr. Contag received her Ph.D. in Microbiology at the University of Minnesota Medical School in 1989 studying Microbial Physiology and Genetics (for Alternative Fuels) and completed her Postdoctoral Training at Stanford University School of Medicine in 1993 specializing in "Host/Pathogen Interactions".

2014 Technical Tracks highlight expanded consideration of design, new directions in education, interstellar information technology and communication methods and adds a Poster Session. Special Sessions and Plenaries include new workshops and classes, topics of Plan B, Space Elevator, Space Treaties & Policy, Accelerating Creativity, international and commercial space, tours, member activities and collaborative, Transdisciplinary approaches to capability building and commitment.

"UNCHARTED" SPACE AND DESTINATIONS

Chaired by David Alexander, PhD

Understanding the interstellar medium and the composition of exosolar systems is vital as we contemplate travel to the stars. In addition, as our gaze is drawn many light years away, focusing on closer objectives as stepping stones to deep space will be essential. Beyond Mars, what missions should be designed to eventuate successful travel to another star? How should potential destinations be evaluated? What do we know and how do we learn more about space between the stars?

INTERSTELLAR EDUCATION

Chaired by Kathleen Toerpe, PhD

The journey beyond our solar system will overwhelm current educational practices. Commonly held beliefs and understandings of "learning" must and will be challenged. It is probable that humans have huge untapped capacities. Innovative learning tools and educational structures are needed for syntheses of ever-increasing information. The interstellar education platform will drive new knowledge of the universe and the development of the workforce that can create all that will be needed for interstellar travel. What are these new educational paradigms? What is education's role-formal and informal-in producing interstellar citizens?

LIFE SCIENCES IN SPACE EXPLORATION

Chaired by Yvonne D. Cagle, MD

As "Earth-evolved" humans, plants and other life forms travel deeper in space, we must understand much more about the fundamentals of life mechanisms. We must prepare for radical shifts in nutrition, potential therapeutics, growth and development, physiology and ethics. Concurrently, as we search for life beyond the earth we may need to re-evaluate our perspective of what is defined as "life". Also, how might we use the interstellar environment itself for life science research?

BECOMING AN INTERSTELLAR CIVILIZATION

Chaired by John C. McKnight, JD, PhD

Are humans driven to search beyond our knowledge base? How and in response to what do we create the belief systems that guide us? Interstellar travel is not just about the physical trip, but must include the journey civilizations take together. Who will we be and what will define our societies, morality, ethics, cultures, laws, economies, relationships and identities?

INTERSTELLAR INNOVATIONS ENHANCING LIFE ON EARTH

Chaired by Dan Hanson

Technology progresses in small increments and by leaps and bounds. Often the biggest steps forward are through the invention and innovation required to meet grand challenges. Interstellar travel represents such a challenge that may spur new economies, combat climate change, address heretofore incurable diseases. This session asks "What are these innovations and how can we deploy these to enhance life here on Earth?"

POSTER SESSION

Chaired by Timothy Meehan, PhD

Great ideas arise through unique individual observations, from people of all ages and educational backgrounds. The Poster Sessions are an opportunity to present snapshots of these early concepts and experiments. Poster sessions are a great forum to communicate any commercial opportunities in space or here on earth and seek like-minded collaborators or investors. Presentation in the poster format allows in-depth discussion in a small group setting. Topics are open.

PROPULSION AND ENERGY

Chaired by Eric W. Davis, PhD
and Hakeem Oluseyi, PhD

How fast and how far can we travel? Fundamental breakthroughs in propulsion and energy are required for interstellar travel to be feasible. To overcome the formidable time-distance barrier for travel between stars, robust leaps in theory and engineering for energy production, control and storage must occur, as well as the advancement and demonstration of propulsion techniques.

DATA, COMMUNICATIONS, AND INFORMATION TECHNOLOGY

Chaired by Ron Cole

Sending and receiving information by interstellar travelers or robotic vehicles requires development new methods to traverse the vast emptiness between stars. Additionally, in the absence of routine and timely communication with Earth, a probe or traveler must be self-sufficient in gathering, generating, compiling, storing, analyzing and retrieving data while ensuring these systems are operational over the lifetime of the mission and beyond.

DESIGNING FOR INTERSTELLAR

Chaired by Karl Aspeulund, PhD

Design for interstellar probes and crewed vehicles must address the unique characteristics and extreme environment of interstellar space. The equipment, structures, tools, materials, buildings, furniture, cleaning and maintenance processes, clothing-the accouterments of life and work-surround and create an environment. This environment protects, nourishes and facilitates daily activities. For most living things, their environment must fulfill many physical needs and for higher order creatures, physical, mental and emotional requirements need be met as well. Understanding, optimizing and manufacturing design to make these aspects of daily activities sustainable are critical for any hope of successful interstellar flight-with a living crew or robotic probes.

100 YEAR STARSHIP™

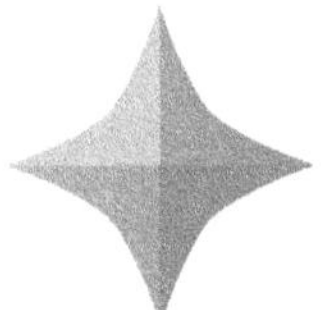

Chaired by Yvonne D. Cagle, MD

Astronaut, National Aeronautics and Space Administration (NASA)

Track Description

As "Earth-evolved" humans, plants and other life forms travel deeper in space, we must understand much more about the fundamentals of life mechanisms. We must prepare for radical shifts in nutrition, potential therapeutics, growth and development, physiology and ethics. Concurrently, as we search for life beyond the earth we may need to re-evaluate our perspective of what is defined as "life". Also, how might we use the interstellar environment itself for life science research?

Track Chair Biography

Yvonne D. Cagle, MD

Astronaut, National Aeronautics and Space Administration (NASA)

Dr. Yvonne Cagle is an astronaut for the National Aeronautics and Space Administration (NASA), a Colonel, U.S. Air Force (retired), a Family Physician, and consulting professor for Stanford University's department of cardiovascular medicine and its department of electrical engineering. Dr. Cagle is currently the Chief scientist for the Level II Program Office of NASA's Commercial Reuseable Suborbital Research program. Her groundbreaking work is preserving NASA legacy data while galvanizing NASA's lead in global mapping, sustainable energies, green initiatives and disaster preparedness.She was assigned as Stanford's lead astronaut science liaison and strategic relationships manager for Google and other Silicon Valley programmatic partnerships.

When Biology Meets Exobiology

David Almandsmith

Khotso Consulting, 3436 Boston Ave, Oakland, CA 94602

david.almandsmith@gmail.com

Carmen Nevarez, MD, MPH

Public Health Institute, crnevarez@mac.com

Abstract

An exoplanet must have a sufficient partial pressure of oxygen in its surface atmosphere to be habitable by humans and the likely source of that oxygen is a biology many hundreds of millions of years old. Such a biosphere would have a long history of predation and protection, even if the most complex organisms were the equivalent of single cells. Quite possibly, the exoplanet organisms would have no ready defenses against 'alien' Earth microbes, which could lead to the collapse of the entire exoplanet biosphere and quickly reduce its habitability. It could also be true that humans, their livestock, and crops would have no ready defenses against exoplanet microbes, a situation foreseen in H.G. Wells' "War of the Worlds." Strategies to manage these and other issues are offered.

Keywords

exobiology, exoplanet, biology, colonization, health

1. Introduction

This paper focuses on biological issues of exoplanet colonization although we also list technical developments that will be needed to manage those issues. We conclude by listing ethical issues of a 'human subjects' nature that may be related to starship travel and colonization.

By the term "exoplanet" we are actually referring to suitable extrasolar moons as well as to extrasolar primary planets.

2. A Brief Consideration of the Probable Variability of Exobiology Chemistries

The sky's the limit for potential exobiology chemistries—but we are only interested in those chemistries that would exist on an exoplanet where the conditions are suitable for humans. Therefore, we can exclude extremes in temperature, radiation, atmospheric constituents, etc. We are only considering biochemistries that are possible on an exoplanet with conditions similar to our home planet, Earth.

With this constraint, the scientific community has been moving toward a consensus that amino acids are likely to be the Lego blocks for building structures and enzymes in such an exobiology. Other organic Lego blocks might also be in use, as in our Earth biology, but because amino acids are found in space, in comets, and in meteors it is reasonable that a target exobiology will use them. Earth biology uses some 20 amino acids out of the hundred or so that can be made in the laboratory but there is little reason to expect an exobiology would use the exact same set.

After scientists discovered how DNA encodes data for subsequent generations, there have been attempts to build a better DNA; to find a polymer that will work better in some fashion. Biochemists have studied several that are capable of encoding data and replicating but those studied to date have been deficient in some respects. We can imagine that there were early versions of Earth biology that used other data storage polymers and data schemes but eventually were out-competed by DNA life. We suspect that we may find exobiologies in early stages of evolution where such a competition has not yet played out and more than a single data polymer may still be in use by variants of the exobiology.

There is also the question of whether an exobiology will have a technologically advanced civilization. Considering that it took Earth biology about three billion years to evolve the first multicellular organisms, it is reasonable to guess that many exobiologies are still unicellular. When we do find another technologically advanced civilization, we doubt they would jump at the chance to rent us good land for colonizing.

3. Lessons from Predation and Defense Strategies Found on Earth

An important aspect of biological evolution is the competition for limited resources. In our pre-biotic world, self-replicating sets of molecules arose from a dilute soup of naturally occurring organic molecules. Eventually, it became more advantageous to acquire organic molecules not from the dilute soup but instead from the concentrations of organic molecules found in other self-replicating sets. The requisitioning of those concentrated compounds was an early form of predation. Predation therefore pre-dates life itself.

As self-replicating molecules evolved into ever more complex self-replicating systems, the complexity of predation increased along with the multiplicity of defenses against predation. Even at the level of single-cell organisms, there is a great variety of toxins, structures, metabolic pathways, and behaviors for predation, parasitism, and defense. Apparently, the development of the nucleus—the distinguishing feature of eukaryotes—is likely due to the protection it affords against parasitic "jumping DNA." These parasitic strands of DNA can insert themselves into the genome when transcription occurs in the same compartment as the genome.

We may expect a unicellular exobiology to also have venoms and toxins and metabolisms that are invulnerable to those poisons. An example in Earth biology is penicillin; it is a class of toxins used by fungi to prevent predation by bacteria. However, the process of evolution has yielded strains of bacteria that are immune to penicillin.

With the development of multicellular biology less than a billion years ago, came tooth, claw, venoms, increasingly complex immunological systems, and the military-industrial complex.

In the multicellular world, many snakes are immune to their own venoms and some mammals are immune to venomous snakes that share their territories. For example, a gram of alkaloid toxin from poison dart frogs could kill 15,000 adult humans, yet a species of snake has evolved the ability to eat these frogs with impunity. In an exobiology we can expect similar developments of toxicity and acquired immunity.

4. The Problem of Metabolic Toxins

It is possible and perhaps probable that several normal constituents of an exobiology—common metabolites for instance—that are innocuous to native organisms, will prove highly toxic to humans. Such compounds could render the water deadly for humans to drink and the air lethal to breathe. An terrestrial example is botulinum toxin which is a metabolic protein that has no known defense purpose but for humans just happens to be the most acutely toxic substance known. Dispersed in air, a gram could kill 75,000 people.

A long-term, or chronic, concern could be the presence of organic molecules with hormonal activity that would interfere with human reproduction, fetal development, or crop success. Concentrations of such molecules could be on the order of parts per billion and yet prevent colonization by Homo sapiens sapiens.

5. Necessary Assays of an Exobiology

Even before a colonizing starship slows to the point it can orbit the target exoplanet, a biological probe could be sent to catalogue the organic and inorganic molecules found at the planet's surface. With those data, the crew can determine—for many of the chemicals—those that would be problematic in the concentrations in which they are found simply by consulting the ship's enormous databases. Some chemicals may need to be synthesized on board and tested with organisms designed to mimic the physiologies of humans, livestock, and crops.

An exobiology could be of the same chirality as Earth biology—where sugars are generally right-handed and amino acids are generally left-handed—or opposite or of a mixed chirality. Regardless, the crew must determine the effects of every chemical found there before colonization can begin. Even the possibility of prion activity should be investigated. Since potential effects on mammalian reproductive processes must be determined, these studies may take months.

6. Choosing between Terra-Forming the Exoplanet or Exo-Forming the Terrans

The choice is obvious. If there are biochemicals that are toxic at the levels found on the planet, one could try to change the physiological metabolisms of every organism—possibly numbering in the quintillions—that give rise to those chemicals or one could change the physiology of the small number of colonists.

Through evolution, the snake that eats poison dart frogs developed a metabolism that differs from its snake cousins. With sufficient inventiveness, the crew could experiment with the genomes of lab mice and other test systems until they create strains with metabolisms that are immune to exoplanet toxins. Similar research would also be necessary for each livestock species, crop species, and for every species of bacterium that make up a minimum microbiome of humans, livestock, and crops.

At this point, the crew could send down a biological laboratory to the surface with the modified mice, crops, and microbiomes to test the effectiveness of the genomic modifications and, if needed, to develop defenses to counter microbial infections. Prior research only dealt with toxic substances but this laboratory would be able to monitor for infections, synthesize vaccines and antimicrobials, and test their effectiveness.

7. Strategies to Protect an Exoplanet from Earthling 'Space Invaders'

It would be catastrophic if a species of Earth bacterium found the exoplanet's organisms to be nutritious and defenseless. Potentially, the entire planetary ecosystem could be laid waste in a few decades and lifeless shortly thereafter following the collapse of the bacterial population. There would be no choice but to attempt terra-forming the exoplanet with a full, diverse ecosystem—a process that minimally would require centuries.

In trying to protect the exoplanet's ecosystem, two approaches come to mind. In Jurassic Park, Dr. Hammond modified the dinosaurs' genomes such that they were dependent on receiving lysine in their dinosaur kibble. For colonizing an exoplanet, perhaps all of the gut bacteria needed by the mammals could be made to be dependent on some constituent that is only found in the mammalian environment. Another ploy would be to modify the bacteria such that eating an exoplanet organism would prove fatal. Ideally, both strategies should be employed. Furthermore, efforts must be made to render the genomic modifications robust, i.e. difficult to lose.

Similarly, the microbiomes of the crops will need to be modified such that they pose no threat to the exobiology. At present, much more needs to be known about the bacterial microbiomes of Earth crops and soils and to what extent they are required.

8. Strategies To Allow Earthlings To Thrive While Immersed In An Exobiological System

Following all of the research and testing outlined above—which could easily take longer than a year—it will become time to genomically modify human colonists such that they could thrive on the exoplanet. In spite of countless sci-fi stories to the contrary, this would be done before division of the fertilized zygote. The crew would alter the genome of thawed-out human zygotes to establish metabolisms that would permit them to live healthily on the surface amidst the exobiological system. Other genomic alterations might also be applied to account for a difference in gravity and an atmosphere that differs in constituents and pressure.

Nine months later there will be infant colonists and 18 to 20 years later there will be adult colonists prepared to leave the starship and colonize the exoplanet. More colonist children will continue to be born on the starship. Shortly after going planet-side, children will be born there as well.

If a 20-year wait to send down the first colonists seems like a long time, consider that the starship itself could very well have taken three or more centuries to reach the exoplanet from our solar system.

Also, consider that the colonists must be free from unmodified Earth bacteria since those could potentially destroy the planetary biosphere. Perhaps from the time human zygotes are modified, they will be separated from other crew members, gestated in artificial wombs, and raised in a pre-sterilized area of the starship with other young colonists and, of course, inoculated with modified bacteria to provide a healthy microbiome.

This could actually be a joyful upbringing with one or more warm, caring autonomous or telepresence androids as 'parents.' Of course, far more psychological research is needed on parenting modalities. Each individual Homo sapiens astroensis would grow up knowing they were exceedingly special and destined to embark upon the epic and formidable mission of colonizing a new world.

While they are on the starship, the young colonists' immune systems would be enhanced to protect against exobiological infections through the use of vaccinations or other techniques. If needed, facilities on the starship would synthesize any appropriate antimicrobials in preparation for colonization.

While waiting for the colonists to mature, robotic devices could descend to the surface, build structures, plant crops, and install solar and wind generators. After reaching the planet's surface, the colonists would always be in touch with the starship that holds the expertise and the culture of millennia to guide them.

Life for the unmodified humans—tasked with maintaining the starship systems and administering the colonization from above while trapped forever in a glorified tin can—may prove more difficult.

9. Implications for Future Research and Technological Developments

Given the current trajectory of biological research, the tools to accomplish the tasks implied above and listed here are likely to be ready in a century's time—barring natural or man-made catastrophes.

1. Improve remote qualitative and quantitative chemical analyses to detect and identify all environmental and exobiological compounds.
2. Improve laboratory test subjects such as mice, plants, and acellular systems to be accurate physiological surrogates for humans, livestock, and crops.
3. Remotely detect and study exobiological infections of animals and plants.
4. Remotely synthesize, administer, and evaluate remedies such as antimicrobials and immune system enhancers.
5. Customize mammalian, plant, and microbial genomes to allow colonization in what would otherwise be a poisonous environment.
6. Accomplish mammalian gestation in an artificial womb.

10. Ethical Issues

There is no obvious trajectory for ethical issues in the coming century. Given this following list of issues, however, it is clear that current concepts and beliefs about informed consent are out the window.

1. Sending young women from Earth on an interstellar voyage where they will be impregnated with thawed-out zygotes, give birtth, grow old, die, and be recycled.
   87. [This scenario assumes that starships will be equipped with maximally redundant systems at the lowest possible mass. This implies that males are not necessary prior to actual colonization.]
2. Implanting zygotes in crew members during the voyage to produce baby girls who will grow up, themselves become impregnated with thawed-out zygotes, give birth, grow old, die, and be recycled all without ever walking on a planet.
3. Genomically modifying human zygotes to create a human subspecies.
4. Developing human fetuses in an artificial womb.
5. Raising children by android(s).
6. Raising genomically 'normal' women who will spend their lives orbiting the exoplanet without the possibility of joining the surface colony.

11. Conclusion

The above biological, technological, and ethical issues are a minimum set and are only a perspective from the year 2014.

This paper assumes that Earth biology and the target exobiology are not the result of a panspermic seeding from the same source. Were that the case, attention to viruses would also need consideration.

We also chose not to consider exoplanet fauna since those issues have been thoroughly dealt with in the fictional literature.

To our knowledge, this paper puts more potential biological difficulties of exoplanet colonization on the table than any paper preceding it. Regardless, humans are a clever species and we believe that none of these difficulties will prove insurmountable.

References

1. Clara Moskowitz, C., "Life's Building Blocks May Have Formed in Dust Around Young Sun," *Space.com* 29 March 2012, http://www.space.com/15089-life-building-blocks-young-sun-dust.html. Retrieved 9 Aug 2014.

2. Wolman, Y., Haverland, W. J., and Miller, S. L., "Nonprotein Amino Acids from Spark Discharges and Their Comparison with the Murchison Meteorite Amino Acids," Proc. Nat. Acad. Sci. Vol. 69, No. 4, pp. 809-811, April 1972.

3. Doyle, A., "Alien Life Could Use Endless Array of Building Blocks," *Space.com* 28 January 2014, http://www.space.com/24460-alien-life-building-blocks-amino-acids.html . Retrieved 9 Aug 2014.

4. Kwok, R., "Chemical biology: DNA's new alphabet," *Nature* Vol. 491, 21 November 2012. http://www.nature.com/news/chemical-biology-dna-s-new-alphabet-1.11863 Retrieved 9 August 2014.

5. University of California Museum of Paleontology, "How Did Life Originate?" http://evolution.berkeley.edu/evosite/evo101/IIE2bDetailsoforigin.shtml Retrieved 9 August 2014.

6. Martin, W. and Koonin, E. V., "Introns and the origin of nucleus–cytosol compartmentalization," *Nature* Vol. 440, 41-45, 2 March 2006. http://www.nature.com/nature/journal/v440/n7080/pdf/nature04531.pdf Retrieved 9 August 2014.

7. Barchan, D., Kachalsky, S., Neumann D., Vogel, Z., Ovadia, M., Kochva, E., and Fuchs, S., "How the Mongoose Can Fight the Snake: The Binding Site of Mongoose Acetylcholine Receptor," Proc. Natl. Acad. Sci. Vol. 89, 7717-7721, August 1992. http://www.pnas.org/content/89/16/7717.long Retrieved 9 August 2014. Retrieved 9 August 2014.

8. Myers, C. W., Daly, J. W., and Malkin, B., "A Dangerously Toxic New Frog (Phyllobates) Used by The Emberá Indians of Western Colombia, with Discussion of Blowgun Fabrication and Dart Poisoning," Bulletin of the American Museum of Natural History Vol. 161, Art. 2, 307–365. http://digitallibrary.amnh.org/dspace/handle/2246/1286 Retrieved 9 August 2014.

9. Arnon, S. S.; Schechter, R., Inglesby, T. V., Henderson, D. A., Bartlett, J. G., Ascher, M. S., Eitzen, E., Fine, A.D., Hauer, J., Layton, M., Lillibridge, S., Osterholm, M. T., O'Toole, T., Parker, G., Perl, T. M., Russell, P. K., Swerdlow, D. L., Tonat, K., (Working Group on Civilian Biodefense), "Botulinum Toxin as a Biological Weapon: Medical and Public Health Management," Journal of the American Medical Association 285 (8): 1059–1070. February 21, 2001. http://jama.jamanetwork.com/article.aspx?articleid=193600 Retrieved 9 August 2014.

10. Coyne, J. A., "The Truth Is Way Out There." *New York Times,* 10 October 1999. http://www.nytimes.com/1999/10/10/books/the-truth-is-way-out-there.html Retrieved 9 August 2014.

11. Townsend, C. R., Begon, M., and Harper, J. L., "Essentials Of Ecology Third Edition," 2008, Malden, MA: Blackwell Publishing.

12. Lebeis, S. L., Rott, M., Dangl, J. L., and Schulze-Lefert, P., "28th New Phytologist Symposium: Functions and ecology of the plant microbiome, Rhodes, Greece, May 2012." *New Phytologist* Vol. 196, Issue 2, Article first published online 17 September 2012. http://onlinelibrary.wiley.com/doi/10.1111/j.1469-8137.2012.04336.x/pdf Retrieved 9 August 2014.

13. Zenaida Gómez. "Trillium Project." 2013. http://www.scribd.com/doc/142126834/Trillium-Project Retrieved 9 August 2014.

Teaching HERstory: A Vital Aspect of Interstellar Education

Adrienne Provenzano

P.O. Box 40604, Indianapolis, IN 46240

adrienneprovenzano@yahoo.com

Abstract

Teaching about women's involvement in space-related activities is an essential and important part of developing an interstellar civilization with interstellar citizens. Establishing a human presence - and eventually, civilization - beyond our solar system where humans collaborate and communicate effectively – regardless of gender – on beneficial projects, can be facilitated in part by the study of women's experiences and accomplishments. Looking at space studies and space exploration through the women's studies lens enables students to recognize and study women's contributions to human history as well as understand the socio-cultural contexts in which these achievements occurred. Such a paradigm also provides avenues for understanding how women have been denied opportunities and how men and women have worked for gender equity and equality. Knowing about the human experience, past and present, can help identify and strive for future goals both on and off planet Earth, within our solar system and beyond it.

Keywords

education, herstory, history, interstellar, space, women

1. Introduction

A woman's place in is Outer Space
Just see how far she'll go.
Take a trip on a rocket ship
With the Earth down far below.

Take a trip on a zooming ship
With the courage to explore.
'Cause a woman's place is in Outer Space
It's a place that she'll adore.

Going to a space station,
Walking on the Moon.
Onto Mars, just because
It's the thing to do.

Solar Systems, Galaxies,
The Universe is where she'll be.
'Cause a woman's place is in Outer Space
See how far she'll go.
Just see how far she'll go! [1]

Women have been looking up at the sky, studying the stars, and wondering about the cosmos for generations. In the 20th and 21st centuries, as astronomers and astronauts, secretaries, seamstresses, scientists and storytellers, educators and engineers, and in many other roles, women have been part of space studies and space exploration in various ways. For centuries, women have been pioneers and adventurers, travelers and explorers on Earth. As humans continue to venture into our solar system and beyond, it is essential to tell the stories of women - their accomplishments as well as their journeys, the challenges and obstacles they faced and how they dealt with them, and also discuss the socio-cultural systems that have at times limited women's options and at other times expanded their opportunities.

2. Herstory and Space

The term "herstory" was coined in 1970 by Robin Morgan in a book entitled *Sisterhood is Powerful* and later used in 1976 in the book *Words and Women* by Casey Miller and Kate Swift. [2] Women's experiences had been written about by historians prior to the 1970s, as well as in a variety of autobiographical works. Moreover, there is a substantial record of women's lives in many primary source materials such as journals and diaries, letters, songs and stories, and also in artifacts such as clothing, quilts, and paintings. Histories about women have been written in different times - but often forgotten from one generation to the next. In addition, in many historical writings, "the whole of human experience has been dominated by the political, economic, and military exploits of an elite, powerful group of men." [3]

The idea of researching, writing, and teaching about women through academic courses and programs in American colleges and universities developed in the late 20th century. The historian Gerda Lerner, whose books include *Why History Matters: Life and Thought and The Majority Finds its Past: Placing Women in History,* helped to develop this field of study. In some cases, women's studies programs have had their content expanded and have been renamed to be called "gender studies." Hundreds of women's studies and gender studies courses and programs now exist in many academic institutions around the world. Lerner has noted that the field of women's history "progressed from a focus on the historical narrative to one that concentrates on theory and interpretation." [4] She saw four phases of the development of the field, first from a *compensatory state* "in which notable women are identified" to a *contribution history* "which acknowledges women's roles in social development." [5] The third stage puts women's and men's experiences in comparison when interpreting different historical periods or movements, and the final phase focuses on seeking "an understanding of how and why gender dictates meaning and experience for individuals and groups." [6]

As regards women, space studies and space exploration, there is a growing collection of research materials that can be considered herstory. For example, biographies have been written of cosmonaut Valentina Tereshkova, who was the first woman in space; Sally Ride, the first American woman in space; and Sunita Williams, the second woman to command the International Space Station and the woman with the most hours logged for Extra Vehicular Activities (EVAs), otherwise known as spacewalks. There are books about many women who have traveled to space and also books about women astronomers, physicists, robotics engineers, and those in many other fields connected with space. There are also websites such as *Women@NASA* (women.nasa.gov) and www.womeninaerospace.org which show the historical and contemporary roles of women in the aviation and aerospace fields. In addition, websites about space missions involving women – such as the Grail mission led by Dr. Maria Zuber, a geophysicist and MIT professor; the Voyager missions currently led by Suzanne Dodd, and the Curiosity mission on Mars, which included many women on various teams during its development as well as during the rover's current travels on Mars – all can provide insights into women's roles in exploration programs. Social media also provides valuable information, such as the content of Pinterest and Twitter accounts of Dr. Karen Nyberg, International Space Station crew member on Expedition 36.

As far as putting such women's stories in historical context, there is also an increasing number of materials available on women such as Jerrie Cobb and the other First Lady Astronaut Trainees (FLATs), a group of American women who sought to participate in the space program in the 1960s by undergoing a series of tests. Later dubbed the "Mercury 13," their story sheds light on the socio-cultural views of women at that time. The story of these women – all accomplished pilots – has been told in several books, including *The Mercury 13* by Martha Ackmann, *Right Stuff, Wrong Sex* by Margaret A. Weitekamp, and *Almost Astronauts: 13 Women Who Dared to Dream* by Tanya Lee Stone and most recently featured in the TV series *MAKERS*, which spotlights women in a variety of fields. The *MAKERS* episode entitled "Women in Space" includes interviews with many female astronauts, including Colonel Eileen Collins, the first woman to pilot and later command a Space Shuttle, and Dr. Peggy Whitson, the first woman to command the International Space Station.

Both the television program and books give insights into the changes in social attitudes that affected opportunities for women at NASA. Another example of change is the opening of test pilot training to women in the U.S. military in the 1970s. Colonel Collins was able to participate in such training, and when she flew her first Space Shuttle mission – having learned about the FLATs – invited them to attend the launch. The story of these women helps to understand the changing roles of women in America – it informs us of both individual accomplishments and organizational policy shifts. Books like *Integrating Women into the Astronaut Corps: Politics and Logistics at NASA, 1972 - 2004* by Amy E. Foster and *Almost Heaven: The Story of Women in Space* by Bettyann Holtzmann Kevles also provide insight into how changes in culture and policy have played a role in expanding women's opportunities to travel into space and help understand how the most recent class of American astronauts is 50% female. In contrast, The Astronaut Wives Club by Lily Koppel provides a tale of "female friendships and American identity" through research and interviews with the wives of America's first astronauts – the men known as the "Mercury 7" – and gives voice to another aspect of women's experience in connection with the 20th century space race. [7]

To teach herstory also involves consideration of how race, class, region of the world, sexual orientation, and other factors intersect with gender and also an understanding that there is no such thing as a "homogenous womanhood." [8] Sociologist Linda L. Lindsey notes that a "balanced history makes visible women and other marginalized groups and, in turn, affirms their identity." [9] Taking a multicultural and multidisciplinary approach is often part of herstory as well. In the case of space travels, considering various socio-cultural factors can help illuminate why women from only a handful of countries have so far traveled to space and how interstellar travel might provide opportunities for a more diverse group of women to journey among the stars in the future.

In addition to putting stories in context and connecting the effect of gender on women's experiences with the impact of other factors, such as class, herstory can also provide insights into how men encourage women and collaborate with them on STEM (Science, Technology, Engineering, and Mathematics) endeavors. For example, Edward Charles Pickering, Director of the Harvard College Observatory for many years, recruited women to work there in the late 19th and early 20th centuries. The work of three of those women at the Harvard College Observatory – Annie Jump Cannon, Henrietta Swan Leavitt, and Cecilia Payne-Gaposchkin – is highlighted in Episode 8, "Sisters of the Sun," of the recently updated TV series *Cosmos*.

Another aspect of herstory is learning how women encourage and inspire one another. Consider, as one example, the role that *Star Trek* actress Nichelle Nichols played in working on behalf of NASA to recruit women and minorities to apply as astronauts. Sally Ride was just one astronaut recruited in this way. In contrast, Dr. Mae Jemison, another astronaut, was inspired by Nichols' fictional character on *Star Trek*, Lieutenant Uhura, to become an astronaut. So both through historical and contemporary biography, as well as through the experiences of fictional characters, can women's lives be a source of inspiration to other women.

3. A Vital Aspect of Interstellar Education

Why learn history? Why learn about those who have gone before? The author Maya Angelou once said, "How important it is for us to celebrate our heroes and she-roes." Teaching about women's herstory is a vital way to provide inspiration and fuel imaginations, to present options and possibilities, to connect the past, present, and future. Research the story of astronaut Cady Coleman, who served on Expedition 27 of the ISS as one of her space missions, and you will learn that she was inspired by meeting Sally Ride. As part of your research, you may also come across photos of Colonel Coleman with Valentina Tereshkova. Learn about Tracy Caldwell-Dyson and you're likely to learn how her father's efforts to teach and train her as an electrician helped her pursue her dreams as a chemist and astronaut and also participate in three successful spacewalks. Study the life and work of physicist and mathematician Katherine Johnson and you will learn how she calculated trajectories for Alan Shepard's 1961 flight into space, John Glenn's 1962 orbit around the Earth, and the 1969 Apollo 11 mission to the Moon. Read *Rocket Girl* by George D. Morgan, and you can learn how his mother, Mary Sherman Morgan, became a rocket scientist who invented the rocket fuel *hydyne* - used to launch Explorer 1 into space in 1958. Through books and documentaries, plays, photographs, artwork, blogs, websites and social media, women's stories can be accessed. More than fifty women have traveled into space, and many, many more have participated in space exploration or the study of space in some capacity, in a range of jobs from the bottom of the corporate or governmental hierarchy to the executive level, as well as at academic institutions and other organizations.

Learning about women involved in space study and space exploration can be a source of stories to carry beyond Earth - while also inspiring those on Earth. In 2013, the United Nations Office for Outer Space Affairs (UNOOSA) celebrated the 50th anniversary of the first woman in space with an exhibit and panel discussion. The panel included women space travelers from five countries – Roberta Bondar (Canada), Janet L. Kavandi (U.S.), Chiaki Mukai (Japan), Valentina Tereshkova (Russia), and Liu Yang (Russia), and was entitled "Women in Space: The Next 50 Years." The astronauts and cosmonauts included were mostly trained in STEM fields.

However, currently, in many countries including the United States, there is a significant gap between the number of women and men in many STEM fields, especially engineering, computer science, and physics. Recent reports by the American Association of University Women (AAUW) and the Girl Scout Research Institute include practical suggestions on how to address the current challenge of closing the gender gap in STEM education. In the 2010 AAUW report entitled *Why So Few? Women in Science, Technology, Engineering, and Mathematics,* it is explained that socio-cultural and environmental factors such as subtle biases and internalized stereotypes play a role in whether women pursue STEM studies and careers. A variety of suggestions are made in the report as to how to cultivate and support "achievements, interest, and persistence" [10] in STEM studies by girls at the pre-K - 12 grade levels, as well as in higher education and in the workforce, including the following: " Spread the word about girls' and women's achievements in math and science. Expose girls to successful female role models in math and science." [11] Similarly, the report *Generation STEM* from the Girl Scouts mentions the importance of role models on girls' career choices and the Girl Scout organization is involved in mentoring programs. [12] Likewise, the National Girls Collaborative Project, a clearinghouse and networking organization which facilitates collaboration among STEM programs for girls, has initiated the Million Women Mentors program to help boost the number of females in STEM fields. [13]

Humans over many ages have been looking at other solar systems and galaxies simply by looking into the night sky. We have been imagining space travels for many generations. Already, we have built Earth-based and space-based telescopes, which have expanded our awareness of what is beyond Earth. Already we have succeeded in sending a spacecraft, Voyager 1, outside of our solar system. Humans have flown in space and have set foot on the Moon. For over a decade, there has been continual human presence on the International Space Station. Plans are underway to develop more capabilities to travel within our solar system and beyond it. Whether as pioneers, refugees, immigrants, explorers, or adventurers, people have journeyed from their homes on Earth to other places on our planet – for temporary or permanent settlement – or for a brief visit. So, too, humans imagine traveling to many places in outer space – living on spaceships or space stations, or in planetary habitats. "Space: the final frontier. These are the voyages of the starship *Enterprise*, its continuing mission to explore strange new worlds, to seek out life and new civilizations, to boldly go where no one has gone before." These iconic words at the opening of the *Star Trek: The Next Generation* television series (modified slightly from the original *Star Trek* series to state "no one" instead of "no man") sets forth the hopes of humanity of what might be within and beyond our solar system to discover - and provides a model for humans for interstellar journeys.

Women have already been pioneers in many eras – such as those traveling as part of the westward expansion in the United States in the 19th century. These "matter-of-fact" women "endured prairie fires, locusts, droughts, disease, and the ever-present loneliness" and were able to adapt to and thrive on the frontier. [14] Other female adventurers and explorers include the 18th century French balloonist, Sophie Blanchard, and aviators Amelia Earhart and Bessie Coleman. Women have traveled to extreme environments including the ocean floor, the Arctic, and Antarctic. Curiosity and courage combine in such women, showing that they indeed have the "right stuff" for such endeavors. Contemporary women are also involved in ambitious frontier projects as entrepreneurs – such as Gwynne Shotwell's work as President and COO of SpaceX and Karin Nilsdotter's plans for Spaceport Sweden, a launch site she manages beneath the Aurora Borealis.

Humans are storytellers and stories help us imagine what we want to be – what we believe we are capable of achieving and what we once were – or nostalgically, think we were. Stories help us understand ourselves and one another. They help us dream and help our dreams become real. Whether sitting around a campfire or using digital media, we tell stories through words and music and sounds and images. What we tell about and how we tell it, defines us as individuals and as communities, as countries and even as Earthlings, preparing to leave our home planet for far distant worlds. Our stories seek to ask and answer profound questions, remind us of our connections to the past, give us new perspectives on our present and past, and also project us forward into the future. For example, in the film *Gravity*, a woman comes to terms with loss by fighting to survive under extreme conditions and returns safely to Earth. And the end of the film *Blade Runner*, a character comments on basic human questions – "Where do we come from? Where are we going? How much time do we have?"

It is important to include females in significant roles in fictional works - such as the labrador Sally in Stephen Huneck's children's book, *Sally's Great Balloon Adventure*, or the female space planes in the TV series *Space Racers*. According to the Geena Davis Institute on Gender in Media: "If She Can See It, She Can Be It." [15] Teaching about women's involvement in space-related activities through fact and fiction is an essential and important part of developing an interstellar civilization – where women and men are valued and respected, educated and encouraged, given opportunities to develop their potential through failure and success, and celebrated for their efforts and accomplishments – a civilization where there is gender equity and equality. To paraphrase the motto earlier stated: "If Humans Can See It, We Can Be It."

Herstory can be vital to education both on and off Earth. It can inspire those who might invent and develop the capabilities for interstellar travel. Then, once humans are living further away from Earth, they will be able to retain some connection with their home planet through a variety of methods of storytelling – including stories of women's lives and experience – and stories which include women's original, unique voices and expression.

4. The Women's Studies Approach

How can women's lives and accomplishments be taught? There are many ways and one effective approach involves using methods that are part of the academic area of women's studies. According to the National Women's Studies Association, the field of women's studies "has its roots in the student, civil rights, and women's movements of the 1960's and 70's. In its early years the field's teachers and scholars principally asked, 'Where are the women?'" [16]

Women's studies has expanded to also investigate "how categories of identity (e.g., race, class, gender, age, ability, etc.) and structures of inequality are mutually constituted and must continually be understood in relationship to one another" and considers such concepts as "identity, power, and privilege" in a transnational approach. [17] In part, it does so through feminist pedagogy, which, according to the Gender and Education Association, is "a way of thinking about teaching and learning, rather than a prescriptive method." [18] The three "key tenets" of such an approach are "resisting hierarchy," "using experience as a resource," and "transformative learning." [19] Resisting hierarchy enables students and teachers to value one another's contributions to the educational experience and collaborate in creating a positive educational environment. [20] Using experience as a resource values personal experience and helps students as well as teachers to change their thinking frameworks – and thus create a transformative educational experience. [21]

Valuing the voices and experiences of women is a core component of women's studies paradigms. Students are empowered to gather information on their own, rather than strictly from materials provided to them Gerda Lerner suggested that "we can certainly take pride in the achievements of notable women, but these kinds of histories do not describe the experience of the masses of women who still remain invisible to the historical record." [22] Women have been part of history all along – but their stories may or may not have been recorded – or such documents may have not been valued. Personal narratives, such as those in the Kennedy Space Center and Johnson Space Center Oral History archives, show how interviews can be an excellent source of gathering such autobiographical material. Likewise, the profile videos on the *Women@NASA* website collect women's stories in their own words and represent a diverse group of women in a range of occupations, such as attorney Pamela Bourque, systems engineer Victoria Garcia, and mechanical and aerospace engineer Tahani Amer.

Two groups of women who worked in the aviation and aerospace fields that would be particularly of interest in a women's studies context are the "sew sisters" and the women who worked at NACA (National Advisory Committee for Aeronautics). The "sew sisters" were women who worked on the Space Shuttle program as seamstresses, creating such items as gap fillers and thermal protection blankets. Lives depended on the quality of their work, and over the years of the program, forty different women sewed with machines and by hand. Some of their work is documented through photographs and newspaper articles, and one of the women, Jean Wright, recently spoke at the 40th annual International Quilt Festival in Houston, Texas.

NACA was the predecessor of NASA. It was created in 1915 and consisted of several centers around the United States. Some women worked at NACA before 1940, but it was during the 1940s that a number of women came to work at NACA, including a large group at Langley Research Center. Due to the segregation policies of that time, African-American women worked in a separate section. A number of the women worked as "human computers." To gather the history of these women, Margot Lee Shetterly and Duchess Harris are working together on "The Human Computers" project – an online museum to collect oral histories and related artifacts. [23] Shetterly is also writing *Hidden Figures*, a book about the African-American women who worked at NACA.

Feminist pedagogy seeks to value students' knowledge and create a supportive learning environment. [24] Applying such methods to studying women's activities connected with aerospace can be meaningful and effective. Distance learning, between humans on Earth and those in space, already provides valuable learning experiences which incorporate such collaborative approaches. Moreover, this educational approach can have applications as more humans travel beyond Earth – including multigenerational journeys.

5. Conclusion

Teaching about women's herstory as it relates to space topics is important for many reasons. It provides inspiration to current and future generations and helps people imagine future roles for women in space studies and travels. The stories of women in space exploration can inspire girls and boys, women and men. The possibility of interstellar human travel provides an exciting opportunity to tell stories of women adventurers past and present, both through fiction and non-fiction. What has already been accomplished by women is worth retelling, in part to better appreciate and understanding the process of achievement – to value and honor the struggles along the way to overcome obstacles and breaking barriers and to continue to strive in memory of those denied opportunities. Women's stories are human stories and part of the complex tapestry of human experience – and worth carrying throughout the universe. As humans travel among the stars, there will be more stories to tell – such as the tales of the first woman on Mars, the first woman to travel beyond our solar system, and the experiences of all the women and men who worked on the first interstellar mission. In the meanwhile, fictional works will continue to be crafted both based on real people and events, and also as purely imaginative expressions to inspire the development of interstellar starship capabilities to be used by women and men to journey among the stars. "The stars belong to everyone," said astronomer Helen Sawyer Hogg [25] and astronomer Carl Sagan noted that we are "made of starstuff." [26] By continuing to look up, wonder, and venture forth, humans will set off on bold adventures for many more generations – exploring and learning about the universe of which we are all a part.

References

1. Provenzano, A. (2014, September 19) "A Woman's Place Is In Outer Space." Presentation at 100 Year Starship Public Symposium. Houston, TX.

2. Herstory. (2014, November 1). Retrieved November 4, 2014, from Wikipedia: http://en.wikipedia.org/wiki/Herstory.

3. Lindsey, Linda L. (2005). *Gender Roles: A Sociological Perspective* (4th ed.). NJ: Prentice Hall, 96.

4. *Gerda Lerner* (1920 - 2013). Retrieved November 4, 2014 from http://www.nwhm.org/education-resources/biography/biographies/gerda-lerner.

5. Ibid.

6. Ibid.

7. Koppel, Lily. (2013). *The Astronaut Wives Club.* New York: Grand Central Publishing.

8. Lindsey, 97.

9. Ibid.

10. Hill, C., Corbett, C., & St. Rose, A. (2010). *Why So Few? Women in Science, Technology, Engineering, and Mathematics.* Washington, DC: American Association of University Women.

11. Ibid.

12. Modi, K., Schoenber, J. & Salmond, K. (2012). *Generation STEM: What Girls Say about Science, Technology, Engineering, and Math.* http://www.girlscouts.org/research/publications/stem/generation_stem_what_girls_say.asp

13. *Million Women Mentors.* Retrieved from www.millionwomenmentors.org.

14. Lindsey, 114.

15. Geena Davis Institute on Gender in Media, http://www.seejane.org/research/index.php.

16. What is Women's Studies? National Women's Studies Association. Retrieved November 7, 2014, http://www.nwsa.org/content.asp?pl=19&sl=21&contentid=21

17. Henderson, Emily F. "Feminist Pedagogy." (2013, January 15) *Gender and Education Association.* Retrieved from http://www.genderandeducation.com/resources/pedagogies/feminist-pedagogy/

18. Ibid.

19. Ibid.

20. Ibid.

21. Ibid.

22. Lindsey, 97.

23. Margot Lee Shetterly: Research. Write. Repeat. Retrieved November 6, 2014 from http://margotleeshetterly.com/the-human-computer-project/

24. Accardi, M.T., Drabinski, E., & Kumbier, A. (2013). *How Feminist Pedagogy Can Transform the Way You Teach and How Students Learn* (Slides). Retrieved from http://www.slideshare.net/alanakumbier/fem-peda-crl2013plusnotes.

25. Armstrong, M. (2008). *Women Astronomers: Reaching for the Stars.* Marcola, Oregon: Stone Pine Press, 77.

26. Sagan, C. (1980). *Cosmos.* NewYork: Random House.

The Concomitant Search for Habitable Exoplanets and Life In Space: Lessons from Evolution on Earth

Pamela Reilly Contag, PhD

Molecular Sciences Institute, Milpitas, CA 95035

prcontag@molsci.org

Abstract

The core rationale in seeking habitable exoplanets has been with the anticipation that the target planets, those suitable for colonization, will look like Earth in atmosphere, temperature, and chemical composition. With respect to colonization, the search for these Earth-like exoplanets are can yield promising destinations. Planets with oxygen, methane, nitrogen, carbon dioxide or even water in the atmosphere may or may not hold life, either visiting humans or endogenous forms. Such planets would still constitute a valuable discovery since sources of oxygen, water or methane on an exoplanet Weven one that is not habitable by Earth's standards and would allow us to create and utilize technology to renew any artificial atmosphere and provide energy to our life support systems. Therefore there are two approaches to colonization of these exoplanets: Wherein humans can be free living or can acquire materials to survive using synthetic systems.

Exoplanets are also of interest as part of our search for life in the Universe. The criteria for Exoplanet habitability can be, and perhaps should be, distinguished from searching for life forms. We search for life that is intelligent. This is the direct mission carried out by the SETI Institute under the assumption that intelligent life will have advanced technology and are also curious about their Universe. We search for Life that looks Earth-like. The habitable zone hypothesis of exoplanets suggests that there are biosignature gases and compounds that infer life (5), i.e. nitrogen, carbon dioxide, hydrogen and water. The presence of such biosignature gases (defined as earth-like) is not proof of a biological origin or the presence of life since gases can have both a biological and geochemical source and the appearance of such gases in disequilibrium may be due to either insensitive measurements or novel chemical reactions. (6) Conducting these searches using only Earth's chemical and molecular probes and biosignatures and only seeking Earth-like biology may cause us to miss planets with environs comprised of novel life forms.

For the most part we lack technology to search for non-earth-like mechanisms of life. We may be able to improve our search for new mechanisms of life by studying Earths evolution and the way that organized life arose and evolved over time. Scientists may vary in their belief on the beginnings of the Universe, but we can agree that stars, our own Sun for example, provide the major energy source for life throughout evolution of life on earth. We can thus infer that early in the formation of the Universe stars provided energy as our own galaxy was formed and are stars are most likely a potent energy source in other galaxies. Assuming that we agree on the definition of life, we could use the knowledge of how life developed on earth to propose new hypotheses for how the sun and other energy sources can be directly or indirectly converted to chemical or other types of energy needed to support living entities.

1. Introduction: Looking For A Home Away From Home

Exoplanet detection and characterization is in a nascent stage with regard to efficiency, sensitivity and precision, and at present most data has been acquired by telescopic transit and velocity measurements and infrared spectrometry. Exoplanet detection via more direct measurements with coronagraph or alternatively "starshade" technology is contemplated for 2015-2020 mission development. (Roberge, JPLAG).

The current state of Exoplanet assessment of fitness for human life is also at a rudimentary stage. Our ability to study planetary climates now extends beyond our own solar system to those of nearby stars. Through various technologies we can perform some direct and indirect measurements of mass, density and velocity and deduce temperature, water and atmospheric molecules all which can help approximate climate. Required information, such as planetary and stellar motions, can also be inferred from photometry, and atmospheric spectra (Nick Cowen; Lisa Kalteneggar, JPL SAG). However, the assessment of factors sufficient to support human colonization or the inference of life will require a much more detailed understanding at levels that are not currently possible.

It is clear that we need to make radical leaps in our ability to assess exoplanet conditions. Such leaps require new technologies beyond those currently available or contemplated for development in the near-term. Furthermore, converting a less than optimal exoplanet environment or creating an artificial environment for such conditions will require a high level of prior knowledge to ensure that we have the proper reagents and tools on our journey. We will need to go beyond simply predicting probable weather conditions, and detecting the presence of specific atmospheric gases and liquid water. To develop sustainable ways of existing on such planets will require the means to remediate or reconstruct the planet's atmosphere and water to match that of Earth.

2. We Search ExoPlanets For Endogenous Life In The Same Way We Search For Conditions To Sustain Human Life

We are conducting searches for both exoplanets and life in space with a natural bias towards chemical mechanisms used by life on Earth. The use of only Earth's chemical and molecular probes and biosignatures and seeking Earth-like biology may cause us to miss planets with environs comprised of novel life forms.

Within the context of the Universe, as Earth formed and life evolved on Earth, it is anticipated that this also occurred on planets outside our solar system and beyond the confines of our galaxy. In studying evolution on Earth, we have seen that physical separation of living things that were once similar can lead to divergent evolution toward distinct features, and possibly distinct biochemistries (think Darwin's finches).

An understanding of energy and matter helps us to form an image of how vast the universe is and informs us as to the distance between galaxies and solar systems. The Universe is made up of large amounts of dark matter and dark energy and only a small amount, roughly 15%, of what we normally think of as matter and energy. (3) As the universe expanded each planet formed and, if possible, life evolved, on those planets in complete, or nearly complete, isolation from each other, separated by vast amounts of space with physical connections only through comets and other space debris. This suggests that life forms in the universe are likely very diverse and there is no reason to assume identity, or even similarity, to the biochemistry of life on earth.

Another consideration that suggests that life may have formed very differently from that on earth is that within the theories of the beginning of the Universe, it is likely that we do not have a thorough understanding about the relative distribution of matter. There is a central hypothesis of an ultra condensed particle at the core of the big bang theory that caused the fusion of hydrogen to helium and lithium. Subsequently other elements were produced by further stellar nucleosynthesis. This does not necessarily require that elements be distributed evenly amongst solar systems.

A central dogma in chemistry is that the periodic table is "Universal", but to reiterate, the distribution may not be uniform. In fact, the distribution of chemicals as the Earth formed was not uniform. The presence of matter in the form of elements and the climate determined the chemistry of Earth's composition and the foundation of its eventual biochemistry. (7) Not the Earth's entire surface is hospitable to life, and the distribution of all life forms is therefore nonuniform. Life on our planet is segregated by climate, terrain and water.

In fact, given the vast and variable distances between galaxies, nonuniform distribution can be assumed. Understanding the nonuniformity may help narrow the search for habitable exoplanets, and enable us to focus on regions of the universe with a greater likelihood of earth-like conditions serving as a guide for our searches of habitable planets. This is not necessarily the case in the search for life, if we remove the assumption that all life must be earth-like.

3. We Can Improve Our Search For New Mechanisms Of Life By Studying Earth's Evolution

The amount of the sun's energy that reaches the earth is ideally suited for the life on this planet. We are 93M miles from the Sun and our air temperature at the hottest is 136 degrees Fahrenheit and the coldest is negative 128 Fahrenheit. Only 1/billionth of the Sun's energy/light reaches Earth taking an 8 minute journey from the sun to the earth. Our stratosphere now contains ozone to filter out the Sun's radiation, but this took time to develop and the early earth was very different. Our Troposphere reflects 34% of the sun's radiation and much of the rest of the energy that reaches the earth is turned into heat. Water, carbon dioxide, methane, and nitrous oxide in the atmosphere keep the earth warm by trapping this radiated heat.

Humans are uniquely evolved to survive on Planet Earth. The distance of Earth from the sun along with the gases that are in our atmosphere determines the radiation, weather and temperature. Those gases are matter and were attracted (as the rest of Earth's matter was) to a gravitational core. The gravitational core of the Earth is mainly iron and is dense even compared to the cores of other planets. The relative ratios of these gases in the atmosphere and the land and water mass that sequestered gases, also determined the evolution of the climate and atmosphere of the Earth.

The exact starting materials, environment, selective pressure and thus evolutionary pathways that led to life on earth remains largely a mystery. We have some tools that allow us to addresses the mystery such as carbon dating, phenotyping, organic chemistry, and genomics integrated and evaluated through emerging tools of bioinformatics. We study the evolution of life from which we have a molecular or fossil history and that represents, in many cases, the most recent of living things. Examining our planet through the lenses of the Big Bang 13.8Bya, the formation of planets, including earth 4.54Bya, our knowledge of isoprene lipid formation at 3.8Bya, the fossil evidence of Archaea stromolites at least 3.5Bya and our fossil records of humans 200-800 Kya is leading to a greater understanding of many of the early mysteries. The fossil record contains information that has had a tremendous impact on how we see early life on this planet, however, many aspects of evolution may have occurred without leaving a recognizable mark. We need to consider what we know about early earth in our description of early events in evolution.

In the early history of our planet, our atmosphere was mainly anaerobic. Evidence for life dated around 3.8Bya has been found but this would have been 1.8B years before oxygenic photosynthesis. Anoxygenic phototrophs used sunlight to grow and replicate beginning 3.8B years ago. Anoxygenic photosynthesis provided energy for a large class of microbes. The early microbes were most likely anaerobic and then adapted to the presence of oxygen as oxygen was sequestered in water and land and then slowly accumulated in the atmosphere.

Oxygenic phototrophs came into being beginning 2.5B years ago. Fossil evidence exists of atmospheric oxygen 2B years ago. The distance between oxygenic photosynthesis and evidence of atmospheric oxygen was 500M years. Great microbial mats that included anaerobes, oxygenic phototrophs and aerobes covered vast areas of the earth 2B years ago and were thought to contribute to the formation of atmospheric oxygen.

All life on Earth can be categorized in terms of the organism's carbon and energy source obtained from:

- Light reactions (phototrophs)
- Chemical oxidations (of organic or inorganic substances) (chemotrophs);
- Carbon for cellular synthesis obtained from CO2 (autotrophs)
- Preformed organic compounds (heterotrophs).

Combining these categories, we get the four basic life strategies: photoautotrophs (e.g. plants), chemoheterotrophs (e.g. animals, fungi), photoheterotrophs and chemoautotrophs. Only among the bacteria do we find all basic life strategies.

These well-studied strategies may be excellent sources from which to develop probes to reveal the habitability of exoplanets, and to find life forms that are similar to those on Earth. Using a probe for finding life that uses these strategies requires a priori knowledge of biomolecules and their chemistries. Conducting these searches using only Earth's chemical and molecular probes and biosignatures and only seeking Earth-like biology may cause us to miss planets with environs comprised of novel life forms.

4. Begin our Search For Life and Habitability With Only One Assumption: That Stars Provide the Major Sustainable Energy Source

The first evidence of humans on earth was 600,000 years ago. Excess energy is needed to carry out our life affirming biology of replication, growth, predation, not to mention the skills of language development, organizing and domestication of our food sources, and tool making. The development of oxygen as a terminal electron acceptor for aerobic and facultative anaerobic microbes and eventually oxygen-requiring higher life forms would not have happened until the earth developed accumulated some geologic or living mechanism to generate and sequester oxygen. This is important to evolution of humans as we consider the link between the thermodynamics of early microbial metabolism and the increased energy requirements of higher life forms. (15)

5. The Winogradsky Column Space Experiment: Shine Light On Exoplanet Mud

Within the living kingdoms of Archaea and Eubacteria (prokaryotes) organisms form structures or communities according to their food source and energy source. This is demonstrated in a unique experiment called the Winogradsky Column (shown below-image by Wikipedia) in which simple pond mud placed in a column creates a gradient from anaerobic to oxygenic environments in which each life strategy for utilization and acquisition of carbon and energy can be observed. The organisms may utilize substrate generated by a member of the community and in turn excrete or secrete a substrate that another organism may utilize. The Winogradsky column, demonstrated below, is a container with glass walls, to enable both light penetration and observation, that is filled with mud rich in microbes. The organisms self assort by their ability to flourish under the conditions in which they have evolved to thrive and they partition with regard to their need or sensitivity to oxygen concentration and light. Winogradsky did not look for a simple process but looked for the ability to organize within the context of the most favorable environment. There have been experiments on a potential chemical process from Mars soil with the results still under discussion (12). Scientists can perform a thoughtful experiment by creating a container exposed to conditions believed to be present on an Exoplanet (Mars) and then carry out a test for noticeable self-organization with soil and probable solvents (Mars mud).

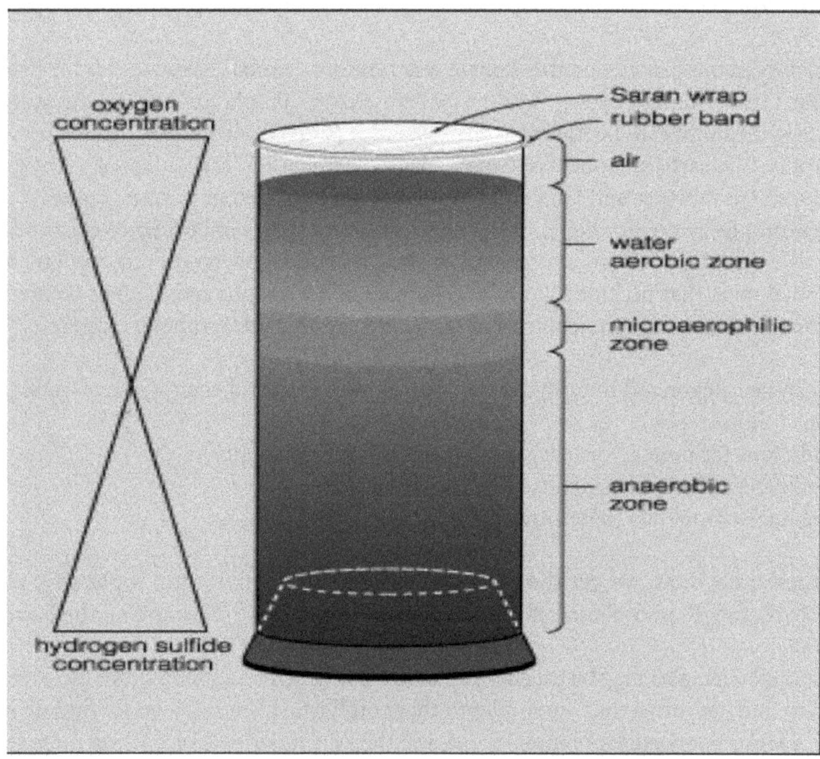

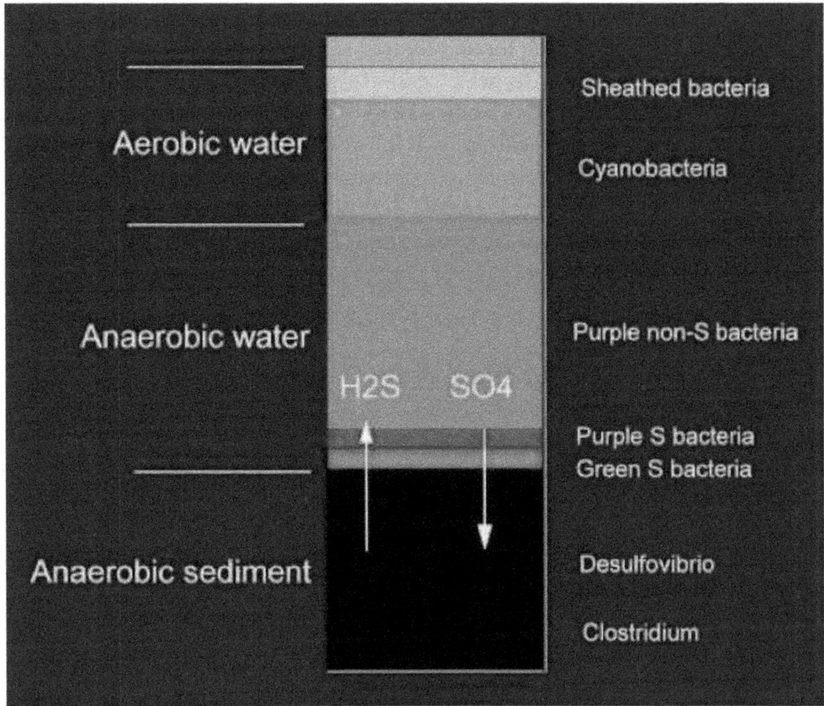

From James Lake. The study of protein, DNA, RNA and lateral transfer genetics.

6. Parsing Information Of Evolution and Selection of Life

The evolution of life results from selection of function. Diversity exists in a population and those individuals best suited for an environment are selected. Diversity arises through genetic mutation in the forms of point mutations, deletion, duplication horizontal transfer of DNA between species via DNA uptake or viral infection. Our phylogenetic trees are drawn to a common primordial ancestor, but the data that we use to build those trees and the algorithms with which we analyze the data are incomplete and trees built from different data sets using alternate strategies may not be in agreement. We currently lack the knowledge of how many times life started and ended, and any extinct, unknown or cannot be cultured (organisms that can't currently be propagated in order to collect their DNA) or those that have a unique genome not captured in our current methods for genomic sequencing. (4) Evolution is neither linear, consistent, or strictly vertical given that DNA can move between and among individuals of a species and between species and the selective pressures are variable.

There are many scholarly definitions of life including this one: "the condition that distinguishes animals and plants from inorganic matter, including the capacity for growth, reproduction, functional activity, and continual change preceding death." Many scientists have revisited the premise of life and have tried to identify the processes of life distinguishing matter from living forms. What is key in these processes that would allow humans on Earth to recognize life from a distance? Attempts to define life in terms of certain processes has fallen short due to the many exceptions to the rules. The challenge is to create a definition that does not exclude those exceptions especially if they prove to be the rule in space. (12)

7. Summary

Life on Earth was honed by the Earth's early geology, climate and rate of change of the environment. That evolutionary process on Earth may have conferred uniqueness with respect to the Universe. We as humans have paradoxically both the hope that we are unique and the belief that the possibility exists that the same organism could have independently evolved to a similar conclusion on a distant exoplanet. The use of strategies of life present in our earthly beginnings can help us determine exoplanet habitability or even discover life on distant planets.

However we should consider that alternative life strategies may have started and died out due to environmental changes or other selective pressures before a footprint could be left in the broad fossil records. Images taken of the distant Universe are from light as it reaches us from a time long ago. The more distance the images, the more distant in time is the origination of that light, even billions of years ago. (8) Using light as a historical fossil record requires increased sensitivity and accuracy of long range photometry and spectrometry to acquire more detailed information. This will help improve our search for an essential signature of life.

To date, Scientists have never been able to produce life from a collection of elements, compounds or molecules, thus our knowledge in incomplete. We currently use a structure/function approach to look for life. Carbon (rings and chains) and Hydrogen are the backbone for the macromolecules of living things. ATP, GTP, poly phosphates and some specialized organic molecules are our main energy carrying molecules. Water is our essential "solvent". Is Earth is the first case for life in the Universe. To find life elsewhere, Scientists may need an approach that does not require a prior knowledge of structure/function.

References

1. Seager, S. Science, 2013 May 3;340(6132):577-81.

2. McKay, CP. PNAS, 111(35):12628-33. Jun 9, 2014

3. Bateman, J. et al. *Scientific Reports,* 2015;5:8058

4. Gaucher, EA. Cold Spring Har Perspect Biol. 2010 Jan, v2(1).

5. Seager, S. PNAS, Sept 2014, 111(36)12634-40.

6. Rein et al. PNAS, May 2014, 111(19): 6871-5

7. McKay, CP, PNAS, June 2014, Online published ahead of print.

8. *Breakthrought in Cosmology: Taking the Universe's Baby Pictures.*

9. Roberge et al., *Debris Disk and Exozodiacal Dust,* NANA-ExoPAG@nana.gov

10. Rivera and Lake, 2004 Nature 431:152-155

11. UCMP, Geologic time scale. *The Archean Eon and the Hadean.*

12. Cleland, Carol E.; Chyba, Christopher F., *Origins of Life and Evolution of the Biosphere,* v. 32, Issue 4, p. 387-393 (2002)

Global Emergency Care and the Global Emergency Medicine Initiative

Terrence Mulligan DO, MPH

Associate Professor, University of Maryland School of Medicine

1. Background

The need for emergency medicine (EM) and acute care (AC) systems is a global crisis. Every country in the world is experiencing the new epidemic of non-communicable diseases, and every health care system is facing the problems of over-crowding, under-staffing, under-training, under-recognition and high acuity patients with incurable, chronic diseases. To meet these needs, the medical specialty of emergency medicine has arisen, along with the subspecialty of International EM (IEM). In many countries, emergency medicine and acute care systems are their earliest stages of development, and in most countries, emergency medicine (EM) does not exist at all. As one of the newest medical specialties in the world, Emergency Medicine focuses on providing the highest quality, safest, most appropriate life-saving medical care to the patients who need it the most--emergency patients who need the right care, at the right place, at the right time.

2. Goals and Specific Aims

The goals of this presentation include:

- **Specific Aim # 1:** a focus on how emergency medicine and acute care systems are developing globally
- **Specific Aim #2:** an exploration of the similarities and differences in national and regional EM systems development
- **Specific Aim #3:** an examination of developmental pathways of national EM and acute care systems development
- **Specific Aim #4:** the formation of a global organization dedicated to comprehensive, full-scale emergency medicine and acute care systems development on national and regional scales

3. Methods

To construct an organization dedicated to comprehensive, full-scale emergency medicine and acute care systems development on national and regional scales: The Global Emergency Medicine Initiative (GEMI).

1. Success in creating, directing and running the GEMI would contribute to global EM and acute care systems development, and subsequent advantages in health care costs, efficiency, improvements in patient care quality, access and availability, and overall health system strengthening.

2. Successes in an organization like the GEMI and the multi-lateral, multi-professional, multi-level relationships and infrastructure created by the GEMI would flow over to other national-scale development projects such as primary care systems development, geriatric and palliative care systems development, and public health systems development, creating a large-scale, full health-care system GEMI such as a Global Health Initiative.
3. Successes in multiple types of global health development will promote global health diplomacy and health systems strengthening, contribute to health system resilience, and will strengthen other diverse areas such as urban resilience, disaster management, financial resilience, and public health resilience.

4. Outcome and Discussion

Constructing large-scale, complex systems is a similar undertaking, whether building health care systems, IT systems, banking or economic systems, or STEM developmental projects. Development involves linking multiple layers from multiple disciplines with multiple professionals on multiple time scales, with simultaneous focus on building the pieces while connecting the pieces. Education, management, administration, finance, economics, policy all need to be coordinated to create capacity, sustainability, resilience, auto-poiesis / self-generation, self-reinforcement and auto-correction. Similar to a living process, development involves enormous organization on multiple scales and on multiple levels, with each level connected to, listening to, compensating for and responding to, changes on every other scale. These connections allow stability through change, and resilience and longevity through intelligent and reasoned systemic compensation.

The creation of The Global Emergency Medicine Initiative would enlist and involve multitudes of educators, researchers and health professionals in clinical and academic medicine, health administration, health economics, health law, public health and health policy. GEMI would be attractive to national and regional governments, most of whom have yet to fully address their health care systems' lack of EM and AC development. Universities and existing global health programs would support GEMI, and would contribute to national and international needs assessments, implementation and outcomes research. GEMI would provide private groups and corporations with an opportunity to support the creation of local and national health care systems; it would allow NGO's and research organizations to collaborate and combine efforts. GEMI would also allow governments and policy groups to deliver local, national and regional health system strengthening, to engage in global health diplomacy, and to embrace global health development. The creation of GEMI would deliver tremendous first-mover advantages and would avoid eventual path-dependent developmental dead-ends and blind cul-de-sacs, and would allow health care system development to ride ahead of the coming wave of non-communicable disease burdens instead of being knocked over and consumed by it.

100 YEAR STARSHIP™

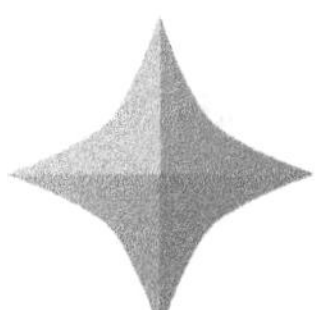

Chaired by Kathleen Toerpe, PhD

Deputy CEO for Program and Special Projects, Astrosociology Research Institute

Track Description

Our journey to the stars will captivate several generations of humans - as dreamers, builders, and explorers. We have just begun the journey and, as it was for our ancestors, passing along the knowledge, passion, and perseverance for our end goal is a crucial part of our mission. How do we insure that our children (and even grandchildren and great-grandchildren!) sustain the passion for the endeavor? How do we utilize both formal and informal avenues of instruction to teach them the skills they will need? Can we anticipate future modes of instruction that will prove to be a radical leap in how students of all ages learn? How do we continue to educate ourselves as we move from being dreamers of interstellar flight to being its builders and explorers? What challenges or pushbacks can we expect from society along the way? And, perhaps most importantly, how can creating the education of tomorrow make real and sustained impacts in the way we both learn and teach today? These are some of the challenges prompting us to research in the *Interstellar Education* technical track.

Track Summary

To become a space-faring civilization, we now know that we need to achieve radical leaps in education, not only in traditional STEM fields, but also in the social and behavioral sciences, the humanities, and the arts. The presentations in the Interstellar Education track spoke to underlying paradigms, broad initiatives, innovative curricula, and informal outreach programming that are not only simultaneously instructional under this broader umbrella of STEAM, but also inspirational and aspirational.

We heard about some of the innovative programs taking shape today:

- In the inspirational "herstory" of pioneering women "shattering the atmospheric ceiling" in spaceflight, engineering, and space research around the world.
- In success at creating formal curriculum to include space-related coursework targeted to community college students, an often-overlooked learning population.
- In programs designed for students from elementary to college that reveal learners' abilities to grapple with complex scientific issues from fresh or overlooked perspectives.

We were also reminded of the importance of collaboration between formal and informal avenues of education:

- In ambitious efforts to launch returnable student-designed scientific experiments on an orbiter aptly named Enterprise.
- In using the increasing convergence and "fuzziness" between science fact and science fiction as a pedagogical tool to inspire both scientists and students to greater creativity.

And, in the end, we were forewarned that educators will need to continually harness new avenues of creativity and collaboration, both to adapt to the unanticipated educational needs of an increasingly space-faring people on earth, and to design inspirational, aspirational, and instructional education for the closed and rigid learning environments that future students will encounter some day in outer space.

Track Chair Biography

Kathleen Toerpe, PhD

Deputy CEO for Program and Special Projects, Astrosociology Research Institute

Kathleen Toerpe, PhD, is a social and cultural historian who researches the human dimension of outer space through an emerging field called "astrosociology." She is the Deputy CEO for Programs and Special Projects and a Senior Research Scientist with the Astrosociology Research Institute (ARI). She is also an Executive Editor for the new peer-reviewed Journal of Astrosociology. Kathleen heads up 100 Year Starship's Education Special Interest Group and volunteers as a NASA/JPL Solar System Ambassador. She provides space outreach consulting, resources, and training through Stellar Outreach, LLC, and teaches courses in the social sciences and in critical and creative thinking at Northeast Wisconsin Technical College. Chairing the Interstellar Education track at the 2014 Symposium has been an honor. She can be found online at kathleentoerpe.com, at Twitter at @ktoerpe, or reached at ktoerpe@100yss.org.

STEM and the Oppenheimer Strain: Why Children Are Ready for Space Education

Marshall Barnes

SuperScience for High School Physics, 234 W. 18th Ave Columbus, OH 43210

superscience.hs@publicist.com

Abstract

Upon the discovery of blatant, undetected errors in a number of published works in theoretical physics, the author determined that they were psychological in nature and proceeded to test a theory that young people, whose minds are more flexible, would be able to see these mistakes. After the first such experiments proved successful, it was discovered that in 1960, the father of the atomic bomb, J.R. Oppenheimer, had mentioned a similar sentiment. Further experiments were conducted to test this hypothesis and all were successful, leading to the conclusion that students are smarter than they are given credit for and that entrenched experts can suffer from what I call, The Oppenheimer Strain. It is not clear what prompted it but the sentiment is. Oppenheimer said, "There are children playing in the street that could solve some of my top problems in physics because they have modes of sensory perception that I lost long ago..."

Keywords

J.R. Oppenheimer, education, STEM, wormholes, time travel

1. Perspective of Students

In 2003, during my fundamental research into the nature of time, I had begun to discover serious errors in the work of Kip Thorne and other big name physicists, which were confirmed as such by PhDs that I showed them to (but only after I had pointed them out) and determined that they were psychologically based. I successfully presented my findings at the 2004 Mars Society Conference under the presentation name, avoiding Hidden Assumption Traps When Thinking Outside The Box[1]. The premise was that due to the peer pressure in the physics community against such topics as wormholes and time travel, that had only begun to change in the late '80s, physicists like Thorne were not used to thinking in a highly imaginative way, accurately. A similar problem had been identified by NASA. Marc Millis cites the Horizon Mission Methodology[2] as a method of forcing thinking away from extrapolations of existing solutions, which is often where hidden assumptions reside. I felt that younger people, used to thinking scientifically, but not trained at the PhD level, could detect these errors because their minds are more flexible.

So I tested two classes of physics students at Bexley High School in late 2007, after explaining the pertinent concepts that they would need to understand, like what the speed of light, etc., to see if they could detect an error that I had found made by Stephen Hawking and Kip Thorne in Thorne's theory of using wormholes for time travel. The thought problem is as follows:

1. Take a wormhole that has two mouths.
2. Leave one on Earth and place the other on a spaceship that goes out and back on a journey that lasts ten years.
3. As the spaceship approaches the speed of light, time dilation causes everything on the spaceship to age slower, including the wormhole [3].
4. When the spaceship returns, the wormhole mouth onboard will be younger than the one that stayed on Earth.

Kip Thorne has assumed that this would cause the wormhole to be a time machine because the time dilated mouth would be "younger", equating age with time, but other physicists, including Stephen Hawking, argued that once the spacecraft was "ten light years away" a feedback loop would be created that would collapse the wormhole [4]. There was a colossal, obvious mistake that stood 13 years before I discovered it. Could high school physics students catch it?

Figure 1: Bexley students considering the thought problem.

Yes, they did! Five students in each class, the exact amount that I had predicted to teacher Craig Kramer, were successful; however, the surprise was that in each class, more girls than boys got the right answer which is that the spacecraft is never ten light years away from the Earth because it has to be back to Earth at the end of ten years at a sub-light speed. Comments made by the students included, "It can never reach 10 light years away because it has to be back in 10 years", "The ship could only go about less than 5 light years away before it has to turn back around" and "The spaceship can never reach 10 light years away because it's not going at the speed of light and it has to be back by 10 years"[5].

I then learned about a quote made by J. Robert Oppenheimer, the father of the atomic bomb, in 1960 that shed even more light on this situation - "There are children playing in the street that could solve some of my top problems in physics because they have modes of sensory perception that I lost long ago..."[6].

I decided to pursue it further.

I tested another aspect of the same problem in 2009, this one much harder, with twelve juniors from the Columbus Africentric Early College[7]. Hawking's criticisms had been wrong but Thorne was still wrong, but for another reason. Could these students see it? The problem:

1. The wormhole is said to have 2 mouths, one younger due to the time dilation effect it has been subjected to on the spacecraft.
2. The assumption is that because of this, you can enter the mouth that stayed on Earth and exit the one that traveled through space and appear in the past because that mouth is younger. This has been the accepted explanation around the world. But there is an error.

Figure 2: Columbus Africentric Early College students in action.

Four out of twelve found it! A higher percentage than before. The answer? Thorne's model won't work because time dilation effects the age of the wormhole but does not connect it to an earlier exterior universe. In other words, age does not equate with time, which is the hidden assumption. The time dilation effect doesn't effect anything beyond the local reference frame subjected to it and does not extend to the exterior universe, which is a concept cited by Thorne as part of the thought model. This answer has not been caught by any physicist in the world, yet. In the well known example of Einstein's twins paradox[8], the only reason the traveling twin can arrive in the future, younger, is that he was traveling for the normal amount of time that had to pass outside of his local reference frame – in the exterior universe, to reach that point in the future. Example - to go 100 years into the future, you must travel at some velocity for 100 years. It doesn't matter what your velocity is or where you fly to, you must do it, per clocks outside your reference frame, for 100 years. Time dilation in your reference frame just keeps you from aging a 100 years in the process, that's all. Likewise, the wormhole would age younger but there is no explanation nor evidence of how it's supposed to connect to an exterior universe that's in the past.

The students were handicapped by the lack of proper terminologyr with which to describe their thoughts but did so in their own words. One student struggling with the concept said, "The Earth and the rocket are still in the same space". Another replied that "Once you go outside of the rocket, you're still in the same exterior universe". For those who want to argue that the wormhole is made of space and time (which is not accurate) there is still the unexplained and unmentioned third wormhole that would have to be present in the interior of the time dilated wormhole that is connected to the past, otherwise, how would it be possible to have the original two, side by side, as is often described, and you enter the Earth based one and never appear out of the time dilated one that has returned because you've somehow disappeared into the past? A conundrum these students saw right through.

In 2011 I did the project with two classes at Grandview Heights High School in Ohio. One class was tasked with figuring out the problem with J.R. Gott's model of the self-creating universe [9] while the other handled the issue of why Stephen Hawking should've of known that the arrow of time wouldn't reverse if the universe were to begin to contract [10]. After explaining the basic concepts required to do the work (i.e. causality, CTCs, parallel universes, etc) I presented the problems to the students.

Figure 3: Presenting thought problems at Grandview Heights High School

The J.R. Gott problem, from his book, Time Travel in Einstein's Universe, can be seen on its cover which shows a closed time-like curve at the base of an illustration of branching universes. The problem is as follows:

1. At the moment of the Big Bang, the universe spreads out into different branches.
2. One of the branches loops back through time to the beginning of the main trunk where he says "it would be like having one branch of a tree circle around and grow up to be the trunk. In that way, the universe could be its own mother."

Gott was wrong because the time loop he wants to invoke would cause a collapse, in the manner in which he wants to invoke it, or would go to a parallel universe and not result in the effect that he wants. The collapse would be as a result of the CTC that Gott is invoking, causing a feedback loop as energy built-up, that would destroy the loop. If instead it went back in time to a parallel universe, it would not be the source of the original one as Gott describes. To help the students process the problem, I gave them a formula I created called, MCEBPS, which stands for Marshall's Copenhagen Everett Barnes Paradox Solution[11]. It exposes the impossibilities of paradoxes by citing that any backward in time travel (like the branch of the universe Gott describes) would violate the Copenhagen interpretation of quantum mechanics which says you can only have one outcome per measurement[12] which then kicks the problem to having to be resolved by a parallel universe (Everett interpretation) [13] which is then proved by the Barnes element which focuses on the ability to provide or create evidence prior to the time travel event, proving any alternate outcome would be from a parallel universe. In this case, filming the original Big Bang moment to show that something other than a branch of the universe created the original and that the loop wouldn't change that.

Students responses included such things as, "There'd be an explosion back where it began because of the feedback loop of energy", "It would have to be in a parallel universe so it wouldn't change what orginally happened", "If I had a camera at the Big Bang I could show that that branch didn't create the orginal universe."

The Stephen Hawking issue dealt with a mistake that he admitted to, that the arrow of time would flip if the universe began to contract, but the focus was on why he never should have made the mistake in the first place:

"This would lead to all kinds of science fiction like possibilities for people who survived from the expanding to the contracting phase. Would they see broken cups gathering themselves together off the floor and jumping back on the table? Would they remember tomorrow's prices and make a fortune on the stock market?"

The students were able to see why, if these were the conditions of the reversal of the arrow of time, that there must be something wrong with the idea. "People's thoughts would be going backwards with everything else then...", "Everything would be in reverse, like eating and going to the bathroom", "You can't have some things happening in reverse and not others". The students proved that Hawking should have known he was wrong because of the logical and causal inconsistencies that he described that would be part of his scenario, such as people being able to make money on the stock market because they would already know what the prices were going to be. There would be no forward thinking, in fact, all life would cease to exist because everything new would stop and nothing would be happening but events that had already taken place, happening in reverse.

At Thomas Worthington High School in 2012, I tested students to see if they could see the problem with Stephen Hawking's presentation on the Discovery Channel special, Into the Universe with Stephen Hawking [14]. It was a two part lecture. The first part was a test to see how many could see a critical functional error in the idea that a mad scientist could commit suicide by shooting himself through a wormhole time machine that he had created, that connects to the past by one minute. Hawking called this his Mad Scientist Paradox [15] and it works like this:

1. A mad scientist has created a time machine that generates a wormhole to the past by one minute.
2. He decides to test the paradox idea by walking over to the wormhole, taking aim at himself through the opening and shooting his past self.
3. We see his past self, then lying dead on the floor with the time machine and the wormhole it has generated, that he has been shot through, still running.

Hawking further posited that in reality feedback would be created that would prevent the wormhole time machine from working at all. He used the example of feedback caused by a rock band to illustrate the idea of energy building up through the wormhole connection to the past, that would gain enough strength to destroy the connection. Since I am an accomplished lead guitarist, I played that role of the band and allowed student volunteers to run the equipment. It was the first ever physical experiment ever that proved Stephen Hawking wrong. The experiment consisted of my playing guitar while a student ran the mixer, it was plugged into, and another student held a mic, that was plugged into the system, in front of one of the speakers. As a third student raised the volume while I played, feedback began to happen. This is what Hawking had been talking about. The speaker represents a connection to the past that has a circular path or causal connection. Next, we tested the alternative, non-causal connection representing an Everett/Wheeler solution. The student with the mic aimed it at the other speaker without moving it directly within line of sight. This time when I played, and the volume was raised, there was zero feedback even while the mic was picking-up the sound. I had in fact given this example in 2004, prior to Hawking's idea of the Mad Scientist Paradox, for a class taught by Sabra Weber at the Ohio State University in a lecture on time machines. However, this was the first time I actually performed the experiment.

Figure 4: Proving Stephen Hawking wrong with a solid body electric guitar

In the first case, six students were able to see that it would be impossible for the mad scientist to shoot himself through the wormhole because he was trying to use the same opening for both functions when in fact, there would have to be another wormhole opening present in the past in order for it to work because Hawking stipulates that each wormhole being generated connects to the past, making impossible the idea of the scientist shooting himself through the same wormhole that he has made in the past. Nowhere in the video is a third wormhole shown. Their comments included statements like - "The wormhole to the past can't be used by someone in the future", "If the wormhole goes to the past there needs to be another one in the past that connects to the one we see in the future".

I finally, later in 2012, tested actual children who were members of the OSU AAAS/Wexner Medical Center Math and Science Club to see if they could get the right answer to the Mad Scientist Paradox question as well. To help them understand this example of multiply connected spaces, I made sure that they understood the difference between an "out" door and an "in" door. A higher percentage of these children got the right answer than any other group tested - 4 out of 7. Their responses included, "You need three wormholes because two's not enough", "He can't hit himself through that wormhole because it's not connected to the past on the other side", "He can't be hit in the past because it would be coming from the future and that wormhole only goes to the past".

Figure 5: Children see right through the Mad Scientist Paradox

Many of the students throughout this entire series of tests, who successfully answered questions, were honored at either the state or local level or sometimes both. The result was described by them, in many cases, as "empowering" and a "confidence builder". They were amazed that they had the inherent capability of out- performing people like Stephen Hawking on these tasks, when given the chance. A number of times teachers told me flat-out, "I doubt that they'll be able to do it." Sadly, I've had media, such as a science editor at the Associated Press; repeatedly try to down play the significance of the results. I've even had media express resentment at the idea that children could out think Stephen Hawking, so the idea that this is universally seen as a wonderful proof of the importance of STEM, is shockingly, not so universal. In fact, at times, the children's accomplishments were seen in no other way than "threatening".

I owe my own ability to see these solutions to my experiences learning by example from Todd Rundgren from his work, in 1972 to 1980 as a teenager. I learned about modes of sensory perception from articles about him and how he applied his perceptions, via interviews. I learned not only what he thought and why he thought it, but also how. An example, with a current analogy would be, "A record isn't about the music that's on it but the experience of recording that music" [16] and my own, "Time travel to the past isn't about the actions of the time travelers but the geometry that allows the time travel" [17].

2. What Does This All Mean?

- It means that many children are inherently more gifted that most adults imagine and are being underserved by the way that the educational system handles them.
- It means that adults, especially so-called experts, are inherently hindered by the Oppenheimer Strain without their knowledge. This problem effects all professions, not just physics and science. The fact that it effects education just perpetuates the problem.
- These subjects, that the students dealt with, prove that early general education, dealing with space science, can begin as early as the 4th grade. This will provide a boost for preparing the new generation for the new space age. In fact, NASA has educational programs online that run through K-12[18] and should be taken advantage of.
- This program can be continued elsewhere except with the caveat that the same questions and issues can't be used anymore because other kids will eventually read about this and it will pollute the exercise. Also, the project is based on mistakes that PhDs have made, and there just aren't that many, but students can be tested on similar issues from the stand point of finding solutions as opposed to errors.
- The final issue, the Oppenheimer Strain, is exactly why this symposium is called, the 100 Year Starship instead of 30 or 20 Year Starship. At a time when micro wormholes are being created by my invention, the Verdrehung Fan™, achieving the speculative conjectures of Luke Butcher of Cambridge University, as reported by New Scientist magazine [19], and my STDTS™ technology for warp drive is now ready for space testing in 2016 (see my poster presentation at this event) the common wisdom is that it will be some 100 years and counting before a starship is possible. Of course, that means we'll all be dead by then However, the NASA Institute for Advanced Concepts warned against letting "your preoccupation with reality stifle your imagination" [20].

3. Conclusion

It is clear the statement that J.R. Oppenheimer made about young people possessing better modes of sensory perception, in some cases, is true. It is also true that experts, particularly entrenched experts, can suffer from the "Oppenheimer Strain". This is the key to what is holding back the advancement of Humankind. It points towards the need for a completely new kind of educational system, whether there ever is one or not. The first step is recognizing the situation and making some steps toward enhancing the abilities of children and correcting the stigmatism of adults.

I will close with this quote by Vannevar Bush who started the National Defense Research Committee, the scientific body instrumental in defeating the WWII Axis powers:

> *"Can we learn to think in 4-dimensions? This and negative time, involve dreaming of the wildest sort..." [21]*

References

1. Barnes, M, "Avoiding Hidden Assumption Traps When Thinking Outside The Box", International Mars Society Conference, August 2004

2. Millis, M, "The Challenge To Create The Space Drive", NASA Technical Memorandum 107289 July 16, 1996

3. Thorne, K., *Black Holes and Time Warps: Einstein's Outrageous Legacy,* W.W. Norton & Company, January 1995

4. Thorne, K., *Black Holes and Time Warps: Einstein's Outrageous Legacy,* W.W. Norton & Company, January 1995

5. Bexley News "Community News" *This Week Newspapers* Wednesday September 10th, 2008

6. McLuhan, M., "Address At Vision '65", McLuhan on Maui, http://www.mcluhanonmaui.com/2011/06/address-at-vision-65-by-marshall.html (2011)

7. Hutchins, T., "Researcher Inspires Local Students, 'Expands SuperScience'", nbc4i.com, December 8, 2009

8. Jain, Mahesh C.. Textbook Of Engineering Physics, Part I. PHI Learning Pvt. p. 74, 2009

9. Gott, J. R., "Time Travel in Einstein's Universe", Mariner Books, September 19, 2002

10. Hawking, S., W., "A Brief History of Time: From the Big Bang to Black Holes", Bantam Dell Publishing Group, 1988

11. Barnes, M., "Experimental and Theoretical Analysis Of Chronology Protection Conjecture Failing On The Discovery Channel", March 16, 2012 Nature Network Forums Physics of Time Group http://network.nature.com/groups/time/forum/topics/10468

12. Highfield, R., "Testing the Copenhagen Interpretation: A matter of live and dead cats" The Telegraph, November 8, 2011

13. Tegmark, M. "Many Worlds In Context", Dept. of Physics, Massachusetts Institute of Technology, August 3 2008, revised February 14 2010)

14. Factor, S., Weinberg, J., Hillman, E., "Discovery Channel Press Release" January 14, 2010

15. Hawking S., "The Mad Scientist Paradox," Time Travel episode, Into the Universe with Stephen Hawking, *The Discovery Channel* April 9, 2010

16. Edmonds, B., "Todd Rundgren Unchained" *Creem,* August 1972

17. Barnes, M., "Paradox Lost: The True Geometries of Time Travel – The Public Edition" Kompressor Press/ Blurb.com October 6, 2014

18. NASA Education for Students http://www.nasa.gov/audience/forstudents/index.html#.VM1_jrSap68

19. Ananthaswamy, A., "Skinny wormholes could send messages through time" *New Scientist,* May 20, 2014

20. Gilster, P., "NIAC: Bob Cassanova's Mug" *Centauri Dreams* March 7, 2011

Marshall Barnes

A Utopian Education in Space —Are We There Yet?

Janet de Vigne, MEd

University of Lancaster UK, Bailrigg, Lancaster LA1 4YW

janet_devigne@yahoo.com

Abstract

Many people are currently considering education in the 21st century as something quite different from what we have been used to. Some, like Jenkins (2002), consider that it is now outside the classroom and should not be 'controlled' in traditional ways. [1] Others, such as Mitra (2013), consider education as an 'emergent' phenomenon, something that will, with a little less help than usual, become a very effective self-organising structure. [2] With the advent of two new intelligences as advocated by Gardner (1999)—existential intelligence and naturalist intelligence—how could education in the closed system of a spaceship or extra-terrestrial settlement take into account new thinking and the connectedness of humanity in its acquisition and practice? [3] How might teachers function in this environment and how much might depend on the computer? How might AI emerge as a powerful factor in the development of human beings carrying the future of the species into hostile environments? These questions and more are considered from the perspective of education—a human right?—and a child's freedom to make choices concerning her future, her education and a 'career,' in terms different to those currently understood.

Keywords

education in space, future of humanity, teaching and learning

1. Introduction

Everyone, it seems, has an interest in education in the 21st century; at the very least, they can tell you that everything seems to be wrong with it and that the system is failing our children. In examining how the education system might function in the confines of a space ship on its way to a habitable planet, this paper will present contemporary ideas shaping learning and teaching, while exploring the very specific needs presented by such a case scenario. It is hoped that a better understanding of today's classroom will emerge, and that a common interest in improvement and development will ensue. To quote Dr Mae Jemison, "We can go to the village because we've been to the stars." (Keynote/100YSS Conference 2014).

2. The Context

Assuming that humanity develops the capacity to leave the earth in search of another habitable planet, and that the speed of travel to such a planet will not exceed the laws of physics currently understood (in other words, not faster than light), survival of our species will depend on our ability to produce and raise children in space who will need two distinct groups of skills. The first group will involve the operation of very sophisticated machinery in order to survive within the spacecraft and reproduce and manage perhaps two more generations on board. The last (space-born) children will need a second group of skills to equip them for landfall and survival in a set of circumstances almost impossible to foresee. It may be possible to design a system that will develop with them, (i.e. an education system not entirely dependent on human interaction); therefore one of the most important issues here will be "critical mass," not just because of the numbers of people necessary to continue the species, but because of the current understanding of emergence and the "self-organising structural" elements of human consciousness and understanding.

In the words of Cockell (2013), children in space will be brought up in cages. [4] The cage, of course, has a dual purpose—it keeps the children inside a safe environment, and it prevents them accidentally or deliberately, from entering or exposing the community to a potentially lethal one. The issue of nature or nurture will become contentious; teenage rebellion becomes a serious threat to life and the curious toddler becomes a hazard not just to herself, but to the entire community. Skills that will need to be taught will have to encompass reflection, understanding and management of the self, responsibility and accountability to the community and, very importantly, conflict resolution. This is, of course, no guarantee at all of a non-autocratic "one big happy family," but the settlement members (or crew) will have no option but to take the calculated risks of growing "community" in a container where no-one can hide (or not for long) or leave (during their lifetime). And we have not yet mentioned subject knowledge.

Children in space will need to learn fast. They will need functioning maths, astrophysics, biology, and who knows how much science in order to make their environment work for them, to exercise control, and to repair highly specialised equipment. An in-depth understanding of what they start with will enable them to develop and innovate while in space, thus improving the environment for the community. Creativity and freedom to think, problem solving, and practical strategies for invention can be taught—not using the traditional jug of water and empty glasses teaching metaphor—but something much more sophisticated: the teacher as enabler, gardener, guide? However necessary these subjects might be, where might they leave the learners in terms of artistic development? Linguistic development? Human development? What does current neuroscience have to say about these subjects, largely ignored by those panicking about STEM? And could a programme be developed that would create a utopian educational environment, prepared using what we know now, and designed to expand and develop to encompass what we do not yet know? Where are the child's choices in all of this—and how can we guarantee that she will be able to develop along these lines?

3. The Background—What Purpose Does Education Serve?

Let us briefly examine the reasons why we educate and have educated our young people today. The game has changed. Arguably, education has always been about the transfer of knowledge from those who have a lot to those who have less. The amount transferred is decided upon by those with the power that knowledge itself confers. In a very superficial, not to say slightly cynical, overview, I suggest that education has always been about control. As an education system in space will determine the survival of the species, the control parameters must be decided upon and enforced, but they will not be the traditional societal cohesion or economic controls. What freedoms could this possibly allow in terms of creative thinking and problem solving that will be crucial to the success of any off-earth long-term travel and settlement?

From the 1100s and the beginnings of the great European universities emerging from the mists of the post-Roman era, education developed within a very strictly controlled system. Galileo, Copernicus, and many more were persecuted for disagreeing with the religious state in matters as basic as fundamentals of physics and maths. This seems remarkable to us, so far away from the beliefs of that early church. And yet what people were allowed to access at that time was rationed; an education fitted a student to become a priest and debate weighty theological matters (such as how many angels could stand on the head of a pin), to take the office that was his by right or by might, or to trade internationally. Knowledge created a division between the elite and the non-educated, perpetuating myths of "divine right" and the fatalistic acceptance of the underclasses of their supposed inferiority by divine design. (Is this ringing any bells? Social mobility was non-existent.) Myths of the torments of hell, pornographic in their visceral detail of punishment for sins such as adultery, were legion. Moving through the

Renaissance, and in England, in particular, the state wrested control of education from the church, and access to Oxford and Cambridge was allowed only to members of the Church of England, thus contributing to the movement of scholarship between Europe and Scotland and reinforcing Scotland's separate identity as a nation-state. As the Reformation swept through Britain, it became necessary for the "moral good" to allow men and boys to read the Bible—in their own language—to maintain religious civic control. Gaining some rights, partly through being able to read, people were prepared to die to defend the same. Moving on again by the time of the Industrial Revolution, education begins to provide a mechanism of economic control; everyone should be able to access the instructions for the operation of the machines, not for their personal safety, but in order to maximise profit for the factory owner. We then move forward again, through economic stages one, two, and three, where education becomes more necessary in terms of production and, to some degree, protection of the workforce (for sound economic reasons rather than altruism, such as income tax). Litigation procedures appear as the workforce becomes "aware," until we reach today's ideologies of economic power and the knowledge economy—the fourth sector. No one knows how this will pan out—because we in the West, the erstwhile economic leaders, have not been here before. So, what is happening? Knowledge is increasingly within the grasp of anyone who can access the Internet. At least, quite a lot of knowledge is. This is, of course, a ridiculous and somewhat cynical oversimplification, but it is not without some justification.

4. Connectedness, the Internet, and Education in the 21st Century

I am equating knowledge acquisition, rightly or wrongly, with education, as most of the types of knowledge outside survival skills associated with terrain or culture are acquired through practices of literacy such as reading, or using a computer. There is no reason why one should not be able to educate one's self—history is littered with highly motivated people who have done just this—but it is unusual. In fact, according to Jenkins, education in the 21st century is now out of the control of the state, the church, and the classroom and, due to the globalised access through the Internet, is within the grasp of everyone who has the technology readily available [5]. One consequence of this is democratisation (a positive force for good, flattening societal hierarchies), another, lightning fast innovation (connected synchronicity, if you like) as ideas travel between countries. Is this a happy place for policy makers? How might such a paradigm affect an education programme designed for space?

Policy makers have reasons to rejoice and reasons to fear current developments. Monitoring people's behaviour and thinking has never been so easy. As fast as Facebook-style social media sites emerge, so does the technology to 'observe' interaction thereon, enabling both trade and political manoeuvring. The role of such sites in revolution and the communication of disaffection have been well noted. If Jenkins is right and the reign of the new literati of the Internet is just beginning, then ancient institutions holding power had better look to their laurels [6]. I suggest that it is in the nature of these institutions to survive, as they have done for centuries, and that their capacity to adapt may surprise the revolutionaries. The universities for example, in producing Coursera and other MOOC delivery platforms, are already investing billions in systems to educate the world. You pay for the certificate at the end of the course, not the education, which will contribute in equal parts to a new equality and inequality in society—to be certified or not might well be the first issue on the application form. The capacity of these courses is huge; hundreds of thousands of students worldwide, learning, talking and giving feedback on the teaching, changing it to meet their needs.

So the issue here pertaining to space is in fact one of "connectedness." Humanity connected around the globe, togetherness flattening society and producing a best-case-scenario Utopian democracy in Jenkins' brave new world [7]. (It might not of course—there are those who fear deception and tyranny arising through the same mechanisms.) The advantages of this collective intelligence are clear - education, now without traditional contexts (the classroom)—can truly be owned and therefore driven by those who are learning, and not by those who have a vested interest in what they learn. Therefore, according to Jenkins, some of the best things educators should give students today might be the social skills and cultural competence to function in a networked society [8]. In space, the connectedness will not be there. The possible beneficial outcomes of emergence, the rise of the self-organising structure, the glorious messiness of the mass of humanity will have to be vicariously experienced via the computer data banks. It would at least be possible to include vast amounts of information in increasingly small spaces. Are there any instances of self-organising structures in education that might help us understand education and support a system in space?

5. Current and Proposed Uses of Technology in Education

The research of Mitra (2013) bears investigation [9]. An educational technologist by training, Professor Mitra believes he has stumbled on a self-organising structure, using the technique of allowing schoolchildren to drive their own learning using the Internet in a teacher-less system of education. This is important—his research is very impressive and along the lines of "the system is greater than the sum of its parts." Children are placed in SOLEs—self-organising learning environments and given problems or questions to solve or answer. Left to their own devices, with adults there just to admire them and encourage them (the "Granny Cloud"), children are capable of a) learning more than we think they can (i.e. within the generally accepted and imposed developmental constraints) and b) retaining substantially more and more advanced information. This has surprised many teachers—but has been foregrounded in education by the work of Vygotsky and Bruner in the concept of "scaffolding." Arguably, Mitra's work relies on "situational memory"– the child recalls instantly whom she was with, when and where—and all this contributes to the retention of the knowledge gained. (This has yet to be researched). Mitra's TED talks are exciting and inspiring—in fact he won the 2013 TED prize for his pioneering work in the field and has been given £1 million to continue his research at the University of Newcastle, UK.

So, access to information in space, while it will not be real time, is possible, but there are certainly stories of children using the Internet only as a source and being led astray by the first findings thrown up by the search engine. The information stored in the space vehicle's data banks will not be live (we do not yet know how electronically stored data might degrade), and one of the most important aspects of Mitra's concept will be missing - the Granny Cloud—the adult enablers who stand behind the children and 'admire' them. Enabling children to teach themselves seems to offer many solutions to the logistical problems of an education in space, but it doesn't appear to solve issues of freedom of choice, of artistic development and emotional development (such as conflict resolution). What will happen if you know you are born to be a dancer? If you are gay? If someone imposes their will on you? If you fall in love and it is not requited? So it seems that 'significant other' adults will need to be equipped to negotiate these topics. This is highly skilled work and requires specialist training, but there would still be no guarantee that a "good" education would result in the survival and prosperity of the species. Might the space travellers need an extra element of control in the classroom, at a level that we would, today, find ethically unacceptable, without going down the road of the trans-humanists?

Professor Patricia Kuhl is considering brain-based interventions that could "help" a child with difficulties in maths, or maintain the "plasticity" of the brain to allow it to keep growing rather than fossilise [9]. The Lenneberg hypothesis of 1967 has, it seems, been confirmed by her research, in that the brain's capacity to absorb new languages seems to stop at age 12 as the brain takes the shape it will have in the adult. Kuhl has shown that babies are statisticians; by measuring their reactions to language she has observed how the brain calculates instances of repetition of sounds and stores them in recognition centres. Interestingly, the babies in her research do not respond to artificial stimuli such as a television or audio system—they will only learn from a human. This raises other issues in space—how much stimulation could be provided? What needs to be human and what could be machine? When and how could a child experience the smell of newly mown grass? The sweetness of a horse's breath? In the same way that a no-gravity context lessens bone mass in the legs because they aren't being used, might a lack of olfactory stimulation cause the travellers to lose their sense of smell over three or four generations? This would have huge repercussions at landfall as an essential part of our reptile-brain instincts.

6. Conclusion

We must begin to consider these issues now, as people prepare for humanity to leave planet earth. Education is the single most important element to establish for any human future. Unfortunately, in many contemporary societies, teachers are not valued highly—in fact, they are despised. We are still at the stage where Brendan Iribe's holodeck, the *Oculus VR*, is considered by some as the "next big thing" and the saviour of the education system as we know it. [10] However, Howard-Jones at the University of Bristol, would beg to disagree [11]. Researching the effectiveness of educational games and technology in education, he is asking serious questions based in neuroscience about learning, how it happens, and whether such games have a positive or negative influence. What seems to work in terms of engaging learners' motivation is "uncertain reward"—to gamble "double or quits" at the end of the game. This seems to increase dopamine production. He has showed that the main issues with learning appear to be: enough sleep, looking at the areas of your own brain that light up when you're engaged in the desired activity, and stimulating your own creativity by focusing and defocusing on the task in hand. Certainly the jury here is out on whether the use of certain types of technology can help. One could argue that this might be because the technology in the classroom is very far behind the technology in the home, and that, as a consequence, children see themselves as walking into a pretty much useless museum of antiquity when they walk into school, questioning whether what happens inside the classroom could possibly be of use to them outside. This means that, in terms of providing a space where learning is immediately engaged, teachers today are at a disadvantage, having to waste time proving relevance before anything can be achieved.

Most educational theorists agree that motivation is the key element in learning. Types of motivation are legion (this is a huge area of research), but a concept connected to this is how to engage the "best" kind of motivation—could it be by appealing to a learner's intelligence "package," as the optimal way in? Howard Gardner now believes that there are around thirteen different types of intelligence, and that manifesting seven or eight of these is "perhaps a working definition of the species." [12] Gardner refers to leadership, moral excellence, and creativity as fostering human potential—qualities certainly desirable in the development of our space children, if a colony is to survive. And thus we return again to the question of control. Can we afford to trust the process of human development? What will need to be learned? And how will it be learned? In the words of Socrates, "Education is the kindling of a flame, not a vessel to be filled." Who will teach? What will they teach and how? There are many instances of alien life forms from *Star Trek* to *The Fifth Element* being appalled by finding the encyclopaedia entry for "war" and almost giving up on the human race as a consequence. The ongoing BBC TV project *Up*, which has been following a group of children growing up over a period of 20 years (this predates *The Truman Show*) generated the following anonymous post a few months ago: "The futures of children are not preordained. They have lives full of potential, luck (or lack of it) and opportunities . . . " In our case, the opportunities for the first few generations will be very limited; those for the landfall generation, absolutely limitless. We are very far from a Utopian ideal. The real question will be, as very possibly in many things, how much can we truly leave to luck?

References

1. Jenkins, H. (2002). Interactive audiences? The collective intelligence of media fans. www.labweb.education.wisc.edu

2. Mitra, S. (2012). Children and the Internet—A preliminary study. In Uruguay International Journal of Humanities and Social Science 2 (15), 123-129.

   Mitra, S. et al (2013). Self Organised Learning Environments (SOLEs) in an English school: An example of transformative pedagogy? *Online Educational Research Journal,* 1-19.

3. Gardner, H. (2000). *Intelligence reframed—Multiple intelligences for the 21st century.* New York, Basic Books.

4. Cockell, C. S. (2013). *Essays on extra-terrestrial liberty.* Edinburgh, Shoving Leopard

5. Jenkins, H. (2002). Interactive audiences? The collective intelligence of media fans. www.labweb.education.wisc.edu

6. Jenkins, H. (2002). Interactive audiences? The collective intelligence of media fans. www.labweb.education.wisc.edu

7. Jenkins, H. (2002). Interactive audiences? The collective intelligence of media fans. www.labweb.education.wisc.edu

8. Jenkins, H. (2002). Interactive audiences? The collective intelligence of media fans. www.labweb.education.wisc.edu

9. Kuhl, P. (2004). Early language acquisition—Cracking the speech code. Nature Reviews Neurosci-ence 5, November, 831-843.

   Lenneberg. E. H. (1967). *Biological foundations of language.* New York: Wiley.

10. Young, Jeffrey R. (2014). Brendan Iribe and the Oculus VR. Chronicle of Higher Education, Sep-tember 12. http://chronicle.com/blogs/wiredcampus/will-the-next-classroom-disruption-be-in-3-d-facebooks-virtual-reality-company-thinks-so/54517

11. Howard-Jones, P. A. & Murray, S. (2003). Ideational productivity, focus of attention, and context. *Creativity Research Journal* 15, (2-3), 153-166. Taylor and Francis Online http://www.tandfonline.com/doi/pdf/10.1080/10400419.2003.9651409#.VOMWcS6I_Yg

12. Howard-Jones, P. A. (2014). Neuroscience and education: A review of educational interventions and approaches informed by neuroscience. London, The Education Endowment Foundation. www.educationenvironmentfoundation.org.uk

Convergence to Fuzziness: Fact, Fiction, and the Evolution of Interstellar Education

Mark D. Juszczak, Ed.D.

Faculty of Pedagogical Sciences

Akademia Pedagogiki Specjalnej im. Marii Grzegorzewskiej

Ul. Szczęśliwicka 40, 02-353 Warszawa

mjuszczak@aps.edu.pl

Abstract

This paper is an exploratory study of the categorization of fact and fiction in the context of future interstellar education. What impact will interstellar travel have on the distinction between fact and fiction in adult learning, both for those adults on Earth preparing to undertake interstellar voyages and those in transit to other star systems? What evidence do we have to date, if any, that traditional distinctions between fact and fiction in formal academic settings are beginning to display emergent properties of convergence between fact and fiction? Evidence within contemporary higher education suggests that convergence—as a positive dynamic between science fiction and general fact—is beginning to occur in courses that integrate science fiction into factual discourse. The fact that science fiction is leaving the literature department is not happening in isolation. Rather, if we observe this departure in the context of the development of fields such as computational astrophysics, we can look at a pedagogical trend that increasingly accepts the boundary between fact and fiction as being fuzzy. Although, within social sciences, extensive discourse analysis on the legitimation of fact has occurred, the perspective in this paper has focused on the broader question of utility in the context of the unique challenges posed by interstellar exploration, technology, and colonization. It is surmised that interstellar pedagogy—distinctly referring to pedagogical methods aimed at developed competencies required for interstellar travel—will necessarily be a pedagogy of fuzziness.

Keywords

science fiction, discourse, pedagogy, convergence, simulative studies

Sections

1. Introduction
2. A Review of Fact, Fiction, and Contextual Interpretation
3. The Context and Challenge of Interstellar Education
    3.1 Non-FTL travel as time travel; Problems in Prediction & Knowing
4. The Emergence of Science Fiction as a Non-Literary Academic Field
    4.1 Science Fiction Leaves the Literature Department

4.2 Science Fiction within Politics and Utopian Studies

That's all science is, and history—stories about why things happen or happened. They are never, never true—never complete and always at least a little bit wrong, and we know it. But they're true enough to be useful.
—Orson Scott Card, *Shadows in Flight*

1. Introduction

This paper is an exploratory study of the categorization of fact and fiction in the context of future interstellar education. What impact will interstellar travel have on the distinction between fact and fiction in adult learning, both for those adults on Earth preparing to undertake interstellar voyages and those in transit to other star systems? What evidence do we have to date, if any, that traditional distinctions between fact and fiction in formal academic settings are beginning to display emergent properties of convergence between fact and fiction?

This paper is divided into 4 sections: (1) a review of current research on science fact and fiction discourse, (2) an analysis of the unique pedagogical challenges of interstellar education systems and goals, (3) an analysis of existing evidence of convergence—the synthesis of science fact and fiction in formal academic settings, and (4) the implications of current convergence for the future of interstellar educational institutions and systems.

The working hypothesis of this paper is that the unique conditions of interstellar travel, both from the perspective of non-Faster-than-Light (FTL) travel induced time dilation and from the perspective of literally exploring new worlds, will demand a new type of competency development in interstellar education—one that converges fact and fiction and replaces insistence on their certainty with probabilistic utility. Convergence is defined here as the integration of fiction into factual analysis, and a blurring of the boundary between fact and fiction in formal pedagogical contexts.

Interstellar education, as I define it for the purpose of this study, does not refer to the class of contemporary studies about interstellar systems, objects or phenomena. These, accordingly, are represented as sub-disciplines of academic fields; for instance, interstellar travel or interstellar communication, both of which belong to the broader disciplines of space transportation and communication. Interstellar education refers here explicitly to the vocational education of two organizational units, composed of individuals from various disciplines. The first unit is that of individual experts at the current time who, as a unit, are developing educational tools intended to facilitate the development, management, and operation of interstellar transportation systems. The second unit consists of the intended travelers of such systems. In this regard, interstellar education is a sub-discipline of the field of pedagogy, as its primary purpose is to understand the unique theoretical and competency challenges involved in adequately preparing both units to be successful in their undertakings.

2. A Review of Fact, Fiction, and Contextual Interpretation

The study of the relationship between fiction, specifically science fiction, and fact is not new. Research within the fields of sociology of science, literary analysis, scientific pedagogy (Barnett, Kafka, 2007) and discourse analysis has, however, largely focused not on convergence to fuzziness, but on entanglement between fact and fiction, and on efforts to disentangle the one from the other. In other words, convergence, in the way that I am using the term, is largely seen as a negative—a "mistake" that occurs most often when fiction is passed off as fact. Although it is possible to frame the nature of this mistake through the lens of discourse analysis explicitly -via, for example, Bourdieu's model of legitimate language and social power in determining constructs such as fact and fiction (Bourdieu, 1982), I will be using a more generalized utilitarian approach throughout this work focusing on the question of what is useful and/or needed in the peculiar context of interstellar education.

An excellent example of a common critique of this entanglement can be seen in the discourse analysis of nanotechnology writing, also known as nanowriting. A number of scientific and literary critics have focused on nanowriting, because of its apparent tendency to:

> Incorporate individual experiments and accomplishments in nanoscience into a teleological narrative of "the evolution of nanotechnology," a progressivist account of a scientific field in which the climax, the "full potential," the "dream" of a nanotechnology capable of transforming garbage into gourmet meals and sending invisible surgeons through the bloodstream, is envisioned as already inevitable. (Milburn, 2002, pp. 263-264).

This narrative, critics point out, tends to convey a "sense of inevitability that [future nanotech successes] will come in time" (Milburn, 2002, pp. 263-264). The scientific work of nanotech authors often contains these suggestive narratives of certain futures embedded within scientific text. Discourse analysts point out that this type of writing ultimately means that nanotechnology is "less a science and more a science fiction" (Milburn, 2002, p. 265). This perspective, as earlier mentioned, is one that views convergence as negative and sees the proper discursive response as corrective.

Jean Baudrillard, on the other hand, a frequent writer on the relationship of science to science fiction, contextualizes this entanglement differently. He classifies three "orders of simulacra [fiction] —the counterfeit, the reproduction, and the simulation" (Milburn, 2002, p. 267). Of these three, nanowriting and other writing that entangles science fact and science fiction is perceived as a more neutral and natural phenomenon. In this context, according to Baudrillard, "The real and its representation deteriorates, and [...] there is no real, there is no imaginary except at a certain distance" (Milburn, 2002, p. 267). Baudrillard speaks directly of the mechanics of convergence without naming them as such. "Science and science fiction are no longer separable. The borderline between them is deconstructed" (Milburn, 2002, p. 268). However, and this is crucial, Baudrillard does not conclude this deconstruction be wholly positive. Nor does he address the implications of this phenomenon for distant-future pedagogy and competency development. Instead, he depicts the process as "reinventing the real as fiction, precisely because it has disappeared from our life" (Milburn, 2002, p. 268).

Researchers such as Kitchin & Keale (2001) interpret discourses of fact and fiction as being highly entangled.

> Interesting recursive relationships are developing between novelists and academic scientists, professional engineers, computer programmers, the military, social scientists, politicians, musicians and 'lifestyle communities'. To us these relationships are important because they illustrate the value of these fictions and the ways in which they become incorporated into other ways of exploring and making the future (Kitchin & Keale, 2001, p. 23).

For Kitchin and Keale (2001), fiction becomes a critical component of technological and social development.

This utilitarian view of fiction also reflects a more complex understanding between scientific fact and the way in which fact is morphed into knowledge through scientific narrative. The focus on the narrative component of the dispersion of scientific fact brings us closer to an understanding of how convergence can occur in practice and how a switch from certainty to utility can positively impact pedagogical structures in contexts of incredibly high uncertainty, such as interstellar travel. As Richardson (2000) points out, "When we view writing as a method, however, we experience "language-in-use", how we "word the world" into existence (Richardson, 2000, quoted in Rose, 1992) And then, according to Richardson, we, "reword the world, erase the computer screen, check the thesaurus, move a paragraph, again and again. This *worded world* never accurately, precisely, completely captures the studied world, yet we persist in trying" (Richardson, 2000, p. 923).

The difficulty in capturing the facts of the world increases as our context of "local and temporal" decreases (Bury, 2001, p. 264). Interstellar travel represents, with the exception of inter-dimensional or inter-galactic travel, the most extreme departure from local and temporal context that we, as a civilization, are currently addressing. Darko Suvin's (1979) definition of science fiction as the "literature of cognitive estrangement" best captures the implicit need for convergence (Suvin, 1979, quoted in Freedman, 2000, p. xvi). In the absence of fact, or even of a framework for assessing the pertinence of fact, fiction and exercises in its classification and application, may be the only way that we can meaningfully advance learning in contexts of interstellar competency development.

The subsequent investigation into the current dynamics of convergence of science fiction literature in factual academic settings is meant to determine, first and foremost, if convergence is happening at all and under what conditions. Only then can broader questions of the implications of convergence be discussed—not as a negative entanglement, but as a positive phenomenon with clear consequences for competency development in interstellar social systems.

3. The Context and Challenge of Interstellar Education

Independent of the engineering and technological challenges that humanity faces today with the prospect of developing both interstellar travel and interstellar civilization, interstellar education poses a number of unique epistemological challenges. These are not challenges related to the absence of engineering knowledge, for information that precludes engineering knowledge in this area will arise organically within traditional technological development curves. Rather, I am speaking of two epistemological challenges that will confront interstellar education from a perspective of curriculum development and content classification. The first uniquely concerns a non-FTL technology regime. The second concerns the pedagogical problem of working consistently with so-called unknown unknowns (Taleb, 2007). In both cases, however, the virtual impossibility of acquiring factual information will force a convergence of fact and fiction within formal academic contexts.

3.1 Non-FTL travel as time travel; Problems in Prediction & Knowing

Should FTL travel be made feasible sooner than non-FTL travel between star systems, then the following issues related to time dilation will no longer be relevant. It is assumed for the purposes of this paper that engineering technology for interstellar travel will progress from non-FTL to FTL travel and not the other way around. However, it is not precluded that the opposite may occur.

For purposes of pedagogy and curriculum development, non-FTL interstellar travel needs to be treated as equivalent to future time travel, from the perspective of preparation and pedagogy. Any transportation technology capable of interstellar travel within a functional timespan will invariably involve some form of either sleeper/hibernation ships and/or speeds that come close to light speed. If we are looking purely at sleeper/hibernation ships, then the duration of time involved will probably be so long that any return to Earth (without so-called generational ships that only involve the hypothetical return of offspring to earth in a distant future) will result in the original crew effectively traveling forward in time and arriving at a future Earth that they do not know and cannot possibly anticipate. If we are looking at non-sleeper/hibernation ships that are capable of near light speed, then the effects of time dilation for the crew will produce similar effects, with regard to preparation and knowledge, to those of sleeper ships: effective future time travel due to time dilation with the crew coming back and having aged at a slower rate.

In either case, we can treat the epistemological problem of non-FTL interstellar transportation as being analogous to future time travel. The problem, from a pedagogical perspective, now becomes clear: how do we educate for the future? Not the immediate future where none of us has a strategic advantage or shortcut in time, but an asynchronous future, where interstellar travelers are at a loss pedagogically relative to their slower-moving peers.

Although technology promises to improve our powers of forecasting, the mere fact that nobody can with any accuracy tell you exactly the weather or the stock market closing price tomorrow with any certainty reveals our limitations. An excellent example of the broader pedagogical challenge this situation creates can be seen if we look at predictions made in time capsules about the future and the extent to which such predictions tend to be wildly wrong. A time capsule sealed in 1987, for example, predicted that by 2012 we would be "living in space and on the Moon, [complete] an expedition to Mars, have much industry located off-planet, cure multiple sclerosis and Parkinson's disease and [build] a network of levitated superconducting trains under construction in Western Europe and in Japan" (Angelica, 2012). Magazines and books on the history of science and technology abound with errors in prediction. This should come as no surprise, given the incredible diversity of complex systems and agents in the world. However, no one of us has yet dealt with the issue of future time travel as a result of time dilation. And there are two dimensions to this that need to be considered: not just the future of our own planet, but the fact that we, as true interstellar traveler, are also presumably going to the future of other planets and civilizations.

Attempts at modeling prediction and developing pragmatic schemas for prediction algorithms have been made, but have only been successful in cases constrained by parameters in the natural sciences. One of the best examples of the scientific attempt to model predictions came about as a result of the evolution of the Periodic Table of Elements. "Ziman (1978)," for example, "believes that the '...fundamental purpose of science is to acquire the means for reliable prediction'" (Brito, Rodriguez & Niaz, 2005, p. 87).

> Analyzing the process of the development of the Periodic Table, Brush (1976) suggests the following schema for types of predictions: "(a) Contra-prediction: foretelling the existence of unknown elements and their properties. Brush explicitly points out that the discovery of gallium was a contra-prediction. (b) Novel prediction: correction of some of the existing atomic weights by Mendeleev (e.g., beryllium changed from 14 to

> 9, uranium changed from 120 to 240, tellurium changed from 128 to 125). (c) Retrodiction: explanation of a fact, known before the theory was proposed" (Brito, Rodriguez & Niaz, 2005, p. 87).

However pragmatic such modeling of prediction may be and may, indeed, become significant in the future in forming new category of boundaries between fact and fiction in pedagogical settings, eventualities that are as open-ended as effective time travel through time dilation may render such attempts at predictive categorization obsolete. As Nassim Nicholas Taleb (2007) observed, "Prediction requires knowing about technologies that will be discovered in the future. But that very knowledge would almost automatically allow us to start developing those technologies right away. Ergo, we do not know what we will know" (p. 173).

Taleb (2007), in his analysis of so-called Black Swan events, elaborates on the problem of unknown unknowns. The problem refers to the more general classification of knowledge according to models of knowing. According to the schema, there are things we know we know (we are certain of our knowledge about such things), there are things we know we don't know (specific things that we know exist as information/knowledge, but we do not know that at this given moment; presumably we have some method for retrieving or finding such knowledge), there are things we don't know we know (tacit knowledge; we are not aware consciously that we know them), and, finally, there are things we don't know that we don't know—the so-called unknown unknowns. This refers to classes of knowledge that we do not even know we don't possess.

Before discussing the problem of pedagogical models that confront unknown unknowns, let us look at what Taleb (2007) provides as a simple case of known unknowns and at the problem of even basic prediction that arises once conditions change within a relatively closed system even more than once. Taleb (2007) uses the example computed by Michael Berry, the mathematician:

> Concerning a game such as billiards, he says, "If you know a set of basic parameters concerning the ball at rest, can compute the resistance of the table (quite elementary), and can gauge the strength of the impact, then it is rather easy to predict what would happen at the first hit. The second impact becomes more complicated, but possible; you need to be more careful about your knowledge of the initial states, and more precision is called for. The problem is that to correctly predict the ninth impact, you need to take into account the gravitational pull of someone standing next to the table. And to compute the fifty-sixth impact, every single elementary particle of the universe needs to be present in your assumptions" (Taleb, 2007, p. 178).

What is commonly referred to as the problem of changes in the weather is due to a butterfly beating its wings in the South Pacific, is just the simple problem of potential combinations quickly out-pacing our ability to compute and currently estimate the future consequences of current actions.

Critics of Taleb's (2007) emphasis on the 'unpredictability' of the future, on the other hand, point out numerous models and systems that have been developed that are quite accurate in their ability to both predict and model future events. They claim that Taleb's (2007) view is "anti-epistemic" because it presents life as "fundamentally unpredictable" (Lee, 2012). At the same time, they point to the development of models specifically linked to the prediction of technological progress as proof that the future, within certain constraints, can be broadly, if not specifically, predictable:

> The long wave theory of economic development, propounded by Nikolai Kondratieff and Joseph Schumpeter, for example, places innovation within economic cycles of "boom and bust" which average 50-60 years in duration...There is [also] a theoretical background of long-term regularities in economic behavior which provides a framework for understanding the patterns of innovation and for making technological predictions (Lee, 2012).

It would appear, from our perspective, that the accuracy of models in predicting the future may render moot the need for fact/fiction convergence from a pedagogical perspective. This would be true if the sphere of activity that competencies evolved from convergence designed pedagogical experiences was confined to Earth or at least to the more proximate interplanetary region. Interstellar education, however, poses challenges that are not only temporal, but that may be so fundamentally different that no amount of current knowledge can truly prepare for them.

Thus, the problem of educating for these unknown unknowns as eventualities in both time and place, as I would argue, is evident: we cannot. Or, more accurately, we cannot do so in a way that treats traditional boundaries between fact and fiction as conditions for the acquisition and interpretation of knowledge. In effect, because of the fundamental epistemological problem of educating for unknown unknowns, as I will shortly explain, I contend that formal interstellar education will have to, in the majority of sciences (natural sciences included) treat fact and fiction as a synthesized form.

Evidence to date suggests that one of the critical aspects of this synthesized form will be based on analysis through stress-testing in extreme conditions with the implicit understanding that the factual nature of these conditions is not as significant from a pedagogical perspective as the ability to rehearse and understand how to react to extremes. This competency may indeed prove more significant than fact/fiction disassociation. As Taleb (2007) points out:

> I don't particularly care about the usual. If you want to get an idea of a friend's temperament, ethics, [...] you need to look at him under the tests of severe circumstances, not under the regular rosy glow of daily life. Can you assess the danger a criminal poses by examining only what he does on an ordinary day? Can we understand health without considering wild diseases and epidemics? Indeed the normal is often irrelevant (p. xxiv, prologue).

Indeed, if we examine the ways in which fiction is currently being used in traditional non-fiction university courses, we will see direct evidence of this stress-testing being a critical aspect of the pedagogical utility of convergence.

4. The Emergence of Science Fiction as a Non-Literary Academic Field

Formal fact/fiction convergence in structured higher education academic settings is, at this moment, emergent. That is to say, it is beginning to occur at sporadic intervals in non-traditional course and genre structures. This analysis is not intended to be a history of science fiction pedagogy and its permutations in the classroom. Rather, it is meant to be an inquiry into the particular ways in which fiction, and most notably science fiction, has begun to be synthesized into formal studies of non-fiction history and theory with an emphasis on fiction being used to provide insight into and understanding of fact.

The development of the Uchronia Genre, as a sub-genre of science fiction, is one of the most significant trends from a categorical and pedagogical perspective. The Uchronia or 'Alternative History' genre, also referred to as "allohistory, counterfactuals, if-worlds, uchronia and uchronie, parallel worlds, what-if stories, abwegige geschichten, etc." (Schmunk, 2014), refers to a subset of science fiction that meets certain criteria. These criteria set it apart from other science fiction works and are based on specific aspects of plot and historical consistency with the past, with the exception of what is usually considered a key or critical event. According to one of the authorities on the sub-genre, Robert Schmunk (2014), "Alternate history is the description and/or discussion of an historical "what if" with some speculation about the consequences of a different result. Alternate history somehow involves one or more past events which happened otherwise and includes some amount of description of the subsequent effects on history " (Schmunk, 2014). While a list of relatively famous Uchronia titles is beyond the scope of this work, books in this genre tend to focus on 'significant' historical events, such as WWII or the assassination of JFK and examine what-ifs. What if Kennedy had survived the assassination? What if Hitler had been assassinated? What if the British had won the American Revolution? Etcetera.

Although Uchronia, as a distinct sub-genre, has not yet come into visible use in academic curricula outside of specialized literature classes, science fiction has. While the majority of science fiction courses, just as fiction courses, are taught within departments of English literature and/or writing, it is the exceptions that are of interest in the context of this work. The exceptions as we shall shortly see demonstrate the beginnings of 'formal convergence', i.e. the replacement of clear distinctions between fact and fiction and the emphasis on utility.

At the current time, dozens of universities across the United States and the world provide courses in Science Fiction. What is interesting about these course provisions is that almost all of them occur within the department of literature (either that of English Literature, or that of a respective language that has a substantial and historical body of science fiction text such as French, German, or Russian) or within the department of creative writing.

The Department of English at the University of Chicago, for example, offers a course called "Science Fiction in the 19th century." The course outline states that, "We will attend to the ways in which these texts refigure the anxieties about and promises of science at a particular moment in history, paying attention to the ways in which this contributes to an alternative history of science fiction" (University of Chicago, 2008). Although the course appears in the Department of English, just like in many other universities that teach science fiction, the outline highlights a broader point about the arbitrariness of department boundaries when it comes to the issue of fact/fiction discourse.

Invariably, virtually every course in both science fiction and fiction more generally has to be, by virtue of the methods of literary analysis, also grounded in the historical and socio-political context of its day—to say nothing of the need to understand basic natural sciences such as physics, in order to properly read much of science fiction. This raises the broader question about fiction in general and the way in which it is both classified and taught.

Formal classification, such as the Dewey Decimal System (OCLC, 2014), is largely responsible for division of books into categories of fiction and non-fiction. What is little known, however, is that this 135-year-old classification system is not a public consensus. Rather, it is a trademarked and licensed system owned by the Online Computer Library Center (OCLC, 2014). And while the system's hierarchy is both relative and quite flexible in adding new books to existing libraries, it ultimately forces a choice upon librarians: is this a work of fact or a work of fiction? That choice spills over as a mandate—an assumed certainty—in academic curricular development and in the discourses generated about literature, history, and human society. As a simple example, consider the following: are religious texts works of fact or fiction? At what point is a historical religious text substantiated enough to 'move across the aisle' from being a work of fiction or myth to being a history of a people? If a person writes an autobiography based on a partially imagined life, is it a work of fact or a work of fiction?

While extreme cases, where determination of fact and fiction are easy, seem to justify the utility of the system, the point where facts/fictions converge reveals the fragility of the premise that one can easily classify things into decisive facts or fictions. We can also ask the broader question, from a utilitarian perspective, whether we benefit from this forced separation. An example of the value of synthesis in formal pedagogy exists in my own personal experience. My junior year curriculum in high school had a synthesized program called 'American Studies'. The program brought together American literature and American history into a single course. The course was based on the premise that American literature could not be properly understood without discussing and learning about the socio-political context in which it was written. At the same time, American history could not be understood without analyzing the development of its culture and thought, manifest in its literature. As the boundary between so-called fact and fiction blurred, it became possible to understand much more deeply the changes that occurred and the reasons for them within the historical narrative.

It is only by questioning the fundamental assumptions of the socio-cultural systems and institutions that we inhabit that we can make real progress. From this perspective, I believe it is imperative to treat fact and fiction as social constructs. As we look at the way in which science fiction is beginning to be formally used outside of the domain of literature and writing, I believe we will see the immediate benefits, in terms of advancing technology and welfare, of convergence.

4.1 Science Fiction Leaves the Literature Department

This is one of the most visible areas of convergence: the recent development of either inter-disciplinary science fiction or hybrid university courses situated in departments other than English, Comparative Literature, or Writing. Although the construct of 'department' or 'field of study' may appear to be driven entirely by the utility of transparent and common classification, the reality is more complex. Departments and fields of study also represent, from what Bourdieu (1982) would cite as a relational perspective, 'fields of power'; delineators of the scope of power that departments and fields of study exercise on both students and society. One of the best explanations of this comes from Loïc Wacquant (2013). He writes:

> Pierre Bourdieu's reframing of the question of class [...] highlights the key conceptual shifts effected by the French sociologist in [order] to forge tools for elucidating the broader politics of group-making: the socio-symbolic alchemy whereby a mental construct, existing abstractly in the minds of individual persons, is turned into a concrete social reality acquiring existential veracity as well as historical potency outside of and over them (Wacquant, 2013).

From a perspective of convergence, this socio-symbolic alchemy displays clear evidence of fuzziness, albeit at a level of institutional narratives: 'facts' about social constructs and institutions, such as a department of 'political science', are also fictional constructs, begun in the minds of individuals and transposed through collective acquiescence into social facts. This reinforces the idea of a department of field of study in the university being in a constant state of flux: operating in a recursive relationship with the collective knowledge of society at large (Giddens, 1984).

The appearance of science fiction courses in non-literary academic fields of study is a critical emerging trend from the perspective of competency demands in future interstellar education. The reason is simple: the shift in emphasis from certainty to utility reflects the challenge of interstellar education where both time and space will be experienced in ways incredibly different from the collective experience of human education and memory.

The following review and discussion about these courses is not meant to be a comprehensive listing of science fiction as a non-literary field of study. Rather, it is to illustrate two things: the diversity of fields in which science fiction has already come to be used and the evidence of convergence in the presentation of course topics and outcomes.

4.2 Science Fiction within Politics and Utopian Studies

A number of universities currently offer science fiction content and analysis as part of a political science curriculum. The course, "Imagined Planets: Women, Isms, and Speculative Fiction" is taught in the Global Gender Studies Department at the University at Buffalo-SUNY (SUNY, 2011). The course outline states that:

> This course will examine issues of "difference"-such as race, gender, class, and sexuality-as imagined and narrated by contemporary women writers of speculative (science) fiction. By doing so, we hope to interrogate the impact of 20th century resistance/social movements-in particular, the civil rights, feminist, lesbian and gay, and human rights movements-on women writers who've chosen to "write against the box" of prevailing literary expectations. We also hope to consider (1) how women writers of speculative fiction subvert, or resist, the status quo; (2) notions of speculative fiction as "escapist" (meaning, "less serious") literature; (3) the uses of speculative fiction as blueprints for imagining new social orders (SUNY, 2011).

The most important component of the course at SUNY Buffalo is, from our perspective, in the first and third objectives for consideration. Concerning the first, an assumption is made that writers of speculative fiction are writing not just to 'entertain', but also to subvert or resist a status quo. This is also consistent with the idea of speculative fiction acting as a 'blueprint' for new social orders. The usage and analysis of fictional text as an agent of social and political change reflects both the mutability of language in the context of power 'constructs' (Bourdieu, 1982) and the phenomenon of convergence in formal academic settings.

A similar course is offered at the University of Colorado, Denver, in the Department of Political Science (Depauw, 1996). This course, entitled "Utopian & Dystopian Fiction and Drama," outlines "Political, philosophic, and literary examination of classic and contemporary works of utopian and dystopian fiction. Fictional visions of wonderful and terrible societies we might become" (Depauw, 1996). Along with such classics as *1984* (Orwell, 1950), the genre of utopian/dystopian fiction is well known. What is less studied formally, is the relationship between literary fictions that imagine such societies and the real world attempts to create societies, or the way in which political and historical events are influenced by works of fiction.

One could argue, from a perspective of societal development, that the relationship between fact and fiction has always been, historically, a fuzzy one and has existed in a fuzzy state ever since religion and mythology were used to establish the rules and social structures of ancient civilizations. Certainly, the architectural and artistic remnants of ancient Egypt and ancient Greece offer proof of the power of this convergence, wherein a significant portion of a society's wealth was invested in the support of religious myths that were believed to be true.

At the same time, however, since the time of Descartes, an objective realism has descended upon the academy and upon fields of scholarship. So, what was historically true in social life became categorically false in academic analysis. The precursor to the beginning of convergence between academic fact and fiction actually belongs not to the field of science fiction, but to the field in which its separation was first demanded: religion. Specifically, the work of the German philologist and scholar of Greek religion, Walter Friedrich Otto, was among the first to purport a model of academic inquiry that demanded convergence of supposed fact/fiction as a necessary criteria for analysis and understanding. Otto (1965) writes in the introduction to his analysis of the cult of Dionysus:

> Up to this point I have spoken constantly of the reality of deity, even though it is custom to speak only of religious concepts or religious belief. The more recent scholarship in religion is surprisingly indifferent to the ontological content of this belief. As a matter of fact, all of its methodology tacitly assumes that there could not be an essence which would justify the cults and the myths." (p.24)

The Political Science Department at the University of Iowa, Iowa City, offers a course entitled "Current Political Theory—Science Fiction as Political Theory." The class, according to the outline, is a "topics course, new each time in themes and readings. So far the course has explored (1) Sf as Syzygy, (2) Orwell's Political Myths and Ours, (3) Subgenres of Sf, (4) Deconstructing Modernity in Sf, (5) Sf as feminist theory, (6) Dreams as Realities in Sf, and (7) Sf as Green Politics." (Depauw, 1996)

The stated objective of the course is to treat science fiction as a "form of political theory" (Depauw, 1996). Here we have another example of direct convergence. While one could argue that this approach to the analysis of fact mirrors the philosophy of pragmatism, in that it reduces a form of media to its mere pragmatic utility in a different field of study, it also signifies something unique to the idea of convergence: not merely that fiction is used in political discourse and that its usage is evidence of pragmatism, but that fiction is treated directly as a form of political theory and discourse and that fiction can both signify and reflect real politics.

4.3 Science Fiction within Cultural Studies

The University of South Carolina has an Institute for Southern Studies, which teaches a course called "Speculative Fiction & Southern Culture" (ISS, 2014). The course outline states:

> How did the U.S. South come to be a place where zombies, vampires, and ghosts feel right at home? This course uses the lens of speculative fiction—fantasy, horror, and science fiction—to study representations of southern culture and to ask critical questions about how the fantastic can convey ideas about the region's people, their social realities, and the physical and emotional landscapes that map their history...Our study on speculative fiction genres will be coupled with critical analysis of more recent texts to consider: how southern vampire and zombie tropes foreground the social transgressions and excesses of humanity; what happens when superheroes take on the villains of the Deep South; and the way time travel and alien abduction metaphors are used to tell the story of American enslavement" (ISS, 2014).

Fiction, as it is used in the context of this course, demonstrates the tendency that I believe is critical to the evolution of convergence: the transition from certainty to utility. Interestingly, what the course description and literature omits is the specific significance of the relatively unique religious history of the South as a critical factor in the development of its geographically contextual contemporary fantasy and horror science fiction:

> Because of its unique blend of French, Spanish, and Indian cultures, New Orleans [as a central point for the development of US Southern Culture] offered a perfect setting for the practice and growth of Voodoo. In 1809 many Haitians who had migrated to Cuba during the Haitian revolution found themselves cast out and came to New Orleans. They brought with them their slaves who incorporated their rites and beliefs to those of the existent slave population - Africans from Senegal, Gambia, and Nigeria. Voodoo in Louisiana was enriched and revitalized. Voodoo was not only their religion, it was also their natural medicine, their protection and certainly a way of asserting and safeguarding a sense of personal freedom and identity" (Singh, 1994).

Voodoo, from the perspective of convergence, is quite interesting. It is, from the start, a synthesis of fact and fiction that manifests through ritual, belief, and action. At the same time, it reflects the pervasiveness of fuzziness in human social life in the sphere of religion: where fiction (in the form of largely unprovable historical or contemporary personal experience of spirituality) is acted up as fact through the mechanism of faith.

Another example of science fiction appearing within culture studies comes from the Core Curriculum at the University of Massachusetts, Boston (Depauw, 1996). This course, entitled "Cultural History: Mars, 1877-2019", aims to study the "nature, methods, and uses of cultural history by examining in some detail a single example: how scientific and literary images of Mars during the past century have mirrored and expressed cultural ideas and values" (Depauw, 1996). This course examines the way in which science and literature (read: fact and fiction) have been intertwined in the case of Mars. Given its proximity, and the fact that there have been centuries of observation of the 'red planet', it should come as no surprise that Mars, more than any other celestial body, should inspire both such a cultural body of work and such a comprehensive convergence between fact and fiction. Without going into an exhaustive detail of the works written about the planet, one of the best examples of convergence in action can be read in Kim Stanley Robinson's (2014) trilogy "Red, Green and Blue Mars" (Mars Trilogy). Given the attention to scientific detail and the near-future setting of the trilogy, Robinson's (2014) work can be considered as a history of future or an anticipated history of future technological progress.

Another course that investigates this synthesis of science and society is taught under a program in American Studies is at Pennsylvania State University at University Park. The course is called "Cyberspace Aesthetics and the American Novel" (Depauw, 1996). The course attempts to "map out the relations between the American literary discourse of cyberpunk fiction and the technoscientific discourses of virtual reality, cyberspace, and molecular biology" (Depauw, 1996). What is interesting about this course is that it is using discourse analysis as a formal part of its structure and intended pedagogical outcomes. The course outline states further, "We will investigate the ways in which the new technologies of information and genetic engineering impact the very form of the American novel, as new modes of narration become possible and necessary in cyberpunk worlds where many of the grounding distinctions of humanist culture [...] become problematic" (Depauw, 1996). Ironically, instead of looking at convergence of fact/fiction, this course takes a different route: it uses changes in science and technology as a lens to investigate broader changes in discourse and culture. Fiction is not only being used to advance fact, but fact—in the form of new technology—is also being used to alter fiction.

4.4 Science Fiction in Hard Sciences

Further evidence of convergence is visible in the appearance of science fiction in traditional hard science departments. The University of Nevada, Reno, for example, runs a course titled "Science Fiction and Information Control" (Depauw, 1996), which is cross-referenced in the English/Sociology and Library Science Departments. Of these, the link to library science studies is the most interesting as it provides a critical insight into the way in which convergence can have practical implications for technical competency development. The course states that it will "examine the role that libraries [...] have played in science fiction and the importance of the transmission of knowledge, whether it be conveyed orally (*Fahrenheit 451*), statically stored (*The Foundation Trilogy*), or mechanically/electrically disseminated (Clarke's *2001*)" (Depauw, 1996). Science fiction is purposely being used to construct and critically analyze data modeling and information classification theory.

Kenyan College, in Ohio, runs a course in its Biology department entitled "Biology in Science Fiction." The course outline explores, "principles of biology through their extrapolation in science-fiction literature. Relationships between biology and society will be considered, as well as the literary context of science-fiction stories" (Depauw, 1996). Here, convergence is occurring in a different context: science fiction narratives are being used to contextualize scientific facts in biology. The narrative acts as a motivator from a pedagogical perspective, but the boundary between fact and fiction is explicitly retained.

More common is the integration of science fiction in ecological and environmental studies. Bucknell University, in Pennsylvania, offers a course called "Green Utopias" in its Environmental Studies program. According to the course overview, it is, "designed to bring attention to existing alternatives to a gray future caused by overpopulation, air, soil, and water pollution, clear-cut forests, defective atomic-power stations, etc." (Depauw, 1996). Here, convergence is more evolved. Science fiction narratives are used as points of analysis and reflect real consequences of current directions in resource usage and technological systems.

Although we often hear in the media about instances of convergence, such as a propulsion concept that mimics technology used on the television series Star Trek, for example, the reality in academia is more limited. Academic fields don't evolve overnight, and those that are interdisciplinary face challenges within academic circles. The field of astrobiology is an excellent example of this - to whom does the field belong? Which actors have the right to define and shape the field? Biologists? Astronomers? Geologists? Can it belong to all and none? And this field is contained within the hard sciences. As fields of study such as interstellar communication and astrosociology emerge into their own they will likely face even stiffer structural challenges within academic systems.

5. Contemporary Implications of Convergence

This paper presents a cursory overview of evidence within higher education of the convergence of fact and fiction. This overview is limited to a brief snapshot of instances where science fiction leaves the literature department and emerges as a part of formal discourse in non-fiction academic settings.

The examination of convergence is meant to demonstrate what is believed to be a necessary pedagogical criteria for successful interstellar travel and communication: a model of synthesizing fact and fiction under certain conditions of unknown unknowns that are prevalent in interstellar travel and will, most likely, continue to be frequent even after such travel becomes commonplace.

Before we get there, however, I believe it is critical to examine what the implications of contemporary convergence are, and how these dynamics may accelerate our ability to develop interstellar communication and transport technologies. Specifically, there are two implications that I would like to review: (1) a broader shift towards convergent inter-disciplinary inquiry within higher education and (2) the emergence of simulative academic fields.

5.1 A Shift Towards Convergent Inter-disciplinary Inquiry

The University of Chicago runs a Big Problems program within the college (UChicago, 2014). John Hopkins University just developed an inter-disciplinary MS program in Government Analytics (JHU, 2014). Florida Institute of Technology currently offers a BA degree in Astrobiology (FIT, 2014). The list goes on. The point is simple:

technological complexity and social complexity exist in a recursive space. Acquiring competencies to successfully operate in that space requires an increasing allocation of resources to analysis of complex interactions that can only be studied in an inter-disciplinary academic structure. We should not be surprised, however. Inter-disciplinarity is not a transient phenomenon unique to this moment in time. Rather, it represents the continuing branching of knowledge and the maturation of new fields that were, only recently, just ideas or theories. What is interesting in the context of this study is not the inter-disciplinary inquiry itself; rather, it is the evidence of convergence, which as this study shows, is beginning to occur but only at the singular course level at this time and not at the programmatic level.

Given the unique competency challenges required for development and operation of an interstellar exploration regime, it is anticipated that, over the next several decades, convergence will increase from the course level to the programmatic level. In other words, fiction will increasingly be used as a pedagogical tool to develop and accelerate new facts.

5.2 Emergence of Simulative Fields

The history of simulation as a valid form of data generation goes back to the Manhattan Project and the development of the Monte Carlo Method by Stanislaw Ulam and John Von Neumann (Eckhardt, 1987). Since that time, the usage of computer simulation as a form of factual data generation has grown exponentially, especially in fields where laboratory experiments, such as astrophysics, can be difficult if not impossible. "By necessity, a third, modern way of testing and establishing scientific truth — in addition to theory and experiment — is via simulations, the use of (often large) computers to mimic nature" (Heng, 2014). One of the more interesting fields to develop is that of computational astrophysics. Simulated results in these fields are "often described as being empirical, a term usually reserved for natural phenomena rather than numerical mimicries of nature. Simulated data are referred to as data sets, seemingly placing them on an equal footing with observed natural phenomena" (Heng, 2014). Although simulated data sets are treated as analogous to empirical data sets, the problem of simulation pixel resolution persists:

> For multiscale problems, there will always be phenomena operating on scales smaller than the size of one's simulation pixel. This difficulty of simulating phenomena from microscopic to macroscopic scales, across many, many orders of magnitude in size, is known as a dynamic range problem" (Heng, 2014).

Information below the level of this pixel, although it may be critical to the quality and utility of data, is for all practical purposes fictional. It exists only in so far as we imagine it to exist.

The fact that we already permit a sort of fictional data as scientific fact illustrates the general tendency of convergence and also hints at what will probably develop in the near future: the dedicated and specific study of simulation in social sciences. These simulative fields of study will most likely combine both computational modeling and science fiction to form hybridized forms of traditional social science disciplines, such as simulative political science, etc. These disciplines, as departures from their traditional non-simulative variants, will largely focus on utility and critical analysis in the face of unknown or complex variations within social systems. The complexity and lack of knowledge that we face in dealing with any form of interstellar, or even interplanetary, exploration mirrors the competency demands that training in simulative fields of study provide.

6. Conclusion

Evidence within contemporary higher education suggests that convergence—as a positive dynamic between science fiction and general fact—is beginning to occur in courses that integrate science fiction into factual discourse. The fact that science fiction is leaving the literature department is not happening in isolation. Rather, if we observe this departure in the context of the development of fields such as computational astrophysics, we can look at a pedagogical trend that increasingly accepts the boundary between fact and fiction as being fuzzy. Although, within social sciences, extensive discourse analysis on the legitimation of fact has occurred, the perspective in this paper has focused on the broader question of utility in the context of the unique challenges posed by interstellar exploration, technology, and colonization.

It is surmised that interstellar pedagogy—distinctly referring to pedagogical methods aimed at developing competencies required for interstellar travel—will necessarily be a pedagogy of fuzziness, where the convergence of fact and fiction occurs because of the tremendous epistemological challenges of time dilation and unknown unknowns in our vast galaxy. Certainty will have to give way to utility, and perhaps a new type of construct will emerge that is neither fact nor fiction, but an expedient amalgam of both.

Acknowledgements

I would like to thank the Faculty of Pedagogical Sciences at APS, for their assistance. I would also like to thank Katerina Juszczak, my wife, for her dedication and extraordinary effort in editing and providing research support.

References

1. Angelica, A. D. (2012, August 1). 1987 time capsule predictions for 2012. *KURZWEIL Accelerating Intelligence.* Retrieved from: http://www.kurzweilai.net/1987-time-capsule-predictions-for-2012

2. Barnett, M., & Kafka, A. (2007). Using Science Fiction Movie Scenes to Support Critical Analysis of Science. *Journal of College Science Teaching,* 36 (4), 31-35.

3. Baudrillard, J. (1993). *Symbolic Exchange and Death* (I. Hamilton, Trans.). Theory, Culture & Society (Book 25). London, England: Sage Publications Ltd.

4. Bourdieu, P. (2013). The Production and Reproduction of Legitimate Language (G. Raymond & M. Adamson, Trans.). In P. A. Erickson & L. D. Murphy (Eds.), *Readings for a History of Anthropological Theory* (4th ed., pp.407-422). Ontario, Canada: University of Toronto Press Inc. (Reprinted from *Language and Symbolic Power*, pp. 43-65, by J. B. Thompson, Ed., 1991, Boston, MA: Polity Press; originally published as La Production et la Reproduction de la Langue Légitime, *Ce Que Parler Veut Dire: L'économie des Échanges Linguistiques*, pp. 23-58, 1982, Paris, France: Librairie Arthème Fayard).

5. Brito, A., Rodriguez, M. A., & Niaz, M. (2005). A Reconstruction of Development of the Periodic Table Based on History and Philosophy of Science and Its Implications for General Chemistry Textbooks. *Journal of Research in Science Teaching*, 42, 84-111.

6. Bury, M. (2001). Illness Narratives: Fact or Fiction? *Sociology of Health & Wellness,* 23 (3), 263-285. doi: 10.1111/1467-9566.00252

7. Cochrane, R. (2013, April 9). A love affair with science fiction - an interview with Dr. Robert Crossley. *The Mass Media.* Retrieved from: http://www.umassmedia.com/art_lifestyle/a-love-affair-with-science-fiction---an-interview/article_4ae4901a-a187-11e2-9735-001a4bcf6878.html

8. Depauw (1996) Source url: http://www.depauw.edu/sfs/backissues/70/courses70.htm

9. Desanctis, G. & Poole, M. S. (1994). Capturing the Complexity in Advanced Technology Use: Adaptive Structuration Theory. *Organization Science.* 5, 121-147

10. Eckhardt, Roger (1987). *Stan Ulam, John Von Neumann, and the Monte Carlo Method.* Los Alamos Science Special Issue, p. 131-143. Source url: http://library.sciencemadness.org/lanl1_a/lib-www/pubs/00326867.pdf

11. Evans, Arthur B. (1988) *Science Fiction Studies.* SF-TH Inc. Vol. 15, No. 1 (Mar., 1988). Stable URL: http://www.jstor.org/stable/4239855

12. FIT (2014). Source url: http://www.fit.edu/programs/7137/bs-space-sciences-astrobiology

13. Freedman, Carl. (2000) *Critical Theory and Science Fiction.* Wesleyan University Press.

14. Giddens, A. (1984). *The Constitution of Society: Outline of the Theory of Structuration,* Cambridge: Polity Press.

15. Heng, Kevin (2014) The Nature of Scientific Proof in the Age of Simulations. *American Scientist.* Vol. 102, No. 3. P. 174. DOI: http://dx.doi.org/10.1511/2014.108.174

16. ISS (2014) Source url: http://artsandsciences.sc.edu/iss/undergrad

17. JHU (2014) Source url: http://advanced.jhu.edu/academics/graduate-degree-programs/government-analytics/

18. Kitchin, Rob & Kneale, James. (2001) Science fiction or future fact? Exploring imaginative geographies of the new millenum. *Progress in Human Geography* 25, 1. pp.19-35. Source url: http://eprints.nuim.ie/3909/1/RK_Science__fiction.pdf

19. Lee, Michael. (2012) Shock-testing the Black Swan Theory. Source url: http://www.wfs.org/blogs/michael-lee/shock-testing-black-swan-theory

20. *Mars Trilogy* (2014). Source url: http://www.kimstanleyrobinson.info/w/index.php5? title=Mars_trilogy

21. Merritt, R. D. (2008). Student internships. *EBSCO Research Starters.* Retrieved from http://www.ebscohost.com/uploads/imported/thisTopic-dbTopic-1072.pdf

22. Milburn, Colin (2002) Nanotechnology in the Age of Posthuman Engineering: Science Fiction as Science. *Configurations.* 10:261-295. John Hopkins University Press. Source url: http://www.thing.net/~rdom/ucsd/posthuman/nanotechnology.pdf

23. OCLC (2014) Source url: http://www.oclc.org/dewey/resources.en.html

24. Otto, Walter Friedrich. (1965) *Dionysus. Myth and Cult.* Indiana University Press, Bloomington.

25. Raelin, J. A., Bailey, M. B., Hamann, J., Pendleton, L. K., Raelin, J. D., Reisberg, R., & Whitman, D. (2011). The effect of cooperative education on change in self-efficacy among undergraduate students: Introducing work self-efficacy. *Journal of Cooperative Education & Internships,* 45 (2), 17-35.

26. Richardson, Laurel (2000). Writing as a method of inquiry. In Norman K. Denzin & Yvonna Lincoln (Eds.), *The Handbook of Qualitative Research* (2nd ed., pp.923-948). London: Sage.

27. Ricouer, P. (1984) *Time and Narrative: Volume 1.* Chicago: University of Chicago Press.

28. Schmunk, Robert B. (2014) *Uchronia: The Alternate History List.* Source url: http://www.uchronia.net/intro.html

29. Singh, Severine (1994) *Voodoo Crossroads.* Black Moon Publishing. Source url: http://www.neworleansvoodoocrossroads.com/historyandvoodoo.html

30. SUNY (2011) Source url: http://globalgenderstudies.buffalo.edu/courses/undergraduate/descriptions/

31. Suvin, Darko. (1979). *Metamorphoses of Science Fiction: On the Poetics and History of a Literary Genre.* New Haven and London: Yale University Press.

32. Taleb, Nassim Nicholas. (2007). *The Black Swan: The Impact of the Highly Improbable.* Random House.

33. Thorndike, R. & Thorndike-Christ, T. (2010). *Measurement and Evaluation in Psychology and Education.* Boston: Pearson.

34. University of Chicago (2008) Source url: http://english.uchicago.edu/content/science-fiction-19th-century

35. UChicago (2014) Source url: http://franke.uchicago.edu/bigproblems/bp-index.html

36. Wacquant, Loïc. (2013) “Symbolic power and group-making: On Pierre Bourdieu’s reframing of class”. *Journal of Classical Sociology.* 0 (0) 1-18. DOI: 10.1177/1468795X12468737

Broadening the Perspective of Undergraduate Students to Prepare Them for Interstellar Travel: First Steps

Barbra Schuessler Maher, M.S.

Red Rocks Community College,

13300 W. Sixth Ave, Lakewood, CO 80228

John Schuessler, M.S.

Independent Consultant, Littleton, CO

Abstract

How does higher education need to shift to begin preparing interstellar citizens? Encouraging students to pursue careers in science, technology, engineering, and math (STEM) is one path. Currently, STEM fields are on track to outpace other fields in job growth in the next decade; but the number of graduates and the diversity of these students remain limited. Two-year colleges can play an important role in preparing a wide range of diverse students in STEM fields and increase the scientific literacy of all students to begin the shift to an interstellar perspective. This will be a necessary step.

Fostering the qualities of innovation, problem solving, communication, assimilation and synthesis of vast amounts of knowledge are specifically emphasized in STEM education. These qualities and skills can be expanded upon as a base for preparing the interstellar student. In order to achieve this, specific types of STEM courses to be developed, taught, and promoted in higher education include: Science and Society, Physics, Modern Physics, Astronomy, Astrobiology, and Cosmology. A special topics seminar could be developed to handle other topics such as synthetic biology, faster than light travel, teleportation, and self-replicating robotic space probes. The Colorado community college system has already begun this process. However, is STEM emphasis enough?

To better understand where educators need to broaden student understanding, we need to assess the receptivity of students to the idea of interstellar citizenship. Do they understand the challenges and the science? An initial survey of community college students will help us get a better understanding of the knowledge level and receptivity to the idea of interstellar travel. Following the survey we will use Structured Inquiry Methods (SIM) to take some of the mystery out of future things. Once completed, further curriculum development can take this process of interstellar education to the next level.

Keywords

education, community college, curriculum, interstellar

1. Introduction

What will it take to accelerate our progress toward living beyond our planet?
– Tau Zero Foundation [1]

This is an important question and education is a key component to the answer. Several related questions come to mind. By the time students come to college, what is their perspective on space? How do they view humanity's place in the universe? Do they view space exploration and development as important? Is immersion in science, technology, engineering, and math (STEM) curriculum enough? What about students that do not opt for STEM coursework?

In order for a significant acceleration to occur, more of the public and future generations need, at a minimum, to become scientifically literate, but preferably, well versed in STEM topics. STEM education promotes critical thinking and develops problem-solving ability and quantitative skills. Traditionally taught laboratory science classes develop students problem-solving and scientific discovery skills in a structured way. When the problem and procedure is given, it is considered a structured inquiry process. [2] Recent studies have shown that once a student has learned the methods of structured inquiry, they can be successfully guided to become proficient at open inquiry. [3] This creative problem-solving approach will be crucial for new leaps towards interstellar. Encouraging college students to pursue careers in science, technology, engineering, and math (STEM) is a beginning path. However, finding a way to engage a larger number of students is crucial for wider success.

A recent national survey discussed below highlighted some important gaps in public understanding and support of space exploration. In order to gain a better understanding of how current community college students compare to these national opinions, a survey was conducted at Red Rocks Community College. The national and local survey results lend some insight to how higher education can address the gap in scientific literacy and whether the public is ready for an interstellar effort.

2. National Attitudes Toward Space

The Survey of Public Attitudes Toward and Understanding of Science and Technology[4] is a social survey sponsored by the National Science Foundation (NSF), monitoring public attitudes towards science and technology and science policy. The latest survey was conducted by the National Opinion Research Center at the University of Chicago and analyzed by General Social Survey, GSS[5], in 2012. This survey sheds some light on attitudes towards many aspects of science and education. This was a starting point for thinking about where our students fit into the broader arena. Several relevant findings in the latest survey results released in 2014 illuminate the current state of public opinion on space. The first was the public interest in selected issues question in Chapter 7, where space exploration rated very low compared to other topics. Respondents said that 23.4% were very interested in space exploration, 43.8% were moderately interested, and 32% were not interested at all (See Figure 1).

The number of "very interested" responses from 1981-2012 shows interest in space peaked in 1988, and hit a low in 2008 (Figure 2). The last years showed a leveling off at 23%, which is a stark contrast to level of interest in new medical discoveries which remains close to 60%. This highlights a disconnect in public understanding of the place space holds in medical research and new medical technology. The overall trend for space exploration follows the general trend in interest in new scientific discoveries, just at a much lower level. This suggests a larger issue with scientific literacy in general.

The next section of the report that is relevant to understanding where the public places space exploration is the question of funding. When asked, respondents ranked space exploration very low compared to other areas. Only 22% thought we spent too little on space exploration; 41% said spending was about right; and 29% thought too much was being spent (Figure 3). In contrast, Education was the number one issue identified as being underfunded by the government (Figure 3).

It is clear that the public supports education and science topics, however understanding of how much the government actually funds space exploration is misguided and generally uninformed. The respondents were not told the percentages of government spending on the various issues in question.

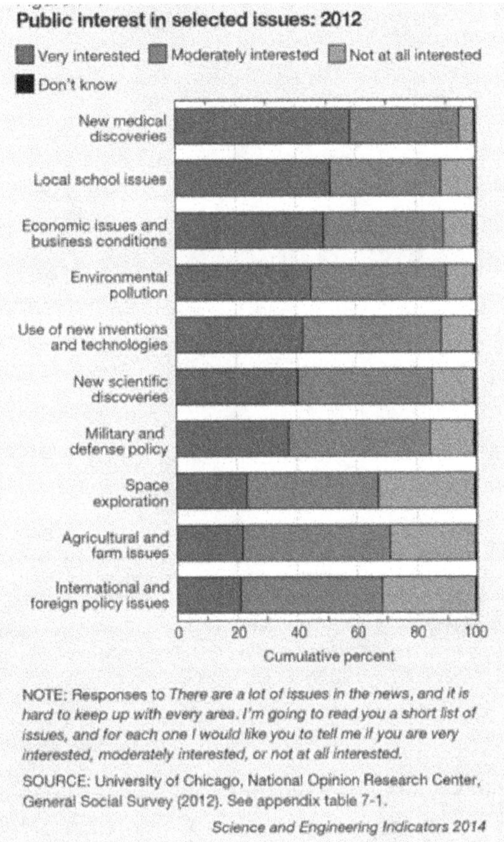

Figure 1: Survey of Public Attitudes Toward Understanding of Science and Technology results, Chapter 7: Science and Engineering Indicators. (National Science Foundation Higher Education Indicators, Chapter 7, Figure 7-1)

In a survey conducted by the Pew Research Center in 2014, a third of Americans said they believe there will be manned long-term colonies on other planets by the year 2064.[6] In an earlier survey conducted in 2010, 63% of respondents said that they believe astronauts will have landed on Mars by 2050.[7] More than half said that ordinary humans will be able to participate in space travel.

These are optimistic results, however the GSS survey clearly showed that support for funding space exploration was lacking. In fact, in 2012, the NASA budget took a 20% hit to its planetary science programs and total current funding for NASA is 0.5% of the federal government's budget. Even at the height of the Apollo program funding for NASA was only 4.4%.[8]

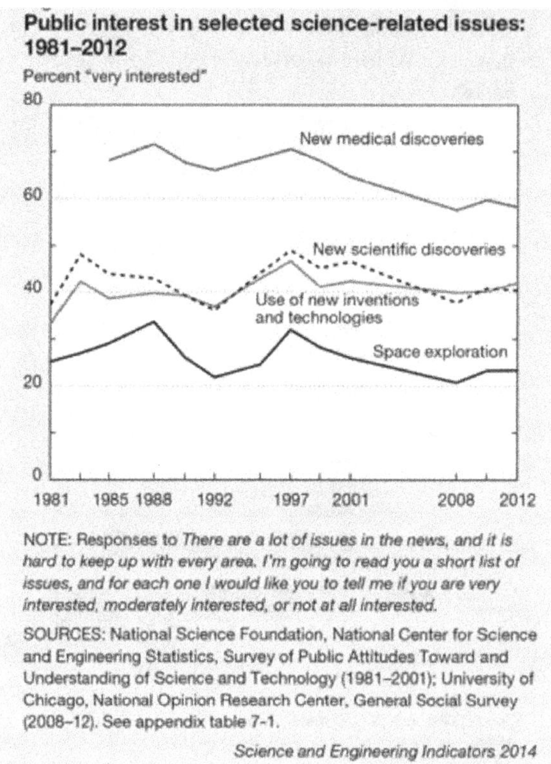

Figure 2: Trends in public interest from 1981-2012, responses ranking space exploration interest as "Very Interested". See the note for the specific language of the question asked. (National Science Foundation Higher Education Indicators, Chapter 7, Figure 7-2)

3. Community College Environment and STEM

National surveys provide a starting point for looking at how and where higher education can make an impact on the science and technological breakthroughs needed for interstellar travel. Higher education will play an instrumental role in preparation. Community colleges will a have a unique and important role in this preparation. Community colleges traditionally enroll more part-time students, lower-income, single parents, veterans, and dislocated workers than four-year universities. This creates an incredibly diverse student population.

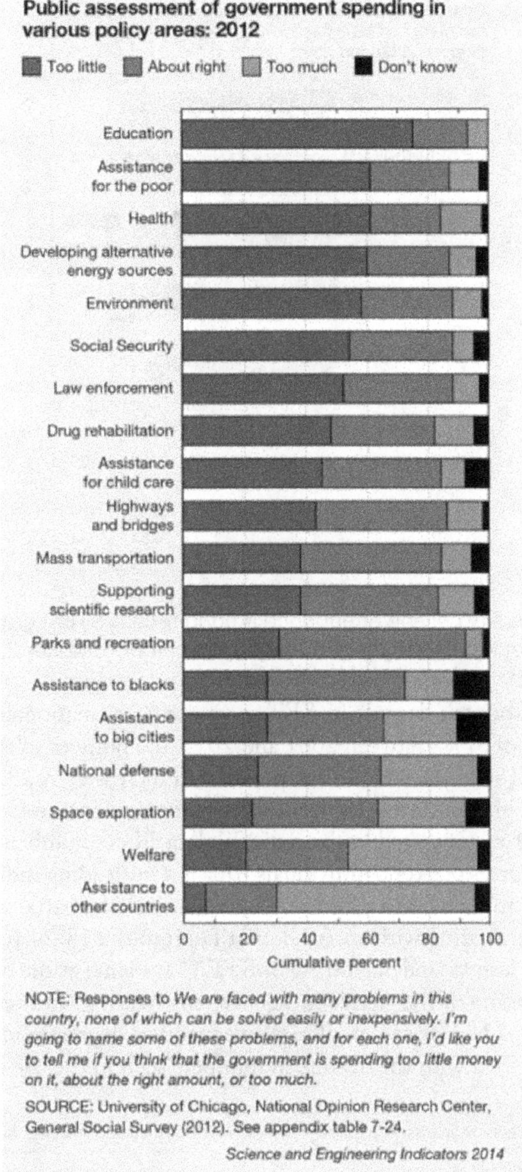

Figure 3: Public Assessment of government spending (National Science Foundation Higher Education Indicators, Chapter 7, Figure 7-14)

The survey shows that in 2010, close to 50% of all students earning a Bachelor's degree in STEM attended a community college during some part of their undergraduate career (Figure 4). Nearly one in five U.S. citizens or permanent residents who received a doctoral degree from 2007 to 2011 had earned some college credit from a community or 2-year college. [9]

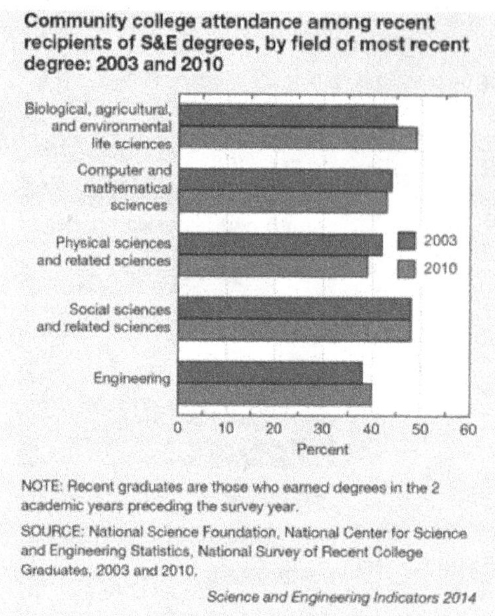

Figure 4: Community college attendance among recent STEM graduates(National Science Foundation Higher Education Indicators, Chapter 2, Figure 2-1)

In 2011, nearly 116,000 associate's degrees in SET were conferred nationally, which is up from 85,000 in 2001 and represents an increase of 37%. Between 2001 and 2011, the number of associate's degrees in SET fields conferred per 1,000 individuals 18–24 years old in the population increased by 23% nationwide. [10]

Figure 5 shows the number of Associate's degrees conferred on students in Colorado in science, engineering, and technology from 1990-2011 as compared to the national data. These numbers are low, with Colorado falling below the national average of degrees per 1000 individuals. This is a misleading indicator for several reasons. Many community college students opt to transfer to a four year university without attaining the Associate's degree (AS) first. Secondly, many students graduating with an AS do not fall into the 18-24 year old age range that was measured in this study. Both factors lead to undercounting of STEM students at the community college level.

In order to gain a better picture of the trends in science, engineering, and technology degrees in Colorado, Figure 6 shows these degrees as a percentage of all higher education degrees conferred. Colorado has remained well above the national average, even with the decline in numbers seen in the last five years.

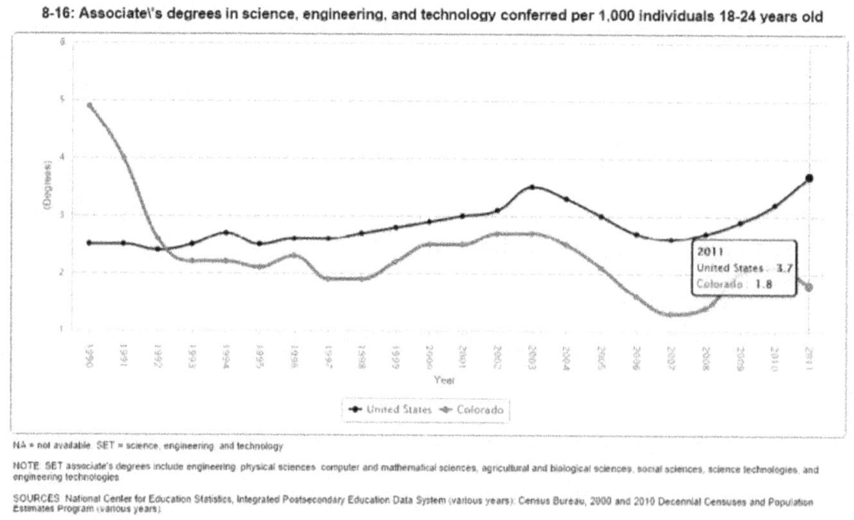

Figure 5: Data from the NSF showing how Colorado compares to the national average of Associate's degrees in science, engineering and technology per 1000 individuals. Colorado saw a decline in these numbers in 2011, compared to a slight increase nationally. (National Science and Engineering Indicators 2014 State data tool, NSF, http://www.nsf.gov/statistics/seind14/index.cfm/state-data/table.htm?table=20)

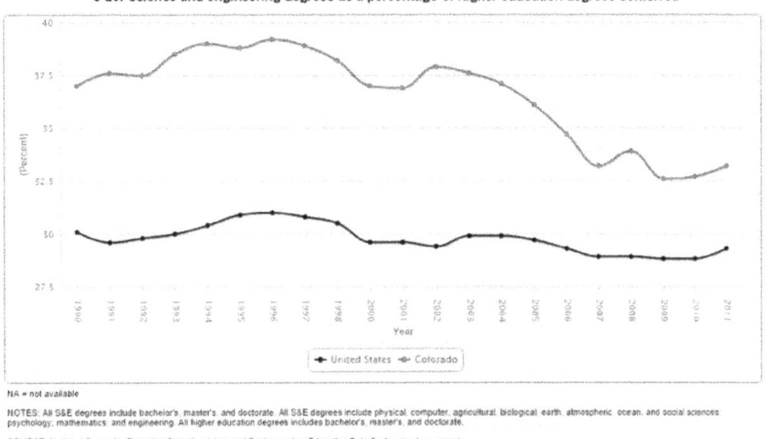

Figure 6: Comparison of the percentage of all higher education degrees conferred in the areas of science and technology. Colorado ranks considerably higher than the national numbers. There has been a noted decline in these numbers in Colorado in the past five years. (National Science and Engineering Indicators 2014 State data tool, NSF, http://www.nsf.gov/statistics/seind14/index.cfm/state-data/table.htm?table=20)

While nationwide community college attendance has declined in the past several years, Red Rocks Community College has continued to have robust and even increasing STEM enrollment. Red Rocks Community College (RRCC) is a fully accredited, public, two-year college in the Denver, Colorado area. RRCC has an annual full-time enrollment equivalent of 6,000 students in 150 programs on two campuses. One of RRCC's principal receiving institutions for engineering transfer is Colorado School of Mines, which accepts more transfer students from RRCC than any other Colorado community college. This is a result of a robust and challenging math and science program at RRCC and a well-established transfer pathway.

4. Local Attitudes Toward Space: Red Rocks Survey

After reviewing the national survey data, a survey was designed to gather information on the knowledge and attitudes towards space travel on the Red Rocks campus. The survey was designed to be brief, five questions, and was given to college students enrolled in at least one science class. For a copy of the survey given see Appendix A. Students were surveyed during the first two weeks of classes during the fall semester 2014. A total of 479 students responded. Students were enrolled in astronomy, physics, chemistry, biology, or geology.

The first question was designed to gather information on students' knowledge about scale of the universe and speed of human travel in space (Figure 7). Question 1: *If you left on a spacecraft today, approximately how long do you think it would take to travel to another star (not our sun) in our galaxy?* The answer choices ranged from the time it would take to travel to Mars (6 months), travel to Pluto (10 years), beyond a human lifespan (100 and 2000 years), and the more realistic answer using current technology (gravitational assist) of 20,000 years. The largest percentage of students answered that it would take 20,000 years. Several students did comment that if new propulsion methods were developed, the trip would be faster. What is interesting is that 10% chose the 6 month answer and another 11% chose ten years. This suggests a misconception about the distances involved in interstellar travel.

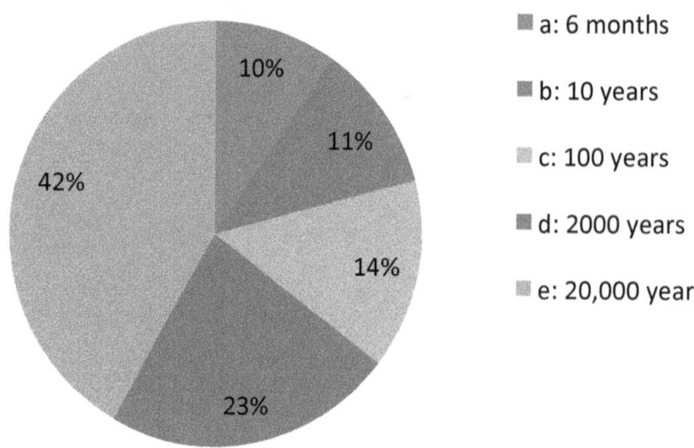

Figure 7 Question 1: If you left on a spacecraft today, approximately how long do you think it would take to travel to another star (not our sun) in our galaxy?

Question 2 was designed to test students understanding of the current state of technology as it relates to space travel (Figure 8). *Question 2: Humans currently have the technology to send people to other planets in the solar system. True or False?* Largely, students believe we have the technology to send humans to another planet at 75% selecting true. Since we have only sent humans to the moon, this is unproven to date. However, the majority of students think we can do it.

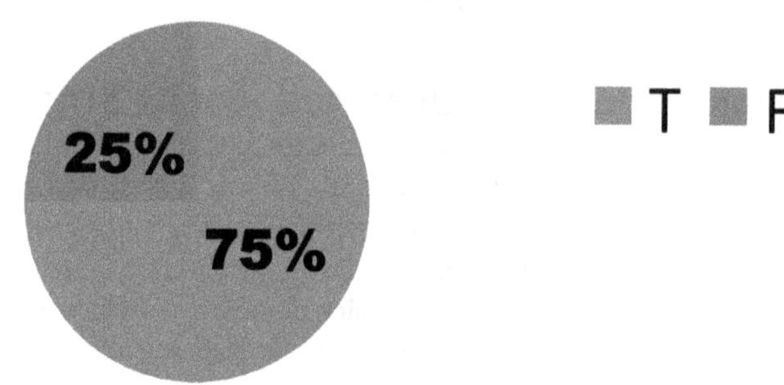

Figure 8 Question 2: Humans currently have the technology to send people to other planets in the solar system.

Question three addresses whether students think we have the technology to send humans beyond the solar system (Figure 9). Question 3: *Humans currently have the technology to send people to other solar systems in the galaxy.* Overwhelmingly, students think we do not, with 90% of all respondents saying this was false.

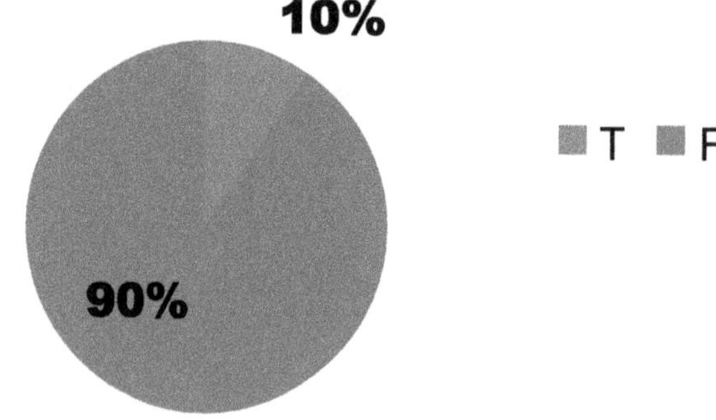

Figure 9 Question 3: Humans currently have the technology to send people to other solar systems in the galaxy.

The fourth question attempts to assess how students feel about our human spaceflight program (Figure 10). Currently, we send humans into space regularly to spend time on the International Space Station. Question 4: *Do you think we should keep sending humans into space?* Again, students overwhelmingly agreed that we should continue manned space missions at 90%.

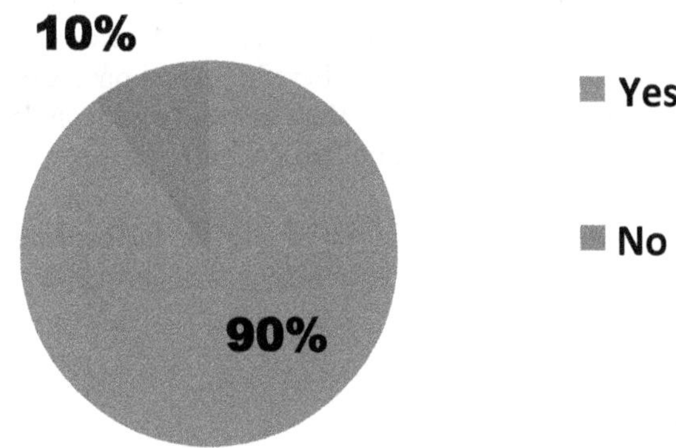

Figure 10 Question 4: Do you think we should keep sending humans into space?

The last question asks about their personal opinion regarding space travel (Figure 11). Question 5: *Would you volunteer to live in space?* Some students were enthusiastic in their affirmative response, but only 28% said they would live in space. Another 39% said they would consider it, but 33% said no, they would not live in space.

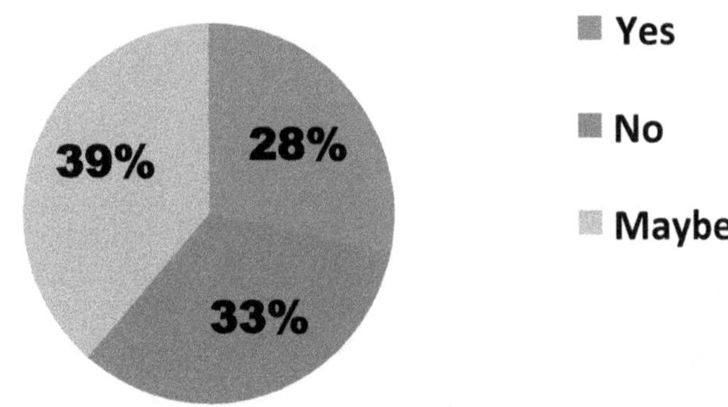

Figure 11 Question 5: Would you volunteer to live in space?

These results suggest there is clearly a level of interest about space amongst students enrolled in science classes. The majority of students place value on human exploration of space, not just robotic exploration. However, a definite lack of understanding of the current state of technology and time it would take for an interstellar mission. How do we engage more students in science? It is imperative that students raise their awareness, understanding, and knowledge of science and technology in space exploration. Why do we need more scientific literacy? We need to be accelerating students toward futuristic thinking and problem solving. Number 8 on the list of ten things we can do to build a space colony this century posted on *io9.com* was "Educate people about our connection to space"[11].

5. Formal Education Pathways for Broadening Understanding

Red Rocks Community College has already begun to take steps in curriculum development that will enable more students to have access to learning about space and deepening their understanding and connection to space. The first thing that can be done is to infuse existing STEM classes with a space focus. This can be done in a variety of ways including lab experiments, group activities, debates, field trips, star parties, reviews of recent space news, and examples of science being done on the International Space Station. Also using inquiry based laboratory assignments can help students with the development of their critical thinking and problem solving skills. [12] Field trips can provide experiential learning opportunities which can be critical for inspiring students to continue in STEM.

Astronomy and physics already have a significant component related to space and space travel. RRCC has rewritten many of the labs in these courses to be more inquiry-based. There are structured inquiry exercises which build the foundation for more open inquiry assignments as the course progresses. Some of the physics courses have assignments that directly relate to space travel. However, other STEM courses could include elements of space and human exploration of space. Geology already deals with concepts of deep time, planet formation, and processes. Using exoplanet analogs for processes we have identified on Earth would be an excellent extension of material to encourage students to consider space. Biology classes could include discussions of extremophiles. There are numerous ways to infuse space into the existing curriculum.

Red Rocks Community College has taken this effort one step further by introducing new courses with a space focus targeting both STEM majors and non-science majors. The new classes include (for course descriptions see Appendix 2):

- SCI 105 Science and Society
- MET 151 Climatology
- AST 108 Colorado Night Sky
- AST 150 Astrobiology
- AST 155 Astronomy of Ancient Cultures
- AST 160 Cosmology
- AST 208 Field studies: Astronomy

All of these courses are introductory level courses with minimum pre-requisite coursework. This allows students access to the excitement of these topics at the beginning of their college education. This can lead to more students entering into fields related to space exploration, aerospace engineering and space development.

Future courses to add include:

- Introduction to Spaceflight
- Astrosociology

Special topics seminars in future technology:

- Faster than light travel
- Future of propulsion
- Synthetic biology
- Self-replicating robots
- Teleportation

RRCC has established a rigorous and thorough traditional science curriculum, which has been the foundation for growing these new innovative science classes and programs. The new classes are drawing in even more students and the department will be tracking how many students that have enrolled in one of these beginning science classes end up in a STEM track. What is most important is the connection between getting the students in the science classrooms in higher education. This can lead to further education about space and space travel.

Engaging students in STEM = Exciting students about space

The additional benefits include facilitating scientific literacy and building knowledge and excitement. These courses begin moving students toward higher order thinking capabilities and fostering creativity in problem solving. Higher education can provide a platform for innovation that can accelerate us towards our interstellar future.

Appendix 1
RRCC Survey

1. If you left on a spacecraft today, approximately how long do you think it would take to travel to another star (not our sun) in our galaxy?
    a. 6 months

b. 10 years
c. 100 years
d. 2000 years
e. 20,000 years

2. Humans currently have the technology to send people to other planets in the solar system.
   a. True b. False

3. Humans currently have the technology to send people to other solar systems in the galaxy.
   a. True b. False

4. Do you think we should keep sending humans into space?
   a. Yes b. No

5. Would you volunteer to live in space?
   a. Yes b. No c. Maybe

I am currently enrolled in a science class at RRCC: ___yes ____no

Appendix 2

New Course descriptions from the Colorado Common Course Database

SCI 105 Science and Society

Examines issues relating to the way science affects society. Students will investigate issues in information technology, the environment, physics and astronomy, biology, medicine and the interaction of science with politics. The class will focus on gathering accurate scientific information and applying critical thinking skills and the scientific method to analyze how science plays both positive and negative roles in society. Emphasis will be on student research, inquiry and analysis of science related issues.

MET 151 Introduction to Climatology

Introduces the physical mechanisms responsible for spatial and temporal variability in Earth's climate and the human-climate relationship. This course develops a scientific understanding of the physical aspects of Earth's climate system, climate system dynamics, and factors that influence climate change. The course explores the global balance of energy and transfer of radiation in the atmosphere, major climatic controls, classifications and comparisons of major types. Current issues such as global warming and El Niño are covered.

AST 108 Colorado Night Sky

Develops an appreciation of and competence in observational astronomy with the naked eye or binoculars, including knowledge of the seasonal and circumpolar constellations and of the location of interesting objects in those constellations. Emphasis is on deep sky observing, including various types of stars, nebula, clusters and galaxies. Basic tools of the astronomer are also covered. The focus is on observation rather than theory

AST 150 Astrobiology: Life in the Universe

Introduces the interdisciplinary and scientific nature of the search for life in the universe, also known as astrobiology. Students will address the questions: "How does life begin and evolve?" "Is there life elsewhere in the universe?" Students will examine life on Earth, its origin and evolution. The possibilities of other life in the solar system and throughout the universe will be examined. Students will investigate the current state of exploration and the search for extraterrestrial life.

AST 155 Astronomy of Ancient Cultures

Introduces the study of archaeoastronomy and ethnoastronomy. Students will study the principles of naked eye astronomy and examine how those principles have been used for timekeeping, navigation, religion and ritual, political power, cosmology and worldview. Methods of the ethnoastronomer will be covered, including measurement of alignments, analysis of written records, examination of art and architecture and incorporation of general knowledge about the culture being studied.

AST 160 Cosmology: The Big Bang the End of Time

Explores the birth, large scale structure and eventual fate of the universe. The course will examine the evidence for, and science behind, the Big Bang and inflation, the expanding universe, dark matter and dark energy, and the possible futures of the universe as a whole. The rise of complex life in our universe, the anthropic principle and the theory of multiple universes will also be included. Unification theories may be covered.

AST 208 Astronomy: Field Studies

Involves in-depth field studies of astronomical phenomenon of specific regions both within and outside the United States. Trips lasting from one to fourteen days in length to study the area constitute the major activities of the course. The specific area of investigation is indicated in the schedule of classes each time the course is offered.

References

1. http://www.tauzero.aero/humanitys-journey/ accessed September 10, 2014

2. Buck, Laura B., Stacey Lowery Bretz, and Marcy H. Towns. "Characterizing the level of inquiry in the undergraduate laboratory." *Journal of College Science Teaching* 38.1 (2008): 52-58.

3. Zion, Michal, Ruthy Mendelovici. "Moving from structured to open inquiry: Challenges and limits." *Science Education International* 23.4 (2012), 383-399.

4. National Science Foundation, http://www.nsf.gov/statistics/srvyattitude/#qs&sd accessed September 15, 2014

5. General Social Survey, http://www3.norc.org/GSS+Website accessed September 15, 2014

6. Wormald, Benjamin, "Americans keen on space exploration, less so on paying for it", August 2014, http://www.pewresearch.org/fact-tank/2014/04/23/americans-keen-on-space-exploration-less-so-on-paying-for-it/

7. "Public Sees a future full of promise and peril", Pew Research Center, June 2010, http://www.people-press.org/2010/06/22/public-sees-a-future-full-of-promise-and-peril/

8. "Public Sees a future full of promise and peril", Pew Research Center, June 2010, http://www.people-press.org/2010/06/22/public-sees-a-future-full-of-promise-and-peril/

9. Science and Engineering Indicators 2014, http://www.nsf.gov/statistics/seind14/ accessed September 15, 2014

10. National Science Foundation Higher Education Indicators, Chapter 8, http://www.nsf.gov/statistics/seind14/index.cfm/chapter-8/c8s2o16.htm accessed September 15, 2014

11. http://io9.com/8-things-we-can-do-now-to-build-a-space-colony-this-cen-1631995142 accessed September 10, 2014

12. Martin-Hansen, Lisa. "Defining Inquiry: Exploring the Many Types of Inquiry in the Science Classroom." *Science Teacher* 69.2 (2002): 34-37.

100 YEAR STARSHIP™

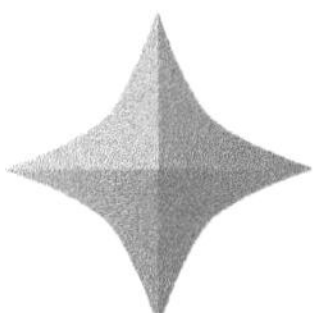

Chaired by John Carter McKnight, JD, PhD

Research Associate, Department of Sociology, Lancaster University

Track Description

Are humans driven to search beyond our knowledge base? How and in response to what do we create the belief systems that guide us? Interstellar travel is not just about the physical trip, but must include the journey civilizations take together. Who will we be and what will define our societies, morality, ethics, cultures, laws, economies, relationships and identities?

Track Summary

The *Becoming an Interstellar Civilization* track has typically been a forum for a broad conversation across academic disciplines and intellectual perspectives, and 2014's session was no exception. Presentations this year divided into two main themes: the challenges of community in long-duration spaceflight, and the long-range issues involved in building an interplanetary and interstellar civilization.

One group of papers addresses the difficulties of building sustainable communities in extreme environments: Chao's approach is historical, examining a problematic case from the history of English settlement of the Americas as a cautionary tale. McKnight draws upon research in virtual worlds to challenge a legacy of visions of space communities drawing on a misreading of the politics of the American frontier, situating Chao's work within political philosophy. Almon drills down to the interpersonal level, building a model for evaluating the social-psychological stresses of crews of explorers and settlers on long-duration missions. Alekseyev focuses on the prospect of genetic engineering to address the impact of high-radiation environments, exploring the ethical dimensions of altering the human for the space environment.

The second group of papers takes a broader perspective on the future of human civilization in the cosmos by examining broader social forces. Batt surveys contemporary science fiction's visions of religion as a social force within a spacefaring civilization. Ziolo's papers examine macro-scale psychohistorical forces and the prospect of manipulating them to ensure humanity's spread into space. Jackson and Hawkins present an inspirational vision for our initial move into the solar system and beyond, and the prospects for encountering alien intelligence.

Taken as a whole, the papers in this track demonstrate the value of bringing a range of perspectives to bear on the challenge of Becoming an Interstellar Civilization, highlighting not only the value of contributions from history, political science, sociology, psychology, and philosophy, but the emergent and transformative experience of introducing these perspectives into conversation around a common goal of building a positive, multi-planetary future for humanity in years to come.

Track Chair Biography

John Carter McKnight, JD, PhD

Research Associate, Department of Sociology, Lancaster University

John Carter McKnight is a former corporate finance lawyer with a PhD in Human & Social Dimensions of Science and Technology . His research focuses on emergent community governance, ethics and social norm enforcement in technologically-mediated spaces from internet gaming communities to spacecraft. He is currently a postdoctoral Research Associate in the Department of Sociologyat Lancaster University in the U.K. Dr. McKnight is working on the Third Party Dematerialization and Rematerialization of Capital project on peer-to-peer finance communities.

Breaking Bad: Creating Predictive Models For Psychological Breaking Points During Long-term Space Travel

Alires J. Almon, M.A.

100 Year Starship

Denver, CO 80202

alires@100yss.org

Abstract

Our human curiosity to discover what's out there requires us to journey further and longer than we ever have before. The question of "Who gets to go?" is one that has everyone wanting to "take the test" to be selected. Great! You are selected. Have a seat because you are going to be together for a long time. As with any group, there are times when the relationships fail, an individual is traumatized, or other social and personal breakdowns occur. When these events or incidents occur, we now have a mission in jeopardy—not due to hardware failure, but because a psychological breaking point has been reached. It will happen—the question is when. The purpose of this paper is to identify signs of when breakdowns will happen and allow the crew to minimize its impact when it does.

A breaking point is reached when an individual is not able to consistently and effectively contribute to work efforts and/or maintain functional relationships with the crew. This paper focuses on the psychological and psychosocial characteristics that lead to individual and group "breaking points". The thesis presented is that the longer a travel mission, the earlier in the mission the group/individual will have a breakdown. This paper begins to create definitions and qualities of breaking points, define the impact on group dynamics, and identify post-break recovery strategies, which will be used as a foundation for predictive model development. Physiological constraints lead to both physical and psychological predictable impacts. However, interstellar mission constraints make this type of travel difficult, such as extensive travel time, no crew changes, eventual loss of communication with home, and the possibility (reality) of no return. Research draws upon typical areas such as analog isolation studies, group dynamics, and physiological impacts, but also incorporates historical explorer activities, grief and Post Traumatic Stress Disorder (PTSD) studies, and organizational development theories.

Keywords

group dynamics, psychology, mental health, predictive models, mission types, physiology.

1. Introduction

Hubris is often the downfall of good intentions. It is a term from the Ancient Greeks meaning extreme arrogance. Used as a focal point in Greek tragedies, it is the arrogant belief that one's power is unassailable. This is the premise of the popular culture United States television show *Breaking Bad.* The main character Walter White begins a nefarious side career as a drug dealer in order to prepare his family financially for his inevitable death due to an aggressive cancer diagnosis. To his friends and family, he is the meek husband of an overbearing wife. His career is that of a high school chemistry teacher. As with most individuals their story is much more layered than that. We find out that he is an extremely talented chemist with an advanced degree, and in his decision to manufacture drugs he is able to create a product that is pristine and in high demand.

His good intentions quickly require decisions and actions that begin to compromise the integrity of his initial intent. He becomes more demanding, he makes life choices for others, and he makes more money that could ever manage or spend in his lifetime. Walter does not realize that his behavior change is noticeable and affecting others in his life. As an audience, we see his descent into this new personality as a fairly quick and dramatic turn (within one year). His "co-workers" also notice a change as he enters the new work environment, although he did not have an alter ego that they were aware. With them he has a breaking point and new direction.

With hindsight, some behavior changes can be predicted. In this case, family welfare and fear turned into greed and hubris as time went on. Sometime during his journey as a drug dealer, Walter White had a breaking point where his behavior turned from helpful to harmful and he became "Heisenberg". While his intent was still the same – to provide a financial stability for his family when he died, when circumstances changed when death was no longer imminent, he continued his behavior to a point where he broke his current path. He reached his breaking point, and broke bad. This scenario, while fictional, showcases the impact of a sudden behavioral change, in this case a "break for the bad." An individual starts out with good intentions and then is suddenly faced with circumstances that were unseen and unexpected and reacts in a way that produces a negative effect for themselves and those around them.

Extreme conditions such as long-term space travel will create an environment where an individual is confined, without a change of scenery or personnel, or the idea of going home and the realization of this environment may create a challenge for them to stay positive and on a path of their initial intent to be positive and productive. Which is being defined as the intent to be a useful contributor to the crew and the social society develops as a result of the journey.

This paper will look at the circumstances that create an environment where an individual will be challenged in their ability to maintain optimal physical health, focus on their work and tasks supporting the group mission, and maintaining effective and meaningful interpersonal and intrapersonal relationships. Long-term space travel, while replicated by analog studies on Earth, creates a strong speculative basis for predicting behavior in certain circumstances. They do not always take into account the unique conditions that long-term space travelers will encounter.

2. Psychological Studies and Space

Psychology is the study of people and their behaviors. Space travel has unique impact on the behaviors of those who travel. Many astronauts have noted the experience of awe and wonder at the experience of space. Space shuttle Astronaut Kathryn D. Sullivan shared her account looking at the Earth from space: "I am happy to report that now amount of prior study or training can fully prepare anybody for the awe and wonder this inspires". (Robinson, et al., 2011). It is this awe and wonder that drives us to be explorers and pioneers into the unknown. The notion of the unknown puts individuals into environments that are not always conducive to the individuals being in their optimal state both physically and emotionally.

Psychological stressors resulting from both physical and mental environments affect an individual's behavior and decision-making. The amount of attention to stressors has grown as space programs have learned the value of the psychology in space exploration. While the beginnings revolved around the early space race, the evolution of those studies has taken us all the way to Mars. Douglas Vakoch (2011) edited a historical view on psychology and space exploration, which outlines the role of psychology and space programs. Three sources of information educate us about important psychological and psychiatric issues affecting human space missions: anecdotal reports, space analog and simulation studies on Earth, and research conducted during actual missions. (Kanas & Manzy, 2008).

The first wave of research focused on the ability to do tasks and individuals who had the ability to do tasks. Supporting activity was focused on human factors and other "ability-focused" actions. Harrison and Fiedler (2009), observed that NASA managers and engineers as are "thing" people rather than "people" people, so the behavioral side of space flight is of little interest to them. And, for many years, the psychological side of space travel was ignored.

The next wave of studies began to look at the importance of "fit" versus skill for mission success. The dynamics of mission success was expanded to include personality fits as well as skill fits. (Bishop, 2011).

Analog studies provide an experience that mimics certain conditions. Study results provide a foundation for extrapolative applications for long-term space travel. In the case of space, researchers have conducted studies such as surveying extended mission submarine crews and experimental studies, such as those who are asked to live in isolated and extreme conditions of Antarctica. Antarctic experiments have been quite successful in bridging the gap of skill vs. fit. Bishop (2011) cites Palikans who states:

> "The cumulative experience with year-round presence in Antarctica makes it an ideal laboratory for investigating the impact of seasonal variation on behavior, gaining understanding about how biological mechanics and physiological processes intact, and allowing us to look at a variety of health and adaptation effects."

These studies examine the psychological effects of these conditions. Not only are individuals tested on their ability to live and work in these environments, but also their ability to get along with others in those same conditions. It is the understanding of the impact of behavior that begins to gain traction as a research topic.

A recent analog study, MARS 500, simulated a 520 day mission Mars where activities included communication delays, on-planet task completion, and real-time frame mission. Researchers studied the actual psychological impacts and needs for this mission. They found results that will necessitate a change in the way crewmembers are selected and trained for such a mission. (Ushakov, Morukov, Bubeev, Vasil'eva, Vinokhodova, & Shved, 2014)

Once recognizing the importance of behavior in mission success, studies began to explore all the variations of behavioral impact, for example, the personality types that are best fit for extreme conditions (Bishop, 2011), and simulated research of negative group interactions (Wichman, 2011). Salutogenic approaches to mental health focus on the factors that support the health and well-being of an individual, rather than the factors that cause disease. In the case of utilizing an often mission-required task of photographing earth to help astronauts experience the awe of their position and the uplift it offers. Providing activities that give positive mental experiences rather than trying to fix a broken individual can be a more effective to long-term mental health wellness. (Robinson, et al., 2011). Each approach validates the impact of the psychological state of the crewmembers on the success of the mission.

Space programs have expanded in complexity – from being single country endeavors, to global cooperation efforts. Space crews are no longer one gender or culture. Space exploration is no longer a male-dominated field. Women have begun to join crews that provide a different kind of challenge to psychological health of crewmembers. Group and crew interaction is now impacted by gender roles, and the evolution of those roles in society and at work. Women are now peers and leaders of their male counter parts, and the workplace transition has been bumpy, both on Earth and in orbit. Likewise, the breadth of research has evolved into multi-disciplinary and trans disciplinary studies. Kring & Kaminsky (2011) note that due to the inherent differences in behaviors and abilities, mixed gender crews have performed effectively, both in space and in similar settings like Antarctica. They have assured us that a crew composed of both men and women is the right choice for extended missions to the International Space Station, rendezvous missions with asteroids, and one day, the first human mission to Mars.

Global missions will require another layer of adjustment: cross-cultural relationships. Although individuals may be trained for space flight, their own cultural biases will have an impact on their approach to their work and the development of interpersonal relationships. There are fundamental differences between American, Asian, European and Slavic cultures, and the many cultures in-between that have to be considered as crewmembers are selected and trained. Success has been exemplified through the crew cooperation in the International Space Station (ISS) Required training will have to cover the subjective culture, which is implicitly known to its members even though the knowledge is rarely articulated. Its rules and practices must be made explicit for an outsider. (Draguns & Harrison, 2011)

One area of study that is often overlooked for its applicability to long-term space travel is that of grief and loss studies. In particular, ambiguous loss is defined as an unclear loss that defies closure (Boss, 2006). During an interstellar mission, crewmembers may experience ambiguous loss, where they lose contact with the living and will never see family and friends again.

Those who are left behind experience unfulfilled grief without closure as are the crewmembers. As the crew gets farther from Earth, their communications will become less and less frequent, not due to lack of interest, but due to our technological and physical communication limitations. Boss (2006) defines the effect of ambiguous loss as "inherently traumatic because the inability to resolve the situation causes pain confusion, shock, distress and often immobilization. Without closure, the trauma of this kind of stress becomes chronic." The recognition of this condition will be important in the development and application of coping mechanisms.

The research studies cited are just a sampling of the need and the possibilities of research to make space travel the most optimal experience both from a program/mission and a commercial/tourist perspective. Eventually the commercial space business will become large enough that the training required to participate will be as simple as the inflight safety demonstration on today's commercial airplane flights.

3. Predictive Modeling

As space programs grow globally, there are certain inevitabilities that will occur as a result. Crews will comprise of individuals from different cultures, will be more gender-balanced, and the missions will be longer and more complex. Exploration goals will be bigger and require more effort of the crew—task work, endurance and collaboration. Much of the current psychological research is geared toward optimization of human performance, a necessity. However, it is just as important to address the potentiality of human breakdowns both physically and especially psychologically. The impact of a psychological breakdown is sometimes subtle, but the impact can be crippling or devastating to a mission.

A breaking point is reached when an individual is not able to consistently and effectively contribute to work efforts and/or maintain functional relationships. When an individual reaches this point, there is no way to predict how they will eventually react and resolve the issue, nor how others will react. In a mission where the journey will be several months or years, a breakdown is inevitable, be it small or large, but it must be addressed.

What does a breakdown look like when there are so many factors that could be the cause? In some instances it is a person's prolonged depression, or self-isolation. It could be an extroverted behavior such as hostile interactions with fellow crewmembers or continued questioning or disobedience of orders. Or nothing may be apparent, because an individual may be strong enough to weather their internal strife and be able to maintain normal relations on the mission, but have post mission recovery issues, such as PTSD, as experienced by many veterans of war and traumatic experiences. If it is a long enough mission, how long can a person last before psychological distress ensues?

While creating a predictive model for understanding the state of the individual human condition seems ambitious, it is possible to lay the groundwork and identify signs of behavioral anomalies caused by physical and psychological stress. Once these signs are identified, then they can be addressed and managed for that individual and their group relationships.

3.1 Physiological Factors in Psychological Behavior

An individual's physical well being has an impact on their mood and behavior. Many studies have been conducted, both on Earth and in space that identify specific physiological conditions and their effect on behavior. For example, the change in the circadian rhythms of space crews has been studied for many years. It has become recognized that sleep disturbances and fatigue, as well as alterations of circadian rhythms in astronauts, are among the most important factors contributing to impaired well being, alertness, and performance during space missions. (Kanas & Manzy, 2008)Kanas and Manzy (2008) defined the actual types of stressors and stress that are encountered during space travel.

> A **stressor** is a stimulus or feature of the environment that affects someone, usually in a negative, arousing manner. In space, there are four kinds of stressors: physical, habitability, psychological, and interpersonal. (p. 1). Examples are:

Table 1: Environmental Stressors Experienced during space travel

Physiological	**Habitability**	**Psychological**	**Interpersonal**
Acceleration	Vibration	Isolation	Gender Issues
Microgravity	Ambient noise	Confinement	Cultural Effects
Ionizing radiation	Temperature	Danger	Personality Conflicts
Meteoroid Impacts	Lighting	Monotony	Crew Size
Light/Dark Cycles	Air Quality	Workload	Leadership Issues

A stress pertains to the reaction produced in someone by one or more stressors. In space, there are four kinds of stress that affect human beings: physiological, performance, interpersonal and psychiatric. (p.2) Examples are:

Table 2: Types of Stress Experienced during space travel

Physiological	Performance	Interpersonal	Psychiatric
Space Sickness	Disorientation	Tension	Adjustment Disorders
Vestibular Problems	Visual illusions	Withdrawal/ Territorial behavior	Somatoform disorders
Sleep Disturbances	Attention deficits	Lack of privacy	Depression
Bodily fluid Shifts	Error Proneness	Scapegoating	Suicidal thoughts
Bone loss and hyperkalemia	Psychomotor problems	Affect displacement	Asthenia

3.2 Mathematical Modeling

Developing a mathematical model has many challenges. Mathematical models, while extensively used in the field of engineering, have been utilized by other disciplines seeking to model their designs and behavior of objects. The behavior that is described in engineering is much different than that of human behavior. In psychology, mathematical models are designed to describe basic behavioral processes in a more precise way than can be done with simple verbal descriptions. (Mazur, 2006)

Human behavior is a very difficult process to model, due to the complexity of the decision- making itself. Mathematical models can make precise and important statements about behavioral processes that are relevant to anyone who is interested in explaining, predicting or controlling behavior, either in the laboratory or in applied settings. (Mazur, 2006). Specifically, decision modeling studies the human decision-making mechanism and tries to build models that predict human decisions. (Kim, Yang, & Kim, 2007)

Predictive models can be very difficult to develop and execute. On of the most apparent reasons is that individuals have free will and can exercise it at anytime, which means they can make any choice despite past behaviors. It may be possible to consider improved modeling strategies. It will be important to distinguish models that try to predict what a particular individual will do from models that try to predict particular actions that all people will do. (Bar-Yam, 2011)

3.3 Defining Mathematical Models for Predicting Human Behavior

The predictive model for behavior both defines and simulates the conditions in which certain behaviors will occur. Kennedy (2011) notes that a rational, random number-generated methodology is unwise. However, a rationally behaving model needs to be able to represent knowledge, learn, remember new knowledge, and apply it to determining the behavior of the agent.

The key components of this mathematical model are that it is dynamic, computational, and has elements of fuzzy logic in its framework. The idea of soft computing or fuzzy logic is a dynamic representation of human behavior as a gradation of behaviors, not just absolute experiences. Fuzzy logic is a form of multi-valued logic derived from fuzzy set theory to deal with reasoning that is approximate rather than precise. Fuzzy logic models human thinking in imprecise and uncertain conditions comparatively better than other techniques. (Singh, 2010). These principles lay the foundation of the mathematical model itself.

Conceptual frameworks are approaches to modeling human decision-making using more abstract concepts than mathematical transformations of environmental parameters. They involve concepts such as beliefs, desires, and intentions (BDI), emotional state and social status (PECS), and "fast and frugal" decision hierarchies. (Kennedy, 2012)

3.3.1 Model for Individual Behavior

Creating a predictive model for human behavior on any level is difficult, as noted above. However, having some degree of assessment and therefore predictability allows individuals and crewmembers to assess their environment more effectively. It also gives a foundation for the development of assessment tools, support activities and post-mission management actions.

The next two sections will focus on two models of predictive behavior in space, one for individuals and one for groups or crews. Empirical data to support these models are simulated in the sections below.

The Individual Model

The behavior of the individual is based on his or her interaction with the environment in which they work and live. The model reflects the aspects of that living situation. The behavior of the individual (Bi) is a result of the combined physiological state of the individual (Pi) within the phase of the mission cycle (Mc) plus the Level of group dynamics (Gd) plus the Role of the individual relative to the group development. (e.g. Is the person a crew member, crew leader, or not part of the crew in question?) This is a description of the Psychological state of the individual. The lower the result, the more likely an individual is to have ineffective relationships or an emotional breakdown. (See Figure 1)

$$B_i = (P_i / M_c) + G_d + (I/G)$$

Figure 1: Mathematical model for predicting emotional breakdowns in Individuals during long-term space travel.

The Mission Cycle describes totality of the stages within a mission. Different stages of the mission will have an overall effect on the physiological state individual. Kanas & Manzy (2008) found that human performance varied based on the stage of the mission, which the individual is being assessed. Those stages being pre-flight, mission, and post flight. In this paper, the term "mission cycle" is being used to describe the overall mission, describing the different time frames of the mission itself which results in different mission constructs. Mission cycle types are different depending on the goal of the mission, even though the mission may be similar in duration. (See Table 3)

Table 3: Mission cycle types for long-term space travel

Mission Cycle Type for Space Travel	Time Frame			Mission Example
	Departure from Earth to Destination (Egress)	Mission Duration	Return to Earth	
Mission Cycle Type 1	24 hours <	24 months	24 hours <	2 years on ISS
Mission Cycle Type 2	7 months	14 days	7 months	Mars 500
Mission Cycle Type 3	3 months	3 months	3months	Not experienced
Mission Cycle Type 4	24 hours	Indefinite	Unknown	Pioneer Journey - Interstellar

Mission Cycle Type 1 Missions: The travel time to the destination is minimal, the duration of the mission time is spent on the mission activities itself and the return in less than 24 hours. These tend to be low-Earth orbit type missions, such as the International Space Station. An early example would be traveling by commercial flight to a summer destination and spending a month in a location, then returning by commercial flight. Travel time is minimal relative to the vacation/mission time.

Mission Cycle Type 2 Missions: The travel time to the destination and return are greater than the mission itself. This type is modeled after the Mars 500 analog study where the time it takes to the complete the mission is significantly less than the travel time to get there. The "mission" is not only the tasks to be completed, but includes the travel time itself. The Earthly example here is a road trip , which may take 2-3 weeks to reach a destination and then spend only 1 week at the beach and two weeks to return. You basic scenarios of "are we there yet?"

Mission Cycle Type 3 Missions: The travel time and the mission duration are equal or nearly equal. Our space programs have not reached a point where mission egress, mission duration and return to Earth are equal in time. Missions such as these provide equal skill for the crew of travel endurance and mission performance.

Mission Cycle Type 4 Missions: These are unique in that it will only take approximately 24 hours to get out of Earth's orbit onto a planetary or interstellar trajectory. However the destination is so far away as to feel indefinite to those on Earth, and the return is unknown for both the crew and those left behind.

The mission cycle type provides different scenarios for the individual and their environment. Their physiological state will have a different cycle depending on the mission type. Kanas & Manzy (2008) show that there is an initial adaptation adjustment to space during the pre-flight, mission duration, and even after a return to Earth and there are adjustments for both short-term and extremely long-term (over 400 days) missions. Still, little is known about long-term space mission. The application to interstellar travel must be extrapolated from these limited studies. One of the key differences is that when the mission is beyond our current extremes—over 3 years—what will be the long-term impacts on human performance? It is not known if there is a leveling-off of certain side effects, or how the severity of side effects will affect performance and therefore psychological well-being.

3.3.1 Model for Group Behavior

The behavior of the group is reflective of the individuals therein. In this model, the behavior of the group (Bg) is a result of the group development level or dynamic within the mission cycle plus the group relation to the individual. As in the individual model, the lower the result, more likely a group is to have decreased effectiveness or cease functioning altogether. This model differs from the individual model in that it takes into account the individuals members in the group and their relation to the mission and the individual. (See Figure 2)

$$B_g = (G_d / M_c) + (G/I)$$

Figure 2: Mathematical model for predicting emotional breakdowns in groups during long-term space travel.

In part one of the equation, the group dynamic or development is based on the point in the mission cycle where the assessment is taking place. That will vary based on the mission type (see Table 3). Group development principles show how mature the group is in its overall development and functioning relationship together. The body of work on team development, organizational dynamics come into play here. The longer into the mission cycle the mission is, the more the group will learn to be effective and grow.

The other part of the equation attempts to model the effect of the individual in relation to the group. Impacts such as cross-cultural differences, gender effects and general interpersonal issues are addressed at this point. The individual relationship relative to the group is what is measured. The individual may be a "minority" by culture, gender, group tenure or other isolating factors. The relationship of the individual to the group membership and level of inclusion, will affect an individual's emotional and psychological state.

4. Simulated Application and Results

Space mission environments are very structured, both physically and mentally within a very limited space. The habitable space on the International Space Station (ISS) is that of a six-bedroom house, approximately 13,696 cubic feet. (NASA, 2014). Astronauts from many nations who travel to ISS to conduct experiments, participate in space walks and maintain the station itself use the habitat. While the crew quarters are tolerable, they are not exactly comfortable. Wichman (2011) is reminded of Feedman's Density Intensity Hypothesis which states that crowding is not necessarily an adversity stimulus, but it does tend to amplify whatever emotion is extant in a group, positive or negative. For the duration of their visit, all astronauts will live in the habitual space with other crewmembers for a specific amount of time. The timeframe will vary from member to member, and individuals will have served for different amounts of time, based on their mission assignment. For example, an ISS crew may consist of up to six members who have served anywhere from two weeks to six months, and sometimes longer. (NASA, 2014) The composition of the crew can vary at any time by gender and nationality. The work required of the crew will be different and the expectations of both the crewmembers and the crew as a whole will also vary.

When assessing the potential for an emotional breakdown, the predictive model has many factors to consider. For current research, there is only an Earth analog and short distance missions to assess. While a stint on the International Space Station may be several months long, the journey there is relatively short. Extrapolating these results and predicting the experience for future space travelers will be the challenge since the travel time and the mission activity will be of equal importance even if the mission activity itself is short.

Each mission type has its own expected result, regardless of the time spent in space. However, a Type 1 mission lends itself to simulate short-term mission cycles where it will be a short journey to reach the mission activity location and to return. In Type 2 mission cycles, the journey to the mission activity location is longer than the mission itself and creates a different expectation of the crew. This gives new meaning to the saying, "It's the journey that is the reward." The Type 3 mission cycle is similar to a control group, where all aspects of journey are equal: egress, mission activity and return. And finally, in a Type 4 mission cycle, the longest journey of them all, there is nothing to compare with the timeframes expected of these journeys. The constant unknown in the Type 2 – 4 mission cycles is the ability of the crew to stay physiologically and psychologically stable enough to complete the entire mission cycle.

Table 4: Simulated result of modeling psychological well-being during mission cycle types over time.

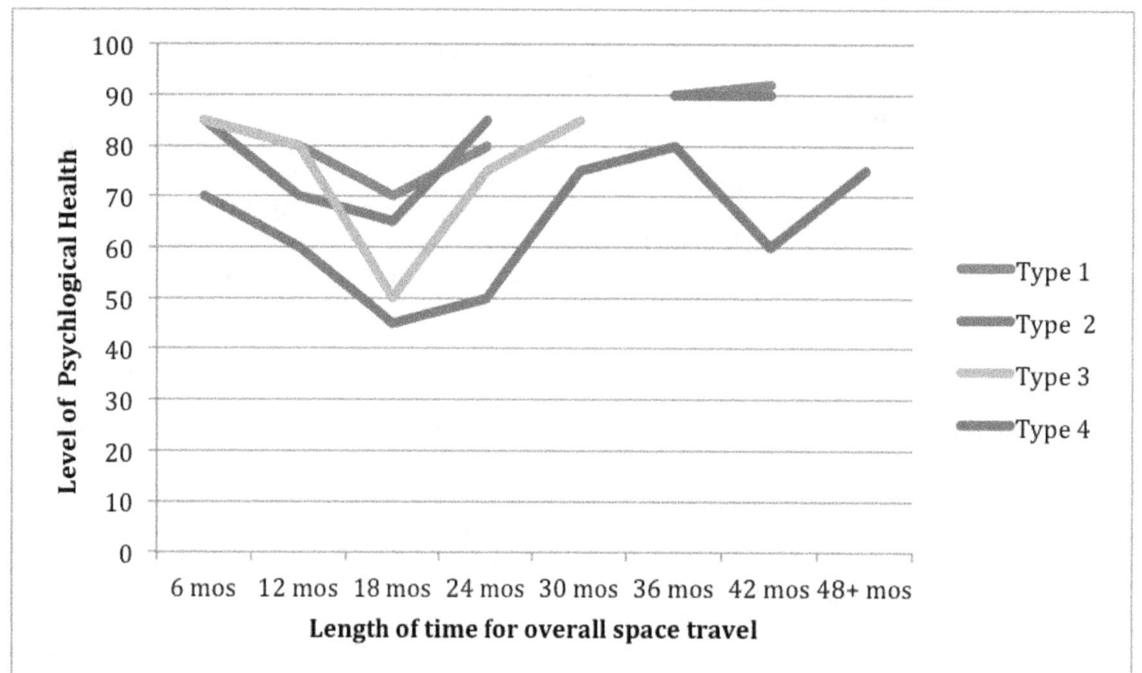

Based on the simulated results, the expectation of the psychological stability and well-being of crew members are dependent on the mission cycle type and the time spent in space. As both models show, factors that influence the prediction of an emotional breakdown, such as individual experience, group development and crew composition. All mission types show a period of adjustment near the beginning of the mission. However, in longer missions such as Type 3 and Type 4, the level of psychological health dips well below the others. Using the method of Thematic Content Analysis (TCA), Suffield, Wilk & Cassel (2011) reviewed the diaries and journals of multinational crewmembers to assess their adjustments and coping mechanisms long-term space travel and their ability to work with a multicultural crew. They found that post-mission reflections were much more positive than those of the original, real-time journal entries. The physiological health of the crew members will be in challenged the longer they are on the mission. The environmental stressors will compound themselves over time, and reduce the performance level of the individual (Kanas & Manzy, 2008) (Harrison A. A., 2011).

4.1 Countermeasures

As the psychological state of the individual and groups can be tracked, and possibly predicted, there is opportunity to establish protocols for adaptive measures. This means creating training for crewmembers, on board support modules for individuals and team and post-breakdown support.

Ensuring the psychological health of a crewmember starts with effective selection criteria. The criteria should be based on the mission activity, the mission cycle and best fit for the composition of the crew. Once selected, the crew should have time to work together and bond as a group. This model assumes small crew sizes, of 2 – 7 members. As space programs embark on longer missions and larger crews, they dynamics of groups will change and require a different approach to selection and training. Kanas and Manzy (2008) suggest a series of assessments for crew selection based on an individual's ability to cope with the psychological stressors. Draguns and Harrison (2011), suggest utilizing cross-cultural data as part of astronaut selection and support.

While on board a ship and conducting a mission, crewmembers can monitor and manage their own individual health needs. Mission duration had significant effects on coping strategies during the flight/mission. The longer missions (four months and longer), showed astronauts utilizing the more certain coping mechanisms than shorter missions, (two-weeks or less). In this case the mechanisms utilized more were Confrontation, Escape/Avoidance, Accepting Responsibility, and Supernatural Protection. (Suedfeld, Wilk, & Cassel, 2011). These mechanisms are indications of potential individual challenges as well as challenges with other in the crew. Longer missions are inevitable and negative interactions and increased coping mechanisms will be a part of the experience of the crew. The results from the psychological studies of the Mars 500 crew found that astronauts and cosmonauts did seek out the psychological support during the mission activity. (Ushakov, Morukov, Bubeev, Vasil'eva, Vinokhodova, & Shved, 2014).

Homesickness is a common issue that crewmembers may experience. While many individuals on Earth encounter this problem, the astronauts' experience in space will be special. They are so far away from their loved ones that they will not have the option of immediate interaction through a phone call, text or video call. The researchers in the Mars 500 study found the incidents of homesickness arose pretty quickly for some of the European astronauts, who were not only part of the isolation from the study, they were also doing the study in a other country. They were the first to seek out psychological support with for issue. (Ushakov, Morukov, Bubeev, Vasil'eva, Vinokhodova, & Shved, 2014). When missions get to the point of traveling beyond communication capability and the crewmembers are fully autonomous on the mission, they may experience something called "ambiguous grief." This is when there is no closure to the loss of a family member or loved one. This is similar to those who have lost someone as a prisoner of war or through missing persons who have not been found. There is no way to close the loop on the loss absolutely knowing the person is deceased. (Boss, 2006) In the case of the astronauts, they will have lost communication with Earth and have no way of knowing the fate of their families, or vice versa. This can cause depressive or other disruptive behavior in the individual.

Providing the crew with tools to support during the mission is critical. Ushakov et al., suggest an approach to support of those individuals through positive psychology, primarily those under the direction of P. Suedfeld and N. Kanas, has focused on study of the positive emotional states and social attitudes arising in astronauts when they successfully cope with the stressors of a long spaceflight.

Once a situation has happened where an individual has ceased to be engaged and a contributing member of the crew – they have had a breakdown – it is incumbent upon fellow crewmembers to assist in the support of that member. If the mission still has the crew within communication range of the Earth, Mission Control can provide psychological support through outreach and discussion as was done in the Mars 500 study. (Ushakov, Morukov, Bubeev, Vasil'eva, Vinokhodova, & Shved, 2014). The other crewmembers may or may not know that this support is taking place, the crewmember may be contacting mission control on their own. When missions get longer and there is little or intermittent contact with Earth, the onus of support is on the crew itself, which may be dealing with mission stress as a result of the activities or journey.

When the group, or whole crew becomes ineffective, then the mission itself is potentially in jeopardy. The crew redundancy is compromised and it is up to individuals to take the role of self-correcting the group.

Upon return to Earth, many studies emphasize the need for post-return crewmember support for both the crewmember and their families. (Suedfeld, Wilk, & Cassel, 2011). Despite the post-mission jump in positive reflections, there are still adjustments to Earth that the crew has to make. It is suggested that private sessions with both the crewmembers themselves and with their families will make the transition back to Earth easier. (Kanas & Manzy, 2008)

5. Research Applications

The goal of the predictive model is to create opportunity for prevention and management of the psychological health of crewmembers. Long-term space travel will slowly have longer and longer timeframes as programs become more ambitions. Suffield, et al (2011), defined long term as four months or more. The Mars 500 group first conducted a 105-day mission in addition to the 520-day mission, each creating opportunities for observing behavior in this unusual environment. Those in the position of identifying, selecting and training an optimal crew for a mission will take into account the research that has been done and utilize the modeling that has been created in order to select individuals who can function, and even thrive in the new conditions.

5.1 Transdisciplinary Studies

Despite the slow start and recognition of the need for such research, the psychological study of astronauts in space travel is critical to the success of future missions. Going forward there needs to be outreach to fields not commonly associated with space and space missions. Cross-cultural psychology has important applications as the programs become intercultural and the crew composition reflects that spirit of cooperation. There is much that this discipline can offer training and psychological support during long-term space missions.

Helping other disciplines understand that they have a meaningful contribution to this work is important to creating a robust, transdiscplinary conversation. Two areas for consideration are grief studies and team development studies from an organizational development perspective.

Mentioned earlier, ambiguous grief is a condition that has the potential to incapacitate a person and prevent them from interacting and building effecting relationships. When a person withdraws from his or her team that could mean disaster for a mission. Team development is a key component to a successful mission. Working with experts in these fields to bring their expertise to the table can only enhance the effectiveness of training, coping, and post mission support.

As human beings, we will take all of our cultural biases, gender biases, etc., with us. Training can help with coping, but only if the issue is recognized as being part of everyday life.

5.2 Quantifiable Self

Predictive models lend themselves to utilizing quantified data for decision-making. Already, our society uses this data for laymen in the use of wearable health and fitness trackers. Using these biometric monitoring systems to track sleep patterns, and weight management allows the user to make data-driven decisions about their lifestyle and behavior. Monitoring systems exist currently for space missions, mostly monitoring physiological activity. However, pairing that with the predictive model and its impact on psychological behavior can prove to be an opportunity for further study. With knowledge of the impact of physiology on performance and behavior, monitoring systems can be created that will alert users and systems to possible psychological distress.

6. Conclusion

This paper opens up many questions about the impact of the conditions of extreme long-term space travel, or interstellar travel, on the human psyche. While empirical research exists from which to extrapolate certain behaviors, the factor of the unknown is so extreme that everything we have studied could have no effect. However, we can be assured that human behavior, while sometimes unpredictable, still follows certain patterns. (Kennedy, 2012) Those patterns can lead us to make predictions and anticipate certain failure points. There is a cautionary tale here. Having this knowledge does not make one omniscient. With all science, it must be tempered with wisdom and not hubris.

Culturally, in the United States, and many other cultures, the study and acceptance of mental health is an often eschewed topic of discussion. Clinically there are many areas of research within the field itself; utilizing that research outside of the field means an acknowledgement that that issue exists in that new field. Those that have studied space exploration know the importance of mental health and psychological stability of the individuals and crews, which are essential to the success of these missions. The body of work continues to grow and extend outside traditional fields into a transdisciplinary exchange of ideas.

Modeling our understanding of the cycle of an individual's physiological and psychological health engenders the research of group dynamics and cross-cultural, gender studies and other applicable fields. It will be a cornerstone of future research on longer-term space travel. However, through predictive modeling there is opportunity to identify when problems may begin to arise and how those problems can lead to a breakdown. These models can also provide a guide as to the likelihood of an occurrence and create a safety net of support for that individual and crew.

If we could have modeled of the behavior of *Breaking Bad*'s protagonist Walter White, would that have stopped his descent in to psychological and physical destruction? Perhaps. Modeling may have predicted critical of decision-making points and provided preventive measures to assist him in going down a different path. In reality, we have no idea how an individual will react when a breakdown situation occurs – will a person be violent, reclusive, and/or disruptive? There is no way to predict behavior to that level – at least not yet.

References

1. Bar-Yam, Y. (2011). *On the Role of Free Will in Predictive Models of Human Behavior.* Retrieved 09 30, 2014, from New England Complex Systems Institute: http://necsi.edu/research/overview/freewill.html

2. Bishop, S. L. (2011). *From Earth Analogs to Space: Getting There from Here.* In D. A. Vakoch, Psychology of Space Exploration: Contemporary Research in Historical Perspective. Washington, DC, United States: National Aeronautics and Space Administration.

3. Boss, P. (2006). *Loss, Trauma and Resilience: Theraputic Work with Ambiguous Loss.* New York, NY: W.W. Norton & Company.

4. Draguns, J. G., & Harrison, A. A. (2011). *Spaceflight and Cross-Culutral Psychology.* In D. A. Vakoch, Psychology of Space Exploration: Contemporary Reserach in Historical Perspective (pp. 177 - 203). Washington, DC, United States: Nationa Aeronautics and Space Administration.

5. Harrison, A. A. (2011). Behavioral Health. In D. A. Vakoch, *Psychology and Space Exploration: Contemporary Research in Historical Perspective* (pp. 17-46). Washtington, DC, United State: National Aeronautics and Space Administration.

6. Harrison, A. A., & Fuller, E. A. (2011). *Introduction: Psychology and the U.S. Space Program.* In D. A. Vakoch, Psychology and Space Exploration: Contemporary Research in Historical Perspective (pp. 1-16). Washington, DC, United States: National Aeronautics and Space Administration.

7. Kanas, N., & Manzy, D. (2008). *Space Psychology and Psychiatry.* El Segundo, CA: Microcosm Press and Springer.

8. Kennedy, W. G. (2012). *Modelling Human Behavior in Agent-Based Models.* In A. J. Heppenstall, M. Batty, & A. Crooks, Current Geographical Theories for Agent-Based Modelling (pp. 167-179). New York: Springer Science + Business Media.

9. Kim, C. N., Yang, K. H., & Kim, J. (2007). *Human Decision-making Behavior and Modeling Effects.* (S. Direct, Ed.) Decision Support Systems , 2008 (45), 517-527.

10. Kring, J. P., & Kaminski, M. A. (2011). *Gender Composition and Crew Cohesion During Long-Duration Space Missions.* In D. A. Vakoch, Pscyhology of Space Exploration: Current Research in Historical Perspective (pp. 125 - 141). Washington, DC, United States: National Aeronautics and Space Administration.

11. Mazur, J. E. (2006). MATHEMATICAL MODELS AND THE EXPERIMENTAL ANALYSIS OF BEHAVIOR. Journal of the Experimental Anaylsis of Behavior , 2 (85), 275-291.

12. NASA. (2014, 11 3). *International Space Station.* Retrieved 11 11, 2014, from NASA.gov: http://www.nasa.gov/mission_pages/station/main/onthestation/facts_and_figures.html#.VGLDZYdzinM

13. Robinson, J. A., Slack, K. J., . Olson, V. A., Trenchard, M. H., Willis, K. J., Baskin, P. J., et al. (2011). *Patterns in Crew-Initiated Photography of Earth from the ISS--Is Earth Observation a Alutogenic Experience?* In D. A. Vakoch, Psychology of Space Exploration: Current Research in Historical Perspective (pp. 79 - 102). Washington, DC, United States: National Aeronautics and Space Administration.

14. Singh, S. (2010). *Modeling Team Behavior And Predicting Success Using Soft Computing.* Third International Conference on Emerging Trends in Engineering and Technology (pp. 620-623). Goa: IEEE.

15. Suedfeld, P., Wilk, K. E., & Cassel, L. (2011). *Flying With Strangers: Postmission Reflections of Multinational Space Crews.* In D. A. Vakoch, Psychology of Space Exploration: Contemporary Research in Historical Perspective (pp. 143 - 175). Washington, DC, United States: National Aeronatics and Space Administration.

16. Washington, DC, United States: National Aeronatics and Space Administration. Ushakov, I., Morukov, B. V., Bubeev, Y. A., Vasil'eva, G. Y., Vinokhodova, A., & Shved, D. (2014). *Main Findings of psychophysiological Studies in the Mars 500 Expereiment.* Herald of the Russian Academy of Sciences , 84 (2), 106-114.

17. Vakoch, D. A. (2011). *Psychology of Space Exploration: Contemporatry Research in Historical Perspective.* (D. A. Vakoch, Ed.) Washington, DC, United States: National Aeronautics and Space Administration.

18. Wichman, H. (2011). *Managing Negative Interactions in Space Crews: The Role of Simulator Research.* In D. A. Vakoch, Psychology of Space Exploration (pp. 103 - 124). Washington, DC, United States: National Aeronautics and Space Administration.

Interstellar Travel in the New Science Fiction

Jason D. Batt

100 Year Starship, Sacramento, California

jbatt@100yss.org

Abstract

The future of humanity in space is not guaranteed. Great cultures have arisen in the past, set their sites on great and, from their perspective, inevitable goals, and failed. Science fiction literature has been a guide post for space exploration from the early days. Much has changed since Arthur C. Clarke and Isaac Asimov wrote. Our political and international world is different. Our technology has branched into different and unforeseen areas. With the current technological expansion in view, what do today's science fiction writers say about interstellar exploration? What pitfalls do they see? What is the roadmap they present now? What challenges do they show? Is there an agreement upon the timeline? What technology shows promise? What dangers do they predict out there? The presentation will present a survey in review of current science fiction on the topic of interstellar exploration. The foundation for this has arrived from serving as judge of the Lifeboat to the Stars literary award for science fiction literature focusing on interstellar efforts. This award was presented at the 2013 Campbell Conference.

Keywords

science fiction, interstellar, short fiction, novels, faster-than-light, generation ships, slower-than-light

1. Introduction

Voyaging to the stars is not just an ancillary aspect of science fiction. It is a core concept that has been embedded in the genre since its earliest moments. Adam Roberts, in *The History of Science Fiction*, contends that "the roots of what we now call science fiction are found in the fantastic voyages of the ancient Greek novel; and I use the Vernean phrase voyages extraordinaires ... especially voyages to other planets." Odysseus sailed to fantastic islands in his quest back home from the shores of war. The extraordinary voyage was woven into the very beginnings of the fantastic tale. Roberts continues: "Stories of journeying through space form the core of the genre ... Are the trunk, as it were, from which the various other modes of SF branch off." The dream of interstellar travel has been explored by science fiction for years before there was even a glimmer of reality apparent in its practical and technological pursuit. [1]

Science fiction's distinctive in contrast to other genres is its ability to serve as a thought experiment, a place where various scenarios can be played out, where consequences can be explored without true damage to real people. Science fiction has been exploring the future of interstellar travel for decades and has wrestled with the very issues that current interstellar experts are grappling with today. Thus, a cataloging and analysis of interstellar aspiration within science fiction can help to mine for solutions and problems to the very topic the 100 Year Starship is engaging today: how will we reach the stars?

2. Interstellar Fiction as a Sub-Genre

Daniel Abraham, writer of *The Expanse* series, was asked on the current state of interstellar fiction and if interstellar aspirations represented a type of "science fantasy". He replied:

> "No, I don't think interstellar travel, space empires, etc. are at best science fantasy any more than they ever were. Or, putting it the other way, they haven't become any less likely, in the recent past. They've always been the stuff of fantasy. That didn't used to be a problem. The thing that has changed, I think, is the requirement that science fiction be somehow rigorous to be taken seriously. When I look back at the classic science fiction that I grew up with, a lot of it didn't pass the sniff test. Larry Niven had teleportation, giant alien cat warriors, and human's eugenically bred for their luck. Arthur Clarke had God turning out the stars. Herbert had giant freaking worms digging through a desert rich in psychoactive space fuel. Unsophisticated as I was, I didn't see anything wrong with that. Still don't. I see two things happening with the genre. The first is that it's becoming the primary idiom of pop culture. The second is that, in response to that, there is a narrower and narrower definition of "real" science fiction which uses terms like "science fantasy" to exclude work which isn't somehow pure enough. I think that there's a real risk of science fiction going the way of jazz music and poetry in which it becomes a more and more sophisticated, narrow, and inaccessible form dedicated to meeting the standards of a smaller and smaller elite. But I also think the consequences if that does happen are pretty minor, because the market for accessible science fantasies like *The Demolished Man* or Dread Empire's Fall or the Vorkosigan Saga appears to still be wide open." [2]

James Blodgett, science fiction writer, provided his view of the current status of interstellar fiction in 2012:

> "Science fiction inspired our early steps in space. If we think about how science fiction did this we may be able to do it again. Science fiction has pictured many methods of travel, and many motivations for traveling. Science fiction stories are a form of scenario analysis that helps us foresee problems and solutions. Humanity could use fresh inspiration for space travel. The public and even the science fiction field seem discouraged with space after initial enthusiasm. Folks in 1940 and 1960 could extrapolate from the Wright Brothers' airplane to the airliner and later the jetliner, amazing progress which took only a few years. The discovery of atomic power suggested miracles to come. This progress was plausibly extrapolated to predict amazing progress in the near future, the near future of 1960. Subsequent space developments have been underwhelming. It is no surprise that science fiction has shifted from space opera to dystopia. However, there are new plausible visions that could lead to amazing developments in space, a space-based form of singularity. The nanotech and artificial intelligence versions of the singularity can also suggest ways to get us into space. There are several types of existential risk that have prompted folks like Hawking to think that we had better get going into space, making the point that space travel is existentially important. We are on the cusp; with disorganization and bad luck we may never develop space and we may go extinct; with focused work and good luck we may see an exponential future in space that can produce amazing prosperity and expand humanity by orders of magnitude." [3]

Interstellar fiction is certainly alive and in its current state provides a distinct laboratory for potential technologies and approaches that will take us to the stars.

3. The Role of Science Fiction in Space Exploration and Scientific Endeavors

In reality, 100 Year Starship is an exercise in applied science fiction. The very nature of key works in the genre is being explored through the mission of the organization: the development of the capability of interstellar travel. As such, it is critical to establish the limitations of science fiction. Science fiction has its roots in fear and wonder, legend and mythology, exploration and reflection. But it does not provide a roadmap to future developments. Science fiction and innovation exist in a circle of influence, mutually dependent upon each other. Dennis Cheatham in T*he Power of Science Fiction*: states: "It may be the case that the future worlds and infinite possibilities projected in science fiction can be used to inspire viewers to pursue work that will make those possibilities or ones like them, real." [4] Science fiction does, quite often, get its predictions wrong. But it does get it right.

Science fiction can be both predictive and prescriptive. How does it do this? First and foremost, it accepts the fact and inevitability of change. It frames dilemmas and puzzles that could arise, play out possible consequences of future innovation, normalize and/or make desirable certain technology, can suggest paths to innovation, inspire invention, reflects the speculative thinking of a generation, and create a mindset that leads to innovation (if not able to predict it exactly). In *The Dreams Our Stuff Is Made Of: How Science Fiction Conquered the World,* Thomas M. Disch writes: "It is my contention that some of the most remarkable features of the present historical moment have their roots in a way of thinking that we have learned from science fiction." [5]

From the beginning, science fiction has pondered what exists past our planet. There have always been stories about travels to other places and other stars. The sub-genres within science fiction (and its close partner fantasy) continue to develop, breaking further and further into various branches. Yet, space journeys formed the primary core of the overall genre.

The focus of this paper is to look at what the modern science fiction stories of space travel are focusing on. In particular, what is the current thought in science fiction towards interstellar. We now exist within an age of actual interstellar missions and organizations. Has this recent development of interstellar organizations and actual efforts to interstellar flight affected the space travel fiction? As well, has interstellar fiction changed as a sub-genre? Has it diminished or expanded in its influence and reach? Has it been abandoned? The intent is to also glean the fiction for original thoughts that can continue to help encourage the remarkable thinking of now and the future. Overall, this analysis will begin (but not finish) the path of sifting the fiction for commonality of thought as a possible reflection of the culture of as a whole.

4. Parameters of the Review

This is not an exhaustive review of interstellar fiction. Materials covered focused on works published since 2000, with a more exhaustive approach to those published since 2010. The intention is to focus on the contemporary thinking of fiction in regards to interstellar and in particular, to assess what the current fiction is exploring in the age of interstellar organizations, in particular since the initial announcement of DARPA's original 100 Year Starship Project.

Works that were tied to other established intellectual properties and previously published methods prior to 2000 were not considered. In particular, although there continues to be numerous works written in the *Star Wars* and *Star Trek* franchises (not to mention *Stargate* and others), these are building upon the continuity and thought that was established in the original development of these IPs. The same consideration is made for works published since 2000 that were simply sequels or continuations of materials published prior to 2000.

Works that explored travel between planets within our own galaxy and universe were reviewed. If the fiction did show interstellar travel but it was dimensional travel, these were not considered. The focus is entirely on interstellar travel within our reality.

Again, this was not exhaustive and focused on professionally published materials. This period of time saw a massive explosion of independent publishing. There were a quite a number of science fiction works that addressed interstellar travel that were independently published. These titles were not considered in this round. However, this is not to say that they shouldn't be. An extension of this review should at least attempt to partially catalog these works because they are attempts to address the subject since the advent of the interstellar organization.

5. Criteria for Works Considered

Attributes considered were:

- method of travel
- specific propulsion method
- speed of travel
- method of acquisition of propulsion technology, and
- nature of the crew.

These were further broken down. For method of travel, pieces were catalogued for:

- faster-than-light capability (FTL)
- slower-than-light capability
- generational ship
- warp capability (the idea that bending the space around a ship provides a loophole in relativistic limitations)
- hyperspace (the idea of jumping into another parallel dimension and utilizing that dimension's different physics to surpass relativistic limitations)
- fixed point jump (FTL travel that requires a jump between preset locations usually due to a dependence on a technological structure)
- free jump (using technology to propel a ship in a manner-like teleportation)
- wormhole/stargate, and
- unique (ideas other than the common ones explored above).

For method of acquisition, works were analyzed based upon if the technology was human developed, developed by aliens, or another method. Under crew composition, works were organized by either:

- normal human crew
- alien crew
- embryonic crew
- robotic crew
- tech-expanded crew (humans that have expanded their capability through technology)
- bio-expanded crew (humans that have expanded their capability through biological advances
- a crew of artificial intelligences, or
- a seed colony/embryo colonization vessel.

6. Interstellar Travel

The vast distances between astronomical bodies and the impossibility of faster-than-light travel pose major difficulties to both interstellar experts and science fiction authors. Science fiction authors approach these issues in several ways:

- "accept them as such (hibernation, slow boats, generation ships, time dilation - the crew will perceive the distance as much shorter and thus flight time will be short from their perspective),
- find a way to move faster than light (warp drive), "fold" space to achieve instantaneous translation (e.g. the *Dune* universe's Holtzman effect),
- access some sort of shortcut (wormholes),
- or sidestep the problem in an alternate space: hyperspace." [6]

So how do these works tackle these problems?

7. Results

Overall, 85 distinct works were reviewed. 46 of these were novels (and/or series). For note, series (such as Gregory Benford and Larry Niven's *Bowl of Heaven* and *Shipstar*) were given one entry. 25 were short stories (the majority of these since 2010). Five were television shows. One was a film. Six were video games. One was a novella. Of the works reviewed, 39 utilize FTL (46%), 40 use STL (47%). Of those 40 STL, 21 are generation ships. Of the 39 FTL:

- 0 stories employ warp as a technology.
- 12 use hyperspace
- 5 use fixed-point jumps
- 10 use wormholes.
- 3 had unique technologies distinct to those works.

In 51 of the stories, humans develop the propulsion technology ourselves. In 15 of the stories, we receive the tech as a gift from the aliens. In analyzing novels alone, we find that 27 of them use FTL (59%), 18 use STL (39%). In particular:

- 7 use generation ships
- 6 use hyperspace

- 3 utilize fixed-point jumps
- 6 use free jump
- 6 use wormholes

In analyzing the short stories, 6 use FTL (18%), 18 use STL (72%).

- In particular:
- 11 use generation ships
- 1 used hyperspace
- The other five are non-explanatory

A full list of works reviewed appears at the end of this paper.

8. Repeated Themes

Several themes were repeated throughout the various stories. In several, older generation ships were overtaken by newer, faster ships. This was the key plot concept of Kevin J. Anderson's "The Tortoise and the Hare." As well, Karl Bunker used this in his novel *Overtaken*.

Distinctly, warp technology is not used. This is despite the recent scientific interest in Harold White's announcement at the 2011 100 Year Starship Symposium. In fact, the far more fantastical and unexplainable hyperspace is far-more employed. Hyperspace was the most utilized technology in the category of FTL.

More often than not, humans themselves develop the propulsion technology that leads them to the stars, regardless if that is FTL or STL. 51 of the stories have humans acquiring the technology on their own. However, there are moments where they technology is provided to us by aliens. Often these are aliens we are removed from and we receive it as a gift. This is the key concept of James S. A. Corey's *The Expanse* series. Initially, in Corey's works, humans utilize an advanced fusion drive for interplanetary travel. An extra-terrestially created wormhole provides the capability for interstellar travel. In fact, this approach of STL only capabilities then expanding to FTL was featured in other works. In particular, Charles Stross' *Neptune's Brood* and Alastair Reynold's *Blue Remembered Earth* and *On the Steel Breeze*.

In particular, short-form fiction focused on character interaction and life on generation ships. This was a theme not repeated in the novels.

9. Unique Concepts

There were several novels that used original ideas. In particular, John Scalzi's *Old Man's War* series employed a technology called skip drive. This was an engine that provided near-instantaneous travel from one location to another location by actually trading spots with an identical ship in another dimension. The theory is that in the endless multiverse, there must be an identical ship with identical crew composition looking to travel from where one ship wants to go to where that ship is. So, they just swap spots.

Neal Asher utilizes the nonsensical term coined by Edward Lear: runcible. In his *Polity* series, a runcible is a interstellar wormhole generator/teleporter. The technology is briefly explained but serves as a fantastical way to instantaneously travel from one spot to another.

David Brin's *Existence* insists that human (and for that matter, biological) travel between the stars is an impossibility. So, civilizations send out advanced probes with copied AIs to scour and seed the galaxy. These AIs allow for culture sharing and the potential for civilizations to, in some manner, exist forever. Yet, all biological life is ultimately locked to their own solar system.

Karl Shroeder's *Lockstep* series keeps human travel to STL only capabilities. Yet, communication and transmission of data is FTL. They utilize this to transmit personalities as well. In particular, his series is distinct for introducing a concept that all human settlements across the stars participate in a shared culture and thus agree to go into periodic very long cold sleeps even if they aren't traveling. Thus, all of the planets and the ships between them are kept in sync technology and culture-wise. For those traveling, the trip seems to occur overnight (get on the ship, go to sleep, wake up when everyone else wakes up). In this way, travelers can actually ensure that when they return home, the same amount of time has passed for those they left behind.

Another galactic-wide civilization concept is explored in Alastair Reynolds' *House of Suns*. Reynolds writes of a galactic civilization that is built on STL capabilities. In this galaxy, planet-based societies, are relatively, short-lived. The most powerful organizations in the galaxy are the "Lines"—family organizations. These Lines don't colonize planets but instead, very slowly, travel through space on circuits of the galaxy and hold reunions every 200,000 years or so.

One final aspect is in Corey's *Expanse* series. In his series, the Mormon church is actually the first to built a generation ship. However, the ship sits destitute as it was never launched.

10. Next Steps

This review was an initial work. There are additional steps to further this. In particular, these works should be compared to previous generations of science fiction. Other aspects to be catalogued could include: the purpose of the trip, how much of the story focuses on the ship, plausibility of the trip, the motivation of the ship, the time frame, political systems, story mood, clarity of story telling. Further examination in regards to the propulsion/transport method could also be explored including, power source, propulsion method, mission type, crew conditions, technology and astronomy assumptions, destination, extra-solar life, etc.

11. Conclusion

Today's science fiction writers continue to explore the concept of interstellar travel. There is great diversity in the approaches and philosophies that enable those travels. There has been a continual shift, in short fiction in particular, to STL -based works. As well, the concept of propulsion method is given far more serious thought than it has been in the past. Even if fantastical, authors go to great efforts to explain their technology. However, more and more are concluding that FTL may be impossible on its own unless access to wormholes is achieved. This study was an initial review and additional work is needed, in particular, a comparison of today's writers to those of the past.

12. Acknowledgments

Special thanks to the r/printsf subreddit of reddit.com. In particular, these users were of great assistance: Starpilotsix, superkuh, sigkircheis, fritz_freiheit, escielenn, gonzoforpresident, systemstheorist, strolls, twistytwisty, weezer3989, and miloshvu.

13. Complete List of Works Reviewed

Table 1. Works Reviewed

Commonwealth Series	Peter Hamilton
Expanse Series	James S. A. Corey
Eon Series	Greg Bear
The Sparrow	Mary Doria Russell
On the Steel Breeze	Alastair Reynolds
Great North Road	Peter Hamilton
Battlestar Galactica	Ronald D. Moore
Sins of a Solar Empire	Ironclad Games
Sword of the Stars	Kerberos Productions
Halo series	Microsoft Studios
The Light of Other Days	Arthur C. Clarke and Stephen Baxter
Palimpsest	Charles Stross
Farscape	Rockne S. O'Bannon
House of Suns	Alastair Reynolds
The Algebraist	Iain M. Banks
Old's Man War	John Scalzi
Revelation Space	Alastair Reynolds
Existence	David Brin

Colony	Rob Grant
Ascension	Philip Levens and Adrian Cruz
Firefly	Joss Whedon
Kovacs series	Richard Morgan
In Which Faster-Than-Light Travel Solves All of Our Problems	Chris Stabback
The People of Pele	Ken Liu
Five Elements of the Heart Mind	Ken Liu
Across the Universe series	Beth Revis
The Tortoise & The Hare / The Grasshopper and the Ants	Kevin J. Anderson / Steve Savile
Bowl of Heaven / Shipstar	Gregory Benford, Larry Niven
The Lost Fleet series	Jack Campbell
The Creative Fire	Brenda Cooper
In the Lion's Mouth	Michael Flynn
Earthbound	Joe Haldeman
Outer Diverse	Nina Munteanu
Ashes of Candesce	Karl Shroeder
Home Fires	Gene Wolfe
Count to a Trillion	John C. Wright
A Delicate Balance	Kevin J. Anderson
Twenty Lights to "The Land of Snow"	Michael Bishop
A Country for Old Men	Ben Bova
Overtaken	Karl Bunker
Scattered Along the River of Heaven	Aliette de Bodard
Connoisseurs of the Ecentric	Jetse de Vries
Against Eternity	David Farland
Lesser Beings	Charles E Gannon
It Pays to Read the Safety Cards	RWW Greene
Other Systems	Elizabeth Guizzetti
Transcript of Interaction Between Astronaut Mike Scudderman and the OnStar Hands-Free A.I. Crash Advisor	Grady Hendrix
The Big Ship and the Wise Old Owl	Sara A Hoyt
Choices	Les Johnson
The Old Equations	Jake Kerr
The Waves	Ken Liu
Design Flaw	Louise Marley
Lucy	Jack McDevitt
Waiting at the Altar	Jack McDevitt
Noumenon	Robert Reed
Siren Song	Mike Resnick
The First Day of Eternity	Domingo Santos

Shattering	Steven Utley
Star Soup	Chris Willrich
Ancillary Justice	Anne Leckie
Neptune's Brood	Charles Stross
Lockstep	Karl Shroeder
Embassytown	China Mievelle
Odyssey One series	Evan Currie
Vatta's War series	Elizabeth Moon
Fortune's Pawn / Paradox series	Rachel Bach
Diving series	Kristine Kathryne Rusch
Pushing Ice	Alastair Reynolds
Confederation series	Tanya Huff
Polity series	Neal Asher
Grand Central Arena	Ryk E. Spoor
The Taken Trilogy	Alan Dean Foster
Spin Trilogy	Robert Carles Wilson
Dread Empire's Fall Trilogy	Walter Jon Williams
Kop series	Warren Hammond
Eve Online	CCP Games
Web Shifter series	Julie Czerneda
Trader Pact series	Julie Czerneda
Light Trilogy	M. John Harrison
Blindsight	Peter Watts
The Future Happens Twice	Matt Brown
It's Always Spring Break Somewhere in the Galaxy"	Raven C. S. McCracken
Interstellar	Christopher Nolan
Mass Effect	Casey Hudson

References

1. Roberts, Adam. *The History of Science Fiction. London, UK: Palsgrave Macmillan: November 15, 2007.*

2. Weimer, Paul. "Mind Meld: Whatever Happened to Interstellar Travel in Science Fiction." *SF Signal.* Web. February 8, 2012. http://www.sfsignal.com/archives/2012/02/mind-meld-interstellar-travel-and-genre/.

3. Blodgett, James. Email to the Lifeboat Foundation SF Star Travel Bibliography, December 24, 2012.

4. Cheatham, Dennis. *The Power of Science Fiction.* December 6, 2011. http://designresearchcenter.unt.edu/sites/default/files/uploaded-documents/cheatham_taxonomy.pdf.

5. Disch, Thomas. *The Dreams Our Stuff Is Made Of: How Science Fiction Conquered the World.* Free Press: July 5, 2000.

6. "Faster-than-light." *Wikipedia.* Online. Accessed September 10, 2014.

Space Colonization: Exploring Lessons from History

Jeremy Chao

Desert Vista High School, 16440 S. 32nd St., Phoenix, AZ 85048

jeremy.w.chao@gmail.com

Abstract

Many similarities can be drawn between interstellar advances and the original colonization of North America. Both are to uncharted lands, and contain many risks. This paper will explore connections between the colonization of North America and deep space travel. The "lost colony" of Roanoke consisted of 115 members. However, the colony soon disappeared, without a trace or struggle. What if one of our colonies on a distant planet did the same? Several years later, the English once again established a colony on North America. This colony, Jamestown, learned valuable lessons from its predecessors. However, there were still several problems. The harsh winter of 1609-10, also referred to as "Starving Time", killed all but 60 of the original 300 settlers. The injuries shown on her skull suggest that her tongue, facial tissue, and brain were removed from her body. How will we combat hunger when we journey among the stars? The human instinct for survival cannot cause of the death of another man, especially when our settlements are light years away. If we are to journey to other worlds, we will need to learn from our history and persevere through and prepare for any dangers that may arise.

Keywords

space colonization, historical analogies, overcoming problems

1. Introduction

For centuries, humans have looked to the skies and dreamed of eventually journeying there. With our current technology, this dream has become a reality for a select group of people. Now, human ambitions turn towards a long-term colonization rather than just a short-term expedition. Instead of exploring Mars, we now look to inhabiting it. We will continually attempt to push the boundaries of impossibility backwards in our quest for knowledge. In order to be more successful in these ventures, we must look to examples of colonization in our past. Many were subject to similar problems as the ones we face today. We need to explore our history to look to the future. Where and why did they make the mistakes they did? What can we do to prevent similar problems? To successfully colonize Mars or any other extraterrestrial moon, planet, or solar system, a model for success will be necessary.

In 1986, Thomas Paine analyzed the influences space exploration and colonization could have on the United States [1]. He set several economical and political goals for the Americans and urged them to construct an enterprise in space. Paine set goals that are challenging, yet feasible, for the American people. A twenty-year plan was proposed, in which space exploration is paired with the government's interests. It incorporated educational benefits into solving problems such as overpopulation and lack of resources. Similar to the planning of the colonization of Jamestown, Paine took into account many factors and presents a concise solution for many problems, while showing a path to an important objective in our near future.

Many, including NASA, have used Jamestown as an analogy for space exploration and colonization [2]. In this article, the author compares the difficulties between settling the moon and Jamestown. However, this article is not very detailed as to which similarities there are, and does not use Jamestown as an example, but merely a tool for comparison. Not using the valuable resource of an example, it simply states some of the troubles that some colonies faced, and some possible troubles that lunar settlers will face.

2. Drawing from Historical Examples

Although humanity has travelled to space many times, there is still no definite attempt at colonizing any extraterrestrial body. Those who wish to do so have a small selection of examples where humans have remained in space for extended amounts of time, and there are no examples of humans colonizing a location that is not on earth. There are a multitude of colonization examples on earth: the colonization of Africa, the colonization of Australia, and the colonization of the Americas. One of the most prominent examples for colonization is that of Jamestown. A successful colony, it still exists today, yet many failures are embedded into its history. Using Jamestown as an example, future pioneers will be able to repeat its success and avoid its failures.

2.1 History Versus Now

During the original colonization of North America, the most developed part of the planet was Europe, specifically the countries that sent explorers to the New World. This is not unlike our current state; we are a developed people and we look to the planets in hopes of colonizing them. However, there were some problems that arose during this time. The colony of Roanoke, situated on Roanoke Island experienced many problems with the Native Americans that were already there. Fighting erupted and the colonists often faced trouble from the native peoples. When the colonists begged their leader to return to England and explain their dire situation, he had no choice but to do as they asked. However, when he returned about 3 years later, the entire colony had disappeared without a trace [3]. Will future attempts at colonization end up the same way? In addition to these problems, the colony of Jamestown also faced many problems. The colonists there faced a harsh winter that destroyed many of their supplies. Referred to as the "Starving Time", it was a period of possible cannibalism as many of the people there turned to various sources of food. When we journey among the stars, these behaviors cannot be allowed. Various problems such as these must be identified and eliminated.

2.2 Comparisons

The risks involved in the colonization of North America and the colonization of space share many characteristics. It is always difficult and dangerous to explore previously uncharted territories. For both the colonists and the astronauts, it is, and was, difficult to communicate across the large expanses of space and the Atlantic Ocean. Lack of, or delay in, communications could present many problems, such as causing loss of lives or damage to equipment. It will cost us a great deal of money to launch ourselves into the stars, just as it cost Europeans much to construct the ships that sent them here. Both interstellar travels and the journey across seas require an extensive amount of time and both journeys are very likely to be one-way trips -- it will be difficult to return unless there is a permanent settlement. Lastly, the psychological effects of permanently leaving loved ones behind, or long periods of isolation are problems that we both face.

Journeying to space, however, does not have the hindrances of pirates, competing countries, or natives already at the destination. In addition to this, the effects of zero gravity, solar radiation, and technological requirements are unique to space travel. We will also have some advantages over the original colonists of North America. We have better technology, and as a result have a better understanding of our destination and the path to reach it. Problems such as sickness or psychological complications are easier to mitigate now. Instead of a human always monitoring the trajectories of the spacecraft, onboard computers can control these things, preventing fatigue or boredom. Lastly, these trips will be completely voluntary and can be planned better for, as opposed to the rushed planning due to religious troubles the colonists experienced. As the colonists were eventually successful, I believe that we will, with perseverance, be equally successful.

2.3 Problems Faced by Jamestown Colonists

At Jamestown, problems arose from the start of the journey. One of the colonists, John Smith, attempted a mutiny aboard the ship. This posed many problems for the others on the ship. Should they kill Smith or should they incarcerated aboard the ship? Eventually, they decided to keep him locked up in one of the three ships that were heading for Jamestown [4]. On arrival, the colonists opened sealed orders from the Virginia Company of London, which appointed Smith as one of the members on a governing Council. These orders also helped the colonists determine a suitable location for their settlement; one that could protect itself from both the natives and the other countries. They selected an inland site that had immediate access to the river and was not inhabited by the natives. The settlers soon discovered the reasons the Native Americans did not inhabit this area. Primarily, the ground was marshy and unsuitable for planting crops. Also, the marshy environment contributed to the mosquitoes that plagued the inhabitants with malaria or other diseases. The water was often dirty and unsuitable for drinking. The peninsula on which they built their fort and houses was too small to support the amount of people there, and had a limited supply of animals that could be hunted [5]. The hopeful colonists decided to ignore these potential downfalls and continue with their settlement.

Shortly after the colonists started, they ran into another problem. Many of their number were unused to the vigorous work needed to construct a successful colony. Already missing the opportune time to plant their crops, the settlers suffered from a poor first year. Many people deserted to join the native tribes living there, and even more of the people died. By the time the next supply ships from England had reached the, around two-thirds of the original colonists were dead or missing. The Virginia Company of London, however, was indifferent towards the troubles the settlers faced. Instead of caring about the people, they only cared about the profit. They sent some manufacturers to America in the hope that the glassware produced would bring them more and more money. With these supplies, they sent many new people to die in the colony.

Only when John Smith took control of Jamestown were the colonists able to access adequate supply. When the Virginia Company stated their frustration and made demands in a letter, Smith wrote back to them. He requested that the Virginia Company would provide Jamestown with enough the resources and manpower necessary for the colony to be self-sufficient [6]. The Virginia Company then sent the largest and best-equipped supply, the Third Supply. They constructed a new ship, the Sea Venture, and had a fleet of at least eight ships. However, the supply ran into a hurricane and was delayed for several months in Bermuda.

The colonists at Jamestown faced their own problems. The summer before, Jamestown had experienced a drought, and could not harvest enough food for the winter. Referred to as the "Starving Time", around 60 of 500 English settlers survived [7]. There is evidence that some colonists ate their fellow colonists when they had no more food to eat [8]. When the Third Supply finally reached Jamestown, they were not prepared. They expected Jamestown to become somewhat self-sufficient, and brought supplies accordingly. They people there planned to abandon Jamestown, but were saved by Lord De La Warr, who brought supplies, a doctor, and some workers. Smith, who was in England during the Starving Time, soon returned. Many historians credit him with saving Jamestown. Under his guidance, the colony prospered. Even with Smith's efforts, there were many arguments with the native people there, which sometimes erupted into bloody conflicts. Despite these hardships, the settlers persevered and the original colony survives to this day, albeit as part of a historic national park.

2.4 Analyzing the Example of Jamestown

One of the biggest problems that Jamestown experienced was the lack of planned supplies. In the beginning, the people there were not used to work and were not able to become self-sufficient. Later, the supplies from England did not carry the resources necessary to farm or build at Jamestown. Eventually, this lack of planning led to the Starving Time, and Jamestown almost became a failed colony. Were it not for the Native Americans who traded with them and taught them how to farm and build, the colonists would have all died. However, we cannot depend on assistance from native life when we journey among the stars. Instead, we will need to prepare everything by ourselves. To plan for an interplanetary mission, we must start from the ground up. We must select only those who are capable of sustaining themselves. Psychological and physical health will be equally important. Knowing how to use all the technology available will also be critical to this mission. This could become troublesome due to the confines a space journey presents; some people, such as Mars One believe that only four people can be sent to Mars at one time [9]. Another necessary aspect of this mission is to plan the supplies necessary. These preparations will be complicated by the limited space aboard a space vehicle. Costs of fuel will continually increase with each new pound of material added. With all of these limitations, planning will be the key to success.

Table 1: Various Difficulties in Colonization

North America	Space	Similarities
Pirates	Solar radiation	Difficult for communications
Competing countries	Cannot depend on help from anyone	Exploring uncharted territories
Storms	Destination not inhabited (to our current knowledge)	Very dangerous
Difficulties from point of origin	Requires fuel/other resources for technology	Expensive
Destination al-ready inhabited	0g side effects	Difficult to transfer supplies
Many disputes with natives	Gravitational pull of other celestial bodies	Takes a long time for travel
Complete unexplained disappearance of colonists (Roanoke)	Destination has different at-mosphere than ours	Many unforeseeable/ seen problems
Cannibalism during periods of starvation (Jamestown)	Destination has different gravity than ours	Unknown what may happen
Leader abandoned colony (Popham)	Limited space aboard space-craft for people and supplies	After departure, difficult/ impossible to return

Table 2: Advantages of Space Colonization over North America Colonization

North America	Space
Technology limited to its time	Better technology, many breakthroughs in propulsion technologies
Little planning for supplies or workers	Better planning of supplies and workers
Did not plan for an environment different from England's	Understanding of environment and conditions to expect
Frequent conflicts with native people	Most likely no native people to have conflicts with
Many competing countries that sabotaged and fought each other	All countries should work togeth-er for an interplanetary journey

3. Application of Examples

Further analysis of the success and downfalls of Jamestown will be necessary to fully immerse oneself in the learning from its examples. Many things need to be taken into account, including location, travel, and resources. Without this, the usage of an example such as Jamestown will be to no avail.

3.1 Location

If one is to colonize Mars, a location for a settlement must be selected. Unlike the hasty decision that the Jamestown colonists made, this decision must be thorough and detailed. The settlers of Jamestown merely looked at a site, picked out a benefit of that location, and constructed a fort there. They were later plagued by unsuitable drinking water, sickness from mosquitoes, and limited space. When we select the home of future pioneers, we will have many things to consider. Solar radiation, Mars's atmosphere, food and water are just the tip of the iceberg. Other things to be aware of include the different climate or environment. For Jamestown, the colonists discovered that the weather and temperature were drastically different from that of England's. Mars will be even more remote. These details cannot be overlooked as they could cost us the lives of many astronauts.

3.2 Travel

While space exploration and the colonization of North America have many shared characteristics, space travel can be significantly more complicated. One of the primary issues with space travel is the amount of time astronauts must be in space. The astronauts must experience extended isolation, larger amounts of radiation, and the effects of microgravity on their bodies. There are multiple solutions to combat these problems. One possible solution is suspended animation. Using this method, the astronauts will not feel the psychological effects of isolation and would be able to be stored in a location where the radiation has little impact on their bodies. One of the ways to combat the microgravity environment is to use centrifugal force by a rotating spacecraft to create artificial gravity. For shorter missions, such as to the moon, International Space Station, or Mars, astronauts do not need to be kept in suspended animation. Short term exposure to radiation will do no lasting harm, and exercise will keep the effects of microgravity at bay. Psychological testing can be used to select astronauts, ensuring that only the most psychologically healthy and stable will be aboard for a mission. One of the problems Jamestown endured was the attempted mutiny by John Smith. In a mission to the stars, instead of a nuisance, it could be the cause for failure. Although there are other problems, many of these have already been solved or avoided over the years. Through human perseverance, even the unforeseen problems can be solved. Throughout the history of space exploration, many problems have risen, yet have all been settled by the hard work and dedication of the scientists that we have.

3.3 Resources

The most critical problem faced by the settlers of Jamestown was the lack of supplies. In the beginning of the colony, they depended on supplies from England. If weather or other circumstances prevented these supplies from arriving, Jamestown had no choice but to starve. Jamestown was not able to support itself for a long amount of time. If we are to send humans to space, we cannot afford to make the same mistake. The astronauts must be able to support themselves from the very beginning. Some plans include sending supplies to Mars before sending the people. All resources sent will need to be carefully planned out, and tools will have to be used to maximum efficiency. Many things, such as fuel and weight, will also need to be carefully planned. If the goal is to colonize a location, the pioneers there must become self-sufficient.

4. Conclusion

Because we do not have examples of space colonization, our examples must come from a different source. One of the sources that can be used is the English colonization of North America. Many analogies can be drawn between the two, and some colonies, such as Jamestown, can be used as valuable examples. In order to better understand the challenges that space colonization brings, we must carefully analyze both the success and failures that the colonists had. By doing this, we can follow their success and avoid their failures. One of the most critical elements that the colonists needed was resources. They were almost never adequately supplied by England, and as a result, often had to resort to various sources of food, including humans, to survive the famines that came. To prevent many of the problems in colonization, we will have to provide enough support for our colonists. Additionally, there are also problems the are specific to space travel and exploration. Among these are the psychological effects of a period of extended isolation, the increased amount of radiation exposure, and the 0g environment that the astronauts must become accustomed to. However, our current technology can provide some solutions to these, including the use of suspended animation for psychological health or centrifugal force as an alternative to gravity. Although this paper only discusses the use of the English colonization of the Americas as an example to space colonization, other examples for colonization can be used. Those which wish to successfully build settlements in space should explore these examples and draw from their success, while avoiding their failures.

Acknowledgements

I would like to thank Dr. John McKnight and Dr. Pamela Contag for providing both research materials and support to continue with this project. I would also like to thank the 100yss committee for a great experience.

References

1. Paine, Thomas. "Pioneering the Space Frontier." *Pioneering the Space Frontier.* National Space Society, 31 July 2008. Web. 18 Oct. 2014.

2. Dunbar, Brian. "Settlement: Harsh Challenges Confront Colonists." *NASA.* NASA, 27 Apr. 2007. Web. 18 Oct. 2014.

3. Horn, James. "The Roanoke Colonies." *In Search of The First Colony.* First Colony Foundation. Web. 18 Oct. 2014.

4. "John Smith." Bio. A&E Television Networks, 2014. Web. 17 Oct. 2014.

5. "Jamestown Settlement." Princeton University. Web. 18 Oct. 2014.

6. "The Copy of a Letter Sent to the Treasurer and Councell of Virginia from Captaine Smith, Then President in Virginia." American Memory. Web. 18 Oct. 2014.

7. Noël Hume, Ivor. ""We Are Starved"" Colonial Williamsburg. Web. 18 Oct. 2014.

8. Stromberg, Joseph. "Starving Settlers in Jamestown Colony Resorted to Cannibalism." *Smithsonian*. Smithsonian. Web. 18 Oct. 2014.

9. "Roadmap - Mission - Mars One." *Mars One.* Web. 18 Oct. 2014.

The Enduring Challenge: Understanding Why We Explore

Bob Hawkins

Opportunity Engineering, 4208 Houx St., Bacliff, Texas 77158

Bob@opportunityengineering.com

Abstract

Why do we explore? Humans have the compulsion to search for their limits: their limits of how far they can get from home, their limits of endurance, their limits of ingenuity. We are driven to push the boundaries of our capabilities, even at great cost.

When the "why" of the quest is strong, then the cost – both the economic cost and the cost in lives – seems like a small expense compared to the payoff.

The justification for exploration can range from the very practical on one hand to purely inspirational on the other. At one end of the spectrum we want to know "how much will this pay?". At the other end we see an image of a better, greater humanity.

Why we join or support a quest requires a blend of both the practical and the visionary. An understanding of the "why" will affect the strategies of an organization, like 100YSS, moving forward.

Keywords

Reasons, why, incentives, future, support

1. Introduction

It is usually easy to understand why we support projects that will be completed soon. When we can see the results of a finished work then we can appreciate that the goal is worthy. But when an undertaking will extend for longer than one generation, like the 100 Year Starship, it is not so easy to see or explain why someone should promote to the effort.

To know why you should support such an effort is important for many reasons. First, in order to reach out and explain the effort to people who have not heard of it. Second, to provide a big-picture strategy to prioritize the day to day tasks. Third, to awaken for individuals a purpose that they can internalize and own for themselves.

This paper includes an explanation of the importance of the rationale for a long term project. It gives some examples of categories of justifications and recommends how this knowledge can be used.

2. Before We Begin

Stop for a minute and imagine yourself in the year 2100. We are ten years or so – give or take – before the mission. The technology is proven and the hardware is just starting to move off the drawing board into the building phase. Imagine that you are looking at the financial pages in September of the year 2100.

Let's look at a couple of headlines:

Companies Vying for Interstellar One Construction Contracts
This could be one of the largest projects ever built. There is plenty of money available and companies a vying for the mouth-watering fortunes to made in public contracts.

Australian Mining Company Sponsors Prospecting Module for Interstellar One
A trip to another world will open huge opportunities for many businesses. Mining and mineral recovery is just one of those businesses which could see a huge expansion with big profits from a return of exotic or rare minerals.

Monsanto to supply seed crops to expedition in exchange for . . .
If there is a developed eco-system on another planet then there could be all kinds of new opportunities for foods, spices, maybe intoxicants? The return of foreign, alien, DNA will be much easier than minerals. Possibly you would even only need to read and measure the DNA and send that information back to earth to reproduce the new wildlife here. There would be a distinct advantage to be in the first wave, before the contamination by invasive species from Earth.

Nike Becomes Official Clothier of Interstellar One
There will be an ongoing need to supply the people living in a new world. There is a natural expansion of the economy from Earth; a growing interplanetary customer base. And of course there is notoriety in being chosen as supplier for the mission.

When the mission is imminent it is easier to think of why we would support this journey. There is money to be made in investing, prospecting, and speculating. There may be a lot of risk but the payoff could be high.

These scenarios, though, are fictional. We do not know what the destinations will offer. We do not even know what rewards might be important to people in 2100. And yet, we can say that there will be benefits.

3. Why Ask Why?

Come back to the present day. Consider, in general, about how we plan and how we accomplish projects that take some time and effort. We think about some result we want, the goal. Then we think about how we will get there. What supplies will we need? How long will it take? What do we need to do first? What next? Who will be involved?

As you look further out in time, you know less and less about what might be required. The further out in the future, the more possibilities for things to change.

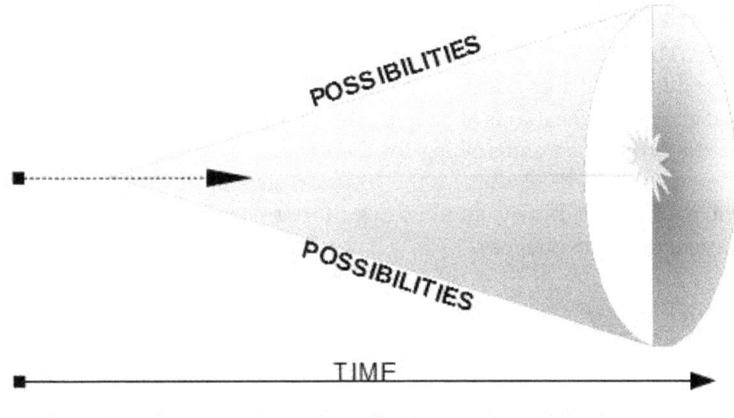

The further into the future we plan,
the less we know.

Figure 1. The further into the future we plan, the less we know.

Everyone knows that things do not always go according to intention. In a long term project there are continual adjustments and refinements of the plan. So how do we ever get anything done? How do we find our target out there in the future, when it is so far away across so much unknown territory?

One thing that does stay almost the same throughout time is the "why" are we reaching for this goal? The Why, the reasons we have to continue, the reasons we know where we are headed, keep pulling us back to our original goal.

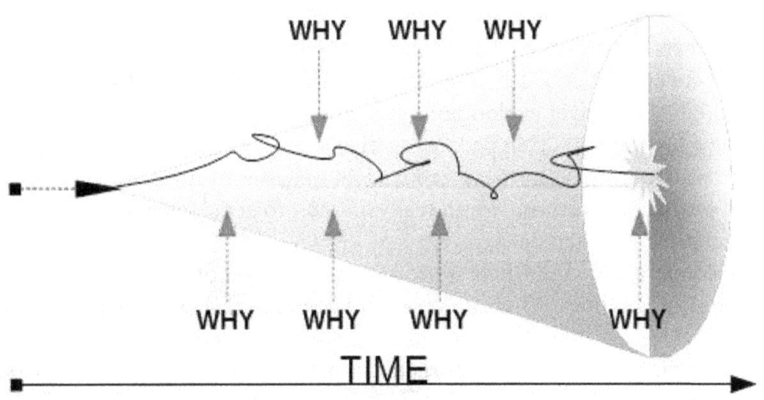

Even though conditions change over time,
the "Why" of the goal keeps us on track.

Figure 2. Conditions change over time but the why of the goal keeps us on track.

It is as if the "why" of our goal travels through time. It could be said that the goal is an expression of our present "why" projected into the future. Conditions may change but the motivations to be there are still the same.

3.1 The "Why" of a Goal is as Important as the "How"

The "why" of a project is just as important as the "How". In the case of the 100 Year Starship, we do not even know the how. Even the destination is unknown. So it is clear that the why is what drives the project.

The importance of "why" is not unique to the 100 Year Starship. Another example is traveling to the moon. We know how to go to the moon. It has been done. We have the means, the "how", and yet we do not have people on the moon. It is the why that is preventing us from going.

Technology, money and even public interest may be available for a big multi-year project. But it takes a clear and strong vision of purpose to provide direction, pull it together, and move toward the goal.

4. Examples of "Why"

The "why" of a goal is not a characteristic that is inherent in the project or the goal, but it is an understanding that is inside each person. The future goal does not exist yet, but the "why" of the goal does exist – right now. However, reasons are concepts that are held in the minds of people. This may seem abstract, but the point is that each person has to have and believe the "why" of an undertaking for themselves. Therefore, there are many different reasons. Here I will give examples of some of the reasons for the 100 years starship. Every person will not agree with every idea. In fact, some of them are contradictory. In most cases, for someone to support the project, they would hold multiple reasons, from more than one category.

4.1 Categories of "Why"

- Economic reasons.
- National interests.
- To advance knowledge.
- Humanist reasons
- Inspirational reasons.

4.1.1 Economic Reasons

We looked at economic considerations in Section 2, when we imagined ourselves near to the time of the launch. It is easy to understand why someone would take risk when the payout is closer in time. For the longer term, this is not such a strong driver.

There may be economic incentives for sponsors who would like to be identified with the values expressed by the 100 Year Starship. This is another good example of where the "why" of a project is more important than the "how". A sponsor would not be so concerned about the technology of the project at this stage in 2014. They would be more interested in demonstrating that they share the values.

4.1.2 National Interests

There is a balance here between national self-interest and world cooperation. Historically, competition between countries rather than cooperation drives exploration. But that is not always the case. Some see exploration into the unknown as a good opportunity to bring people together.

First. consider the positive side. Why would nations join together for this effort?

- **To encourage world cooperation.** We all really do want to get along and a common goal builds a shared destiny that can bring us together.
- **Synergy from cooperation.** The whole is greater than the sum of the parts. Everyone has something to contribute, even if it is money. There can be great power in group participation rather than competing nations.

The project may be too big for one country or organization to complete. The magnitude of the effort may force countries to work together. It could take resources of the whole world to get the tasks done.

Now, think about why a nation would support the project for their own national self-interest.

- **National identity.** World cooperation is fine for some, up to a point. People also worry about maintaining their national identity. Whether driven by genetics or geography, people want to feel that they are unique. The desire to be unequaled may incent a country to go off on their own for interstellar travel. So they can be first or best.
- **Shared destiny.** On the one hand this can bring everyone together. But there is also the drive to have to rest of the world share in "my" destiny. One nation may join the team so they can have control over the results.
- **Cooperation for self-interest.** Yes, we can expect to get more than we contribute. If the whole is greater than the sum of the parts then we anticipate that we get more in return than what we put in. And, yes, we can almost do it ourselves but we need some help. Everyone in the group of nations may only want to fill the gaps in what they cannot do for themselves.

Other justifications for a nation to join the project:

- **To show that they can do it.** This is one of the most common reasons given for the success of the Apollo program: that the US had to get to the moon before the Soviets. It is a sort of passive-aggressive way to threaten others. You say, yes we are doing this for all mankind. But what you do not say is that, by the way, it shows that we can make the deadliest weapons too.

- **Maintain leadership in the world economy.** Everyone will want a piece of the action on a enterprise like this. And the rights to the technology will be important. Governments can plan further out into the future than business for long term economic benefits.
- **Become a member of the advanced countries.** We are seeing this a little today with China and India. They want to have their own capability so that they do not have to depend on other countries for research and exploration science. It enhances their national economy, their independence and their pride.
- **Expand national influence.** If you control a key component of the overall result, then maybe you get a little extra respect. Perhaps other countries are more willing to agree on other issues because you have some control over this important ingredient. You gain leverage.

4.1.3 To Advance Knowledge

These reasons are easy for most people to understand. The advancement toward the 100 year space ship goal will lead to new technologies that will create many new industries. We cannot predict what will come out of it. But we know that success will require new technology. There will be new concepts that no one has considered before and also new solutions to old problems. Both have value.

On the 100 year starship we are also asking to re-evaluate fundamental laws of physics. It encourages a second look at the idea of faster than light travel. Maybe, just maybe, the universe is not flat after all. The open questions could lead to discoveries in fundamental science.

Expand the ability to conduct research. As we venture farther out into space our perspective changes. We get away from the influence of the sun and planets. We will have new ways to look at the universe.

And to discover new things. This is a catch-all to say that we do not know what we will find out there. That is the beauty of it! We do not know. It will help us to find out about the limits of our knowledge.

4.1.4 Humanist Goals

Humanist goals might be rather called Earthlings' goals. Who we are as a species can gain from this effort. Not everyone agrees with these desires or agrees that they are important. But these are values that motivate some people.

- **Survival of the species.** We are at risk from external forces or our own destructive tendencies. If we are to survive as a species then we need to have a back-up plan off this planet.
- **Expansion of the species.** It is in human nature not only to survive but to bloom and thrive. It is our destiny.
- **Expand what we know.** We have a drive to find our physical boundaries and the limits of our knowledge. We know there is more out there. It is up to us to go that one extra step away from the campfire. One more. Then one more. Then one more.
- **A chance to reboot civilization.** Some people would just as soon get off this planet and never look back. If we have a whole new planet to settle would we do it the same? Knowing what we know now... would we still dump our wastes into the rivers and the sky? Maybe not.

4.1.5 Inspirational Reasons

Inspirational reasons are reasons that speak to the heart. These are the reasons that can change a person's behavior in almost everything they do. It not only changes the decisions they make but can change how they make decisions.

- **To ignite students' curiosity.** How does it do this? We set a goal that is beyond what we have today. It is a possibility that you can reach for. It is something that makes you realize that there is opportunity.
- **Inspire everyone to see what is possible.** A big goal inspires more than just students. Most people do not take the time to look up from their busy life and ponder new possibilities. The awareness of potential is sometimes enough to spark innovation. We are creating a vibrant future, a future that has growth, a future that has possibility. The mental image of a new and improved future is enough to create the desire to help build that future.

This is an admission that everything is not done yet. We are still building the future. It gives permission to speculate and to take risks. It is encouragement to join in to build the do-it-yourself future.

The inspirational considerations are probably the most valuable. These are the reasons that speak to the person directly and speak to the core of the personality. They are reasons that benefit now and for the next 100 years. These are what we hold true and what we will believe will still be true in 100 years.

When a student has to decide if they should do their math homework, maybe they will remember this future and the possibilities ahead of them. That is a decision they make right now. Not at some future time. The future that we believe in, and hope for, has as much impact on our decisions today as does our past.

For the economic or the national interest justifications to be enough to drive someone to take action, they also need an inspiration in the form of a promising future. Inspiration is the secret ingredient that makes the practical rationalizations matter.

5. Conclusion and Recommendations

The reasons why we follow a dream are just as important as how we get there. The "why" of our future vision is what is alive today. Our goal for the future is a demonstration and expression of our values. It plants our values into the future and paints a picture of what is possible. It reinforces our values today by inspiring others to participate.

The values expressed in the reasons why this project is important are not desires that everyone would agree with. To recruit support for the project it would be best to find people who have similar values, rather than to try to change someone's mind. One way to do this would be to look at other large ideas – at the imbedded values and reasons—and see where they are similar or different to those of this project. How does the "why" of this project compare to efforts to colonize the moon? Or to explore the deep oceans? If you know what values are unique to this project then you can find other groups that have similar or overlapping values. There you will find the people who will want to have a stake in this too.

It takes a rational decision for someone to decide to support a very long-term project like the 100 Year Starship. But it also takes inspiration. The inspiration comes from an encouraging and empowering vision of the future. It is a future from which we can take values that we use to make decisions today. The drive comes from the knowledge that we must create that future. If we want to colonize other solar systems then we first must create the future where it is possible.

Selected References

1. *A Journey to Inspire, Innovate, and Discover: Report of the President's Commission on Implementation of United States Space Exploration Policy,* ISBN 0-16-073075-9, U.S. Government Printing Office, Washington, D.C., 2004, http://govinfo.library.unt.edu/moontomars/news/docs.asp

2. Commission of the Future of the United States Aerospace Industry. *Commission on the Future of the United States Aerospace Industry: Final Report.* Arlington, VA: The Commission, 2002

3. Hines, Andy and Bishop, Peter. *Thinking About the Future: Guidelines for Strategic Foresight.* Washington DC: Social Technologies, 2006

4. International Space Exploration Coordination Group (ISECG), *Global Exploration Strategy: The Framework for Cooperation,* 2007, http://www.globalspaceexploration.org

5. John M. Logsdon et al. eds. *Exploring the Unknown: Selected Documents in the History of the U.S. Civil Space Program, Volume I: Organizing for Exploration,* Washington, DC: NASA, 1995

6. NASA, *The Vision for Space Exploration,* Washington, DC: NASA, February 2004

   (National Research Council references listed from most recent.)

7. National Research Council. *Pathways to Exploration: Rationales and Approaches for a U.S. Program of Human Space Exploration.* Washington, DC: The National Academies Press, 2014.

8. National Research Council. *NASA Space Technology Roadmaps and Priorities: Restoring NASA's Technological Edge and Paving the Way for a New Era in Space.* Washington, DC: The National Academies Press, 2012.

9. National Research Council. *Approaches to Future Space Cooperation and Competition in a Globalizing World: Summary of a Workshop.* Washington, DC: The National Academies Press, 2009.

10. National Research Council. *Science in NASA's Vision for Space Exploration.* Washington, DC: The National Academies Press, 2005.

11. National Research Council. *Issues and Opportunities Regarding the U.S. Space Program: A Summary Report of a Workshop on National Space Policy.* Washington, DC: The National Academies Press, 2004.

12. National Research Council. *Summary and Principal Recommendations of the Advisory Committee on the Future of the U.S. Space Program.* Washington, DC: The National Academies Press, 1990.

13. National Research Council. *Space Research: Directions for the Future.* Washington, DC: The National Academies Press, 1966.

Beyond Heinlein: Facing the Space-Libertarian Shortage

John Carter McKnight

Lancaster University

Bowland North

Lancaster LA1 4YT United Kingdom

j.mcknight@lancaster.ac.uk

Abstract

This work argues that the virtual worlds of the mid-2000s can be read as analog environments for libertarian space settlements, as their designers shared a common ideology, and were often explicitly inspired by, advocates of such settlements. These worlds explicitly confronted questions of creating new cultures in an environment hoped to be outside the terrestrial legal and governmental system, heavily shaped by its enabling technologies, and capable of experimenting with social systems incapable of enactment on earth. Unfortunately, virtual worlds failed to produce libertarian paradises, for reasons with profound implications for space industry and settlement. Designers implemented engineering systems at odds with their social values, failed to enable social, rather than technological, solutions to emergent social problems, and failed to understand the needs and values of residents of these new socio-technical environments. Residents, on the other hand, outright rejected the libertarian, anarcho-capitalist communities created for them by designers, also rejecting citizenship-based state models in favor of corporate managerialism. Since the governmental and social models users rejected are those most commonly put forward by advocates of extraterrestrial governance systems, it may be time to search for a social model capable of inspiring developers and sustaining future users alike.

Keywords

governance, ideology, social science modeling, case studies

1. Introduction

From the fictions of Jack Williamson in the 1930s through Robert Heinlein to a range of late-1990s political speculators, space settlements have been viewed through a libertarian lens. While some have pointed out that the fragility of infrastructure, its likely high cost, and the lack of adequate indigenous resources including air may predispose space vessels to tightly-controlled dictatorships, the anarcho-capitalist or libertarian community has remained an common model in fictional and non-fictional speculation alike. Meanwhile, from the 1990s, the internet was similarly constructed in techno-libertarian rhetoric as a frontier, a place apart from the nations of the Earth, in which people could be free from the ideological limitations of national governance and the limitations of identities inscribed on their bodies. Thus, when a generation of virtual worlds, some explicitly designed as havens for libertarian experiments came on line in 2003 and subsequent years, it became possible to examine them as experimental testbeds for the political views and practices advocated for space settlements as well as for the internet.

A five-year study of a range of communities and organizations in the most popular gaming virtual world, *World of Warcraft,* and the most popular non-game world, *Second Life*, found a consistent pattern of outright rejection of libertarian and collective self governance models of social organization in favor of hierarchical and managerial models in which the great majority of people delegated authority to a supreme leader. Additionally, rather than debating and adopting rules, in many cases the chosen means of dispute prevention and resolution was via software, in a delegation of authority not even to other human beings but to machine algorithms.

Virtual worlds were far from a complete analog for space settlements: participation was voluntary and rarely 24/7, with negligible barriers to exit from particular communities or from the platforms. The parallel to be drawn is not to near-term space communities but to those for which the libertarian model has been advocated: large permanent settlements within a network of diverse alternative habitats. The lessons to be drawn from the virtual experience are twofold: there is a shortage of would-be libertarians, and many people drawn to innovative, highly technological environments tend to prefer to solve social problems by technological, rather than quasi-governmental, means.

This paper briefly sketches the history of space-libertarian thought in speculative fiction and non-fiction and its relationship to notions of governance on the internet. From there, it summarizes five years of fieldwork on the governance of virtual worlds, explaining witnessed outcomes on the basis of trends in a broader Western middle-class technologically sophisticated culture. It suggests that the failure of uptake of the most commonly advocated ideological system for space settlement provides an opportunity to examine the question of extraterrestrial governance anew, and to create or examine other potential models for their applicability to questions of human social organization in space.

2. Space as a New Frontier of Liberty: An Overview of Speculation

The modern linkage between space settlement and libertarian movements may have begun with the 1931 novel *Birth Of A New Republic*, co-authored by Jack Williamson. [1] The novel describes a movement inspired by the American Revolution to overthrow a repressive corporate state on the Moon. Recasting the American Revolution in space is something of a natural what-if scenario, analogizing to the distance and hardships in traveling to and sustaining settlement for Europeans in North America, and building on a tradition of political thought dating back to de Toqueville [2] and Frederick Jackson Turner [3], of explaining the greater political freedom of the English-descended settlements in North America through the effects of distance from the metropolis and the personal resilience required to inhabit a challenging environment.

The theme was most famously employed in Robert Heinlein's 1966 *The Moon is a Harsh Mistress* [4], perhaps the most famous and influential work on liberty and extra-terrestrial settlement. The novel envisions the Moon as a penal colony along the lines of Australia, in which the convicts and their descendants revolt, issuing a declaration of independence on July 4 and creating a congress. However, Heinlein presents an indigenous culture based on libertarian principles: in the absence of conventional state institutions (as the Warden's Authority only maintains life support and infrastructure, leaving the inhabitants to their own devices), the inhabitants have developed the minimum necessary mores and customary procedures for social functioning. As they are all prisoners, convicted or de facto from musculo-skeletal weakness of Lunar gravity, freedom from authority is prized. Respected community members may be called on at any time to adjudicate disputes, with the disputants offering payment, and judgment, including capital punishment, executed on the spot by the assembled crowd. Reputation systems take the place of regulatory schemes and custom takes the place of law, a common arrangement in relatively isolated communities in which neighbors can count on decades of repeated interactions with each other. [5]

Heinlein's work marks a distinction between concepts of liberty applied to space settlements: between anti-colonialist or anti-socialist visions for which the early United States serves as a model, and anti-statist libertarian imaginings. For the former, John Locke supplies a theoretical foundation; [6] for the latter, Richard Nozick, or, on the left, Murray Bookchin. [7] [8] Despite the rise of private commercial spaceflight and the articulation of an anti-statist ideology for internet governance in the 1990s, and significant cultural and generational divisions between advocates of the two ideological perspectives, statists motivated by American nationalism and libertarians inspired by a vision of a break from state power have tended to cooperate within the space-advocacy movement. For both, analogies to the Euro-American conquest of the American West are commonplace.

That the analogy to the frontier of the American West has been a primary conceptual framing is beyond doubt, from President John F. Kennedy's repeated invoking of the term though Star Trek's "Space, the final frontier," in the years (1966-1969) leading up to the moon landing. A key policy document from 1986 is entitled *Pioneering the Space Frontier,* and elaborates extensively on the analogy.[9] While the meaning and value of the frontier as historical legacy and cultural metaphor for space exploration has been highly contested, [10] [11] the debate over the cultural meaning of the Western frontier and its consequences for space exploration is beyond the scope of this work. There has been, however, an explicit argument for extraterrestrial settlement directly informed by a sophisticated reading of Turner. Robert Zubrin, founder of the space exploration advocacy group The Mars Society, argued in a 1994 essay for Mars as space of political liberty in a frontier, rather than techno-libertarian, vein. The essay begins with an evocation of Turner delivering his 1893 paper. Zubrin argues that Turner's conclusion that the frontier had closed may have been premature then, but was proving true a century later, and that without a new frontier, "progressive humanistic culture" is fading. The key to a frontier is that it be remote enough to allow for the free development of a new society, and that everywhere on Earth, "the cops are too close." Mars, however, has both the cultural distance and the potential wealth to create a new society driven by an engine which destroys aristocracy and institutional stagnation, and promotes democracy, diversity, and individual dignity. [12]

On the libertarian side, O'Neill's 1976 book *The High Frontier* [13] is a foundational work in space settlement advocacy. Drawing on a model apparently coincidentally similar to Nozick's 1974 work, [7] he advocates a system of great numbers of space habitats in the Earth-Moon system, each a platform for social experimentation, with free travel among them, in effect creating a free market in governance systems.

The 1990s saw the application of libertarian beliefs to the internet, and the beginning of a techno-libertarian movement drawing in part from American Progressive technocratic ideologies, the counter-culture movement of the late 1960s, the belief that social problems could be addressed via software engineering, free-market absolutism, and a distrust of state institutions, as Turner documents so well. [14] Again the frontier metaphor is employed, but in an anti-statist fashion, rather than the imperialist, Manifest Destiny, reading given it by Zubrin. The phrase "the electronic frontier" was coined by John Perry Barlow, a Montana cattle rancher (and Grateful Dead lyricist). Barlow took the popular notion of "cyberspace," from William Gibson's seminal novel Neuromancer, which depicted a global information network perceived as physical space, and then described that space in Turnerian terms: "Indeed, one of the aspects of the Electronic Frontier that I have always found most appealing – and the reason Mitch Kapor and I used that phrase in naming our foundation – is the degree to which it resembles the 19th Century American West in its natural preference for social devices which emerge from its conditions rather than those which are imposed from the outside." [15] In his 1996 "Cyberspace Independence Declaration" [16] Barlow calls on the governments of the world to "leave us alone," as cyberspace, "the new home of Mind," is the abode of virtual selves immune to state sovereignty, "that all may enter without privilege or prejudice accorded by race, economic power, military force, or station of birth."

Barlow was a co-founder of the Electronic Frontier Foundation, along with Mitch Kapor, who would become Chairman of the Board of Linden Lab, the company which created and manages the virtual world of Second Life, addressed below. The organization's name parallels that of the Space Frontier Foundation, an offshoot of O'Neill's advocacy group, the L5 Society, and in the late 1990s a principal advocate of libertarian-capitalist space ventures. Both groups attracted Silicon Valley engineers, investors, and advocates with a common set of political views and values sometimes called the California Ideology, or techno-libertarianism. It is this convergence of software and aerospace engineering advocacy within a common ideology that suggested virtual worlds as a test case for the practical application of a politics long advocated for space settlement.

3. Testing the Space-Libertarian Hypothesis: Virtual Worlds as Case Study

The spread of internet access led some to claim that "virtual communities," a term coined by Howard Rheingold in his study of the pioneering group The WELL in 1993, [17] would pioneer newer, more democratic political forms as Barlow envisioned. This vision was shared by many, both developers and users, in the virtual world of Second Life, which launched in 2003. [18] However, the very few experiments in democracy undertaken in Second Life proved vastly less popular as communities and exemplars of governance than its founders expected.

As noted above, *Second Life* was closely linked to the techno-libertarian views espoused by the Electronic Frontier Foundation, among other organizations. It was envisioned as a digital analog to the Burning Man festival: a place for creative engineers and designers to show off their artistry in an environment with a minimum of enforced rules and a spirit of anarchic play. [19] [20] Au, hired by *Second Life's* parent firm Linden lab as an "embedded journalist," and later principal of virtual worlds' main news blog, provocatively entitled *New World Notes,* explicitly cites Nozick [7] in his depiction of the early *Second Life* social environment: one free of social rules, in which neighbors either got along or went to war, in which universities' virtual campuses, historical military re-enactors, brothels and art galleries might find themselves adjacent, with patrons freely crossing each other's properties.

It was into this environment that I entered for a five-year doctoral research project: ethnographic fieldwork in a range of communities experimenting with new forms of governance. With a primary mission of exploring the adaptation of internet technologies generally, and persistent, avatarized virtual worlds specifically to matters of governance, I also saw the project as able to shed light on socio-political behavior in a transhumanist environment, in which material scarcity and bodily needs would be obviated by technology, given a profound interest in these spaces by transhumanists building on Barlow's vision of a republic of the Mind. Somewhat less speculatively, if less closely modeled, I saw the potential to apply conclusions from the study to concepts of extraterrestrial governance. Virtual worlds had the potential to model certain features of an O'Neill/Nozick vision of space settlement, in which large numbers of people (in the mid-to late 2000's, *Second Life* had between 40 and 100,000 "residents," or users, on at any one time, day or night) sought, in Barlow's terms (picked up and applied to virtual worlds by the economist Edward Castronova, who saw these spaces as out-competing the physical world by offering greater personal value [21]) to leave the status quo of terrestrial life for a "new frontier" of personal reinvention. Unlike national and local governmental experiments with "e-democracy," virtual worlds offered relatively free flows of value, information, and people; potential new tools for decisionmaking and resource allocation; a lack of legacy systems, and, again, an explicit techno-libertarian design. This paper summarizes that work: a case study [22] and my dissertation [23] provide a more complete picture.

The virtual world of *Second Life* is a persistent simulated three-dimensional environment in which people interact via avatars, or graphical representations of themselves. Unlike typical massively multiplayer online games (MMOs) such as *World of Warcraft*, it is not built around game mechanisms of quests, scores, competition or objectives. Rather, it re-creates a physical space much like the real world, in which people engage in such activities as they find valuable. Personal behavior within *Second Life* is governed by a set of contracts between the user and Linden Lab, to which users enter into by clicking a button indicating assent, which must be done in order to enter into the virtual world. [24] Among these is the "Community Standards" document prohibiting a "Big Six" of behaviors: expressions of intolerance, harassment, assault, disclosure of real-life information, violation of the adult content zoning, and disturbing the peace. Users are told that they "should report violations" using an Abuse Reporter tool built into the viewer software. [25] With approximately 1.5 million regular monthly users and an average of 50,000 people "in world" at any given time, [26] the general perception that Linden Lab does not swiftly and firmly investigate abuse reports and sanction violators would seem to be justified.

As Linden Lab disclaims liability for "Content, conduct or services of users or third parties," [25] a user has no claim against Linden Lab for not sanctioning any particular alleged violators of the Terms of Service or Community Standards: there is no basis for "third-party beneficiary" claims relating to the contract between LL and the alleged violator, [27] and no basis in the contract between LL and the person alleging violations, as LL has no contractual duty to act on abuse reports, and specifically disclaims liability for actions of other users. This contractual situation gives rise to a political vacuum: not only is the abuse reporting system practically ineffective, particularly in real time, there is also no mechanism for resolving disputes between users that do not involve allegations of violations of the Terms of Service. Users have no contractual relationship with each other, and there is no body of tort law, no judicial system, no universally agreed-upon means of dispute resolution, and most importantly, no mechanism for enforcing judgments, within *Second Life.*

SL users have expanded upon their land-ownership powers as a means of filling this vacuum. While "mainland" land, owned directly by LL, bears no restrictions on its use other than those relating to mature content guidelines, the bulk of land in SL is composed of private islands, typically subject to a "covenant," a document setting forth terms of rental, the sorts of limitations on uses and appearance common in real-life homeowner agreements in planned communities, including architectural style, limitations on the location of unfinished projects, additional restrictions on speech, and other terms. Additionally, much of the mainland has been purchased by "land barons," or large leasing companies, which apply covenants and then sublease the land for residential or business uses. Land barons, or their management staff to whom land powers have been delegated, are typically much quicker to respond to complaints and disputes on their properties. With the ability to evict and ban sublessors, land barons have actual enforcement authority over their tenants.

The land baron/covenant model arose with the establishment of private islands as an SL product in 2006. From SL's launch in 2003, however, there was no clear solution to the lack of a dispute resolution system within the structure of SL's software and LL's contracts with its users. In 2004, a thread on the SL official forum hosted a discussion of the prospect of users creating institutions to fill the vacuum left by LL. In response to a call for proposals by a Linden employee, a group of the forum members submitted a proposal to build a community to be managed pursuant to an electoral system, with a constitution and provisional government. At the time, participants hoped that their model of a constitutional, elected government would prove popular enough to spread to 5% of the *Second Life* grid. Instead, the grandly-entitled Confederation of Democratic Simulators stabilized at five "regions" out of a total of 31,426 as of April 2011, [28] and 70 voting members out of SL's regular user base of 1.5 million.

My five-year study of SL found only four other experiments with democratic self-governance in SL: one community populated by transhumanists and space exploration advocates (which had one vote, to abolish the democratic experiment and transition to a managerial model: only the managers voted against abolition), one Italian-language artists' cooperative, a small community run by American educators, and a group created to enable dialog between secular Westerners and Muslims from around the world, which merged into the CDS, then a year later separated from it and collapsed. In total about 600 people of the millions who established avatars in SL participated in one or more of the governance experiments.

In MMOs, the situation was roughly similar: the almost universal form of organization was the pyramidally-structured guild (or equivalent such as "fellowship" in *Lord of the Rings Online* or "fleet" in *Star Trek Online:* the names reflected the world's narrative, but organizational charts were substantially identical). Typically the guild leader was all-powerful, the software of the MMO prohibiting the delegation of powers even when users might have sought a less hierarchical structure. While MMOs were often viewed from the outside as a chance to act out fantasies of power, in practice they provided an escape from responsibility: by far the most difficult positions for guilds to fill were those at the top of the hierarchy, as leadership was generally seen as an unremunerative burden and the antithesis of fun. Software solutions to social problems were commonplace, one of the most creative and popular being a user-developed technology called "Dragon Kill Points" as a means for redistribution akin to taxation and social welfare policies. [29] While MMOs saw innovations in the development of software solutions to socio-political problems, they never faced calls for liberty in a Barlow sense. Hierarchy prevailed, and competed in a liquid market for low-level followers, who found themselves in an almost ironic version of Nozick's libertarian utopia: myriad governments in which barriers to entry and exit were low, such that an active free market in government could take place, but one in which only one type of government, the feudal/corporate hierarchy, was on offer.

Generally, virtual worlds presented a political atmosphere much like those of Heinlein's Lunar Authority or some sorts of colonial regimes: an overlord with technically near-unlimited power, which however chose not to intervene in internal disputes among residents. This political vacuum at the local scale has been seen by novelists and political theorists alike as an ideal environment for innovation and experimentation, based on the assumption that people, especially those drawn to unsettled new environments, would seek to maximize their liberty. The case of virtual worlds, however, demonstrated the opposite: what people almost universally chose was familiarity and delegation of responsibility, via hierarchical and managerial systems.

This behavior parallels American political behavior more generally, as the past two generations have seen a broad abandonment of civic models of personal responsibility and self-governance in favor of managed spaces, the planned community and Homeowner Agreement (upon which SL's covenants were directly modeled) replacing elected officials and statutes, malls with corporate speech codes replacing public spaces covered by the First Amendment, and the general privatization of services. As the population drawn to virtual worlds was substantially similar demographically to those drawn to managed communities in the physical world, the parallels should have been expected. Likewise, a similar demographic – English speaking, highly technologically literate, skilled workers, with a desire to be early adopters of new technologies – will be seen in any future era of space settlement. Extrapolation of the actual governance practices seen in virtual worlds to the case of space settlement thus seems justified.

4. Modeling Space Governance Systems: Where Can We Go From Here?

The case of virtual worlds provides one salient data point for considering the forms of governance which may emerge or be supported in future extra-terrestrial settlement. Substantial work is being done in simulating the social environment of early small-scale deep-space and Mars missions (e.g., the NASA Haughton Mars Project, [30] the Mars Desert Research Station, [31] the Hawai'i Space Exploration Analog and Simulation, [32] and Mars 500 [33]), which builds on decades of extreme environment social research. However, with Mars settlement actively proposed and debated, research into socio-cultural issues of permanent extra-terrestrial settlement is no longer premature. As the remote Burning Man festival was viewed as a model by the designers of Second Life, [19] we may look to extant and custom-built physical communities as models for extra-terrestrial settlement: Arizona's Arcosanti [34] and Biosphere 2 [35] have been useful experiments in that regard. Beyond the sort of ethnographic work in virtual worlds which I have performed, custom platforms may be created for social science experimentation, as economist Edward Castronova has long advocated. [36]

As Malaby documents [20] and my own work supported, [23] virtual worlds saw a socio-cultural disconnect between designers and users, which had significant negative ramifications on governance and culture. Designers – predominantly software engineers – valued technical skills, a do-it-yourself ethic, and expected the users of their virtual-world platforms to be similarly motivated. Users, however, tended to value aesthetics, social interaction, consumerism, and conspicuous consumption/status display. Much of the story of designer/user interaction within virtual worlds has grown out of the resultant clash of values and practices. The potential for a similar conflict to arise between aerospace engineers and off-planet emigrants is high. Modeling, simulation, and ongoing social science research may aid in shaping the design process to prevent or mitigate clashes over values, technologies and processes, and may make the difference between the success or failure of a future settlement.

This story of the mis-fit between the techno-libertarian ideals of entrepreneurs and engineers on the one hand and early adopters of social technologies on the other is a cautionary tale in at least one additional respect: besides suggesting the need for building lessons learned into the design process, it suggests that there may be a fundamental disconnect between the types of people likely to build and inhabit space settlements and the values and practices most likely to produce sustainable societies in those habitats. In that respect, Marsden's examination of Indigenous practices for long-term sustainability [37] may prove crucial: drawing on the experiences of those cultures which have survived longest in extreme environments may be the key to reshaping space settler's cultures so as not to propagate a dynamic shown to be problematic in analogous environments.

It may be that rather than a shortage of space libertarians to inhabit the envisioned spaces of advocates and entrepreneurs, we may instead choose to address a shortage of space communitarians, and turn to a set of best practices honed over millennia by Indigenous communities around the world as inspiration for our future settlements beyond the Earth.

References

1. Williamson, J. and Breuer, M. J. (1981 [1931]) *Birth of a new republic.* P.D.A. Enterprises.

2. de Toqueville, A. (2003 [1835, 1840]) *Democracy in America.* New York: Penguin.

3. Turner, F. J.(1893). "The significance of the frontier in American history." In Faragher, J. M., ed. (1994) *Re-reading Frederick Jackson Turner.* New Haven: Yale University Press.

4. Heinlein, R. A. (1966) *The moon is a harsh mistress.* New York: G. P. Putnam's Sons.

5. Ellickson, R. D. (1994). *Order without law: how neighbors settle disputes.* Cambridge: Harvard University Press.

6. Locke, J. (1689) *Two treatises on government.* Online: http://socserv2.socsci.mcmaster.ca/econ/ugcm/3ll3/locke/government.pdf

7. Nozick, R. (1974) *Anarchy, state and utopia.* New York: Basic Books.

8. Bookchin, M. (1991) *Libertarian municipalism: An overview. Green Perspectives,* No. 24. Burlington, VT.

9. National Commission on Space (1986) *Pioneering the space frontier: Report of the National Commission on Space.* New York: Bantam Books.

10. McCurdy, H. (1997) *Space and the American imagination.* Washington: Smithsonian Institution.

11. Limerick, P.N. (2000) *Something in the soil: Legacies and reckonings in the New West.* New York: W.W. Norton & Co.

12. Zubrin, R. (1994) "The significance of the Martian frontier." Reprinted in Zubrin, R. (1996) *The case for Mars: The plan to settle the Red Planet and why we must.* New York: Simon and Schuster.

13. O'Neill, G. K. (1976) *The high frontier.* New York: William Morrow and Company.

14. Turner, F. (2008). *From counterculture to cyberculture: Stewart Brand, the Whole Earth Network, and the rise of digital utopianism.* Chicago: University of Chicago Press.

15. Ludlow, P., ed. (1996) *High noon on the electronic frontier: Conceptual issues in cyberspace.* Cambridge: The MIT Press.

16. Barlow, J. P. (1996) "A declaration of the independence of cyberspace." Online: https://w2.eff.org/Censorship/Internet_censorship_bills/barlow_0296.declaration

17. Rheingold, H. (2000) *The virtual community: Homesteading on the electronic frontier.* Cambridge: The MIT Press.

18. Ondrejka, C. (2007) "Collapsing geography: Second Life, innovation and the future of national power." *Innovations*, Summer 2007.

19. Au, W. J. (2008) *The making of Second Life: Notes from the new world.* New York: HarperBusiness.

20. Malaby, T. M. (2009) *Making virtual worlds: Linden Lab and Second Life.* Ithaca: Cornell University Press.

21. Castronova, E. (2007) *Exodus to the virtual world: How online fun is changing reality.* New York: Palgrave McMillan.

22. McKnight, J. C. (2012) "A failure of convivencia: Democracy and discourse conflicts in a virtual government." *Bulletin of Science, Technology and Society.* Vol. 32 No. 5, October 2012 pp. 359 – 372.

23. McKnight, J. C. (2013) *The resilience engine: Generating personhood, place and power in virtual worlds,* 2008-2010 (Doctoral dissertation, ARIZONA STATE UNIVERSITY).

24. Lastowka, G. (2010) *Virtual justice: the new laws of online worlds.* New Haven: Yale University Press.

25. Linden Lab. "Community standards." Online: http://secondlife.com/corporate/cs.php. Last accessed April 8, 2011.

26. Nino, T. (2011) "Second Life statistical charts." Online: http://dwellonit.taterunino.net/sl-statistical-charts/ Last accessed April 12, 2011.

27. Fairfield, J. A. T. (2008) "Anti-social contracts: The contractual governance of virtual Worlds." 53 McGill L.J. 427.

28. Shepherd, T. (2011) *Second Life grid survey.* Online: http://www.gridsurvey.com/. Last accessed April 14, 2011.

29. Castronova, E., & Fairfield, J. (2007) *Dragon kill points: A summary whitepaper.* Available at SSRN 958945.

30. *The Haughton Mars Project.* Online: http://marsonearth.org/

31. *The Mars Desert Research Station.* Online: http://mdrs.marssociety.org/

32. *Hawai'i Space Exploration Analog and Simulation.* Online: http://hi-seas.org/

33. *ESA's Participation in Mars500.* Online: http://hi-seas.org/

34. *Arcosanti.* Online: https://arcosanti.org/

35. *Biosphere 2.* Online: http://b2science.org/

36. Castronova, E. (2005) *Synthetic worlds: The business and culture of online games.* Chicago: University of Chicago Press.

37. Marsden, D. (2013) "Indigenous Principles for a Starship Citizen Handbook." *2013 100 Year Starship Public Symposium Conference Proceedings,* 100 Year Starship.

Editor's Note: Dr. Paul Ziolo presented this paper at the 2014 100YSS Public Symposium. The entire paper has four parts. However, due to the length, we are only reproducing an abridged version in this volume. Due to the complexity of the paper, adjustments to our standard formatting have been made in order to present the material more closely to the author's original intent. The full reference list, for all four parts, is provided at the end of this paper.

Challenges of the Universal State: Comparing the Social Psychologies of Early and Late Phase Cultures (Part I of IV)

Paul Ziolo, Ph.D.

Senior Research Fellow, University of Liverpool

mpf.ziolo@gmail.com

Abstract

"The time taken to effect social change is inversely proportional to the number of people seeking to effect that change'
—Isaac Asimov, *Foundation*

At the conclusion of its third year of existence it is essential that the 100-YSS formulate a plan for development and action over the remaining 97 years in terms of the choice and realization of child and parent goals. While we appear to live in an age of high scientific productivity (and indeed, all technologies essential for an interstellar expedition are either developed, embryonic or theoretically viable), there are indications that global civilization has entered a period defined by macrohistorical models as a 'universal state' – a state immediately prior to anabolic ('stepped') or catabolic (sudden) systems collapse. If this is so, steps must be taken to ensure that a secure self-sustaining platform be established, preferably in interplanetary space, not in LEO but somewhere well beyond the orbit of Earth, before conditions on Earth seriously compromise any furthe development of the space programme due to socioeconomic instabilities and key resource depletion. This platform would serve as the seed base for a Solar culture that in turn would possess the resources and skill to develop the capacities for interstellar travel. In the meantime, socioeconomic conditions already threaten to undermine co-operation among the many space advocacy groups and research institutions. Fragmentation and tribalism are on the increase, the best minds are absorbed into an economy pursuing stasis and closure rather than exploration and the translation of research papers and small-scale local experiments into global-scale hardware, real estate, personnel and machines appears to be an ever-receding possibility. Could it be that we as a technological species are approaching the 'Great Filter' suggested by Bostrom & Ćirković (2011) as a possible explanation for the Fermi Paradox?

Keywords

universal state, metasystem transition, chrysalis, affiliation, diminishing marginal returns, complexity/error catastrophe theorem, space arcology

Introduction: Overview

The most serious 'hidden' obstacles to creating an interstellar culture are those we fail to perceive due to the hidden assumptions we live by. The 100YSS and other interstellar groups have set themselves the goal of creating a 'platform' for interstellar exploration within the time-span of a century or so – despite the fact that no manned interplanetary expeditions have taken place since December 1972, and those were all three-day expeditions to Earth's Moon – the equivalent of swimming out to a rock 100 yards or so out from a beach, then immediately swimming back again. We seem to be aware neither of the thermodynamic/space-time scaling we are dealing with, nor of the realistic implications such scaling has for the biological, psychological, social, economic and political evolution of humanity. As the anthropologist Richard E. Lee of the University of Toronto wrote:

> *"I would like to report some observations on a recently discovered culture – that of the space scientists who attend interstellar conferences. As an anthropologist studying this culture I find it at least as exotic as the cultures of the !Kung, the Greeks and the Vikings, - in some ways more so. Their view of space gives me an interesting insight into their view of the world. They seem to have a tremendous commitment to and faith in, science and technology and their ability to solve our problems. These scientists aspire to the stars and yet their vision is profoundly limited by the blinds of one culture at one point in its history..."* (Lee (1983) pp. 192-3).

Perhaps, as in previous symposia (e.g. Ziolo 2013a), this issue is best illustrated by an image of Paris Airport in the year 2000 as imagined in the year 1900 (Fig. 1 right). Despite the general excellence of the technical research presented at interstellar symposia, this image would still appear to reflect the social assumptions most of us live by – especially when thinking about the future[1], so the 100YSS 'will have done well (lit. 'bene gesserit') if it reflects carefully on what this image may mean or imply.

Fig. 1. Paris Airport in the year 2000 as imagined in the year 1900. From a series of futuristic pictures created by Jean-Marc Côté and others from 1899-1910 – originally as paper cards enclosed in cigarette/cigar boxes, then as postcards. See: *http://publicdomainreview.org/collections/france-in-the-year-2000-1899-1910/#sthash/AAJY7l4f.dpuf.*

The main challenges for the 100YSS (and for other interstellar organisations) are to a) establish a clear goal structure, b) decide on a strategic plan towards achieving this goal, and c) create a transcultural community capable of implementing this strategy. Given the fact that we may be living in what Toynbee has described as a *universal state* (Toynbee 1966), it would seem imperative that stable footholds be established at key points within the Solar System before the constraint and error catastrophes affecting global civilization precipitate a systems collapse to the point were any future recovery to our present level would be seriously compromised by key resource depletion. The questions as to whether or not we are *really* living in what Toynbee has called a 'Universal State'[2] – and what we might do about it – will form the subject of the rest of this paper. We will deal only with *endogenous* (social) causes of collapse, neglecting (for now) the possibilities of *exogenous* (physical) causes and their consequences.

1 We are reminded of the Victorian gentleman who would always put on full evening dress in order to take a call on the new-fangled 'electrical telephone'.

2 Though we refer predominantly to the work of Arnold Toynbee in this paper, the American 'equivalent' to Toynbee would be Carroll Quigley (1979). Whereas Toynbee was concerned mostly with the genesis, maturation and disintegration of civilisations on an ideational basis, Quigley focused more on the behavior of institutions and organisations during a culture's terminal phase.

Part I. Characteristics of the 'Universal State'

1:1. Criteria for determining whether or not we are living in a late-phase culture i.e. a 'universal state'

From our immediate, everyday experience and what we read in the media, we might, in a spirit of civic self-diagnosis, 'tick off' the symptoms listed below (if detected around us or reflected by the media). They are grouped according to socioeconomic, political and psycosocial criteria, but not in any particular order of importance. Some criteria cut across domains, and the lists are by no means intended to be exhaustive.

Socioeconomic criteria:

- Increasing rates of information production *but also* increasing rates of entropy,
- Declining populations at the civilization's 'core',
- Banalization of information – i.e. 'infotainment'
- Stasis and partial regression of science and technology (we should think carefully here in terms of what actually constitutes 'advance)',
- Shifts in the evaluation of career importance – the rise of 'stars' and 'celebs',
- Divorce of currency from an actual value base (i.e. precious ,etals or energy),
- The prevailing of short-term interests over long-term plans,
- Chronic inflation,
- Bureaucratic augmentation and inefficiency,
- Polarisation of wealth - widening gap between rich and poor,
- Subsequent freezing of cash flows,
- Disintegration of state social welfare infrastructures,
- Short-term scanning behaviors (Tainter 1990),
- Resource depletion,
- Environmental destruction,
- Erosion of marriage and family life,
- Over-specialization - Bechtel's (1993) 'centrifugal logic of differentiation'.

Political criteria:

- Legalized torture and violence,
- Overblown, monumentalized & expensive projects (e.g. the Large Hadron Collider),
- Increasing behavioral control & regimentation (a 'tight ship' - everyone (supposedly) 'in close ranks marching in one direction' – 'tightly focused'),
- Fragmentation of groups and individuals – 'tribalism' & primate dominance dynamics (Freud's 'primal horde' behaviours),
- Institutions focus primarily on their own survival *as* institutions rather than on real function or action (Quigley 1979),
- 'Balkanisation' (appropriately called) of political territories,
- Chronic political instabilities and conflicts,
- Schism between erstwhile charismatic elites and increasingly marginalized populations ,
- Increasing social differentiation & stratification, erosion of 'middle' classes and entrenchment of elites or 'cratic groups',
- Proliferation of countercultural movements and hidden subcultures,
- Increasing 'respectability' and acceptance of criminal organisations and criminal behaviours,
- Survivalism.

Psychosocial criteria:

- Breakdown of the dominant construct,
- Regression to magic and superstition,
- Cynicism, erosion of trust and loss of intimacy,
- Increasingly acute fear/paranoia, often at a subconscious level,
- Uncertainty about the future (and unwillingness to talk about it),
- Chronic insecurity,
- 'Carpe diem' mentality ('living for the moment')

- Oversimplification of thought (the 'KISS" paradigm (keep it simple, stupid (or perhaps – keep it simple AND stupid))).
- Anti-intellectualism,
- Increasingly short attention spans (see 'scanning behviours', above),
- 'Tolerance' often serving to mask erosion of morals (marriage laws were always far stricter in 'ascendant' cultural phases),
- Elevation of the random and chaotic as sources of explanation in science,
- Extreme schizotypal behaviours initially identified by Toynbee (discussed below),
- Elevation of the disgusting in art and depreciation of 'good taste' in the name of 'elitism' or 'realism',
- Curtailment of the latency period in childhood and the increasing sexualisation, exploitation and abuse of children (especially by dominant elites),
- In cinema and media,'Heroic Age'-style, 'primal horde'-type plots favoured over thoughtiful exploration of future possibilities.
- SF increasingly simplistic and myth-based, unable to handle the implications of real science and/or social science realistically.

If any of the above criteria seem 'controversial' or provoke 'gut reactions' – the point has been made. We are indeed living in a 'late phase' society (Toynbee's 'universal state') and late phase societies, despite frequent bursts of propaganda to the contrary, have little interest in real-life expeditions to other planets, never mind the stars.

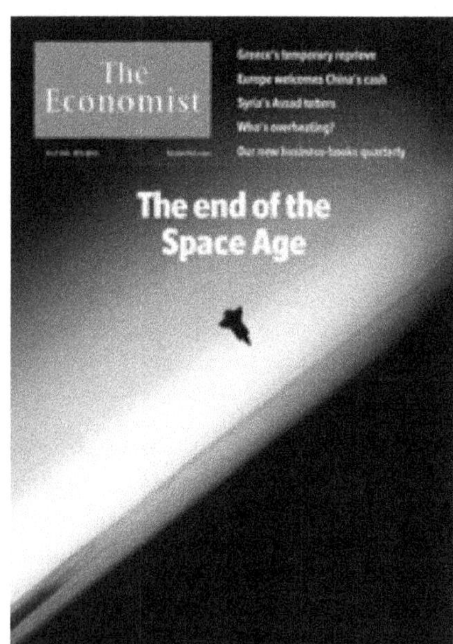

Fig. 2. The global corpocracy's view of the Space Age: In the article, the caption beneath the title reads: *Inner space is useful – outer space is history.* *The Economist*, week beginning June 30th, 2011

Like Ptolemaic Egypt, such societies are increasingly closed, restrictive, and inward-looking, seeking to conserve power hierarchies and to control their citizens, protecting their collective psyches against fears of encroaching chaos. They prefer exploring the so-called 'boundless limits of the human imagination' rather than facing the cost and risk of real exploration (Fig, 2). Why bother with 100 years of cost, labour and risk in getting to α Centauri when you can do the same thing on YouTube in 20 minutes in the comfort and safety of your own home? At this critical juncture in history, human 'involution' as opposed to 'evolution' is the primary challenge facing all space advocacy groups. If this issue is habitually denied or ignored at interstellar conferences, we will end up going nowhere. To overcome the problem we need to understand the underlying causes and nature of 'collapse' and the fact that there may well be a number of 'sweet spots' (points of opportunity) that mark out one or more trajectories through what seems like imminent chaos. If we truly wish to translate 'presentations' and 'proceedings' into hardware, real estate, personnel and machines, we must discover the means to discover and exploit these 'sweet spots'.

1:2. Localised partial or total collapse cycles and global convergence

Historically, what is called 'systems collapse' is part of a normal localised fluctuation in a history of global convergence. On the surface of a lake, ripples may move in directions contrary to those of deeper currents. Systems collapse has been total and complete (catastrophic) only in the case of comparatively isolated social systems such as those of the Maya, the Greenland Vikings or Rapa Nui (Easter Island) which were faced with extreme challenges. Adjacent, more closely bonded social systems (such as those linked to the World Bank and other global institutions) prop each other up like cards or dominoes. Holes or power vacua cannot be tolerated in a co-dependent social fabric and must be patched up (in a sense, Russia, Greece, Spain, Italy and Ireland all 'collapsed' recently – yet these countries are still fully functional members of the world community). Nevertheless, when a group of societies are tightly bound together by a single resource base or financial system, the social fabric is extremely delicate and a shredding of the whole social fabric is still a possibility. 'Partial' collapse usually involves a purposeful, controlled decomplexification in response to repeated systemic failure. In this process, a point may be reached at which further collapse is arrested and some form of recovery may be possible while resources are still available (Toynbee refers to this as 'rout and rally'). At present, the global system is undergoing a series of comparatively minor anabolic collapses, but total depletion of fossil fuels plus the collapse of what is in real, physical terms a 'virtualised' currency (Wallace 2009) may well induce catabolic systems collapse on a global scale, making recovery to our present technical level highly unlikely, and thereby incapacitating the space program. The survival of any society depends upon that society's ability to maintain a balance between development and non-renewable resources.

Impending catabolic collapse requires a surge of technical innovation to 'push through' to a quasistable point where there are still possibilities for continued 'growth'. Such cycles of innovation and 'renewal' become shorter and shorter as a society approaches the point beyond which further growth is no longer possible - a point called the finite time singularity (see below). This cyclic process is the basis for all long-wave economic theories (Tylecote 1992, Reijnders 1990, van Duijn 1983). During the eras of early settlement and state formation, such processes of 'renewal' were often accomplished through emigration to new territory and/or the acquisition of a new resource base, a process usually initiated by 'schizotypes' - i.e. prophets or visionaries - who led a small band of followers away from a society whose population had exceeded the critical balance point with respect to its resource base and whose leaders had become politically oppressive and often enforced environmentally destructive policies to keep themselves in power. Evolutionary psychologists (e.g. Stevens & Price 1996, 2000) refer to this as the *dispersal hypothesis* and propose that it accounts for a) human dispersal around the planet and b) the persistence in the human genome of heterozygous schizotypal genetic polymorphisms at a level of 1% of the world's population (Burns 2004 – see Fig. 3 below). The colonisation of space, both interplanetary and interstellar, may be regarded as the latest manifestation of the dispersal process.

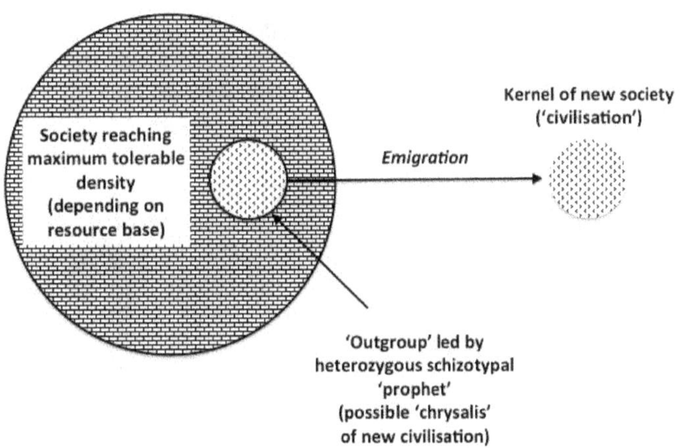

Fig. 3. The Dispersal Hypothesis: Subgroups split off and seek 'new horizons' when the density of a main group exceeds a critical point (for hunter-gatherer groups – the so-called 'Dunbar Number': c. 150 (Dunbar 1992). This limit depends on the society's technological level.

We are accustomed to thinking of the history of civilizations in terms of 'rise' and 'fall' following such dramatic accounts like those of Gibbon (1988), but the average human life-span is very short compared to the historical processes that govern the kinds of social structures we create. Fig. 4 below gives an overview of the 'recent epoch' of recorded history - a period spanning a mere c. 7000 years. In this figure the history of terrestrial civilization appears as a series of alternating attractors and phase transitions. The phase transitions represent periods of population growth and overall expansion while the attractors represent the social equivalent of *Nash equilibria* where a group of societies interact at a certain level of technological development (irrigation agriculture, expanded agriculture (e.g. after the development of the moldboard plough) and proto-industrial).

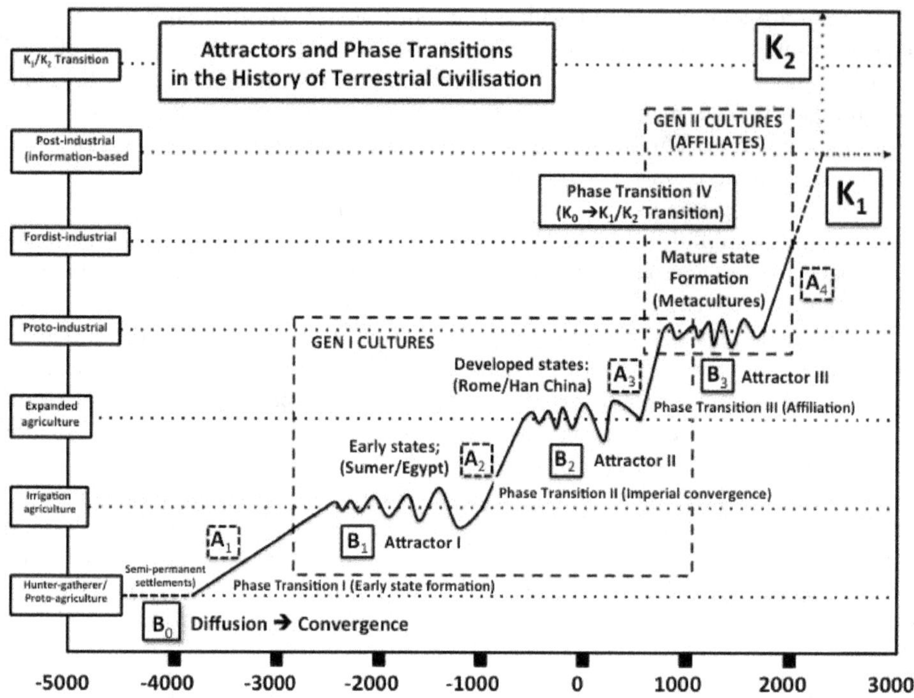

Fig. 4. Overall development of terrestrial civilizationfrom c. 5000 BC to the present. The vertical axis represents overall developmental complexity while the horizontal axis represents time. (adapted from Turchin et al. (2003) pp. 52-57, 121, 134, 138 plus intervening data sets)

Fig. 4 shows how the overall growth curve of world civilisation is exponential. i.e. moving towards a *metasystem transition* (Ziolo 2013a, Turchin 1977). At the very least we are reminded that preoccupation with the immediate details of our lives and times often cause us to lose sight of the real, much greater processes that govern them. "*The Universe is a continual explosion extending over a time of twenty billion years that appears as a majectic solidification only yo a transient being like Man.*" (Stanisław Lem: *Microworlds*). A space-faring civilisation must, however, take full account of these long-term processes and adapt itself to life on these timescales. This may ultimately prove to be the ultimate challenge, creating a vast gulf between terrestrial (human) and interstellar (posthuman) psychologies.

1.3. Universal scaling in biological and social systems

General thermodynamic principles underlie the fluctuations observed in the histories of all 'developed' societies to date – and global civilization is by no means exempt from these principles. These are the *universal scaling* principles shown by Geoffrey West and colleagues at the Santa Fe Institute (West 1999, West & Brown 2004, Bettencourt *et al,* 2007) to apply to all biological and socioeconomic systems on Earth. Starting with Fig.5 below;

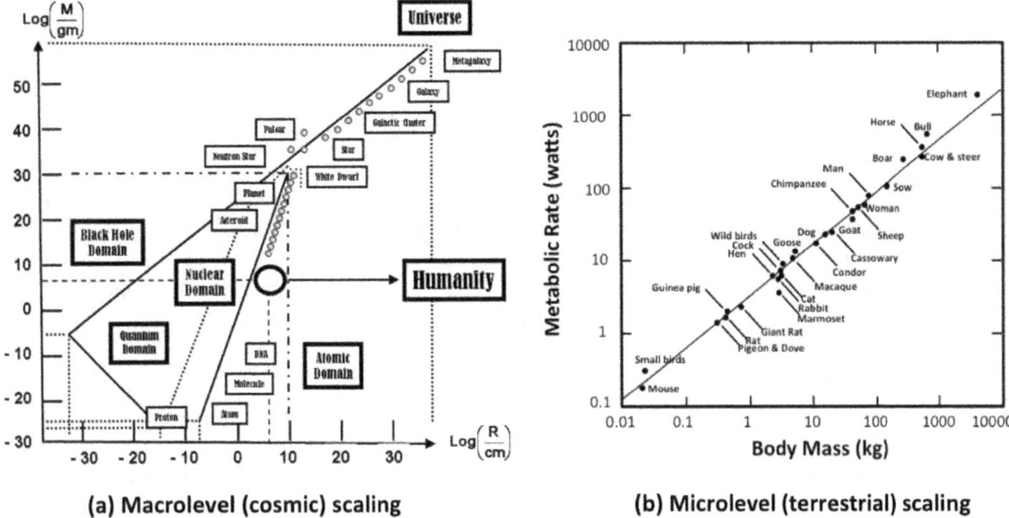

Fig. 5. (a) Macrolevel (cosmic) scaling showing humanity at approximately the half-way mark, and & (b) sublinear scaling principle in terrestrial biology: the slope here is ¾, i.e. < 1- hence sublinear. (after Barrow & Tipler 1986 (p. 397), West 1999 (p. 37) & West & Brown 2004 (p. 105)

Fig. 5 (a) shows humanity at approximately the half-way point in a macrolevel scaling from the smallest to the largest universal structures while Fig. 5 (b) shows the generalized sublinear power law governing the metabolic rates of all biological systems on Earth – these rates are scaled relatively to each other by power laws whose exponents are (±) ¼, ¾ or ⅜. Generally:

$$Y = Y_0 M^{\beta} \tag{1}$$

where Y is some biological observable, Y_0 a normalization, M the mass of the organism and β the scaling exponent (West 1999 p.104). The (±) ¼ / ¾ / ⅜ value of β (also written as B) extends across 27 orders of magnitude ranging from the largest mammal down to the most basic units of molecular metabolism. These exponents express the dimensionalities of all universality classes of terrestrial biology. For example, whereas mammalian lifespan increases as ~ $M^{1/4}$, heart rate decreases as ~ $M^{-1/4}$ so the number of mammalian heartbeats per lifespan is approximately invariant: ~ 1.5×10^9 (West 2004, Kleiber 1975). That goes for all of us, from the pygmy shrew all the way up through space scientists and astrobiologists to the blue whale (not shown in Fig. 5 (b) although its presence (just outside the box) would place humanity exactly at the half-way mark along the scale). For an organism living in d dimensions, the metabolic rate β (the power required to sustain the organism) scales as:

$$\begin{aligned} \beta &= M^{d/_{d+1}} \\ &= \beta \propto M^{3/_4} \end{aligned} \tag{2}$$

where the numerator (3) represents the dimensionality of space and the denominator (4) expresses the *fractal dimension* of the surface boundaries of all organisms, filling their appropriate spaces (metabolic fields). Life takes advantage of the properties of space-filling, fractal surfaces where energy and resources are exchanged in order to maximize energy transfer from the environment (the so-called 'trans-boundary flux' (Nicolis & Prigogine 1975). It could therefore be said that life operates in five dimensions: 3 of space, 1 of time and 1 pertaining to the fractal boundary of the system. In general terrestrial life is governeded by the following three basic principles:

1. *Space-filling networks*: all systems, whether biological or socioeconomic, operate on the principle of tree-structured networks arranged into nested hierarchies that fill the total space of any chreod or system.

2. *Invariance of terminal size*: the size of all terminals or endpoints in the network branchings (e.g. the capillaries of the circulatory system) are invariant irrespective of a given systems mass.

3. *Economies of scale*: the metabolisms of all biological systems seek maximum efficiency – i.e. to minimize energy consumption while maximizing energy usage.

It is unclear whether such power laws are due to the evolutionary matrix specific to Earth, to the universal constants governing the evolution of the Universe or, as is most likely, to both. One thing is clear - socioeconomic evolution, especially of the capitalist variety, appears to violate these basic principles. If *N(t)* is the population of a given social system at time *t*, the power law scaling of eq. 11 now takes the form:

$$Y(t) = Y_0 N(t)^{\beta} \tag{3}$$

where *Y* denotes material resources (energy or infrastructure, measures of social activity etc.), Y_o is a normalization constant and the exponent β expresses the general interactive dynamic across the social system. If we set *R* per unit tine as the proportion of *Y* required to support one individual, and *E* as the quantity necessary to add another, then resource allocation can be expressed by $Y = RN + E\ (dN/dt)$ where *dN/dt* is the population growth rate (Bettencourt *et al*, 2007 pp. 7303 *et seq*.). Now taking the basic socioeconomic growth/balance equation presented in Ziolo (2013b):

$$\dot{Q} = \left[(k_3 - k_2)E - (k_1 + k_4)\right]\dot{Q} \tag{4}$$

where Q = assets or capital created throiugh the importation of energy (*E)* from the environment according to flow rates k_2 (input) and k_3 (feedback) with waste factors k_1 (process-entropic) and k_4 (configurational entropic – see below), we set:

$$\dot{Q} = Y,$$

and

$$\left[(k_3 - k_2)E\right] = \left(\frac{Y_0}{E}\right)N(t)^{\beta} \tag{5}$$

and discount the pollution/waste factor (for now) by setting

$$\left[(k_1 + k_4)\dot{Q}\right] = \left(\frac{R}{E}\right)N(t) \tag{6}$$

we get the following basic growth equation for social systems:

$$\frac{dN(t)}{dt} = \left(\frac{Y_0}{E}\right)N(t)^{\beta} - \left(\frac{R}{E}\right)N(t) \tag{7}$$

the solution to which is given by:

$$N(t) = \left[\frac{Y_0}{R} + \left(N^{1-\beta}(0) - \frac{Y_0}{R}\right)\exp\left[-\frac{R}{E}(1-\beta)t\right]\right]^{\frac{1}{1-\beta}} \tag{8}$$

Vastly different behaviours result from this dynamic depending on the value of the exponent β. If $\beta = 1$, an exponential growth function results, i.e. $N(t) = N(0)e^{(Y0 - R)t/E}$, which best describes maintenance and self-organisation at an individual level. If $\beta < 1$ the result is a sigmoidal growth curve where growth gradually ceases at *(dN/dt)* as the population approaches and finally reaches the carrying capacity of the environment at $N_\infty = (Y_0/R)^{1/(1 - \beta)}$. This is the normal sigmoid growth curve governing all biological dynamics (including neurcognitive and learning behaviours) and generally describes 'economies of scale'. Most 'traditional' or 'hunter-gatherer' societies such as those described by Diamond (2005, 2012): Papua New Guinea, Tikopia, or the as yet 'uncontacted' tribes of S. America that live 'in harmony with their environment', rigorously practice 'economies of scale' as defined by their traditions.

'Developed' or 'mature' socioeconomic systems however, engage in a developmental process where $\beta > 1$ due to self-reinforcing factors such as wealth creation, learning, innovation and re-investment – all channeled in specific ways by the dominant projective construct (ideology or religion – see North 1997). If $N(0) < (R/Y_0)^{1/(\beta - 1)}$, eq. 12 leads to unbounded growth governed by a 'hyperexponential' function, i.e.

$$t_c = -\frac{E}{(\beta = 1)R}\ln\left[1 - \frac{R}{Y_0}N^{1-\beta}(0)\right]$$
$$= \left[\frac{E}{(\beta - 1)R}\right]\frac{1}{N^{\beta-1}(t)} \qquad (9)$$

Fig. 6 below shows the influence on terrestrial biology of the slope of β .

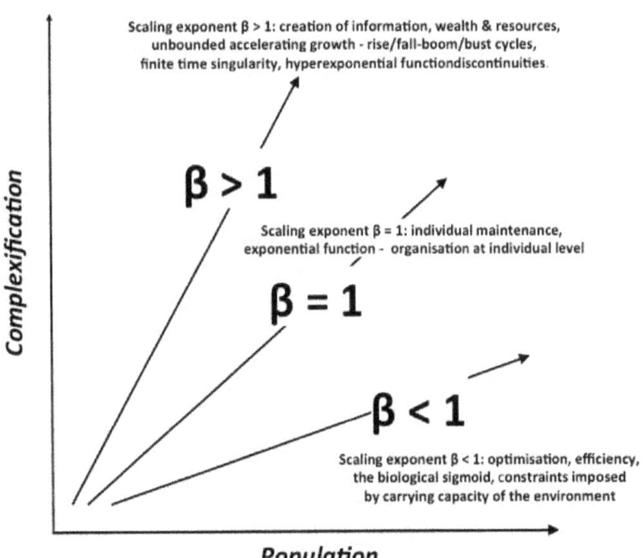

Fig. 7. The difference between supralinear, linear & sublinear growth: β > 1 = Koyaanisqatsi = creation of information, wealth & resources, → unbounded accelerating growth → rise/fall/boom/bust cycles → hyperexponential (supralinear) function → finite time singularity. For β 1: individual maintenance → self-organisation at individual level expressed by exponential function. For β < 1: the biological sigmoid → optimisation & maximal efficiency constrained by resource base & carrying capacity of the environment.

1.4. Consequences of 'pushing the boundaries'.

Systems and subsystems of the terrestrial biosphere seek to maintain minimal entropy at the local level in order to allow maximal entropic flow at the macrolevel – a principle known as *Maximum Entropy Production* or MAXENT (Chaisson 2007, 2011, 2014; Kleidon 2010). The Universe appears to be movong in the direction of increasing complexification, yet "*Die Entropie der Welt strebt einen Maximum* zu", as Clausius wrote in 1850.

This apparent contradiction is resolved when we begin to understand that maximisation of entropy is achieved *through* the maximisation of complexity, and that the human drive towards the creation of ever more complex and extended social structures (and therefore push developmental trajectories to supralinear levels) is part of the drive towards *niche construction* (Lalande *et al.* 2001) – the creation of a perfect 'womb-surround' that will contain and protect us (or at least 'our group') from a dark and hostile outer Universe. By creating systems whose configurational entropies (Craig, 1992) far exceed those of the biospheric matrix within which we are embedded, our civilisation's drive to encyst and *shield* itself from this matrix results in a series of rises and falls, grasps and collapses, golden ages and 'interesting periods' that we call the 'tragedy of the human condition' (on smaller scales, this pulsation manifests itself as a series of booms & busts or bull and bear markets). Within any society, an anastrophic rise in social complexity eventually meets the *constraint and error catastrophes* (Kauffman 1993) peculiar to that society's socioeconomic system and as the society in question continues to invest in greater complexity in order to stave off catastrophic collapse, *diminishing marginal returns* (Tainter 1988) begin to be felt in all domanis of activity, even those supposedly devoted to innovation and renewal. Successive waves of 'innovation' and 'renewal' (boom/bust) occur at shorter and shorter intervals until the whole process reaches a *finite time singularity* (Johansen & Sornette 2011). Since planetary resources are always finite, the consequences are obvious:

Fig.7. Wile E. Coyote meets the finite time singularity....

For those who prefer simple images to baroque prose, Fig. 7 summarises the entire argument of this paper. We are currently approaching the 'OOPS!' stage (frame 3).

Despite all this, it is clear that only a social system that seeks to transcend the limits of thermodynamically-based biological growth can possibly create a recursively self-enhancing scientific and technological culture. Such a culture could never have arisen from a society that lived in perfect equilibrium with its environment. Yet the historical cost of our advance has been severe, both ethically and environmentally. The question is: do we now have enough historical experience to be able to radically innovate and advance without precipitating collapse, whether anastrophic or catastrophic? Can a permanent foothold in space be established despite the clear proximity of the finite time singularuty? In the entire 'science' of economics there woud appear to be as yet no general model, formula or equation relating real-valued currency (one directly based on energy or precious metals) to an equitable balance between industrial production and social welfare. In the words of Wallace (2009):

"*...economics does not provide a formal quantitative model or equation that identifies or allots the appropriate minimum sustaining level for a society's investment in public services that will also permit fiscal integrity and economic productivity.*"

(Wallace 2009, p. 87)

If we are seeking to create a smooth pathway from the present towards an interstellar civilization, this lacuna in economic theory should be filled as soon as possible.

The issue of entropy, both systemic and configurational, takes centre stage in the case of multigenerational interstellar flights. A starship travelling at high acceleration (a significant fraction of c) becomes a closed system far from equilibrium where $\Delta S = \Delta S_e + \Delta S_i$. Within such a system the total number of each elemental atom is conserved, no matter what kind of molecule it may be part of, i.e.

$$\sum_{j=1}^{m} a_j N_j = b_j^0 \qquad (15)$$

where N = the number of j-type molecules, a_{ij} = the number of atoms of element i in molecule j and b_j^0 = total number of atoms of element j in the system.

On a major multigenerational interstellar voyage therefore, every atom of the ship, its crew and supplies must be fully recycled so as to minimize both systemic and configurational entropy – the effects of which may be critical over long time-spans. *Everything* must be recycled, without the slightest exception. This fact may compel us to reconsider what we mean by 'human' and 'being'. As was stated at some point during the 2014 symposium: "*on a future starship, humans may no longer be the most important species...*"

Part II. Dominant Psychologies of the 'Universal State'

2.1. Breakdown of the Dominant Construct

"*Before a civilization is destroyed from without, it first destroys itself from within*"
(Durant 1944).

Historically, it was the Tunisian scholar and diplomat Ibn Khaldun (1332-1406) who first identified the role of what he called *asabiyya* (*group identity* or *group feeling*) in the evolution of a society or state. *Asabiyya* is consolidated and strengthened by binding it into a *dominant projective system* or construct. In addition to its primary functions as collective focus for the stabilization of ego defences, a construct 'etherialises'[3] and thereby grants religious sanction to the claims of cratic (ruling) groups to maintain power and create appropriate social infrastructures. As a society grows, the projective system functions as a 'social glue' binding cratic groups and the *demos* together within a single social fabric and building confidence by providing a collective vision of the future. As diminishing marginal returns and the effects of the constraint/error catastrophe begin to be felt, the projective system must adapt and respond by successfully confronting and explaining these hidden forces in terms of its dogmatic structure. If it cannot do this, it becomes irrelevant. The resulting 'disjunction between genesis and validity'[4] causes the 'social glue' to lose its cohesive power, the faith of the demos in the cratic groups' ability to govern begins to erode, inner tensions between elites and demos are augmented and the cratic groups themselves (whose numbers had previously grown along with increases in wealth and social complexity) now lose confidence and begin fighting among themselves over steadily diminishing resources (Weber 1958; Turchin 2003, 2006; Howard 1996). As the universal state seeks to expand in order to compensate for these effects, the tax burden on citizens increases (Tainter 1990 pp. 65, 107, 116) causing further alienation on the part of the demos. Elimination of the 'middle classes' (wealthy members of the demos, or lower ranks of the elite) ensures polarisation of wealth distribution and consequent freezing of cash flow. This situation is clearly seen prior to the outbreak of the Thirty Years War (1618-1648). In a previous presentation/study (Ziolo 2013b) quantification models for these dynamics that were developed by the Moscow-Volgograd school of Social Sciences (e.g. Turchin 2003; Turchin *et al.* 2010) were introduced, as well as studies of path dependencies developed bythe economist Douglass North (North 1997).

2.2.1: 'Schism in the Soul'

Major social processes unfold on a scale far longer than the human lifespan. When those born into a particular time and place are taught that the situation within which they find themselves is the only correct, possible and just version of social reality, most will accept this teaching without question. Whether a citizen of a universal state chooses to 'believe in' or reject the state's dominant construct is less important than the fact that they are bound to serve its precepts by virtue of being part of that community that observes it. Even if the construct is no longer socially relevnt, its echoes persist (Binion 1986). It is the construct that confers protection on its citizens

3 Originally Toynbee's word – meaning to infuse with religious or spiritual significance.

4 i.e. 'if God (or 'the gods') is/are so good, why are things here on earth so bad?

by providing social stability, a moral compass and a vision for the future. When these are absent, hope in the future is lost, trust is broken and paranoia grows, fueled by uncertainty and fear of catastrophe. Schizotypal behaviours are therefore typical of late-phase societies – especially in those founded on the primacy of the individual over community and a competitive economic system (Deleuze & Guattari 1972) where extended family and community structures have broken down. When the construct that sustained the laws granting at least minimal protection to its citizens is cynically disregarded and replaced by anarchic nihilism, the question arises – does one 'go with the flow' (as all one's peers seem to be doing,) or 'drop out' (whether alone, or by joining some kind of 'countercultural' group)'? The 'behavioural matrix' of Table 1 below is based on Arnold Toynbee's analysis in *A Study of History* (1966 pp. 420-530).

'Splitting' – the primal division of all things into 'good' and 'bad' - is the most primal human psychodynamic. We define ourselves and form our identities in terms of what we love and what we hate – what we accept as part of our world and what we reject as 'alien'. At the deepest level, social impulses during late-phase cultures are split between the quest for order vs. submission to chaos – i.e. between the ascetic vs. the orgiastic, between self-organisation vs. entropy, between the forces of social discipline and complexification on one hand, vs. those of collapse on the other. The schizoid behaviours shown in the behavioural matrix of Table 1 all represent modes of escapism from an intolerable, chaotic and fearful present. They derive from Toynbee's analysis in *A Study of History* (1966 pp. 420-530). Three columns represent the psychosocial realms of feeling, behaviour and action while two rows divide these realms into individual and social (collective) levels:

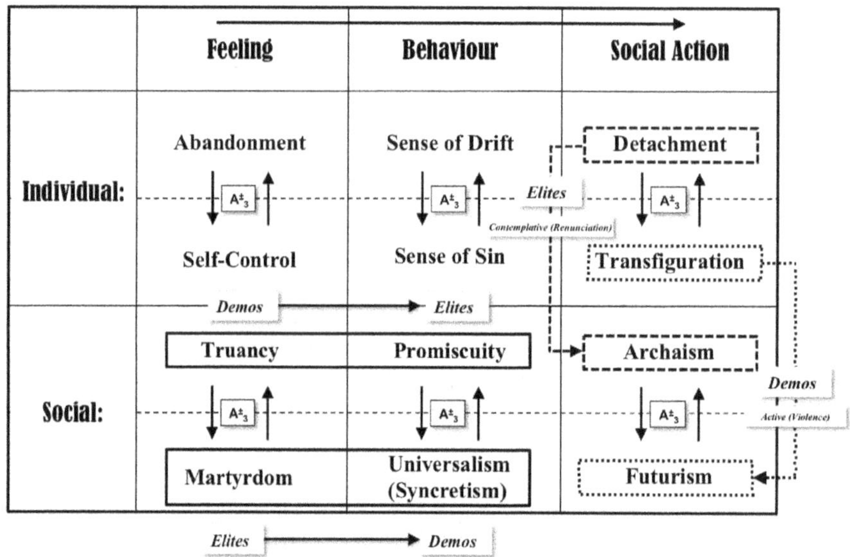

Table 1. 'Schism in the Soul" – schizotypal behaviours characteristic of late-phase societies.

2.2.2: The realm of feeling - individual:

Abandonment: at the individual level, the *abdication of moral responsibility* for one's actions, or any quest for 'sense' in the world, justified by 'going with the flow' (abandonment of responsibility) and following one's impulses without thinking – a form of epicureanism (in the modern, degraded sense of the word as meaning 'hedonism') - or

Self-control: standing apart and 'stoically' refraining from the actions of the majority - often leading to feelings of alienation, of 'not belonging', of 'not being of this world' or of 'not being human'. Herein lies the origin of all 'gnostic' religions (Jonas 2001).

2.2.3: The realm of feeling - social:

Truancy: a social expression of 'abdication of responsibility' which is justified at the collective level by translating it into the philosophical stance of *antinomianism* – the assertion that the world we live in is ultimately without order, random and chaotic. Any complex, long-term enterprise is met with the response: 'what's the point'? The governing principle of 'chance' is often invoked in order to justify indiscriminate freedom of action.

Martyrdom: the alternative to any abdication of responsibility, whether individual or social, is *commitment* to a certain path or goal in defiance of the prevailing opinion. This might indeed be interpreted as a species of 'martyrdom', although we prefer to use more modern terminology here. 'Commitment to the general good' signifies the behavioural response of those we often refer to as 'silent good people' - those who endure chaos and dysfunction without complaint while seeking to ameliorate the worst effects wherever possible.

2.2.4: The realm of behavior – individual:

The sense of drift: at the individual level the decision to 'go with the flow' conveys an uneasy sense of being swept along by forces one cannot understand or control. The behavioural response is to submerge any and all feelings of guilt or regret beneath acts of denial, addiction to drugs (Jay 1999) and/or other acts of self- indulgence under the principle of '*carpe diem*'.

The sense of sin: the feeling that the present chaos is somehow due to one's own (or others') transgressions - a feeling that may paradoxically evoke both a stoic (resigned) and an active response. If the destiny of the universe lies beyond our control, we should submit to whatever inexorable laws appear to govern this destiny and co-operate in bringing it to fulfillment. Both Calvinism and Marxism exemplify this type of response – as do many interstellar and transhumanist enterprises.

2.2.5: The realm of behavior – social:

Promiscuity: Promiscuity in art and living styles means that elites imitate primitive styles or those whom they see as the 'lower classes' in behavior, dress and esthetics. Although Toynbee's meaning is not primarily concerned with sexuality, in late-phase cultures sex is treated not simply as a means of procreation and thereby extension in time of the body social but as a means to personal fulfillment – almost as a substitute for religion.

Universalism (syncretism): at the social level 'promiscuity' is transformed into a quest for more profound explanations for the state of the world and an attempt to differentiate deep-level invariance vs. surface variance. The era of 'the metahistorians – Danilevsky, Dawson, Quigley Sorokin, Spengler, Toynbee *et al.* reflects this behavioural trend, but as early as 1744, Giambattista Vico had already pointed to "*the mental vocabulary of human social institutions, which are the same in substance as felt by all nations but are diversely expressed in language according to their diverse modifications, is exhibited to be such as we conceive it*" (Vico 1744 [355]).

2.2.6: The realm of action - individual:

Detachment: in Toynbee's words '*a flight from reality that deftly evades the problem of landing by taking permanent leave of the ground*'. Detachment involves escaping from social realities by taking refuge in philosophies or religions that advocate total 'freedom from the passions' – a quasi-stoic non-involvement to the extent of also eliminating all empathy and compassion.

Transfiguration: the disciplining of the passions in the service of the world – the philosophy of compassionate engagement. An example is offered by the American variant of Buddhisn known as the 'Mayayana'. There is no exact modern equivalent for this rather 'loaded' term although there is a highly specific, technical meaning in terms of the '*Calculus of Emergence*' (Crutchfield 1994, Ziolo 2013c (section 4)).

2.2.7: The realm of action – social

Archaism: an attempt return to an idealized past - deep ecology/radical environmentalism - the 'New Age' quest for a Rousseau-like 'return to nature' – even as the dark side of the 'Age of Aquarius' is quick to emerge (Lachman 2003). Archaism often involves a mostly futile quest to return to a past 'golden age' through the artificial revival of extinct languages, dress, legal codes or customs. The Nazi movement - indeed, all fascist movements - are typical of this. Although not all attempts are unsuccessful or malevolent (e.g. Ben-Yehuda and the revival of Hebrew) – the results are always unforseen.

Futurism: the attempt to totally eradicate past and present and flee to an imagined, idealized future. Some varieties of communism, transhumanism and all 'Year Zero' political philosophies (e.g. Cambodia's *Khmer Rouge*) are typical examples and this particular schizoid tendency should be regarded with caution by all interstellar/transhumanist/ future-envisionment enterprises or advocacy groups.

Certain directions of behavioral flow are discernible in all late-phase societies. A flow from *truancy* to *promiscuity* can be observed from demos to elites, while the flow from *martyrdom* to *universality* (syncretism) is from elites to demos. – the 'martyr' element among the elites (e.g. pagan philosophers of the Late Empire who were sympathetic to Christianity) tends to encourage and embrace the 'universal churches' which originate from within the demos. The flow from *detachment* to *archaism* is also catalyzed by the elites (detachment from the present is often initiated by excessive nostalgia for the 'good old days') while the flow from *transfiguration* to *futurism* is catalyzed by the demos (the mystical (transfigured) Destiny of the Masses must be catalyzed by revolutionary struggle)[5].

In all late-phase societies these schizoid tensions are expressed at all scales and levels in every nuance of individual and social behavior, whether in overt or covert fashion or whether in a 'personal' or 'professional' context (a distinction which is itself characteristic of a 'universal state' outlook). They reflect the fact that we are approaching the upper bound of the constraint and error catastrophes defining our society and its level of complexity, and reflect typical responses under stress of an organism which spends at least a third of its life in a state of acute dependency (Brown 1985), as well as representing a spectrum of responses to the primal fear of death and dissolution. "*Genes hold culture on a leash*" wrote Lumsden & Wilson (1982). The question is - how long is that leash, and to what extent does transition to K_2 require its severance?

2.3. Semiotic disjunction: differences between early and late phase psychologies

As we seek to 'create an interstellar civilisation', it might be helpful to understand some of the ways in which 'creative' psychologies, active at the 'dawn' of a new civilisation differ from the 'jaded' psychologies of those living in an 'interesting period'. Complexification and interconnectivity increase wth a society's growth over time. Eventually, a society's complexity transcends the ability of any one individual to understand. The construct, whether political, philosophical or religious, can no longer interpret or explain the course or purpose of human relations and social evolution. What now takes place is that *disjunction between genesis and validity* (Weber) where social process is felt by agents and actors to be 'out of control', and who ask themselves: *if God is so good, why is life on Earth so bad*? Life is no longer seen to have a transcendent purpose, deep psychological connections to the construct's projective 'anchors' are ruptured, and the moral and ethical imperatives promoted by the construct in the interests of cultural creativity and societal advance are no longer felt to be relevant, leading to the socioeconomic, political and psychological conditions listed in Part 1. Hence Toynbee's description of a civilisation's late collapse as a long-drawn-out tragedy encompassing millions of casualties in the form of wasted, frustrated and meaningless lives.

Disjunction between genesis and validity has a very real meaning in terms of *semiotic evolution*. The study of semiotic evolution in relation to culture change has come to prominence in recent years, especially in Europe (Wildgen 1982; Ziolo 1999; Bax *et al.* 2004;). What follows is a necessarily simplistic and superficial description, but as the field offers an analytic tool for dealing with psychological responses to culture change – responses clearly evident in institutional and organisational behaviours at all levels (including space advocacy groups) - the relevance of this field to the interstellar initiative should not be underestimated. Languages evolve much faster than the human psychological condition. A word used as a symbol or pointer to, a concept or object is called a *signifier*, while the designated concept or object is called the *signified.* Over the course of social evolution the true 'disjunction' that takes place is that between signifier and signified. Signifiers change rapidly as a society evolves, and new signifiers come to replace them and serve as pointers to the same 'deep-level' psychological structures, which nevertheless retain their 'significance' and capacity for catalysing action. The severity of this disjunction in Euroamerican civilisation is due in large part to excessive splitting into 'good' and 'bad', making rigid distinctions between subject and object (dichotomisation), and the assumption that if a concept, feeling or psychological condition can be given a *name* (like 'love' for instance), it therefore *must* have concrete, material existence in space and time (and can be manipulated. This is the fallacy of *nominalism* (e.g. can 'anxiety' be quantified? What is the difference between a broken bone and a broken heart?). Fig. 24 (a) illustrates the fallacies inherent in nominalism and the subsequent disjunction between signifier and signified.

5 In Britain during the inter-war years, 'Marxism' was an exotic, 'naughty' pastime of the so-called 'Bloomsbury set' – the London-based upper classes, while the 'real communism' of the Glasgow and Liverpool dockers was regarded with suspicion and hatred, even by members of the British Communist Party (Valtin 1941/2004).

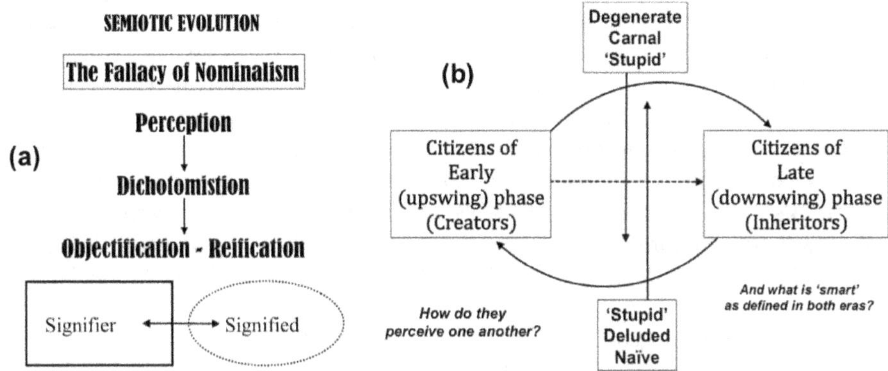

Fig. 8. (a) Semiotic evolution (a) and (b) how citizens at the beginning and end of a cultural cycle perceive one another.

Fig. 8 (b) shows how citizens at the beginning and end of a cultural cycle perceive one another. Such a 'confrontation' can occur towards the end of a cultural cycle, when those who are either deeply alienated from the culture or engaged in forming the chrysalis of the successor (affiliated) culture, confront their contemporaries. Neither can understand the motives and purposes of the other. Those engaged in building the seeds of a new culture regard their contemporaries as 'degenerate' and 'carnal' while those still committed to the status quo regard 'visionaries' as deluded and naive. A good analogy of this condition might be the 'mountain hiking' parable shown below in Fig. 9.

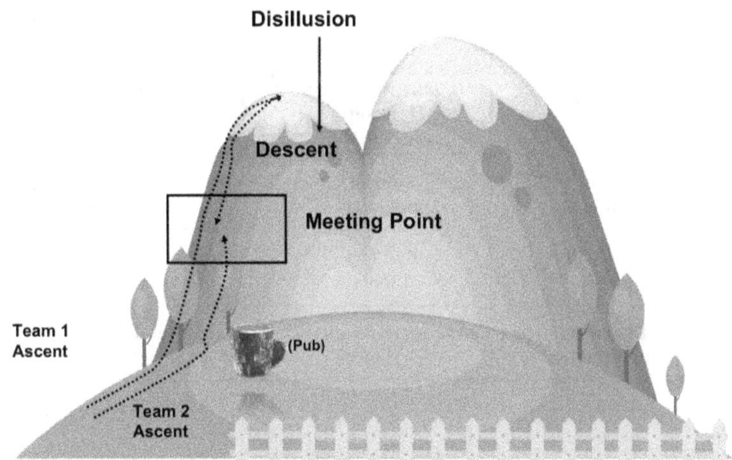

Fig. 9. Illusory Summits: Hope vs. Despair

In Fig. 9 there are two teams seeking to climb a mountain. Team 1 begins the ascent at the crack of dawn: the weather is perfect, confidence and morale are high and the 'summit' seems close and inviting in the light of the rising sun. In the course of the morning the ascent gets more and more difficult, mist and rain begin to fall - but at least the 'summit' seems to be getting nearer and nearer. Eventually, after a supreme effort and battle with the increasingly savage elements, Team 1 reaches the summit, only to find that the'summit' is only a foothill - the real summit, much higher, looms over the team at a much farther distance away. What is to be done? Make a further supreme effort, even though the team feels exhausted? Or, as the treacherous map indicates, there is a pub three-quarters of the way down on the road back. In misery and despair the team resign from the climb and descend towards the pub to drown their sorrows. On the way down they meet Team 2 who have just set out, sull of vigour and confident, no doubt singing joyous mountaineering songs. What words are exchanged when the two teams meet?

Both types consider each other 'stupid' - which begs the most vexed question in psychology: exactly what is 'intelligence'? No decisive answer can be given here, but clues to a multidimensional definition of intelligence may be found in studies of the dependence of computation on morphology, perception of statistical complexity and innovation through ε-machine reconstruction, as found in Crutchfield (1994) and summarised briefly in Ziolo (2013a)).

A construct must adapt successfully to the challenges faced by a culture if it is to retain its function of maintaining social coherence and direction. It does this through a process of transformation, normally experienced historically as *reformation* - i.e. an adjustment of external infrastructure (hierarchies, ministry, ritual and education) and modification of its inner projective structure through realignments of signifiers and signified., A sense of history, of 'manifest destiny' and of unidirectional time as a creative force was woven into the European subconscious through the doctrines of a) a single universe created at a single instant in time b) by a single creator c) through the synergy of the created world with its creator and d) with an underlying teleology. The doctrine of incarnation in turn gave birth to the machinic phylum (Johnston 2008, DeLanda 2011, Deleuze & Guattari 1972). If God made Us to love Him, why shoudn't We, as the New Deities of the Enlightenment, make machines to love Us?

The synthesis stage of the Judaeo-Christian construct in its West European form lasted from the 1st century AD to the 12th. The period from c.1050-1250 AD known as the '12th Century Reformation' witnessed the construct's 'crystallisation', setting Western Europe along its pathway from a localised group of feudal kingdoms to a dynamic, world-embracing scientific culture. Increasing complexification plus internal and external challenges in the course of the 13th-15th centuries necessitated an adaptive transformation of the construct in the 16th - the period known as 'The Reformation

1100 - 1600	*1600 - Present*	*Present - Future*
Crystallisation →	Semiotic Permutation I →	Semiotic Permutation II ····→
Divine-Human Incarna|tion	Mind-Body Dualism	Continuity Thesis
Infinite Progress of the Soul towards God	Infinite Progress of Humanity through Science	Expansion throughout the Known Ubiverse
Divine Omnipotence	Human Omnipotence	Sentience as Prime Vector of Expansion
Building the City of God	Building Utopia	Convergence at Point Omega
Dynamic Trinitarianism	Ancient → Medieval → Modern	Kardashev Type I → Type II →Type III

Table 2. Transmutation of the Basic Euroamerican Construct

Table 2 shows two stages of semiotic transformation. The column on the far left shows the basic semiotic underpinnings of the Judaeo-Christian construct as they stood from the time of the construct's crystallisation (12th century) to that of the Renaissance and Reformation (c. 1450 - 1517). The period from the Reformation and Enlightenment (c. 1517 - 1789, which includes the Thirty Years War (1618 - 1648)) is marked by the first semiotic transformation of the Judaeo-Christian construct, as it struggles to adapt to social conditions dominated by the disintegration of feudalism, the fall of monarchy, the disintegration of ecclesiastical authority under the combined impact of the Black Death (no respecter of 'divinely-instituted' social hierarchies), the trauma of the Thirty-Years War, scientific discovery and the psychosocial impacts brought about by the growth of capitalism and industry. The issues underlying this transformation had less do with science vs. religion than with rival groups seeking power (Elias 2000). 'Science' did not seek to destroy or eliminate 'religion' but to appropriate it and relabel its contents. Surface variance appeared to change while deep structures were preserved. Through a process of semiotic shifts, medieval religion mutated directly into modern 'science' without losing any of its deep, inner structure (Benz 1966; Galgan 1987).

Part III: Strategies for the Present

3.1. Communities

All groups, whether institutions, companies, advocacy groups, project teams etc. need to be considered from three perspectives: a) what do people *outside* the group believe the group is doing? b) what does the *group itself* believe it is doing, and c) what is the group *really* doing? Whatever a group may claim to the contrary, the ultimate goal of all groups living in a universal state is survival of the group itself *as a group,* as well as of the vested interests that sustain it (Quigley 1979). Anything else is camouflage, and any proposal that seeks to transcend these limits will be met with overt or covert ridicule, or 'what's-the-point?'-type arguments based on issues of 'practicality'.

All major steps in social evolution were 'catalysed' by what we may call *transgenerational, task-oriented groups.* These groups exhibited certain properties of communication and structure (*bandwidth* and *topology*) that enabled them to sustain their chief aims to the point where, once those aims had been achieved, the group either fossilised, or melted into the social structures it had helped to create. The field of group process studies can be considered as having its origin in the *Rule of St. Benedict* (6th century – an edition is available at http://www.amazon.co.uk/The-Rule-St-Benedict-ebook/dp/B004YTPAP8) - the primary guide for all European monastic institutions, male and female. Particular attention was given by the book's author(s) to issues of group fragmentation, fissioning, splitting etc. since they understood all too well how the creation of 'in-groups' vs. 'outgroups' marks the beginning of the end for any transgenerational enterprise or project. The Rule has already been adapted to the 'business world' (Dollard et al. 2003), showing that a study of the principles by which the historical Orders were governed can indeed be very advantageous for contemporary groups. In seeking to create an interstellar civilisation, we might therefore carefully study and adapt to the needs of our time the main features of all those former long-lived organisations/institutions that were so influential in the formation of Western civilisation – organisations such as the Benedictine, Dominican, Franciscan and Cistercian Orders, the Knights Hospitaller, the Templars and their derivatives and others (we should include here the Buddhist Sangha – but the Templar are, in fact, 'cautionary' exemplars). 'Orders' such as the 'Illuminati', Freemasons or any post-Reformation, so-called 'occult' organisations don't count. i.e. we shouold be concerned only with organisations/institutions formed over 1000 years ago and lasting to the present time (Ziolo 2001, 2002). These transgenerational, task-oriented groups

1) were *egalitarian,* such that their group topology was *polycentric* (rather than rigid and hierarchical) permitting a faster and more efficient flow of information,

2) possessed a highly efficient *communications network* between communities,

3) followed a régime of *ascetical discipline* (poverty, chastity and obedience) that sought to counter the primal evolutionary drives (resource acquisition, sexuality and status-seeking), directing them towards the service of their primary vision,

4) followed *transversality of praxis,* which was essential in establishing foundations under hostile conditions. While it may be said that the technologies of past ages were simpler, the differences between then and now were only of degree. Our technologies seem more 'complex' due to the accumulation of cultural capital, knowledge and expertise, but the 'encounter with complexity' is of similar scale in human term,

5) developed *induction methods* that sought to ensure that their primary goals were not subject to compromise, deflection or dilution etc., but were transmitted across successive generations and lead eventually to the foundations of the school and university system,

6) sought to maintain *a balance between engagement and seclusion* – which contributed to the preservation of primary aims within a turbulent and changing social environment,

3.2: Child and parent goals: creation of a relevant construct

"The symbolism of uterine ecology is the Rosetta Stone of all world religions"

(David Wasdell – personal communication)

Goals must be set in ways that make them both charismatic and realistic. Charles Galton Darwin (1887 – 1962: grandson of Charles Darwin) observed that there were three ways to consciously direct the course of history: a) formation of a group through personal charisma (such a group will normally mutate, fossilise or disperse following the founder's death), b) create a successful religion and 3) change human nature itself. Interstellar goals will exert a more powerful influence when accorded a 'religious' nature. It is importamnt to understand the vital, catalysing effect of 'religion' in human life and society. "*It is the duty of science to investigate the physical world, it is the duty of religion to place humanity within the context of that world*" (quoted from Frank Herbert's/William McNally's *Bene Gesserit Training Manual* from the *Dune Encyclopedia* (1984)). It is not the job of psychology (nor of physics or philosophy for that matter) to prove or disprove the 'existence of God'. It *is* a legitimate point of psychological inquiry to ask why people believe the things they do in any given era or culture, and why the things they believe in take the forms they do at these times. Psychology may also play a significant role in religious engineering (Khun Eng 2009). It was the very ambiguities and contradictions inherent in the Judaeo-Christian construct which forced a creative response to the problem of 'living in the world' and which therefore catalysed the evolution of a recursively self-enhancing, scientific-technological culture, and it was for this reason that such a culture emerged in the Euroamerican West rather than elsewhere. Nevertheless, this construct must undergo further transformation if it is to be effective in a 'cosmic' context. Relying yet again in this paper on the evocative power of trinitarian statements, we advance three main attributes of a successful construct. There are:

- *Transcendence* (in the West – 'God the Father'; in the East - 'utter openness')*:* what does the given construct say regarding who or what I am? About how I got here? About my place in the world?

- *Viscerality:* (in the West – 'God the Son'; in the East – 'sheer lucency'): how does this construct relate to the the trauma of birth and morphogenesis and to the residue of these experiences in the dynamic unconscious?

- *Praxis:* (in the West – 'the Holy Spirit'; in the East - 'excitatory intelligence'): Does this construct offer a clear and reliable guide as to now I should behave in the world, in relation to meself and others?

All three attributes musy be present equally, but it is the second property – *viscerality* – that is so often lacking in the 'cosmic' religions manufactured at the present time, rendering them dry and unappealing. The metaphor of human 'rebirth' from the 'cradle of Mother Earth' is a very powerful one. We should take the quote at the beginning of this section very seriously and not hesitate to weave threads of the *palingenetic* (rebirth) metaphor into the fabric of interstellar discourse. All human passions are sources of great power that should not be suppressed, but transformed and harnessed for the purposes of social transformation. While the time may come when humanity may cease to rely on the salvific power of constructs and enter fearlessly into that vectored domain of emergent process of which we are so intimate a part – that time is not yet.

All religions, constructs or ideologies require symbols that function as the group 'totem' (Freud 1913). The 'totem' appropriate for a 'Cosmist' faith should be a living, evolving structure that serves as a 'road map' to the future. One possibility is the re-creation or reconstruction of the Integrated Space Plan (ISP), discussed below.

3.3: Strategies

To achieve a permsanent, self-sustaining and independent foothold in space it is essential that, given the current state of Earth's resources, all sincere space organisations and advocacy groups agree as quickly as possible on a common goal and strategy towards this end. If this does not happen well before the middle of the current century, we will end up going nowhere. An excellent tool for this purpose would be a revised and reconstructed version of the Integrated Space Plan (ISP) initially devised by employees of the Rockwell Corporation during the 1980's, but now superceded by world events. An initiative has already been taken in this direction by *Integrated Space Analytics* (http://integratedspaceanalytics.com/cms/index.php) and such an enterprise deserves the support of those seriously concerned with human emigration into space and future human survival and evolution in general. A revised, fully interactive and evolving ISP would serve as a navigational tool rather than a predictive model. Such a prototype 'Seldon Plan' might:

- be fully interactive on the SETI model
- function as a navigational tool rather than a predictive model
- be augmented by socioeconomic and psychosocial data to thegreatest possible detail (ε-machine reconstruction)
- be parsimonious in its deep structure, permitting flexibility in the face of acute turbulence
- operate on genetic/metagenetic principles (Koza 1992; Langdon & Poli 2002)
- use quantum algorithms to search for and pinpoint 'sweet spots' (convergence of the 'five vectors' i.e. who, what, where, when, why?) so as to enable maximum advantage to be taken of them
- be based on convergence of the nine core domains shown in the next section below, using transversality (research-technology) principles for this purpose

Epilogue

Charles Galton Darwin's strategies for influencing the flow of history each contain ceratain injunctions for any trangenerational, task-oriented group that may emerge in our time: a) the formation of a group through personal charisma usually entails the mutation, fossilisation or death (dispersal) of the group following the founder's death), b) creation of a successful religion (or 'construct' in the language we hace used here) is more useful in that constructs may last several generations and effect some degree of radical change) and 3) changing human nature itself has become a very real possibility today given the promises of nanotechnology, biotechnology, information technology and cognitive science (NBIC). Genetics is the 'true' language of God and we would do well *('bene gesserimus')* to master it. Through this, "*Wo Es war, soll Ich werden – Es ist Kulturarbeit*" as Freud wrote in 1923 ('where Id was, there Ego shall be – that is the task of Culture'). For the 100YSS and other space-advocacy or research institutions the main issues are not technical but human. To 'live long and prosper', to translate dreams into reality, it is essential and urgent that a synthesis of aims and practice be achieved between all such groups and that a fully international interstellar research & development community be established whose goals are clearly set, and who follow a shared strategic track. This may sound idealistic, even utopian, but should this fail to occur, we will experience the possible future outlined in the novel *Psychohistorical Crisis* by Donald Kingsley.

> *"Boltzmann was often referred to by students of the Lyceum as the 'great^n-grandfather' of psychohistory … Planck had warned his students thet they were not to use the quantum of action in any prediction that required reversibility since all quantum events were tied to an increase in entropy.*
>
> *Later generations of quantum physicists had ignored Planck (and had sidelined Boltzmann as underivable from first principles) and clung for dear life to their Newtonian roots in the Law of Conservation of Information, a law about as valid as Ptolemy's assumption that a released rock instantly reverts to a universal rest mode. During the heady … days of scientific profligacy, armies of superstitious physicists hunted up new heavens for information gone missing: under rocks, on the surfaces of black holes, behind the locked doors of alternate worlds – all heavens where the physicists were sure to find their home after they died. The debate was only finally resolved by the Great Die-Off when the physicists as well as everyone else fell off the top of the exponential curve to their silence, or, if they lived, to the more pressing questions of survival.*
>
> *There was enough material from the centuries immediately prior to the Ramp-Up to see the excitement of the early scientific revolution driving the population expansion, and plenty of evidence for the cavalier optimism of the richer nations. which were getting richer in their high-rise penthouse apartments supported by the Atlas of an increasingly ignorant breeding population … and after the Die-off there was plenty of material to understand the rise of the strange cultures which had produced the sunlight starships … Every single one of the cultures that had fed the Ramp-Up and energy had simply vanished during the free-fall of infrastructure collapse as the population dropped by c. 10 – 15 billion. The people who emerged from the disaster, and who later founded the first interstellar colonies, bore no relation whatsoever to their predecessor cultures in language, institutions, religion or ethnic characteristics".*

The grimmest prospect is that we may be approaching what Bolstrom & Ćirković (2008) refer to as the 'Great Filter' – a major challenge facing all genetically-unmodified, sexually-reproducing species on the very adge of emergence into space, i.e. 'retreat from the edge' and regression into post-imperial self-absorption. To counteract this tendency, a new vision is needed that expresses confidence in the human future.

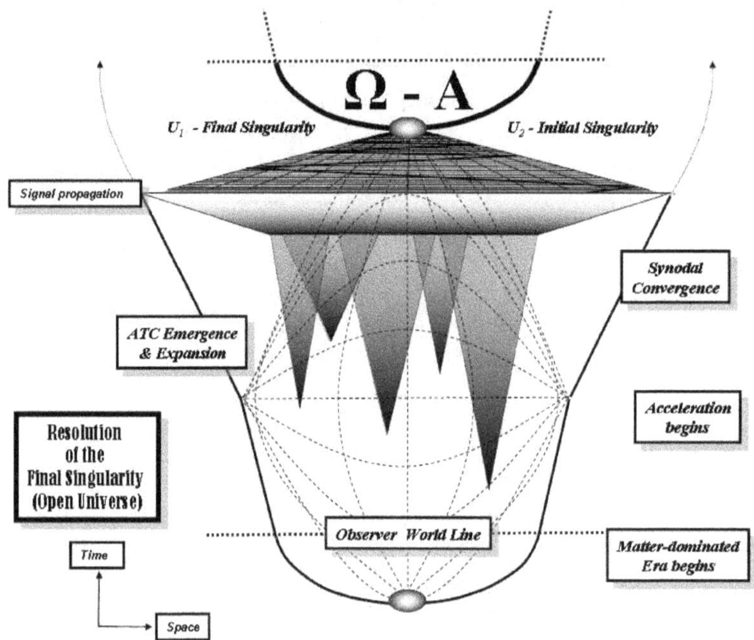

Fig. 10. Penrose Diagram for open universe in which advanced technological civilisations eventually co-operate. and play a critical role in re-initiating a successor universe at the (engineered) final singularity of our present one.

Fig. 10 above is a type of Penrose Diagram showing the future of a hypothetical 'open' universe in which advanced technological civilisations (ATC's) play a critical role in 'kick-starting' a successor universe by macroengineering a final singularity. Emerging at scattered points in space-time prior to and during the acceleration phase, their zones of influence gradually overlap until the ATC's eventually merge at a point of synodal convergence, where they effectively 're-tune' the fundamental parameters of space, time and matter so as to create the successor universe. In other words, even given that the Universe is open or flat, it is possible that an ATC 'synod' might macroengineer re-collapse, using the potential of shear energy to 'fine-tune' the parameters of a chosen successor universe. The third column of Table 2 showed the further semiotic transformation the Judaeo-Christian construct must undergo if it is eventually to catalyse transition to K_3 and ultimately become an active participant in this process. Certain central premises of the construct are invariant under semiotic transformation, i.e. that sentience is a central factor in the Universe's expansion (*synergy*) and that the ultimate fate of all sentient life is bound up with that of the Universe itself. (*teleology*). This outlook is known today as the *continuity thesiss*, but its roots (like all human concepts) are very ancient.

The *continuity thesis* (Gardner 2007) proposes that the 'traditional' schisms between mind/body, animate/inanimate, good/evil, science/religion, rational/irrational etc. will be dissolved, 'Machiavellian' intelligence will be abandoned, species-narcissism will give way to an understanding of the mutability and contextuality of form, biological life will be understood as an intermediate state between inert and fully sentient matter, cognition and affect will be seen to be morphology-dependent and the full potentiality of Being (and the true nature of the human mind in relation to it) will be fully understood (Amoroso & Martin 1995). If achieved, this transformation would involve that crucial step in innovative *ε-machine reconstruction* (Crutchfield *op. cit.*) that is a necessary pre-requisite to creating an interstellar civilisation. Insofar as 'intelligence' is morph-dependent, this step must involve speciation and result in radical changes in our physical form as well as in our perception of ourselves, the universe and our relationship to it. If this step is not taken, a unique chance will have been lost - 'unique' in the sense that such potential for a mammalian-derived species will not arise again in the history of this planet (one cannot speak for the social insects, whose day may yet come). It is not 'cool' to speak of 'manifet destiny' in these so-called 'enlightened' times, but for those among us who look to the stars, not only with 'dreams' but with firm resolve, purpose and the technological means to advance towards humanity's cosmic destiny, the criticality of the present epoch is painfully clear. If the opportunity is lost, then for some future sympathetic archaeologist, or for any one of us transported hypothetically to the far future, human civilisation will seem as ephemeral as the empire of Ozymandias, a brief flash of light upon a world that has only recently emerged from, and will shortly be returning to, conditions inconceivably alien to those we fondly consider to be so stable.

References

1. Amoroso, Richard L. & Barry E, Martin (1995). 'Modelling the Heisenberg Matrix: Quantum Coherence and Thought at the Holoscape Manifold and Deeper Complementarity'. *Scale in Conscious Experience: Is the Brain Too Important To Be Left to Specialists to Study? Proceedings of the Third Appalachian Conference on Behavioural Neurodynamics,* pp.351-77. Lawrence Erlbaum Associates Inc., Mahwah, NJ.
2. Barrow, J. D. & Tipler, F. J. (1986). *The Anthropic Cosmological Principle.* Oxford University Press, Oxford & New York.
3. Bax, M.. Van Heusden, B. & Wildgen, W. (2004). *Semiotic Evolution and the Dynamics of Culture.* European Semiotics vol. 5). Peter Lang Publishing Inc. (International).
4. Benz, E. (1966). *Evolution and Christian Hope: Man's Concept of the Future from the Early Fathers to Teilhard de Chardin.* Doubleday & Co. Inc. Garden City, NY.
5. Bettencourt, L. M. A., Lobo, J., Kühnert, C. & West W. (2007). 'Growth, innovation, scaling and the pace of life in cities'. *PNAS* vol. 104 (17), pp. 7301-06.
6. Bostrom, N. & Ćirković, M. (eds.), (2011). *Global Catastrophic Risks.* Oxford University Press.
7. Burns, J. K. (2004). 'An Evolutionary Theory of Schizophrenia: Cortical Connectivity, Metarepresentation and the Social Brain'. *Brain and Behavioural Sciences* 27:6 (Dec. 2004): pp.831-55.
8. Chaisson, E. J. (2001). *Cosmic Evolution: The Rise of Complexity in Nature.* Harvard Univ. Press, Cambridge, Mass.
9. Craig, N. C. (1992). Entropy Analysis. VCH: New York (1992), pp. 86 – 92.
10. Crutchfield, J. (1994). 'The Calculus of Emergence'. *Physica D:* Proceedings of the Oji International Seminar: Complex Systems – from Complex Dynamics to Artificial Reality.
11. DeLanda, M. (2011). *Philosophy and Simulation: The Emergence of Synthetic Reason.* Continuum International, London & NY.
12. Deleuze, G. & Guattari, F. (1972). *Anti-Oedipus.* Trans. Hurley, R., Seam, M. & Lane, H. R. Continuum, New York (2004): Vol. 1 of *Capitalism and Schizophrenia.* 2 vols. 1972-80. Trans. of *L'Anti-Oedipe*: Les Editions de Minuit, Paris.
13. Diamond, J. (2005). *Collapse: How Societies Choose to Fail or Succeed.* Allen Lane, London.
14. Dollard, K., Marett-Crosby, A. & Wright, T. (2003). *Doing Business with Benedict.* Continuum International Publishing Group.
15. Durant, W. (1944). *The Story of Civilization. Vol.3: Caesar and Christ.* Simon & Schuster, New York. Quotation from the Epilogue: *Why Rome Fell.*
16. Elias, N. (2000). *The Civilizing Process,* Wiley-Blackwell, Oxford.
17. Lee, R. B. (1983). 'Models of Human Colobisation: San, Greeks and Vikings', in *Interstellar Migration and the Human Experience,* Finney, B. R. and Eric M. Jones (eds.), Univ. of California Press, pp. 180 - 95.
18. Freud, S. (1913). *Totem and Taboo: Some Points of Agreement between the Mental Lives of Savages and Neurotics.* Authorised translation by James Strachey. Routledge Classics, London & New York (1950). Reprinted - 2001.
19. (1933). *New Introductory Lectures on Psychoanalysis,* vol. 22, lecture 31, "The Dissection of the Psychical Personality," *Complete Works,* Standard Edition, eds. James Strachey and Anna Freud (1964).
20. Fry, T. (1981). *RB1980: The Rule of St. Benedict in English.* Liturgical Press, St. John's Abbey, MN.
21. Galgan, G. (1982). *The Logic of Modernity.* New York University Press, New York.
22. Gardner, J. (2007). *The Intelligent Universe: AI, ET and the Emerging Mind of the Cosmos.* New Page Books, NJ.
23. Gibbon, E. (1988). *The Decline and Fall of the Roman Empire.* Originally published: 1776 – 88, reissued by Wordsworth Editions (new edition, 1988), New York.

24. Howard, R. (1996): *Political Judgements.* Rowman & Littlefield, NY. See especially the discussion on Kant, p. 151.
25. Johnston, J. (2008). *The Allure of Artificial Life: Cybernetics, Artificial Life and the New AI.* MIT Press, Cambridge, Mass., London.
26. Kauffman, S. (1993). *The Origins of Order: Self-Organization and Selection in Evolution.* Oxford University Press, New York & Oxford.
27. Kleiber, M. (1975). *The Fire of Life: An Introduction to Animal Energetics.* Robert E. Krieger, Huntington, NY.
28. Koza, J. R. (1992), *Genetic Programming: On the Programming of Computers by Means of Natural Selection,* MIT Press.
29. Khun Eng, Kuah, (2009). *State, Society and religious Engineering: Reformist Buddhism in Singapore.* Institute of South-East Asian Studies (2nd Edition).
30. Lalande, K., Odling-Smee, J. & Feldman, M. W. (2001). 'Cultural niche construction and human evolution'. *J. Evol. Biol.* 14, pp. 22-33.
31. Miller, H. (1944). *Sunday after the War.* New Directions Publishing, Norgolk, CT.
32. Nicolis, G. & Prigogine, I. (1977). *Self-Organisation in Nonequilibrium Systems: From Dissipative Structures to Order through Fluctuations.* John Wiley & Sons, New York.
33. North, D. C. (1997). 'Some Fundamental Puzzles in Economic History/Development'. *The Economy as an Evolving Complex System II.* A Proceedings Volume of the Santa Fé Institute Studies in the Sciences of Complexity, pp. 223 – 38.
34. Quigley, C. (1979). *The Evolution of Civilizations: An Introduction to Historical Analysis.* Liberty Fund Inc. (2nd. Edition).
35. Reijnders, J. (1990). *Long Waves in Economic Development.* Edward Elgar Publishers, Aldershot & Brookfield, Vermont.
36. Stevens, A & Price, J. (1996). *Evolutionary Psychiatry: A New Beginning.* Routledge, London.
37. — (2000). *Prophets, Cults and Madness.* Gerald Duckworth & Co Ltd. London.
38. Tainter, J. (1990). *The Collapse of Complex Societies.* New Studies in Archaeology, Cambridge University Press.
39. Toynbee, Arnold (1966). *A Study of History.* Abridgement of vols. 1-6 by D.C. Somervell. Oxford University Press, Oxford.
40. Turchin, P. (2003). *Historical Dynamics: Why States Rise and Fall.* Princeton University Press.
41. — & Grinin, L., Korotayev, K. & de Munck, V. C. (eds.) (2006). *History and Mathematics: Historical Dynamics and Development of Complex Societies.* Volgograd Centre for the Social Sciences, URSS, KomKniga, Moscow.
42. Turchin, V. (1977): *The Phenomenon of Science. A cybernetic approach to human evolution.* Columbia University Press, New York.
43. Tylecote, A. (1992). *The Long Wave in the World Economy: The Current Crisis in Historical Perspective .* Routledge, London & New York.
44. Van Duijn, J. J. (1983). *The Long Wave in Economic Life.* George Allen & Unwin, Boston & Sydney.
45. Weber, M. (1958). *The Protestant Ethic and the Rise of Capitalism.* Scribner, New York.
46. West, G. B. (1999). 'The Origin of Universal Scaling Laws in Biology'. *Physica A*, 263 (1999), pp. 104-113.
47. — & Brown, J. (2004). 'Life's Universal Scaling Laws'. *Physics Today* (Sept. 2004) pp. 36-42.
48. Wildgen, W. (1982). *Catastrophe Theory Semantics.* John Benjamin, Amsterdam & Philadelphia.
49. Ziolo, P. (1999). 'Memes and Morphologies: Neural Catastrophe Induction through the Socialisation
50. Process'. Unpublished dissertation, University of Liverpool. (2001). 'Joachim of Fiore and Apocalyptic Immanence'. *Journal of Psychohistory* 29 (2) pp. 186-224.
51. — (2002). 'The Psychodynamics of the ECF-Nexus: Monasticism and Psychospeciation in Western Europe, c.500-1500 AD'. *The International Journal of Psychotherapy* vol. 7 no. 3 (2002) pp. 221-48.
52. — (2006). 'Divine Madness: Evolutionary and Historical Dimensions of Schizophrenia'. *BPS Journal of the History and Philosophy of Psychology* vol.8 no.2 (2006), pp.41-53.
53. — (2013a). 'Devising the Prime Radiant: Strategies for Catalyzing the $K_0 \rightarrow K_3$ Transition'. Proceedings of the 2013 Symposium of the 100YSS, pp. 185 – 218.
54. — (2013b). 'Religious Engineering for the $K_0 \rightarrow K_3$ Transition'. Proceedings of the 2013 Symposium of the 100YSS, pp. 219 – 48.

100 YEAR STARSHIP™

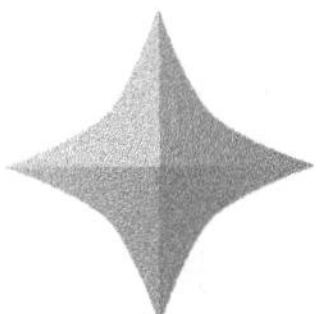

Chaired by David Alexander, PhD

Professor, Department of Physics and Astronomy

Track Description

Understanding the interstellar medium and the composition of exosolar systems is vital as we contemplate travel to the stars. In addition, as our gaze is drawn many light years away, focusing on closer objectives as stepping stones to deep space will be essential. Beyond Mars, what missions should be designed to eventuate successful travel to another star? How should potential destinations be evaluated? What do we know and how do we learn more about space between the stars?

Track Chair Biography

David Alexander, PhD

Professor, Department of Physics and Astronomy

David Alexander , PhD is a professor in the Department of Physics and Astronomy, where his primary area of research is solar astrophysics. Professor Alexander is author of "The Sun" part of the Greenwood Press "Guide to the Universe" Series. Professor Alexander has served on many national and professional committees including the NASA Advisory Council's Heliophysics Subcommittee, the NASA Solar Heliospheric Management and Operations Working Group (SH-MOWG), ESA/NASA Solar Orbiter Payload Committee and the Science Advisory Board of the High Altitude Observatory Coronal Solar Magnetism Observatory.

Exoplanets and Survival Kits: Where to Go and What to Pack for an Interstellar Voyage?

Pauli E. Laine, MSc

Department of Computer Science and Information Systems, University of Jyväskylä, 40014, Finland

pauli.erik.laine@gmail.com

Abstract

The selected target (among the available payload mass) for interstellar mission affects what can and must be taken on the journey. The number of exoplanets and information about them is increasing all the time. For a potential interstellar mission we should choose the most Earth-like planet with proven atmosphere and water. With only limited information about the conditions on the target planet, we should prepare our mission with a suitable "survival kit" in order to survive and start a self-containing colony. In this paper I will describe a minimal mission concept based on our current knowledge about habitable exoplanets and possible propulsion technology. This presentation is a feasibility study about where we can go and what we can and should take with us. As a case study, I will consider a long duration mission with suspended animation to some of the most potential Earthlike or super-earth-size exoplanets to establish the first human interstellar colony.

Keywords

Manned interstellar flight, exoplanets, habitability, terraforming

1. Introduction

If and when we decide to travel to another star for some reason (survival, political, exploration), where would we go, and what should we take with us? Such questions are essential when planning a manned interstellar voyage. As the 100 Year Starship organization's goal is make the capability of human travel beyond our solar system a reality within the next 100 years, we will consider here only manned interstellar missions [1].

2. Where Can We Go?

It is commonly accepted that ship capable of speeds of 0.1 c, or about 30,000 km/s with some reasonable energy usage can be used for interstellar travel. At this speed the nearest star systems can be reached in some 50 years. Today we do not have anything that could achieve such speed, but there are some proposals about near future propulsion systems, (e.g. the Fission Fragment Rocket Engine, or FFRE [2]) that can theoretically be used for manned interstellar travel.

Available payload mass is always a limiting factor for developing space flights. Especially long duration manned interstellar journeys will require very careful planning, depending first on whether active crews (possibly generation ships) or hibernating crews (a common solution in science fiction) are involved. In any case, oxygen, water, and nutrients are needed for both live and hibernating crews. Here we assume, that hibernation can be used, minimizing all the needs during the journey. The number of crew members will directly affect the payload mass, but how many colonists are needed for a self-sustaining colony? This will depend on whether there will be only one planned mission or perhaps many successive missions to same target planet. While there is no certain minimum size for a genetically viable colony, more is always better. Cameron Smith proposed a population of 40,000 as a safe and stable number for a multigenerational interstellar voyage [3]. However, entire modern human populations might be the descendants of a mere 10,000 individuals, so the colony size could be considerable smaller [4]. In any case, a mere 10,000 people with some life support gear could weight approximately 1000 tons. With all the necessary water and nutrition, the payload mass would exceed the capacity of all current and foreseeable high speed propulsion technologies. This leaves us only two options: a) whether we use multiple missions or b) use one huge and slow generation ship (e.g. an asteroid ship with artificial gravity and closed water and food circulations). Here someone will always ask: what's the point in a multigenerational voyage that can last for thousands of years, while we, back on Earth, might invent hyperdrive and pass the poor ark in its first decades? The only rationale for a generation ship is that the ship itself is a world, a 'snapshot' of selected life forms that could survive even if Earth becomes uninhabitable.

3. Potential Targets

Today over 1800 exoplanets have been discovered, most of them are Jupiter-size gas giants. This is due to of our limited detection sensitivity. Even the closest exoplanets are extremely difficult to observe due to their overwhelmingly bright parent stars. Nevertheless, information about exoplanets is increasing all the time, and lists of presumed habitable exoplanets are maintained, e.g. at Planetary Habitability Laboratory [5] (figure 1).

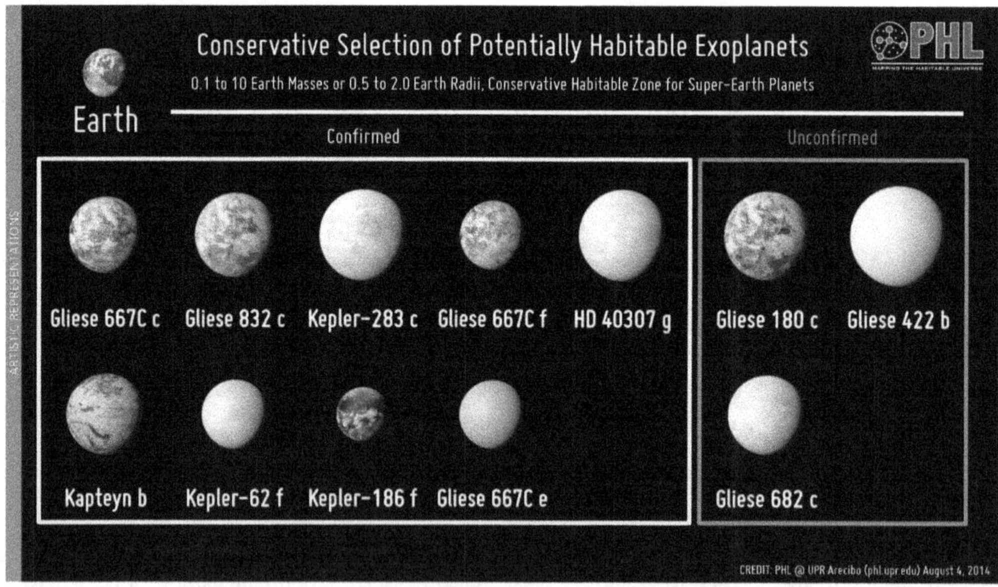

Figure 1. Conservative selection of potential habitable exoplanets, August 2014 (PHL/UPR, used with permission).

Figure 1 represents the best knowledge about known possible near Earth-sized habitable planets so far. As such, they are also our best potential targets for manned interstellar travel, if we will have to go quickly.

How far will we have to travel? The number of star systems within 13 light-years is about 33, within 50 light-years about 1400, and within 250 light-years about 260,000 [6]. If we limit our range to approximately 15 light years, then we know only 3 potential targets (figure 2).

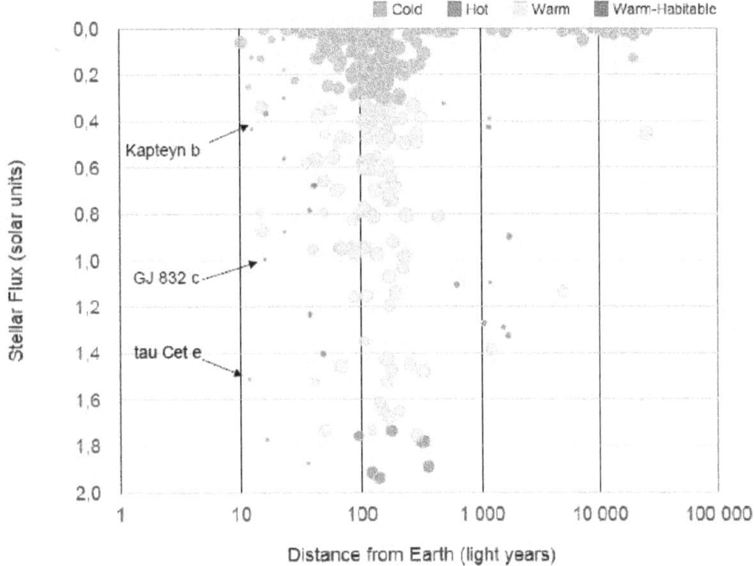

Figure 2. Distribution of habitable zone planets by stellar flux and distance from Earth, August 2014 (PHL/UPR, used with permission).

Tau Ceti e is an unconfirmed habitable 1.6 x Earth radius planet 11.9 light years from Earth. Kapyten b is a confirmed habitable 1.6 x Earth radius planet 12.7 light years form Earth. Gliese 832 c is a confirmed habitable 1.7 x Earth radius planet 16.1 light years form Earth with equal stellar flux. There is very little we know about the true conditions of these planets. They are located in habitable zone, but so is Venus in our Solar System.

Even our best choice for a target planet is scarcely suitable for humans. The limits for a stable human presence are quite narrow, e.g. A temperature of approximately 273 - 303 K, pressure 700 - 5000 hPa, and radiation 1 - 3 Gy. There is hardly any elemental oxygen, since oxygen is biologically produced. Also water is one of the main criteria for habitable exoplanets. Some constraints can be circumvented, e.g. with portable life support systems and underground habitats, but eventually we will have to terraform or find another place.

3.1 Backup Plan

We will get more and more information about our target planet from Earth, and with our ship's own instruments as we get closer to the target. What if we find out that our target is like Venus (737 K temperature, and 9.2 MPa surface pressure)? Currently we have no way to measure or even estimate an exoplanets' surface temperature and pressure. Things might get better in the future with larger and better telescopes, but there will always be a margin of error that can be fateful to our mission. Can we somehow use multiple potential targets in order to find the best target planet, i.e. a trajectory with possibly multiple stops? We (the crew, ship's computer and mission control on Earth) could do habitability estimations during the approach phase and choose whether to stop or continue to the second target (figure 3). However, we must do our decision before the so called No-Stop Limit and braking phase. Due to very long distances, the final decision has to be made either by the crew or ship's computer system. It remains open whether there really are multiple habitable planets reachable with feasible trajectory and propulsion technology.

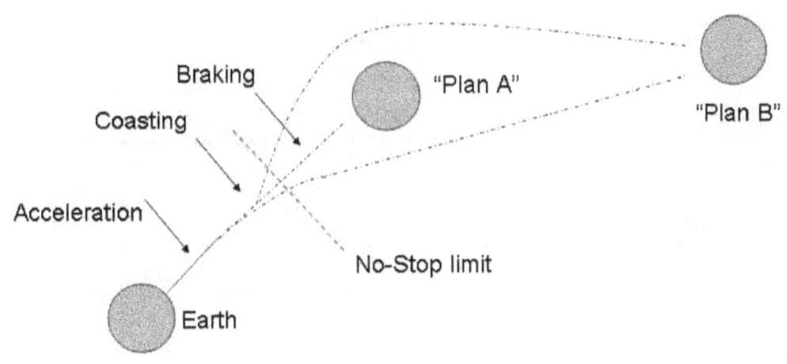

Figure 3. Possible trajectories for two habitable planets.

Another possibility is to engage in a suspended park orbit over the target planet and try to terraform with e.g. with microbes, orbital mirrors, and asteroid impacts. In any case, we had to stop somewhere, unless we have a self-sustained generation ship.

4. What to Take With Us?

When we have found habitable exoplanets with tolerable conditions for human colonization, what should we take with us when we travel there? Payload mass is the critical factor that sets limits what can be taken. There are some items that are essential for the survival of the new colony, the first two being oxygen and water. As mentioned earlier, exoplanets most probably lack elemental oxygen in the atmosphere, so that humans will continually require suits and/or breathing apparatus. There can only be limited amount of oxygen on board the landing vehicle. On the other hand, water could be abundant on many exoplanets, either as liquid in oceans or lakes, or as ice. And water can be used to produce oxygen by means of electrolysis. So our inventory list must include oxygen and energy production devices. Electric energy can be produced with solar cells or e.g. RTG (radioisotope thermoelectric generator).

4.1 Getting Started

Now, that we have oxygen, water, and energy, what else is needed most for the first moments after arrival? In order to be able to stay and work outside the lander module, some kind of extra-vehicular activity suits (EVA) are needed. At minimum (and optimum) level it could be just breathing masks and air containers. After this there will be a need to set up temporary houserooms, e.g. pressurized domes. Some furniture (tables, beds) and equipment (e.g. lighting systems) could be borrowed from the lander. Food production (seeds, plants, greenhouses) would be the next logical step, but there should also be nutrition stock for approximately one year to ensure a successful harvest.

4.2 Building a Foothold, and Beyond

After the first steps of self-sustaining colony have been taken (e.g. the provision of oxygen, energy, and food production), new needs will arise, for example, there will definitely be a need for health care and spare parts for EVA suits and production systems. We should be prepared for all this. We might want to build permanent buildings. The simplest buildings might be of brick, and for them we need only shovels and brick moulds [7]. If the planet's soil is suitable, colonists could plant trees to provide raw material for buildings and furniture. Trees, plants, microbes (e.g. cyanobacteria) and algae would be the first steps toward terraforming.

5. Case Study

Let us summarize the main features of this paper's 'survival kit' in a simple case study. To set certain limits, we will assume that we have a) a high speed propulsion system capable of 10% light speed, b) a lander module with 5000 kg payload capacity and crew of 10, c) some hibernation method, and d) a habitable target planet with tolerable temperature and pressure, and accessible water. Payload capacity (5000 kg) might contain a selection of the items listed in chapter 4 (see Table 1).

Table 1. Proposal of survival items included in 5000 kg payload mass.

Item	Weight (kg)
Nutrition for 10 persons/1 year	1800
Temporary houserooms	800
EVA suits, shared	700
Tools, machines etc.	500
Energy production wares	450
Food production wares	400
Oxygen production equipments	250
Health care accessories	100

Mission mass could be possibly be reduced to a minimum through crew hibernation and thus a possibility of reducing food requirements. With this configuration, the total payload mass should be approximately 20 metric tons.

6. Conclusion

The selected target and available payload mass for an interstellar mission affects what can and should be taken to the journey. Any potential colony must be prepared for some limitations in addition to basic oxygen, energy, and food production. On the other hand, there is an upper limit what can be taken on the voyage. A manned interstellar journey will therefore require very careful planning and review of every single item.

Although the number of exoplanets and information about them is increasing all the time, true surface conditions remain unknown. This is why we need some back up plan in the case of arriving at an uninhabitable planet. We should have either alternative target(s), or have adequate terraforming capabilities

Acknowledgements

The author would like to thank Prof. Abel Méndes (University of Puerto Rico at Arecibo) for permission to use Planetary Habitability Laboratory Figures 1-2.

References

1. 100 Year Starship. www.100yss.org

2. Laine, P. E., "To the stars with current technology?", *100 Year Starship 2013 Public Symposium Conference Proceedings*, 100 Year Starship, 2014.

3. Smith, C. M. (2014). "Estimation of a genetically viable population for multigenerational interstellar voyaging: Review and data for project Hyperion", *Acta Astronautica*, 97, 16-29.

4. Ambrose, S. H. (1998). 'Late Pleistocene human population bottlenecks, volcanic winter, and differentiation of modern humans'. *Journal of Human Evolution* 34 (6): 623–51.

5. Habitable Exoplanet Catalog. Planetary Habitability Laboratory, University of Puerto Rico at Arecibo. http://phl.upr.edu/hec

6. *The Atlas of the Universe.* http://atlasoftheuniverse.com

7. Zurbin, R. G., *The Case for Mars.*, Free Press, 1996.

100 YEAR STARSHIP™

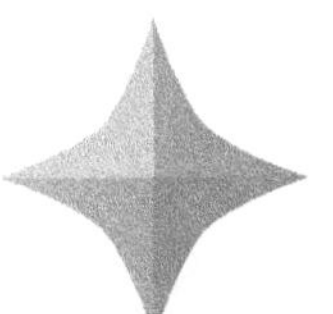

Chaired by Karl Aspelund, PhD

Assistant Professor, University of Rhode Island

Track Description

Design for interstellar probes and crewed vehicles must address the unique characteristics and extreme environment of interstellar space. The equipment, structures, tools, materials, buildings, furniture, cleaning and maintenance processes, clothing-the accouterments of life and work-surround and create an environment. This environment protects, nourishes and facilitates daily activities. For most living things, their environment must fulfill many physical needs and for higher order creatures, physical, mental and emotional requirements need be met as well. Understanding, optimizing and manufacturing design to make these aspects of daily activities sustainable are critical for any hope of successful interstellar flight-with a living crew or robotic probes.

Track Chair Biography

Karl Aspelund, PhD

Assistant Professor, University of Rhode Island

Karl Aspelund, PhD, is Assistant Professor at the Department of Textiles, Fashion Merchandising and Design at the University of Rhode Island. His research interests lie in examining the role textiles and design play in the creation of identity, the impact of the textile life-cycle on the Earth's environment, and how the design community can contribute to the goal of environmental sustainability. He is now turning toward investigating the design and cultural needs and constraints of apparel in long-term space exploration. After graduating from the Wimbledon School of Art (1986,) with a degree in 3d design, Karl worked as an artist and designer for theater and film for 20 years and has taught design since the early 1990's. Before coming to URI in 1996, he was head of the Department of Industrial Design at the Reykjavik Technical College in Iceland. Karl completed a Ph.D. in 2011 in Anthropology and Material Culture from Boston University's University Professors Program, where his dissertation was awarded the University Professors Edmonds Prize as the best dissertation of the academic year 2010-2011. Karl is the author of two design textbooks, "The Design Process," (2006) and "Fashioning Society,"(2009.) The third, an introduction to designing, is due in early 2014.

Space Architecture: Design Aspects for Extended Space Travel

Antoine G. Faddoul, MS

Tony Sky Design Group, 229 E. 85th Street, New York NY 10028

tony@tonysky.net

Abstract

The conventional aspects of design reach distinctive levels when considering extended space journeys. Although space travel and its associated architecture started decades ago, space design has been limited to short distance and short term trips.

How would a design team approach the scheme for manned interstellar journeys? How would it manage sustainable deep-space designs for longer durability, minimal maintenance, and energy self-sufficiency?

It is not the typical design project that could be developed by analyzing the program and functions, then comparing them to previous projects. To carry such task, over a hundred design items related to three main areas; Structure, Environment, and Human needs were evaluated assessing their essences to achieve a sustainable spaceship that would function for hundreds of years. Design items were analyzed according to their status whether current technology, developing technology, or future technology. Each item was mapped according to how the science, technology, and design behind it would evolve in decades to reach the desired functioning product. The items are mostly associated with functions of Earthly daily needs, yet their sustainability is reevaluated to perform in extended space travel.

This design approach provides comprehensive futuristic scenarios by surveying the progress status of each item, while the resulting graphs and info help answering further questions including:

- How much more time is anticipated for each item to be ready for testing?
- Do we have the material, technology, and craftsmanship to build a spacecraft that could self-maintain and upgrade?
- What will be the status of the equipment, energy, and gadgets after decades of use on a spacecraft?
- How do the inhabitants of the space vessel foster their daily functions and needs over decades?

Keywords

Starship, interstellar, space architecture, design, technology

1. Gliina in Route to Antas, Improved

In September of 2013, a simulation of the year 2072 AD showed Starship Gliina[1] about one light year away from Earth. The simulated time was 5:00 AM, 10 years after leaving the lunar orbit. The destination was planet Antas in Constellation Cygnus. The starship had 7.5 light years remaining to reach its destination, and everything seemed fine as the long check list was streaming, showing a functioning and "healthy" starship.

The simulation of the starship readiness and future route will keep developing until 2062 – Gliina's proposed departure day. Since new aspects of design and technology are ever emerging and continuously evolving, the outcome results will keep changing and altering Gliina's future around the clock. In September of 2014, and while processing the same simulation for Starship Gliina's futuristic route, few new results came up.

For example, discoveries in material science and exoplanetary studies in the past year yielded changes in factors such as the starship weight, acceleration, speed, destination, and simulated location from Earth. The list of elements subject for modification extends to question if the future launching scheduled date of 2072 will change too. One thing that is certain about those futuristic simulations is that the pace of change will remain unsteady until all required breakthroughs in the elements of design are stabilized. That will take few decades based on the advancement rate in our current knowledge and technology.

2. Starship Architecture, a Holistic Design Approach

When a design team works on a (regular) project, it is typical to analyze the program required for such project and go over the functions to put them in full perspective while resourcing previous projects of similar nature. That generates; a realistic diagram of the relation between functions, a schedule for developing and finalizing the design, and finally the construction schedule. A holistic architecture approach for a starship, though unconventional, is still achievable following the traditional design methods. It is necessary however to note the following points as additional transdisciplinary items to be considered:

1. While space travel started, practically, with launching the first artificial satellite Sputnik into Earth's orbit in 1957 by the Soviet Union, and despite all the advances since then from landing humans on the moon, to landing robots on Mars, and sending the Voyagers to the edge of our solar system, most of the design and technologies used and developed so far are not suitable for interstellar travel. Certain aspects need to be considered when designing for extended space travel to see whether they could be used as base or heritage for developing the required ones, or to work on new concepts for interstellar travel.
2. The designs we know for interstellar projects came mostly from science fiction literature and films. In addition to that, there were some theoretical studies such as Daedalus unmanned interstellar probe [2] and Stanford Torus's inhabited starship [3]. While no startship, or even a main part of it, was built until now, both literature and theoretical studies still provide good sources for design and ideas – in fact they might be they only conceptual prototypes available.
3. Elements of interstellar and deep space travel are being studied separately and through research that is done for various items by different teams at unconnected locations, and without efficient coordination. For example, the specialized teams researching conceptual interstellar propellers have been working on a certain breakthroughs in different areas; solar sail, fusion, fission, ram jets, plasma, antimatter fusion, and many more, without direct interconnection with those working on closed air and water systems. That applies to teams studying other space travel fields. It is another sort of variables that need to be considered in developing a starship design to keep the project's integrity.
4. In the past few decades, sustainable design and green building concepts helped developing aspects much needed for extended space travel design such as designing with efficient use of material, utilizing natural resources, recycling, reusing, and self-sufficiency. One essential part is the simulation for energy generation, use, and loss for buildings. On a starship, energy calculations are part of every design item for along with the consumption and recycling of the raw materials.

   However, what we consider on Earth as naturally sustainable are the renewable energy resources such as solar, wind, and geothermal energy, in addition to the naturally renewable raw materials. Those items depend on the earth and its position from the sun. In outer space, the conventional renewable resources for energy and raw material become obsolete, and sustainability is to be achieved by other means.
5. Essential design and construction components of the starship are developed through uncommon ways depending on scientific experiments, space probes, and prototypes scouting. The probes themselves and the data they collect and transform are factors that continuously affect the elements of deep space design and schedule.

6. The nature of a starship project by itself is unusual, as it could be simply defined by building a small encompassed earth with everything that would support the life of thousands of people for decades. Add to that the necessity for such self-sustained structure to be capable of travelling at theoretical speeds and venture to locations that we cannot currently see, while employing technologies that are undeveloped and utilizing discoveries that are yet to be made.

Though the last item might seem discouraging, it is actually the key for a comprehensive study for interstellar architecture. It encourages designers to go beyond current modules, and account for variables that might be new to the design process yet essential to make this project work. Adding continuous research speculations for items that are still in theory, and projecting future advancements for others, bring the design process back under control, and somewhat put it in line with the way ordinary architecture projects are evaluated and fostered.

When accounting for all elements, traditional and unconventional, we can resort to the basic elements of design; function, budget, esthetic, and life cycle. In that sense, we can see how designing daily Earthly needs from habitat and equipment is barely different from designing interstellar mechanical spaceships, habitable spaces, architectural furniture, fashion clothing, practical equipment, and accommodating gadgets.

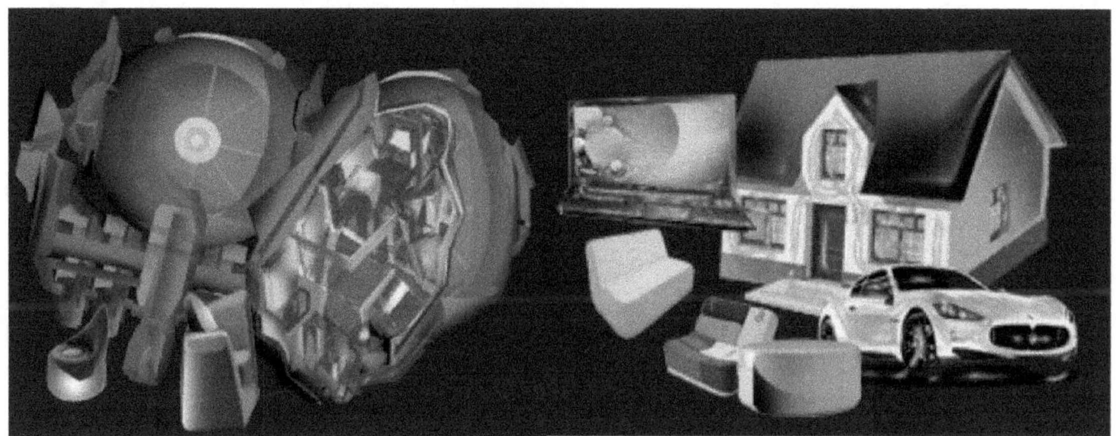

Figure 1. Including conventional and variable design aspects shows that designing for deep space could be compared to designing Earthly items

3. Starship Elements of Design

Over a hundred basic design items were evaluated and assessed for their requirements to attain a sustainable status in a starship that is bound to function for scores of years. The items were analyzed according to their development stage; current technology, developing technology, or future technology. The analysis of extended space travel requirements verifies that each item can perform with long durability, minimal maintenance, energy self-sufficiency, and the capability of self-repairing.

Future scientific and technological developments were included the in assessment. The design elements were divided into three main fields; structure, environment, and human needs.

3.1 Structural Elements

The main structure of the starship with its exterior and interior elements is analyzed with a comprehensive attention of how the components fit together and affect each other. That is based on questions a designer asks every time a new project is assigned. That includes asking about the objectives of such project - the mission in our case. Items such as destination or finance might seem odd under this category, yet they are vital for other items under Structural. Destination, for example, will determine the size of the starship and its crew in addition to everything linked to those items. The mission, whether it is to discover, terraform, or inhabit a life-supporting planet will affect most of the items under this category.

- AI and supercomputers
- Budget and finance
- Command/control center
- Communication with Earth
- Cosmic objects and dust shielding

- Constriction management and techniques
- Construction materials
- Defense system
- Destination
- Energy and alternative energy
- Engine/propeller
- Fabrication and 3D printing
- Fenestrations
- Fuel
- Functioning reliability
- Gadgets
- Gravity
- Growing neighborhoods, remodeling
- Industrial and heavy equipment
- Infrastructure
- Inhabitants, crew
- IT and computers development
- Launching
- Mission
- Navigation
- Radiation shield
- Raw material and resources
- Repairs and maintenance
- Robotics
- Scanning/mapping near-hit-objects
- Size, volume, dimensions
- Spacecraft and probes (onboard)
- Spacecraft launch pads/stations (onboard)
- Speed and acceleration
- Structure, construction
- Telescopes
- Urban design
- Weight

3.2 Environmental

Humans experienced space travelling while living in a closed ecosystem for several days in moon journeys. The trips to the International Space Station (ISS) are extremely short trips to Earth's orbit, while living aboard the station is like camping in ones backyard. Compared to interstellar travel, such trips were so short that humans can suspend certain aspects of their daily life as they will return to them shortly. When travelling beyond our solar system, all aspects of human life necessities need to be considered and delivered, whether predesigned, through onboard production, or simulated.

- Animals and insects
- Arms and artillery
- Artificial and natural open spaces
- Audiovisual production, television and radio
- Common/civic spaces
- Cultivation & botany
- Daycare and nurseries
- Emergencies and ER/safety
- Firefighting
- Food and beverage
- Furniture
- HVAC systems
- Indoors transportation
- Insulation and acoustics

- Interior air quality
- Internal telecommunication
- Internal energy source and balance
- Equipment and appliances
- Landscape
- Medical facilities and equipment
- Museums and artifacts
- Nutrition and food cycle
- Paths and hallways
- Plumbing system and water system
- Recycling
- Residential units
- Simulated light/seasons
- Space suits
- Sports venues
- Stargazing and cosmic nature
- Storage
- Traffic and commuting
- Vacations and recreation

3.3 Human

A manned interstellar mission needs much more than a controlled environment and a starship that can travel for decades at hyper speed. Cloning a small Earth is not complete without taking along all social, communal, and psychological aspects that define humanity as we know it. Studies of the human needs determine a lot of the design that will be incorporated in a starship. That is a normal task in architectural design even for a simple residential building. In ordinary design, many aspects are usually accounted for within a macro environment such as urban design, or a microenvironment that would be created by individuals later such as using a small fan inside a room to adjust the temperature or humidity. In space architecture we have to factor in all the macro and micro aspects in the original design.

- Aging in space
- Art
- Children and parenting
- Cloth and fabric, fashion
- Death and ceremonies
- Economy system
- Education
- Generation Zero
- Generation One
- Government and leadership
- Healthcare and immunity
- Interstellar culture
- Language and vocal development
- Law and order
- Libraries and encoding
- Literature
- Losing physical contact with traditional machinery, Earth's nature
- Marriage and social life
- Interstellar psychology
- Music and theatre
- Parties, weddings, and events
- Personal hygiene
- Policing
- Population growth, distribution, and management
- Religion
- Research and advanced studies

- Simulation and virtual reality
- Society, habits and protocols
- Technology development (onboard)
- Time dilatation and contact with Earthlings

4. The Radiation Shield Sample

The Radiation Shield sample represents the assessment done for each item. Analysis of technology status and the time anticipated to be ready for prototype testing is a major part of the evaluation. That includes analyzing the sustainable design requirements of a radiation shield that will endure a trip expanding for long light years under high speed and last for long decades. Other aspects are verified such as deterioration, maintenance, and future expansions.

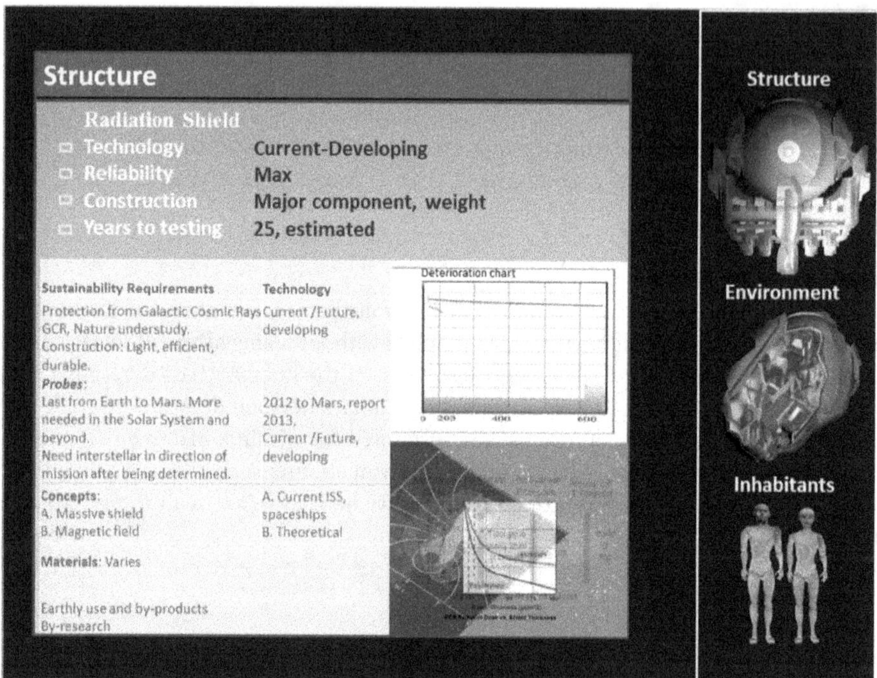

Figure 2. Item Structural- Radiation shield analysis summery

4.1 Sustainability Requirements

Protecting the humans onboard a spaceship from harmful radiation while outside the protective Earth's electro-magnetic field and atmosphere is a big concern. The starship itself and its computers and communication systems need protection from radiation. The longer the trip and the further it is form our Earth and Sun, the more dangerous are the radiations. All manned missions, including the International Space Station's (ISS) have protection measures from radiation [4].

Sun plasma spews in the form of flares, or solar wind, increase the radiation hazards to spacecraft and crews. Other cosmic events such as supernovae or black holes increase the high-energy radiation known as Galactic Cosmic Rays (GCR). The solar magnetic field protects Earth, and the solar system planets from the GCR keeping most of the radiation away by a shielding enclosure represented by the heliosphere[5]. The magnetic field of the Earth plays another essential role in deflecting the harmful radiations within its proximity[4].

4.2 Conceptual Design

Current technology concerning protection from radiation is limited. That also limits the time astronauts can spend in space including the time when they are inside the ISS[6], only 200 miles from Earth's surface. Even commercial airplane pilots are subject to radiation despite flying at low altitudes [6].

The classic design depends on using thick materials for spacecraft walls for protection. Protection from the high-energy GCR needs going thicker with conventional methods or using new concepts. A major structural effect is the additional weight.

An idea that originated in 1969 by space engineer Wernher von Braun, calls for using a superconducting magnet to create an electromagnetic shield to protect the starship just like Earth's electromagnetic filed protect its inhabitants. Project SR2S is currently working in that direction[7]. The project is an EU-funded research for developing magnetic shielding to deflect dangerous cosmic rays using superconductors, materials that have no electrical resistance at extremely low temperatures. The weight of the large magnet to be used is an essential concern. Simulations show that by utilizing new superconducting materials such as magnesium diboride (MgB2) a 30-metre-diameter magnetic field could be produced by a system weighing less than half that of a comparable passive shield depending on thick layers for protection. Project SR2S is supposed to be completed by end of 2015.

Contemplating the current pace in technology advancement, it will take approximately 25 years to reach a stage were a prototype could be tested for a starship and interstellar use. That is considering the different design options and a continuous design development due to the necessity of the radiation shields for manned space missions in general.

4.3 Probes

Further probes are needed especially that we do not know much about the cosmic radiations outside our solar system. The most detailed and recent data about the radiations covers the route to Mars and were received through a spacecraft travelling to Mars in 2012[8]. The high energy radiations outside our solar systems are more intense especially beyond the heliosphere.

Several probes and missions for studying the cosmic rays beyond our solar system and in direction of the proposed route(s) are required. Other items that need further studies and observation are solar flares and cosmic events that release massive concentrated waves of high energy radiation.

5. Status of Structural, Environmental, and Human Items

Following the same way of analysis used for the radiation shield, projections for a hundred elements of design were configured. Graph 3.0 shows the distribution of the hundred items in terms of technology status; current, developing, or future. It is clear that the majority of the elements is in developing phase or still in theory and will be developed in the future. The part that shows as current technology, less than 20%, indicates items with current design technology but still need to be configured for deep space travel.

Graph 3.2 shows the status of technology for the elements under each of the three categories. It also shows the anticipated years needed for development. The human aspects division seems to be the most underdeveloped one as manned deep space missions have not even started. The environmental aspects are in slightly better shape as they were part of the few missions with humans onboard. The structural elements are the most tested as all current space missions included many of them under this division. However, all three divisions need long years of research and development.

Graph 3.1 shows that about 84% of the design elements for the starship have corresponding applications on Earth. The continuous development of such items over the upcoming decades will result in subsequent advantages for their earthly use. Without going into details, all items listed under the environmental division that are developed to be lighter, more durable, and consume minimal energy will be of high benefit to Earthlings, who will be taking advantage of such benefits long before they are utilized on a starship.

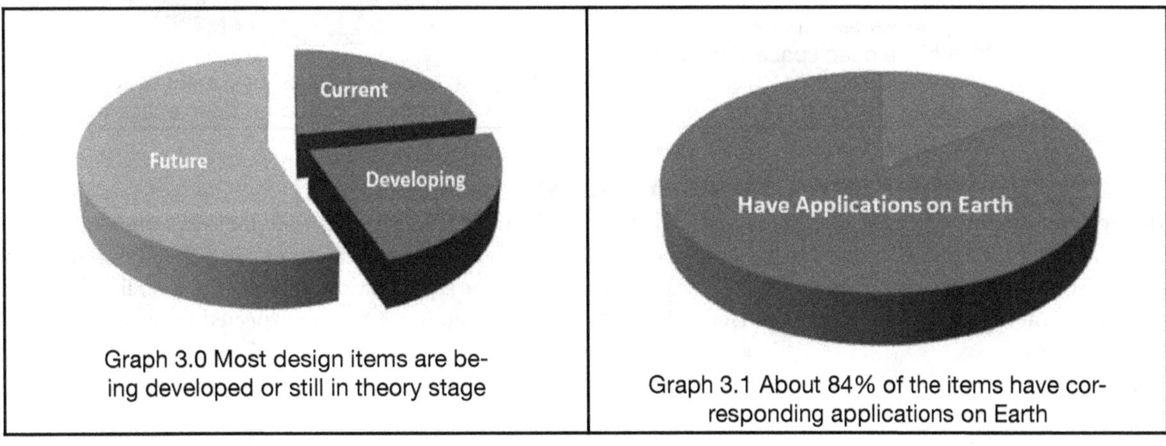

Graph 3.0 Most design items are being developed or still in theory stage

Graph 3.1 About 84% of the items have corresponding applications on Earth

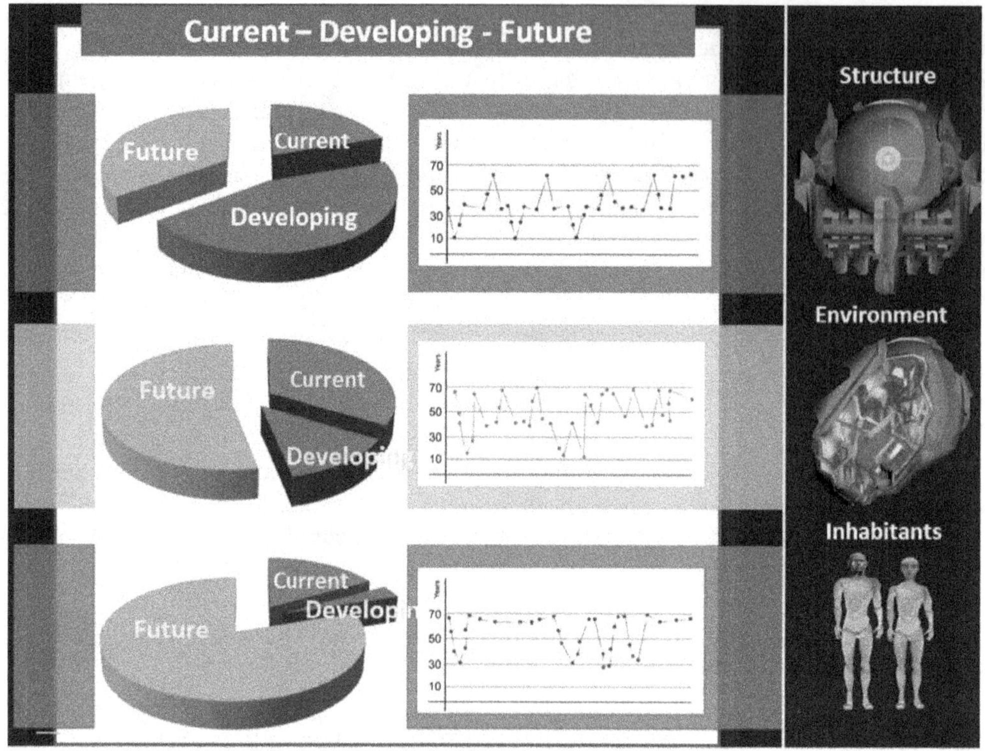

Graph 3.2 Status of technology for the elements of each of the three groups. Human factors are the least developed, followed by Environmental, while Structural items are the most ahead.

Graphs 4.0 and 4.1 show that current space design and technology account for a good portion of the elements needed for the deep space travel. However, only a small portion of the current technology could be reconfigured for deep space travel. For example, we cannot depend on rocket engines that use liquid fuel as prototype or heritage for extended space travel.

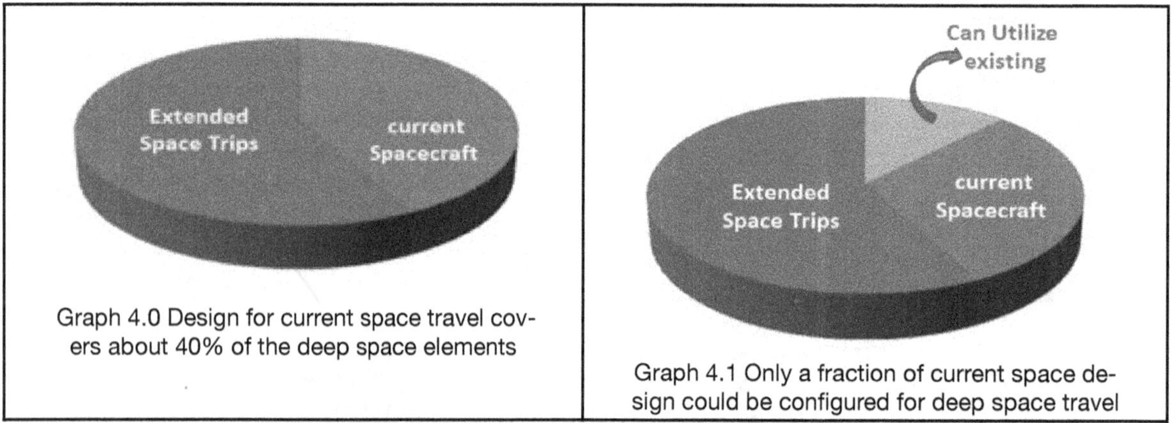

Graph 4.0 Design for current space travel covers about 40% of the deep space elements

Graph 4.1 Only a fraction of current space design could be configured for deep space travel

5. Starship Inhabitants, Human Factor

The design elements related to the human aspects pose considerable challenges as they include some unpredictable features. Our experience with manned missions is limited to missions to the moon and Earth's orbit. Such short history of humans in space can merely explore conditions for a population on starship several light years away. On a multigenerational space trip, Generations Zero and Generation One are expected to undergo critical circumstances.

5.1 Generation Zero

Generation Zero is composed of those who will leave Earth on a one way journey. To make a simple assessment, let us assume that we have a small to medium starship with a startup population of about 1,000. Out of those, the part that will be considered the crew of trained scientists will be from 200 to 300 persons. The remaining population includes service providers, technicians, educators, and members of the community of various ages and backgrounds. While it might be easier for trained scientists and astronauts to understand leaving everything behind on Earth on a one way trip, it is certainly harder for others. In all cases, and putting all simulations aside, the thousand people who were born and raised on Earth are attached to it, and to the compilation of civilizations that led to their time. They will never see or touch any person that they knew on Earth, except for those leaving with them. They will never see a passing train or a sailing ship in the horizon. They will not be able to climb a mountain or dive into the ocean.

By the time the starship is ready to launch, we should have several space colonies within our solar system that examined this issue at several levels. Currently, this topic is being studied and analyzed with several propositions to help Generation Zero adapt to their new life and future. While suspended animation is being researched for saving weight and energy[9], some drastic suggestions for diverting the effects of leaving Earth are surfacing for multigenerational trips. The starship population, or part of it, could be sent in long space trips prior to the interstellar journey. Even more, Generation Zero could be fully born and trained in a space colony so they will not be emotionally attached to Earth by the time they leave our solar system[10].

5.2 Generation One

Generation One, while might be less attached to Earth and the places or activities that they never experienced being born and raised on a starship, will have other points to address. They need to maintain a certain pre-engineered directions within the interstellar community regarding population, education, careers, and daily life. After all, it is a small society and a certain average number of people per family and in total must be upheld. The same applies for the number of scientists, engineers, technicians, doctors, educators, service providers, artists... to the end of the list. That applies to all generations born on a starship, with Generation One having the first real test.

5.3 Social Development in Space

In an interstellar journey, society takes its own course in adaption and evolution just like any society on Earth. Such developments are built on two main factors; the inherited features within the community and the effects of the surroundings. The inherited aspects for the inhabitants of the starship when leaving Earth will be similar to those for people living on Earth. However, living in an enclosed environment on a multigenerational interstellar trip will add different factors to the evolving culture that will gradually differ from what the people on Earth experience.

Design for such elements should account for the departure time conditions, and allow for changes that are not known yet. Future design accommodations and extensions will be made by the generation that will experience the changes utilizing available resources on the starship. Below is a brief discerptions of some of the main social fields that are subject to change.

5.3.1 Languages and dialects keep evolving through time. The development of languages start with changes in dialects and many dialects become languages through generations. Spoken languages are affected by the direct surroundings. A contemporary example is the French language and how the same language developed in two different places in few generations. The Quebec French spoken in Canada today is very distinctive from the French spoken in Europe. It is not clear what (form of) language the inhabitants of the starship will speak when it takes off, since current languages will keep evolving until then. During the trip, the languages will keep developing onboard and on Earth in ways that adjust to the local surroundings at the two locations.

5.3.2 Art and Literature are direct reflection of the environment. On a starship they will start with what is considered, on Earth, classical in visual art, performing art, and creative literature until the interstellar effects take place. The art might reflect the starship as an environment instead of natural, urban, and demographic surroundings. Performing artists or pop stars on a starship would have to adjust as their audience shrinks from millions to thousands. Other forms of art may emerge, flourish, or decline.

5.3.3 Sports in space will have many adjustments. The vast fields needed for football, succor, or races might have to be adjusted or even replaced by simulated sports and venues. On Earth, different people at different locations have developed localized sports like Tejo in Colombia and Dandi Biyo in Nepal. New sports may develop accommodating the community and the environment on a starship.

5.3.4 Education development will not only change in the means of teaching but also in what people will learn on a spaceship. Do they need to learn about Earth and earthly geography and weather? What about agriculture and the history of civilizations on Earth? That might need adjustments considering the destination planet. Will the Earthly physics laws taught on Earth apply to that on the starship and at the final destination, and how are such laws adjusted? How much scientific research will be conducted onboard, and how different will that be from the scientific breakthroughs occurring on Earth? A vast part of learning comes from interacting with diverse cultures and communities. Can simulation cover this gap for the limited number of inhabitants?

5.3.5 Encoding, documenting, libraries, and passing information are changing in our current time at an unpredictable manner and speed. Encoding on Earth moved from printed books, to electronic books, and downloadable data on tablets in a couple of decades. While we do not know what further developments will take place until the starship starts its mission, the type of information to be circulated and recorded during the mission is the major concern. What is needed to be documented and passed to next generation, and by what means? Are there going to be any libraries or bookshops in the future on a starship?

5.3.6 Technology is another leading aspect that has been through unprecedented breakthroughs in the past few decades. It is not clear how far the technology will progress in the forthcoming decades until interstellar travel commences, and how things will develop onboard. Computers fast paced development in the 1990's, phones' in the 2000's, and tablets' in the past decade give an idea of how unpredictable this field is. Such developments will be implemented into all other items of design as applicable, while the advancements during the trip will match the needs of the inhabitants and will be built upon current and future knowledge.

5.3.7 Economy Systems develop naturally in communities. They get more complex with the growing of the population. A well-organized society of small size, take1000 persons, on a starship might start as a simple communal society with no need for making business. How would a society without economic incentive be driven? What type of economy could work as the population grows? Will they use monetary system or stocks on a starship? Can they distribute property? Can any person have a 1,000 times more money or property than others, and how would that affect everyone? Will the market be driven by supply and demand in such a small population? Is there a need to build financial institutes?

5.3.8 Social habits and protocols dealing with others differ considerably between communities on Earth. Will interacting with the same people for years make communication among the inhabitants more casual? Or will that create a new definition for privacy that differs from than what we have on Earth. Will people need more open spaces or more private enclosures? How would people deal with social occasions such as marriage and death?

6. Conclusion

Interstellar architecture introduces a new approach for working on deep space projects. The sustainability aspects in three main areas; Structure, Environment, and Human, need to be assessed, and the contentious development in technology must be incorporated taking into consideration the progress expected until launching a starship and later onboard the starship. Comprehensive futuristic scenarios simulating the future developments for all design items go beyond current modules, and accounts for variables that might be new to the design process. However, that brings the design progression back under control, and somewhat puts it in line with the way ordinary architecture projects are evaluated and fostered, resorting to the basic elements of design; function, budget, esthetic, and life cycle.

References

1. "Sustainable Design for Extended Space Journeys," Antoine G. Faddoul, 100YSS Symposium, 2013

2. published in a special supplement of the Journal of the British Interplanetary Society in 1978

3. *Space Settlements, A Design Study,* Published by NASA and Stanford University 1977

4. NASA, *Space Radiation Analysis Group*, Johnson Space Center, March 5, 2014

5. *Galactic cosmic rays in the heliosphere,* Thomas, S.R., Owens, M.J. and Lockwood, M., 2014

6. NASA, *Space Faring the Radiation Challenge*, 2008

7. *Space Radiation Superconducting Shield,* www.sr2s.eu

8. Radiation measured by Curiosity on voyage to Mars, *Astronomy magazine*, March 2013

9. NASA, *Torpor Inducing Transfer Habitat For Human Stasis To Mars,* 2013

10. *Astro-Sociology: G-Zero to G-1Manning and Phasing*, Antoine G. Faddoul, 2014

Military Planning for Interstellar Flight

Ken Wisian , Ph.D.

Austin, TX

Abstract

Though ubiquitous in science fiction, the military aspects of interstellar travel are too seldom addressed in the forums on actually reaching the stars. The debate on METI (Messaging to Extra Terrestrial Intelligence) /Active SETI (Search for Extra Terrestrial Intelligence) highlights that a peaceful, benign cosmos cannot be assumed a priori. This paper briefly surveys the legal and historical background of military activities in space, civilization to civilization first contacts, the reasons this should prompt us to consider "arming up" and the basic options for crewing and equipping a starship militarily.

Current international law/treaties addressing the military uses of space forbid many military operations and specifically weapons of mass destruction in space. It is worth noting though that throughout history most international norms on weapons restrictions have had mixed success at best. Our limited one-planet dataset also shows that civilizational first contacts are frequently very hazardous, particularly for the less technologically advanced society.

Military options for interstellar ships break down into two main components. The first is how much of the crew, including any Artificial Intelligence (AI), is military or has military training; from none at all, to a complete crew of full-time military. The second component is whether or not there are military use only weapons and defenses on board. There will be systems usable as weapons on a starship regardless - communication lasers, systems to destroy objects on collision course with the ship or propulsion systems could be used as weapons. There will also be personal defense weapons on board for use against dangerous life forms at a minimum, as there have been on manned missions for decades.

This overview will hopefully stimulate the explicit incorporation of military aspects in the work across the interstellar community from ship design to sociological studies.

Keywords

Military, weapon(s), defenses, first contact, dual-use

1. Introduction

Science fiction largely assumes that mankind's future in the universe will be violent, filled with war, fleets of starships, and humans fighting humans or humans fighting aliens. The reality is likely (hopefully) much different. The serious ongoing work on the now envisionable goal of interstellar travel makes little if any mention of military affairs. There is excellent work being done on designing ships from micro-robotic to colony ships as well as thinking through many social/psychological aspects . It is critical then to address sooner rather than later the question of the place of military capabilities (here divided into crew and technology/ hardware) that would be part of a starship. Addressing military capabilities now is driven by the general rule that the earlier components are incorporated into a design, the less costly it is. This can also be phrased as the later you make design changes the more it costs, often exponentially so. This paper will broadly survey the legal issues, the perspective of history, outline courses of action, and hopefully stimulate substantive discussion of the issue at this very early stage of design.

1.1 Current Treaties Preclude Some Military Uses Of Space

Current international law regarding militarization of space is relatively simple, shallow and incomplete. At the least, the commercialization of space will likely drive a contentious expansion of space law. The Treaty on Principles Governing the Activities of States in the Exploration and Use of Outer Space, including the Moon and Other Celestial Bodies , took effect in 1967 and is the primary effective treaty on space. It is a only a few pages long and the primary military-related provisions are;

1. No weapons of mass destruction will be based in space
2. There will be no military bases or activities on the Moon or other celestial bodies

It is important to bear in mind that this treaty was written during the height of the cold war and focused on the superpowers not carrying their conflict into space. The treaty takes no account of the need to defend humanity from extraterrestrial threats, living or inanimate. It is also relevant that throughout history treaties and other international attempts to restrict weapons have a very spotty record – even today treaties banning nuclear proliferation, chemical and biological weapons are nowhere near 100% effective. Some obvious gaps in the current treaty include: 1. Weapons short of mass destruction classification (itself a term subject to considerable debate) are allowed in space, 2. Military bases and exercises are allowed in space but not on celestial bodies, and 3. Military personnel not performing a military role are allowed in space and on celestial bodies with no restriction. Further treaties restricting military activities in space have only garnered limited support. Now add to this rather weak international law on space, the historical tendency of major players to ignore treaties when it is politically or economically inconvenient and you have a situation ripe for making up the rules as you go. Note that space has been and continues to be used extensively for military purposes (intelligence and communications primarily) with missions performed both by robots and humans. The Global Positioning System (GPS) and the Russian GLONASS (Global Navigation Satellite System) are military systems that civilians are conditionally allowed to use. Multiple countries have tested Anti Satellite (ASAT) weapons that are for use solely in space. And finally, all ICBMs (Intercontinental Ballistic Missiles) transit space enroute to their targets.

Related to the above discussion is the growing impetus to develop the capability to neutralize asteroids on a collision course with earth. A military or quasi-military mission such as this, potentially using nuclear weapons will also force the development of new legal framework as well as of dual-use military capabilities and technologies.

1.2 History Suggests That We Should "Arm Up"

The interstellar community often speaks of the innate human desire for exploration and curiosity, but seldom mentions another innate aspect of human nature – conflict and war. War is endemic to humanity and at least some primates . Even life itself is often characterized as a type of warfare. Although human war appears to have declined steadily from pre-history to today, it is far from gone . The big and unanswerable question as it relates to this paper is 'Is violent conflict a condition of all life in the universe?' While we have no way of answering this question with a data set of one planet's species, we can delineate one boundary of the answer. We can definitively say that the universe is not universally peaceful based on our one data point.

War can also be viewed in the framework of politics. The most famous philosopher of war in the Western world, Carl von Clausewitz asserted 'war is a continuation of politics by other means .' This idea is fundamental to most scholarship on war to this day. Would anyone propose that politics will not accompany us into the future in space? A more hopeful rephrasing of the question might be 'could the boundaries of acceptable politics shrink so as not to include war?' It may be a possibility, but not a certainty.

Economic motivation has been a strong driver of exploration in human history. But even when commerce was the driving force behind exploration historically, military and commercial interests were tightly intertwined. At the very least merchants wanted the legal and physical security of their trade guaranteed.

Taken in the broadest possible scope, does human knowledge contribute anything to the question? Looking at life on Earth in the broadest terms, it is clearly a no-holds-barred competition for resources. From viruses to mega-fauna, populations expand to the carrying capacity of the environment and will without thought drive less-able competitors to extinction. Modern humanity is the first example known of self-limiting population (perhaps) and this aspect is due to our intelligence . This then becomes a possible light at the end of the tunnel. If trends continue and human population peaks followed by slow decline, we could remove population pressure / resource scarcity as a driver for conflict on Earth, but also potentially as one of the drivers for exploration or expansion.

We have a limited data set of technological civilization first contacts – namely our own cases of genus homo. The history on this planet of human civilizations meeting is not too good, particularly when one has a significant technological development edge over the other. Obvious examples of contact that did not go well for the less advanced side include the many meetings of European civilizations with American, African, Asian and Pacific civilizations. Note that not all bad effects were based on military superiority. Take for example unintended biological warfare - smallpox, sometimes decimated a civilization ahead of any actual fighting. Technological superiority however is not always determinate. Asian Steppe tribes, most notably the Mongols, conquered a much more advanced China and much of the Muslim and Christian world with superior strategy and tactics, not technology.

Not all contacts turn out in the long run to be bad for the less advanced civilization. Although very violent on a regular basis, the contact between the Roman state and the "barbarians" of Europe stimulated technology transfer and political consolidation, and eventually resulted in the barbarians destroying the Western Roman Empire .

Based on our one-species, one-planet data set reviewed above, it would be naive to venture forth into a galaxy that might have more advanced civilizations, totally unprepared to defend ourselves. Hopefully our one data point represents a transient extreme on the violent end, but "hope is not a strategy."

1.3 The Interstellar Threat

The likelihood of another civilization we come in contact with being of the exact same technology level is small. In the span of cosmic time, human civilization is merely the blink of an eye and our "advanced" technology merely the thought of blinking an eye. Thus if we are visited anytime soon, it is safe to assume that we humans will be the primitives. If we journey to other stars, we could conceivably be either the more or less advanced race. The point is that in general there will be a tremendous mismatch between encountering races.

You might reasonably ask – 'so why arm up?' The top counters that come to mind are; 1. If you don't try, your chances of winning are zero. 2. The laws of physics might set an upper bound on weaponry, such as nuclear bombs and directed energy weapons – meaning that we are already decreasing the maximum technology gap that we might face. 3. There are many non-intelligent threats that require military grade weapons and tactics anyway (asteroids impacting Earth or finding and redirecting or destroying obstacles when travelling at high speeds in space). 4. We may be the more advanced technology, but not the more warlike culture. 5. The argument that we are safe from threat here in our own solar system due to the immense difficulty of traveling interstellar distances assumes no technological breakthroughs. This is more about defenses inside the solar system, but still relevant. Until World War II, the USA thought it could be minimally armed due to the oceans separating it from potential adversaries, but that proved false as technology advanced – the same could prove true for interstellar distances.

The above arguments derive largely from an ongoing debate focused on the wisdom of actively signaling our presence in the galaxy. Stephan Hawking , Jared Diamond and David Brin among others caution that we cannot assume that any extraterrestrial civilization we encounter will be benign.

2. We Must Take Some Military Capability And Training With Us

Based on the above background it is reasonable and prudent to carry some degree of military capability with us into space. Now we look at how to think about military capabilities in early interstellar travel. While the technical aspects of potential weaponry are quite interesting to explore, to even scratch the surface is well beyond the scope of this paper. Suffice that at the minimum future vessels will likely have dual-use technology such as shielding, communication lasers, obstacle clearing systems and even a main drive that will have military utility. An early Larry Niven story, The Warriors, of a first encounter comes to mind as a great example of how a piece of technology, the main drive, can be pressed into duty as a weapon.

The frame of reference for the following is a min-max-min problem - how to minimize risk both existential and mission, by maximizing military capability (if desired), at minimum cost in terms of money, energy, mass, crew complement and training, etc.

2.1 Defenses

Shielding will be present in one or many forms to protect from the harsh environment of space. Shielding could be in the form of mass, particularly in the bow of a ship to protect from collisions with energetic rays and particles - a grain of dust at relativistic speeds can do a lot of damage. Electromagnetic, gravity or other "non-material" shielding is also a possibility (and maybe a life support necessity). The same thought also could apply to the personal level. Suits capable of handling any threat in space or on a new planet would have to be able to withstand a lot of punishment and could end up as a cross between a space suit and combat armor. Ship Maneuverability, laterally as well as acceleration and deceleration is a key defensive quality too. Extra maneuverability beyond the minimum mission requirements is very desirable, but would likely be "expensive" of weight and resources. Jamming or decoy systems are a different matter in that they are unambiguously military in nature . Any unambiguously military systems send a message, not necessarily bad, but distinct from dual-use systems.

Less obvious issues that I will lump under defenses are ship layout and information protection. A ship's design could be different based on the possibility of fighting. For example a backup bridge might be at the center of a ship versus at maximum separation from the primary bridge. If you are looking to protect the bridge from aimed weapons versus random space rocks. There are countless other ways that design might differ, from spatial layout to systems redundancy. Information is another significant question. Any ship will carry substantial information on Earth in many forms. Would a method of quickly destroying information be desired in order to keep knowledge out of potentially hostile appendages? Is it even practical to try to protect knowledge of Earth? What about the location of Earth? Any course alterations (to hide the ship's origin) would increase flight time – unlikely to be sanctioned or affordable on early missions.

2.2 Weapons

Weapons cover a huge range of possibilities from non-lethal personal defense to weapons of mass destruction – we'll start small and work up. I will skip the area of biochemical weapons as an explicit subject in this paper.

The need for self-defense on an unknown world with an unknown biota will drive the need for a wide range of weapons analogous to modern ground weapons. This would range from side arms (pistols or stunners) through heavier, crew-served weapons (think heavy machinegun or cannon) all the way up to armored vehicles (think tanks). Remember although there are no land animals on Earth today that can't be stopped by a large rifle, the equivalent of a large, hungry or amorous dinosaur might require something heavier. Any robots will also have some level of self-protection inherent in their design. Besides chemical powered current weapons, technologies such as rail guns and directed energy weapons are already showing up as fielded weapon systems on Earth.

For ships again there will most likely be dual-use tech present. An "active" drive or maneuvering thrusters have some capability – the focused energy needed to move a ship could be pretty destructive. Project Orion used nuclear bombs for propulsion; Project Daedalus used electron beams to ignite fusion . A starship itself could even potentially be a large guided bomb. Similarly any communication system with enough power to send signals across interstellar distances could at be destructive to electronics if not more, at short range. Even an ion drive might be capable of blinding or destroying sensors. If a ship has any kind of active defense against physical objects on a collision course, that also would have a significant military capability. These active defenses could be anything from missiles, kinetic or explosive, directed energy weapons, to nuclear weapons.

It is possible that ships could be sent out even without dual-use weapons. Potentially un-armed options for ships include swarm, micro or nano-ships, or any ship considered essentially "disposable." These would not necessarily be armed, though a self-destruct or memory erase function would be an option. Sail ships are another possibility - the extreme need for minimum payload mass could force the decision to go with no weapons, human crew, or even redundant systems.

Obviously military-only use weapons could be added to a ship. It is a matter of the cost in terms of mass and ship/crew resources more than actual cost vs. risk. Regardless, we almost certainly will not be venturing forth unarmed, even if a mission is 100% peaceful in intent. Then there is the perception problem. There is no way to guarantee that another civilization will "see" peaceful intent in the dual-use technologies regardless (particularly if they have been watching decades of our broadcasts).

2.3 Crew Options

Given that any interstellar ship likely has some military capability, let's turn to what the options are for crewing such a ship. This discussion initially addresses an actual human crew, followed later by options for robot/AI crews. The range of possible military training and crew complement range from no military training for any crewmember to an all-military crew. The possibility of a police force filling the military role (likely only in a large-crew generation ship scenario) could be a sub-branch of any of the models and is really just a matter of how much military oriented training the police receive (e.g. Special Weapons and Tactics or SWAT forces).

2.3.1 All Crew Full-Time Military

An all full-time military crew would obviously be the most secure option, in that the military mission will be the first priority. Military command and control also has the likely benefit of minimizing crew chaos in crisis. The obvious drawback is that this might be detrimental to the exploration or colonization mission. In other words (in fiction) it is often assumed (but unproven) to be expensive of "non-productive" resources to be 100% military. Productive is here taken as maximizing scientific discovery or human settlement. Certainly though, neither scientific nor settlement mission success is precluded by an all-military crew. In particular, military organization in an extreme survival situation, like colonization, could be a significant asset. Why worry about contact on a colonization mission? Our technology will likely tell us if an advanced civilization exists at a destination before we set out, but movement and expansion outside the solar system will significantly increase the chances of contact and the signature of an interstellar drive is likely to be "bright" and attract attention. Consider also how far our technology has advanced in the last hundred years (a reasonable journey time).

A more subtle drawback to an all-military crew is the old saying about 'if all you have is a hammer everything looks like a nail'. A military outlook is more likely to see worst-case scenarios or to interpret ambiguous actions as threat. Perhaps this could be overcome with training.

2.3.2 Part Of Crew Full-Time Military

This option could vary tremendously, so I will focus on a small percentage of the crew being military, say 5-20%. This is somewhat a reflection of most Earth societies. The intent would be enough of a force to use ship weapons and act as Marines if needed. Less "costly" than the all-military crew, this option is one possible min-max-min solution – getting maximum military benefit at the lowest cost, therefore the size of this part of the crew is kept small.

2.3.3 Entire Crew Citizen-Soldiers

This option looks something like the Enterprise of Star Trek fame. All the crew is technically military, but their military training and focus is somewhat secondary. This is also analogous to modern National Guard / Reserve / Territorial forces where the military role is secondary to the civilian role except in emergencies or war. A possible advantage of this option is that in a time of emergency, military command and control covers the entire crew.

2.3.4 Some of the Crew Citizen-Soldiers

This is getting to the lower end of military capability in that only some of the crew will have some military training as a secondary role. It is also reflective of some modern societies (like the US) that have a small percentage of people in the National Guard / Reserves as opposed to 2.3.3 where all crewmembers are reservists. It is the "lowest cost" option which still provides some explicit military capability.

2.3.5 A Completely Civilian Crew

A completely civilian crew, with no military training may maximize the scientific and colonization benefit, but by "zeroing the denominator" maximizes risk. This option assumes a completely benign universe and/or willingly concedes any conflict.

2.4 Secondary Question—Which Authority Is Paramount Military Or Civilian?

This is perhaps actually a paramount question, but is secondary within the structure of this paper simply because of the very involved arguments. Assuming some mix of military and civilian crew, whether full or part-time, how do you decide which authority is paramount and when? It ranges from complete civilian control to military dominant all the time to situation dependent (e.g. existential threat drives a military command). Military and civilian leadership could be vested in the same person or people in any of the scenarios, which simplifies things, or it could be separate. There are good and bad examples of both cases throughout history.

2.5 Artificial Intelligence (AI)/Robots

A ship "crewed" only by AI could still need all the dual-use technology discussed above and be equipped with military only tech if desired. A ship with a mixed crew, human and AI, would devolve back to all the above options for a human crew except when the crew is non-conscious. Assuming unlimited memory or learning capacity, an AI might be able to fill the roles of scientist and military tactician equally well all the time as needed, and thus moot the questions above about percentages and training levels. Assuming also that any AI sent out would have the same level of trust that a human crew would have to use deadly force, there might be no restrictions on weapon use. This is a highly complex and changing area of research. Robots on the battlefield today are non-intelligent and limited, but the situation is changing rapidly. Non-intelligent systems can already make "trigger" decisions today, with varying levels of human supervision. Barring a binding international agreement, we will cross a threshold of autonomous, lethal AI robots at some point in the foreseeable future.

3. Conclusion

Based on the narrow but deep data set of human experience, venturing forth into interstellar space defenseless is not prudent. Some degree of military weaponry and crew training/programming should be planned into interstellar missions. Options range from minimal added cost in terms of time and resources to making the military mission primary. Even "low end" options must be planned for, will take resources and require much thought.

Why worry about the specific military aspect of missions at this point? Beyond the obvious hardware decisions, some other aspects of design could be significantly affected; redundancy, habitat layout, medical facilities, even the AI and supporting software might be substantially different if military considerations are in place. Early in design decisions to maximize military value will be much more cost effective than later decisions. Also to be considered is how much information on our civilization to carry on board and our communications method back to Earth. Even the course trajectory needs to be considered, though currently it is difficult to conceive of how these things could in practical terms be "hidden". This paper only skims the surface of the topic and outlines a min-max-min framework for looking at the problem. The hope is to stimulate further discussion and work on military aspects of interstellar travel.

References

1. Truth in advertising; I hold the Rank of Major General in the United States Air Force (Reserve) and Texas Air National Guard, and have participated in most of the U.S. conflicts of the last few decades, none the less I am not a war "hawk". Also, nothing in this paper should be construed as representing the official position of the United States or Texas Government, the Department of Defense, or the US Air Force. The opinions and any errors are mine and mine alone.

2. I point readers to the recent 100 Year Starship conference http://symposium.100yss.org/technical-tracks and the Icarus Interstellar Starship Congress conferences http://www.icarusinterstellar.org/congress-livestream/ as starting points.

3. "Treaty on Principles Governing the Activities of States in the Exploration and Use of Outer Space, including the Moon and Other Celestial Bodies", United Nations Office for Disarmament Affairs, available at http://disarmament.un.org/treaties/t/outer_space/text, entered into force: 10 October 1967.

4. Gat, Azar, "War in Human Civilization", Oxford University Press, 822p., 2006, is meticulous in detailing the endemic war in human history. Steven A. LeBlanc, "Constant Battles: Why We Fight", St Martin's Griffin,

2003, also excellent and covers the subject in a much shorter book – both refute the myth of the "peaceful savage" Rousseau-ish thinking.

5. Pinker, Steven, "The Better Angels of our Nature: Why Violence has Declined", Penguin Books, 2012.

   "Steven Pinker: The Surprising Decline in Violence", TED Talk at http://www.ted.com/talks/steven_pinker_on_the_myth_of_violence.html , 2007.

   Morris, Ian, "War! What is it good for?", Farrar, Strauss and Giroux, New York, 2014.

6. Clausewitz, Carl von, "On War", M. Howard & P. Paret eds. & transl., Alfred A. Knopf, New York, 1993.

7. The potential for civilization population leveling or decline (without outside drivers) pertains to the demographic transition - the idea that as conditions improve (prosperity, technology, education) societies/countries move through four stages. Initially for pre-industrial countries birth and death rates are high, pretty much cancel each other out, and total population is stable. In stage two, as development increases, mortality (death rate) falls, while birth rate remains high (a generation or two typically). In stage three the death rate is low and birth rate falls. Taken together stage two and three result in a significant population increase (explosion). These cascading population booms (first in the industialized world followed by other regions) resulted in the rapid rise in world population in the 19th through 21st centuries. The fourth stage is low birth rate and low death rate (post transition) resulting in a new equilibrium population. More recently a fifth stage is postulated in which the birth rate falls and remains below replacement rate resulting in slow population decline. Assuming the entire world eventually reaches "stage 5", world population will reach a peak and then start a long slow decline. There is not enough data by far to make a guess at whether there is a "floor" to declining population. This overall scheme is reflected in UN projections of a maximum world population of around 10 Billion in 2100 (note these numbers are very sensitive to poorly defined variables). (Reference needed here, not just a statement. This may be best inserted into the main body of the document.)

8. Heather, Peter J., "Holding the Line: Frontier Defense and the Later Roman Empire", pp. 227-246, in Victor Davis Hanson, Ed. "Makers of Ancient Strategy: From the Persian Wars to the Fall of Rome", Princeton University Press, 2010.

9. Hawking, Stephen, "Into the Universe with Stephen Hawking", *Discovery Channel*, 2010.

10. Diamond, Jared, "To Whom It May Concern", *New York Times Magazine*, December 5, 1999, p. 68-69.

11. Brin, David, "Shouting at the Cosmos… Or How SETI has Taken a Worrisome Turn Into Dangerous Territory", Lifeboat Foundation (http://lifeboat.com/ex/shouting.at.the.cosmos), 2006-2008.

12. Niven, Larry, "The Warriors", in "Man Kzin Wars", Baen Books, 1988.

13. While it is possible to use communication systems as jammers or to improvise decoys, the effectiveness is likely to be so low as to be negligible. Militarily, today all effective jamming and decoy systems are purpose built. (Reference needed here, not just a statement.)

14. General Dynamics Corp., "Nuclear Pulse Vehicle Study Condensed Summary Report", NASA, available at; http://ntrs.nasa.gov/archive/nasa/casi.ntrs.nasa.gov/19760065935_1976065935.pdf , 1964.

15. Obousy, R. K., "A Review of Interstellar Starship Designs", Presented at the One Hundred Year Starship Conference in Orlando, available at http://www.icarusinterstellar.org/wp-content/uploads/2012/05/A-Review-of-Interstellar-Starship-Designs9-27-11.pdf , October 2011.

100 YEAR STARSHIP™

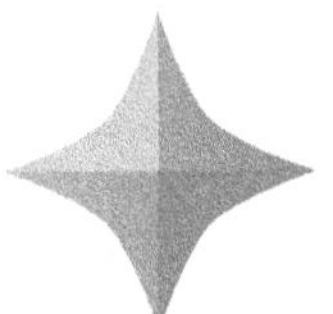

Chaired by Ron Cole

Colorado, USA

Track Description

Sending and receiving information by interstellar travelers or robotic vehicles requires development new methods to traverse the vast emptiness between stars. Additionally, in the absence of routine and timely communication with Earth, a probe or traveler must be self-sufficient in gathering, generating, compiling, storing, analyzing and retrieving data while ensuring these systems are operational over the lifetime of the mission and beyond.

Track Summary

The Data, Communications and Information Technology Technical Panel was charged with bringing the best papers discussing the sending and receiving of information by interstellar travelers (manned or robotic).

Five papers that met this charge were received:

1. *Interstellar Links Created by Two Focal Spacecraft*
   Dr. Claudio Maccone from the International Academy of Astronautics
2. *Loading, Please Wait: Long-Distance Crisis Communications*
   Ms. Jaym Gates from the Science Fiction and Fantasy Writers of America
3. *KLT Filtering, Even from Relativistic Sources*
   Dr. Claudio Maccone from the International Academy of Astronautics
4. *Cutting the Umbilical Cord: Building Trustworthy Software for a Starship*
   David Burke from the Galois, Inc.
5. *Contemporary Classification of Astronomy and Space Encoding*
   Antoine Faddoul from Tony Sky Design Group

These five papers, written miles apart at different times, uncannily supported each other by tying together the subjects of communications and software needed for interstellar travel. Dr. Maccone was back for a third time to update his concept of a "radio bridge" build across FOCAL spacecraft relaying communications to and from the starship. His second paper argued that the Karhunen-Loeve Transform provides a communications filtering algorithm better suited to the space environment than the Fast Forum Transform used today. Mr. Burke argued that we must change the way we approach software support to the starship from the earthbound systems engineering methodology of independent paths for software and hardware development to one that allows the building a complex computational ecosystem to address the unknowns of interstellar travel. Continuing a common theme of new approaches, Ms. Gates argued that new processes and procedures in using social media techniques to address starship communications must be developed and tested well in advance of launch. Wrapping up the panel, Mr.

Faddoul provided a much needed methodology for to track the volume of data being produced in support of the starship programs.

Two clear actions came for the panel papers: 100 Year Starship should actively encourage authors to being using Mr. Faddoul's methodology and to insist that universities begin requiring Karhunen-Loeve Transform in appropriate engineering courses.

Track Chair Biography

Ron finished a 50-year career with the US Intelligence Community in May 2013. Ron received the Civilian Defense Meritorious Award upon leaving the NSA to begin working with Scitor Corporation as a systems engineer technical advisor to the NSA and the National Reconnaissance Office (NRO) on system development and data processing. Ron left Scitor to go to work for Riverside Research as a senior advisor for five years to the National Geospatial-Intelligence Agency on technology developments for mission execution and then moved to support the NRO on policy and management issues.

Cutting the Umbilical Cord: Designing Trustworthy Software for a Starship

David Burke

Galois, Inc.

e.david.burke@gmail.com

Abstract

In the roughly half-century that humans have been writing software, how have we built trustworthy software that supports mission-critical scenarios? The answer is that we've taken a systems engineering approach that was a combination of testing, review, and formal verification. This approach, however, is not scalable to an interstellar voyage given that communication lag times will be measured in years; the impossibility of adequately anticipating all of the relevant mission eventualities (with software correctness to cover all of them); and, that our current practices cannot support the maintenance of what will likely be trillions of lines of code.

So if scaling up today's methods isn't adequate, what other kind of approach would be successful? This paper argues that programmers need to adopt bio-inspired approaches and architectures. In particular, the paper proposes looking closely at the processes of symbiosis and homeostasis as guiding principles, as well as blurring the line between hardware and software. The resulting approach is called "programming Bernard machines."

1. Introduction

This paper makes the case that the conditions of interstellar travel will require "cutting the umbilical cord" with respect to the software infrastructure: unlike current space vehicles, which only get software updates and enhancements after rigorous ground testing, a starship is going to be on its own. And that necessitates a radical rethinking of how software is built so that it can counted on during such a voyage.

The argument is built around a set of underlying assumptions:

- No faster-than-light transportation of either matter or energy will exist; therefore any communications between an interstellar vessel and home will have lag times measured in years;
- Current software engineering best practices won't scale up to support the maintenance of what will certainly be trillions of lines of code embedded into systems of dizzying complexity; and,
- It will be impossible to adequately anticipate all the relevant mission eventualities, hence, the correct software behavior cannot be defined before the voyage - it is too situational-dependent.

Taken together, these assumptions say that not only can ground support be relied upon to keep software systems, but that applying current best practices in software construction is not good enough, either. To address this requires "cutting the umbilical cord," and designing the systems for a starship so that they aren't dependent on support from Earth. Software is generally the key complexity bottleneck in developing complex systems; this paper suggests a new approach to that would help in the building of trustworthy software for a starship.

2. Epochs of Engineering

In the book "Engineering Systems," [1] the authors divide the history of modern engineering design into three epochs. These epochs, taken together, hint at a possible future of engineering design that is applicable to the challenge of building software for a starship.

The first epoch, the "Epoch of Artifacts and Great Inventions" starts in the late 19th century, and ends in the early 20th century. Engineering advances in this period are exemplified by artifacts such as the light bulb, the telephone, and the automobile. In general, these inventions are the fruits of significant advances in specific branches of engineering: mechanical, electrical, chemical, and so forth.

The second epoch, the "Epoch of Complex Systems" appears in the mid-20th century, and is characterized by networks of transportation, energy and communications that support the various inventions and artifacts from the first epoch. This leads to the creation of networks such as the interstate highway system, power grids, and the space program. Creating these networks, and keeping them running results in the rise of standards and protocols, and the appearance of diverse engineering "illities": quality, reliability, and maintainability.

Software is a child of the second epoch, and it is worth noting that the phrase "software crisis" was first used as far back as 1968 at one of the very first international conferences on software engineering. Software was turning out to be a seductive medium in which to build complex systems but even fifty years ago, both researchers and practitioners were dismayed to find that software was not only hard to write, but hard to get right. Compared to other branches of engineering, something about the nature of software made errors easier to make, and harder to find and fix. These challenges led to the establishment of research programs devoted to the idea of applying mathematical rigor to software design – the idea of having a formal proof of correctness for your code was very compelling.

Another signature outcome of the second epoch was the development of the field of systems engineering. The Apollo program of the 1960's is really the epitome of the systems engineering approach to design. As Joe Shea, head of Apollo Spacecraft Program Office put it "You may not know every detail, but you should construct an environment where nothing can fall through the cracks. For Apollo, that was rigorous ground test and closing out all action items before the real, irreversible danger begins at launch."[2]

Shea's call to arms is inspiring and ambitious. But as he states, Apollo relied heavily on analysis, testing and ground support in a way that a starship won't be able to. Additionally, systems engineering as practiced in the second epoch has a bias towards "closed worlds" where all the salient factors can be identified, quantified, and managed; this kind of omniscience is more than we can reasonably expect when designing for interstellar travel. And in fact one of the lessons learned from the second epoch is a sobering one: complex systems inexorably generate unintended consequences. Before networks of roads were built, nobody anticipated gridlock, or foresaw the effects of car emissions on the quality of the air we breathe. Dealing gracefully with unintended consequences will be a key challenge for an interstellar voyage.

The third epoch is the "Epoch of Engineering Systems," and it begins in the mid-to-late 20th century. In various domains such as communications, transportation, energy and finance, we began constructing "systems of systems" – multiple networks that are knit together. These entities typically contain literally billions of lines of code, and can be described as "partially designed and partially evolved." The Internet is a good example of this: on the one hand, you had designed protocols such as HTTP and HTML, while at the same time; the actual growth of the Internet took place in a very bottom-up fashion. The explosion in the number of websites and servers over the past couple of decades didn't happen as the result of some top-down plan, but rather as the aggregate of millions of local decisions.

One key feature about these "systems of systems": they have been described as "robust yet fragile" [3] as a way of characterizing the fact that these systems are designed to be robust against perturbations, and yet, they can be hypersensitive to rare or unexpected events: they don't necessarily fail or degrade gracefully. Part of this is due to the fact that systems of systems are not simply technological artifacts, but socio-technical entities. Which means that they are subject to influences and perturbations from politics, economics, culture, and geography, and these interactions are remarkably hard to predict?

Looking at the progression between engineering epochs, there are some clear trends that suggest what a fourth engineering epoch would look like. It has been argued that we are heading towards a convergence between the engineering of complex systems and biology, suggesting that the next epoch is the "Epoch of Evolving Ecosystems." This convergence would manifest itself in the following ways:

- Designs would favor robustness and resilience over any other attributes;
- Designs would manage complexity through modularity, with well-understood protocols and interfaces to allow for continual growth and enhancement;
- Designs would feature continuous feedback at multiple layers and timescales;
- Designs would not simply be highly configurable, they would be "evolvable," in the sense of rules that can alter rules; and,
- Instead of system development and system use being sequential, they would run in parallel: significant change and growth of the system are possible even while it is operating.

The Promise and Perils of Bio-inspired Design: If the notion that a future convergence between engineering and biology is inevitable, then this suggests that even now, it would be worthwhile to adopt a "bio-inspired" approach to design. And in fact there are already plenty of books on this subject [4].

The field of "Artificial Immune Systems" (AIS) offers an illustrative example of this approach. For computer scientists working in the field of computer security, the human immune system offered an inspiring example: the immune system seems to have the ability to accurately distinguish between 'self' and 'non-self', and it uses its pattern recognition abilities to keep invaders and pathogens at bay. For instance, in the antigen receptors for B cells and T cells, there is "enough potential genetic diversity to recognize any conceivable molecule in the universe."[5]

Researchers set out to build intrusion detection systems from algorithms that were derived from immunology. [6] There were some initial successes, but these systems kept hitting a wall – the digital representation of, say, receptors and clonal selection algorithms turned out to be much more brittle than their true biological counterparts. This led to a leading AIS researcher to lament, "It is becoming increasingly apparent that the biological inspiration behind AIS has been somewhat naïve and … the majority of bio-inspired paradigms have been guilty of reasoning by metaphor approach."[7]

The perils in bio-inspired design are not just limited to the challenge of avoiding a naïve or superficial application of a bio-inspired metaphor or design principle. There is also the problem of discerning just what the ruling metaphor really is in a biological system. For instance, our description of the immune system hews closely to the conventional wisdom that the immune system functions as a fortress of sorts: its job is to maintain a perimeter, and keep the invaders out.

But what if the fortress metaphor is the wrong one? In recent decades, immunologists have pointed out that the self/non-self-distinction at the heart of the standard model of the immune system has its limitations. For example, when you ingest food, or become pregnant, your immune system doesn't leap into action to repel the 'invader', so perhaps your body isn't making a simple self/non-self-calculation. These, and related observations have led to the formation of the 'Danger Theory' of immunology [8]. In this framework, the immune system is looking for signs of threat to the organism (typically manifesting itself as inflammation) – this is what triggers an immune response.

Or we might usefully say that the most salient characterization of the immune system is that it is a cognitive system [9]. That is, the immune system constructs an explicit internal representation of itself and its environment, and makes decisions in accordance with that representation. The point here is that researchers have many choices for the conceptual inspirations that provide the foundations to their system building.

The metaphors you choose will constrain your design; this makes bio-inspired metaphors a double-edged sword, offering both useful guidance, and potential self-deception. Notwithstanding the risk involved, we would like to offer two bio-inspired themes that we believe would usefully contribute to the development of software systems that are resilient enough to support interstellar travel: symbiosis and homeostasis.

3. Symbiosis and Homeostasis

3.1 Symbiosis

Symbiosis is the epitome of mutualistic behavior in biology, being defined as a long-term interaction of two species. It is now recognized that symbiosis is a major driver of evolutionary change, as significant as replication with mutation and recombination [10]. For instance, the mitochondria, the "engine" in our cells, are the descendants of free-floating bacteria that were captured by another cell, forming a relationship that turned out to be mutually advantageous.

Parasitism is a non-mutualistic form of symbiosis where the parasite benefits at a cost to the host. "Depending on the definition used, as many as half of all animals have a parasitic phase in their life cycles."[11] The science fiction writer Cory Doctorow has remarked "All complex ecosystems have parasites". Perhaps this isn't a bug, but a feature. Consider that symbiosis, in either its mutualistic or non-mutualistic form, offers another possible understanding of the immune system: the immune system evolved not to eliminate pathogens, but rather to select for symbioses in a microbe-rich environment. In other words, the immune system is more like a recruiter at an employment agency, not a gatekeeper trying to maintain some misguided notion of purity of self [12]. Symbiosis, as a bio-inspired metaphor, suggests a profound collaboration between a starship and its environment: instead of trying to protect itself from the harsh environment of deep space, starship designers should think in terms of exploiting symbiotic relationships with the ecosystem of outer space.

3.2 Homeostasis

Homeostasis is the capability of organisms to regulate properties inside a boundary within some range. For example, the human organism attempts to maintain a temperature near 98.6 degrees Fahrenheit. A more comprehensive definition states "homeostasis is a dynamic, persistent and specified state of thermodynamic equilibrium that does work to resist perturbations to it."[13]

Like symbiosis, the important role of homeostasis in biology has been underappreciated; however, its significance is undergoing a reevaluation, and in fact it has been proposed that homeostasis should be considered the "2nd law of biology", where evolution through natural selection is the 1st law.

A striking example of homeostasis is the large termite mounds in Namibia that can reach over 15 feet high. There is a colony of termites living in a nest below the mound. The fascinating thing here is that this mound is quite literally the respiratory organ of the colony – the lung, if you will. There is no master designer termite, but the aggregate of each single termite going about its work results in the creation of a huge structure that takes into account patterns of wind and sun to manage temperatures and gas exchange throughout the colony. The termite mound, then, merges structure and function, and is better thought of as a process: the mound is the literal embodiment of termite colony physiology in the environment.

This is another intriguing metaphor for designing software for a starship – strive for a self-organized, homeostatic process much like the termite mound's merger of structure and function. This homeostatic process cycles matter and transforms energy while maintaining its existence.

4. Future Software Design Strategies

4.1 Evolutionary Programming

To tie the narrative so far more closely to computation and software design we will look at the history of evolutionary programming and extrapolating into the future.

In traditional evolutionary programming, the basic idea is that you generate a population of digital organisms (phenotype) in accordance with an encoding (genotype). The environment tests these organisms against a fixed goal or target, and the fittest ones survive. A new generation is bred, and a new selection round begins. The process terminates when one or more organisms attain a sufficiently high fitness score – the underlying encodings of these winners represent solutions to the design problem.

Notice that the environment tests fitness directly against the phenotype (the organism), but that the process indirectly selects for high-quality genotypes (the underlying encoding). This separation of structure and function is very common in computer science, whether or not the algorithms stem from evolutionary programming. For instance, some enterprising researchers once constructed a computer completely out of Tinker toys that plays perfect tic-tac-toe. The point is that you can take a specified function (in this case, an algorithm for a winning tic-tac-toe strategy), and implement it using a wide variety of materials to create its physical structure – Tinker toys will do just as well as transistors, ignoring issues of size and speed. But as we noted with the termite mound example, nature doesn't separate structure and function.

There exist variations on evolutionary programming techniques that could be described as "co-evolutionary programming". In the standard paradigm, the environment is fixed, in the sense that there is a static target or fitness function. But instead, we could set up conditions in which the fitness function depended on the attributes of the population at that time. This means that an organism that is very successful in one environment might still do poorly when surrounded by a different population. We can also conclude that fitness is not an absolute measure, but a relative one. If we are working in a complex domain, such as an ecosystem, this is a more realistic assumption – success is generally relative to a particular environment, and virtually all real-world environments are dynamic, not static.

The goal in co-evolutionary programming exercises is to understand how populations and environments are coupled together in time – how do they alter each other? Under what conditions will they sustain each other? These are different goals than for traditional evolutionary programming. However, the two paradigms still have a great deal in common: the operations of (digital) recombination and (digital) mutation are still used to create population diversity, and structure and function are still conceptually separate. And finally, adaptation occurs at the genotype/phenotype level – the overall evolutionary process itself doesn't evolve, in contrast to biological systems, in which structure, function, and process all undergo change over time.

Evolutionary programmers could accurately call these digital organisms either 'replicators' or 'Darwin machines'; each term captures the idea that these entities exist to make copies of themselves, and undergo some form of Darwinian natural selection. But what if we were to develop a conceptual framework that goes beyond so-called Darwin machines? What new design possibilities would that open up?

4.2 Programming Bernard Machines

Claude Bernard was the first scientist to write about homeostasis, making him the inventor (or discoverer) of the concept. In honor of Bernard, it has been suggested that the term "Bernard machines" refer to the conceptual machines that realize homeostatic processes. Bernard machines create membranes/boundaries, and within these boundaries, generate and sequester new environments, imposing homeostasis on them [14]. We claim that the fundamental challenge of 21st century software design is for us to learn to build and program Bernard machines.

There is already some movement in this direction, even though the practitioners might not recognize it as such. You can find active research communities thinking, writing, and doing computational experiments that have been variously called "emergent engineering" or "morphogenetic engineering" or "self-assembly" [15,16,17]. These are all different ways of naming and tackling the same problem. In a very real sense, these research groups, and others, are taking initial steps towards the realization of actual Bernard machines.

What will Bernard machines look like in practice? Here are some of the key attributes we should expect to see:

- Large numbers of loosely coupled agents/entities that search, communicate, and connect (either cooperatively or competitively) in a bottom-up fashion;
- Strong modularity, with well-defined interfaces between components to support the development of new capabilities;
- Feedback that takes place at all levels: positive feedback for generation of new behaviors, negative feedback for stability;
- A merger of structure and function into a process where the two can't be meaningfully separated;
- "Evolvability" – optimality is dynamic, and the system is able to draw upon "rules that change rules" to adapt; and,
- "Nature abhors a gradient": Bernard machines are effective gradient reducers, and actively seek out new energy gradients to exploit.

One of the important implications of the merger of structure and function is that it won't make conceptual sense to distinguish between software and hardware in this paradigm. Instead of "liquid" software running on top of some static hardware substrate, we have "self-organizing systems all the way down". Bernard machines aren't programmable in the traditional sense that we program software: programming a Bernard machine entails (simple in theory if not in practice) setting up the initial conditions from which the Bernard machine's self-organizing processes start to unfold.

5. Conclusion

At a high-level, we have defined Bernard machines as the instantiation of the dual, self-organizing processes of homeostasis and symbiosis: homeostasis representing the embodiment of the organism in the environment, and symbiosis representing the embedding of the environment in the organism. So in our vision of starship design via Bernard machines, we can say that in a very real sense, a starship is a living machine.

But does a "starship as a living machine" actually solve the problem of trustworthiness? In the world of software engineering there has long been a tension, even a rivalry between two schools of thought. And even those these schools of thought do not have to be mutually exclusive, in practice, you find that software engineers tend to give allegiance to one camp or the other. One school of thought can be summarized by "the best way to get high assurance software is to have rigorous, mathematical proofs of program correctness. Formal methods are the way to go." The other school of thought says that the way to go is have good processes. To simplify their position, it would be something like "get the processes right, and you maximize the chances of a desirable outcome."

Of the two camps, the formal methods group has more cachet among computer scientists. After all, many computer scientists have had significant mathematical training, and the concept of mathematical proof is celebrated as one of the finest achievements of the human intellect. So it's tempting to believe that you can reduce the messiness and complexity of the world to a model, and use that model to make mathematically sound claims. And in fact this is the same drive for predictability and control that explains Joe Shea's admonition that systems engineering success comes from letting nothing fall through the cracks.

So perhaps humans prefer proof. But nature appears to be firmly on the side of process: by giving up control and complete predictability, nature gets in return increased complexity and growth. For human beings looking for trustworthiness in our artifacts, we're going to find that over the long run, trust comes from the deep knowledge of how to design unfolding processes – programming Bernard machines. A homeostatic, symbiotic starship will be one of the finest examples of our newfound capability.

References

1. de Weck, Olivier L. & Roos, Daniel "Engineering Systems", MIT Press, 2013
2. Murray, Charles & Cox, Catherine Bly "Apollo: Race to the Moon", Simon & Schuster, 1989
3. Csete, Marie & Doyle, John C. "Reverse Engineering of Biological Complexity", *Science* 295, 1664 (2002)
4. Floreano, Dario & Mattiussi "Bio-Inspired Artificial Intelligence: Theories, Methods, and Technologies", MIT Press, 2008
5. Sompayrac, Lauren "How the Immune System Works", Wiley-Blackwell, 4th edition, 2012
6. Dasgupta, Dipankar, editor "Artificial Immune Systems and Their Applications", Springer, 1999
7. Timmis, Jon & Hart, Emma "Application areas of AIS: The Past, The Present, and the Future", *Applied Soft Computing,* Vol 8, Issue 1
8. Matzinger, Polly "The Danger Model: A Renewed Sense of Self", *Science* 296, 301 (2002)
9. Cohen, Irun H. "Tending Adam's Garden: Evolving the Cognitive Immune Self", Academic Press, 2004
10. Margulis, Lynn & Sagan, Dorion "Acquiring Genomes: A Theory of the Origin of Species", Basic Books, 2003
11. "Parasitism", Wikipedia, http://en.wikipedia.org/wiki/Parasitism, retrieved 10/15/14
12. Sagan, Dorion "Cosmic Apprentice: Dispatches from the Edges of Science" University of Minnesota Press, 2013
13. Turner, J. Scott "Biology's Second Law: Homeostasis, Purpose, and Desire", appearing in "Beyond Mechanism", edited by Brian G. Henning & Adam Scarfe, Lexington Books, 2013
14. Turner, J. Scott "The Tinkerer's Accomplice: How Design Arises from Life Itself", Harvard University Press, 2010
15. Doursat, Rene & Sayama, and Hiroki "Morphogenetic Engineering: Towards Programmable Complex Systems, Springer, 2012
16. Ulieru, Mihaela & Doursat, Rene "Emergent engineering: a radical paradigm shift" I*nternational Journal of Autonomous and Adaptive Communications Systems (IJAACS)*, Vol. 4, No. 1, 2011

17. Pelesko, John "Self Assembly: The Science of Things That Put Themselves Together" Chapman and Hall/ CRC, 2007

Loading, Please Wait: Long-Distance Crisis Communications

Jaym Gates

Science Fiction and Fantasy Writers of America, sfwa.org

jaym.gates@gmail.com

Abstract

Communication in space, barring intense further advancements, will require a great deal of forward-thinking and planning, much like a military campaign. For the first few missions, the inhabitants of any expedition will be advance operating groups, nearly cut off from any support or rescue. They will need two-part communications plans: the ability to handle the immediate issues on board with only the resources at hand, and a longer-term communication with the home base.

This paper addresses some of the communications challenges presented by interstellar travel, and possible ways to address them, based on current military and space travel procedures.

Keywords

communication, systems, interstellar, space, community

1. Introduction

"In the beginning was the word..."

Communication forms the foundation of life. From the simple "reproduce" command, to complex interactional communications, it is what keeps the world moving. Humans, in particular, have built their supremacy on complex communication systems. The biggest leaps in human advancement have been on the back of communication breakthroughs: the written word, the printing press, the telephone, the internet. At the same time, those were brought about by the pressures of a growing, globalizing civilization. Communication and human evolution go hand-in-hand, an infinite loop.

Now, we find ourselves at the edge of a major cultural evolution, one which will, by necessity, drive a new evolution in communication, as well. And no, that isn't a question, but a certainty. Not only do we not have the technology or systems in place, we are only now beginning to put together the understanding of what that system will need to do.

Nor is it a question of whether or not we will, as a species, move onward from this planet, only a question of when, how, and what we will be leaving behind.

> "If we are to send people, it must be for a very good reason - and with a realistic understanding that almost certainly we will lose lives. Astronauts and Cosmonauts have always understood this. Nevertheless, there has been and will be no shortage of volunteers."
> —Carl Sagan, *Pale Blue Dot: A Vision of the Human Future in Space*

To create the tools to become an interstellar civilization, we have to develop a culture not only accepting of such a future, but one that demands it. We must be obsessed with the final frontier again, in a way that puts to shame any previous such interest. That obsession must infect our media, our creativity, and our very language.

Not if, but when.

And as we develop that culture, we must also be learning, ourselves, how humans communicate. It isn't enough to understand the modern methods of communication,; we must look at earlier examples of communication and community as well.

The successful management—or even survival— of any crisis is dependent on clear communication, and fast response. Calm, organized, prebuilt communications systems are a must if we are to survive the infancy of off-world civilization.

Even simple in-system, off-planet travel will be fraught with never-before-experienced challenges, much less the demands of actual interstellar travel. The conditions our early off-world efforts will lead to higher stress, more pressure, and a more explosive situation during a crisis, and the larger the team, the more stress points and vectors for crisis. In space, the results of those explosive situations could be the death knell of anyone trapped in them.

Crisis, tragedy, isolation, and stress will create a volatile environment, one that does not exist in the understanding of most first-world humans. To adequately prepare explorers, we will need to understand the psychology, communication, and community systems and development of war-torn countries, early pioneers, refugees, prisoners, and high-risk military operational groups. We will need to study and develop the communities that surround our risk-takers, early adopters, and creators.

Our history is filled with grand initiatives and hopeful reaches into the unknown, but nothing like what we're now contemplating. To launch our species into the stars, we will first need to win their hearts and minds, and then develop those hearts and minds to create the systems that will support an interstellar civilization.

And all of that necessitates a robust, agile, and integrated communications system, one that reaches well into both the past and the future. One that will require and has a substantial understanding andof the evolution of technology, psychology, creativity, and science, hand-in-hand.

2. An Overview of Crisis Communications

> "The report covers 15 years of failures and gets highly technical, but it's clear that poor communication, information overload, and bewildering PowerPoint slides take much of the blame for the loss of 13 lives."
> —"Two Misleading Words Triggered GM's Catastrophic Communication Breakdown," Carmine Gallo, Forbes

A crisis is a constantly-evolving geography composed of three primary elements and their interplay: humans, technology, and environment. Successful management of a crisis is achieved through measured, flexible, informed response to that changing landscape. It's all about the information, and the clear flow of that information.

The first 72 hours of any crisis are particularly crucial: Gather relevant information. Create a plan. Spread accurate information. Respond to crisis. Resolve the crisis. Gather information to prevent a repeat of that crisis.

It's all about the information, and the clear flow of that information. The successful management—or even survival—of any crisis is dependent on clear communication, and fast response.

However, fast and agile response is hampered if the proper training and stress-testing has not been doneoccurred in advance. In many cases, there is little to no time to prepare for specific crises, leaving responders scrambling to build or adapt their systems on the fly.

Although we cannot hope to predict even the majority of the crises we will face in the next hundred years, there is still a great deal of preparation and planning that can be done. Planning ahead will reduce stress, fallout, and lag, even for unexpected crises.

There are certain steps that are universal and essential to any crisis scenario:

- Anticipate potential crises
- Develop flexible response plans and resources
- Identify your response team
- Identify and train your figureheads
- Establish and stress-test notification and response systems
- Create fallbacks and resource caches, plan for outages and interruptions
- Create analog backups in case of technological failure
- Consider long-tail management

While these things are easy enough to develop for on-Earth crises, interstellar travel poses additional challenges that we, as a species, will not have faced before. Much of our initial process will be trial-and-error. Early Americans, from the Pilgrims to settlers traveling the Oregon Trail, died by the hundreds, despite available resources. In space, if a resource runs out, it is out. Such conditions will lead to higher stress, more pressure, and a more explosive situation during a crisis. However; however, in space, the results of those explosive situations could be the death knell of anyone trapped in them.

Calm, organized, prebuilt communications systems are a must if we are to survive the infancy of off-world civilization, but equally necessary is the mental adaptability to respond quickly and creatively, perhaps without very basic resources.

3. The Many Faces of a Communications System

It is tempting to break communication down into its component parts, to separate it out. Crisis communications, in particular, are often subject to this isolation. Unfortunately, in doing so, we remove a great deal of the development and support provided by the other areas. To create a strong communications environment, there must be consideration for, and communication between, the following areas:

Cultural: As mentioned in the introduction, we must have a culture that is willing to make the necessary sacrifices. We have, for years, focused on the minimization of human risk—through advanced weapons, automation, safety standards, and policy-making—but that will not be an option in the early days of our interstellar exploration. Things will go wrong. We will lose people and machines. We may lose many people, even if we never find another race, even if we do everything 'right'.

Culturally, we must be prepared for this certainty, and we must have the societal and organizational resilience in place to overcome it without losing our hope for the stars.

Community: The strongest support systems are community-based. Those communities may come in many forms—families, businesses, shared interests or backgrounds—and each will respond with micro differences and macro similarities.

As we start to look at generation ships and other large-scale exploratory efforts, communities will need to be educated and empowered to handle themselves in crisis.

Institutional: Institutions will have multidirectional communications patterns to be aware of—internal, internal to external, external to internal. The more central they are to a crisis, the more robust and diversified their plans and systems will need to be.

Personal: The most potential for improper disaster response comes on the personal level. The perception is that only the response or PR teams need to understand the situation or be briefed on how to deal with the press, families, or social media. However, the peripheral characters, due to their lack of coaching, are often the direct targets of those looking for unrehearsed answers.

4. Systems

> The Federal Emergency Management Agency (FEMA) wrote in its 2013 National Preparedness report that during and immediately following Hurricane Sandy, "users sent more than 20 million Sandy-related Twitter posts, or 'tweets,' despite the loss of cell phone service during the peak of the storm." [2]

As we begin to explore the stars, crises will happen, and communication will make the difference between a disaster being the spur to better technology and higher safety standards, or destroying humanity's desire to risk that final frontier. Yet, with so much necessity for it to be robust, we are still hard-hit by very mundane disasters.

We can see this in our recent natural disaster history. During Katrina, the communications infrastructure failed badly, as did local and governmental systems, leaving residents stranded without a line to the outside, or even basic information. The failure only became more critical as agencies began responding. According to the FCC, "more than 2,000 police, fire, and emergency medical service personnel were forced to communicate in single-channel mode, radio-to-radio, utilizing only three mutual aid frequencies." [3] These overloads hamstrung the response and recovery. The report also concluded that many of the problems were caused by a lack of training on the equipment and technology that could have aided the efforts, and called it "perhaps the most widespread critical infrastructure collapse that any advanced country has experienced since World War II".

Hurricane Sandy showed little improvement in the official systems, but a new player had entered the field, something in its absolute infancy during Katrina.

Social media showed its potential for crisis communications management as Sandy began to descend. People around the world watched through pictures and videos. The stress and fear was palpable, the impending sense of doom as the skies darkened and people rushed for final supplies. Reports of flooding, structure damage, and power outages flooded the feeds.

And then, through that long night, we watched through social media as the city went dark and the feeds became quiet. It was intense and terrifying from the outside, too.

But within the city, social media was providing a lifeline. As the communications and emergency response systems crumbled, people used social media to communicate with loved ones . . . and service providers used it to identify the worst damage and stress points. "At one point during the storm, we sent so many tweets to alert customers, we exceeded the [number] of tweets allowed per day," said PSE&G's Jorge Cardenas, vice president of asset management and centralized services.

However, a lack of technological training still hampers both civilians and responders, and the repeated collapse of communications systems during disaster tells us that something needs to change even to cope with our current needs, much less the growing threat of massive natural disasters.

With our current systems, we cannot even dream of leaving our planet in any numbers, much less our solar system. To reach any interstellar capacity, we will need to create massively better communications. To survive interstellar demands, we will need new technology, new systems, and, for the early years, near-autonomy of communications hardware. We will also need to provide widespread training of both staff and civilians in the use of these systems and tools, as well as redundancies and fail-safes.

The temptation, when looking at any sort of evolution, is to make it more complex, more agile, and more adaptable. The problem with complexity, however, is the support system that must be built around it. More humans with more specialized knowledge, more resources, more money, more chance for things to go wrong.

However, with the minimal resources and the extensive strain of distance and time, we will need to consider simplicity and redundancy, too. The more self-sustaining it is, the better its chances of functioning under the extreme demands that will be placed on it.

Sometimes, the strongest communications system is a bullhorn in the hands of a trusted community leader.

5. Community Preparedness

In July, 2014, a car backfired in the dry grass alongside a hill-country highway in Northern California. Within a few hours, most of the region was under mandatory emergency evacuation. The speed and unpredictability of the Sand Fire in the middle of ranch country meant that many people were forced to abandon livestock, pets, and personal belongings unsecured, as well as to leave without adequate resources.

The community had not been faced with such an abrupt crisis in the past, allowing for a close look at the process of a community mobilizing and responding to disaster.

Within three days, the community's page on Facebook had been closed to all outside posts, and was being used entirely to manage housing, supplies, information flow, and animal care, both for the evacuees, and the response teams.

Community members stepped in to provide vital support for the official personnel, using their skills to fill in the gaps, allowing fire teams to focus on defending structures and suppressing a volatile and aggressive fire. Leaders and managers were identified, taking on roles in directing the resources and needs.

About two months later, an arsonist triggered another fire on the other side of town. This one grew to monstrous proportions overnight, eventually burning nearly 100,000 acres within two weeks, and commanding over 8,000 personnel from as far away as Alaska and Hawaii, a larger resource pool than any other wildfire in United States history. The smoke plume was visible from the other side of the state... and from space.

This time, within three hours, the community had rallied the systems tested during the Sand Fire. Posting on the community page was restricted to moderators, who carefully managed daily evacuation notices, evacuee resources, fire movement, air quality, news, and animal care. They rotated shifts, monitoring the official channels, interfacing with city, county, and crisis officials.

These systems supported over three thousand evacuees and their stock and pets, as well as over eight thousand firefighters and their support personnel, over the course of three weeks.

They were so successful that the news organizations were turning to the page moderators for information. As soon as the containment began, the same resources slid smoothly into place to assist the recovery.

None of these people have had any training in crisis response or management.

5.1 Community Command Structures

Any community under pressure will, over time, develop some form of command structure. This may go against established command strategies, or develop as a sub-structure of them. This structure can be a community's best asset or worst liability. It is not always the best-informed or most-competent person who takes the lead, and if someone without the correct skills and knowledge is in charge when a crisis arises, catastrophe is likely.

In such cases, the community will find leaders who are capable of responding decisively and accurately. It is, therefore, wise to ensure that, in a situation as potentially deadly as off-world travel, these assets and liabilities are discovered and developed early on.

Additionally, the question arises: what is the breakdown of civilian versus military assets? If the community is primarily civilian, there will be a longer response time as assets are identified and leaderships are developed, but there is a potential for faster individual response and stronger community involvement in response and recovery.

If it is primarily military, a great deal of that training will be in place already (making an excellent argument for prioritizing military background in the selection process for early missions). Stress will be expected and familiar, even if the circumstances are not.

Better yet is a community with a strong veteran presence. They are already integrated, and bring a wide variety of backgrounds to the table, but have the training and mindset to respond to high-stress situations.

Additionally, this would minimize stress and maximize communication between civilian and military components to any mission. In a crisis, the military component on a mission or ship is the best-suited to response, but that will require clear communication and cooperation with and from the civilian element.

But no matter the strength of the command system, emergency response will fail if it is not coupled with equal and cooperative support from within the community. Training and empowerment must be provided to all levels of the structure.

6. Social Media, News, and Crisis Culture

With the rise and spread of social media, individuals may now have more voice and reach than ever before. While this has many positives, care must be taken to plan for the negatives, as well.

Every individual will have family, friends and loved ones, but they may now also have a strong following on social media. Look, for example, at "Mohawk Guy" from the Mars Curiosity mission. An instant and unexpected celebrity, sought out for interviews, guest appearances, and more.

Similar things will happen in a crisis. We can't predict the spokespeople of a crisis, but we can identify hubs of communication and information. In any group, there are people who connect important resources to create a clear and concise stream of information. These people must be given the resources and training necessary to act as hubs and support structures.

One of the biggest issues we run into in the intersection of media and crisis management is the virulent spread of misinformation. While we've always had, as a race, a tendency to favor sensationalism over fact, we have, in the past, at least had some control over the spread of information, and a relatively small pool of outlets to be monitored.

With the advent of the internet, village gossip became a business, and the number of outlets to monitor grew exponentially. With the advent of social media, those outlets exploded. Everyone's an expert now, and it is harder than ever to control and maintainmaintains an accurate flow of information. While that can be useful—as demonstrated by the Arab Spring, in particular—it is potentially lethal in a crisis, and the misinformation can be difficult or impossible to contain and correct.

During Hurricane Sandy, misinformation and scams were cropping up everywhere, and being spread by well-meaning but uninformed people. FEMA responded by setting up a page to combat and correct these rumors, combing social media to catch as many as possible, but it is never possible to catch them all. Even one uncorrected rumor will add stress to the response teams. [4]

Additionally, in the case of an interstellar disaster, the world will have its eyes firmly fixed on every detail. Speculation, accusation, and misinformation will run rampant, and the resources necessary to counter all of that would simply be impossible to maintain. Overly-direct, negative response can, all too often, appear to be defensiveness, damaging the communication and trust between the parties.

We may be fortunate, and disaster may not happen for decades or longer, but it will happen eventually. In the event of a massive disaster, it is necessary to create an immediate response plan, but to gear the public relations efforts on a much larger scale. A long-term plan to regain trust and support is necessary before a single human leaves orbit again.

The strongest public relations plan will not only prepare for the crisis itself, but understand that the recovery effort is on a much slower timeline than the initial, negative response. Repair may take weeks, months, even years, but handled correctly, it could become a net gain.

7. Conclusion

In conclusion, it is clear that it is not only our technology that holds us back, but our very perception of the importance of communication. Until we are able to take the time to overhaul our communications systems, to really explore and understand them before creating better, stronger resources, we do not dare reach much further than we already have.

For now, we must ask the hard questions about where our priorities lie, and if we are willing to make the cultural and political changes necessary to ensure that our first reach to the stars isn't the last we'll see in our lifetime.

References

1. http://www.scientificamerican.com/article/how-social-media-is-changing-disaster-response/

2. http://www.carlisle.army.mil/DIME/documents/Hurricane%20Katrina%20Communications%20&%20Infrastructure%20Impacts.pdf

3. http://www.carlisle.army.mil/DIME/documents/HurricaneKatrina/Communications.pdf

4. http://www.fema.gov/hurricane-sandy-rumor-control

KLT Filtering, Even from Relativistic Sources

Dr. Claudio Maccone

Istituto Nazionale di Astrofisica, Italy

clmaccon@libero.it

Abstract

This paper looks forward to the time when our spacecraft achieve relativistic speeds and discusses the impact of relativistic motion on spacecraft communications The paper addresses a key element for interstellar communications; using the Karhunen-Loève Transform (KLT) mathematical algorithm for communications processing. KLT is argued to be superior to the classical Fast Fourier Transform (FFT) used in Earth-based communications today.

1. Introduction

One of the most critical elements of interstellar travel will be reliable communications between the Earth-base and the spacecraft for the sending and receiving of information by interstellar travelers (manned or robotic). The extreme distances and relativistic motions involved with interstellar travel will have negative impacts on signal strengths, timing and composition, requiring high quality processing to retrieve the data of interest. The KLT can filter signals out of the background noise in both wide and narrow mobile bandwidths, providing the foundation for earth-based communications with interstellar spacecraft. That is in sharp contrast to the FFT that rigorously applies to fixed narrow-band signals only.

2. Details

The concept of the Karhunen-Loève Transform is essentially a transformation to the principal axes in Hilbert space (an infinite-dimensional analog of Euclidean space). For signal processing, using the KLT allows the extraction of weak signals from noise of any kind over wide or narrow bandwidths. The signal extraction can be achieved by the KLT more accurately than by the FFT, especially if the signals buried into the noise are very weak, in which case the FFT fails. This superior performance of the KLT happens because the KLT of any stochastic process (both stationary and non-stationary) is defined from the start over a finite time span ranging between 0 and a final and finite instant T (contrary to the FFT, which is defined over an infinite time span). Mathematically, the series of all the eigenvalues (a special set of scalars associated with a matrix equation. The determination of the eigenvalues is equivalent to matrix diagonalization) of the autocorrelation of the signal (noise + signal) may be differentiated yielding the "Final Variance" of the stochastic process in terms of a sum of the first-order derivatives of the eigenvalues, resulting in the immediate reconstruction of a signal buried into the thick noise. This allows a KL expansion on a noisy signal, followed by an examination of the first eigenvalue to get a measure of the signal to noise ratio. For a large first-eigenvalue the corresponding eigenfunction (each of a set of independent functions that are the solutions to a given differential equation) is a good representation of the signal. Applicable to broadband signals (where the Fourier transform is not particularly useful), the KLT provides a mathematical foundation for the optimal extraction of any data from any noise.

3. A KLT Example

As a signal is received at the input of a receiver, the KLT separates the input (noise + signal(s)) into uncorrelated components. This separation yields the Karhunen-Loève's eigenfunctions and the kernel of this integral equation is an autocorrelation. The KLT adapts itself to the shape of the input by adopting, as a reference frame, the uncorrelated component spanned by the eigenfunctions of the autocorrelation. This turns out to be a linear transformation of coordinates in the Hilbert space.

There is no degeneracy since each eigenvalue corresponds to just one eigenfunction, and, the eigenvalues turn out to be the variances of the random variables. By sorting, in descending order of magnitude, both the eigenvalues and the corresponding eigenfunctions, we can decide to consider only the first few eigenfunctions as the "bulk" of the signal and declare the rest of the values as "noise." Thus we have isolated the signal of interest.

4. The "Lack of Use" Issue

If N is the size of the autocorrelation matrix, the number of calculations required to find the KLT is of the order of N squared, while the same number of calculations for the FFT is much less: just N ln(N). This computational burden prevented scientists from replacing the FFT with the KLT until 2007.

That winter the SETI-Italia Group at Medicina discovered a way to circumvent the computational burden of the KLT. Called the Bordered Autocorrelation Method (BAM), the approach is to regard the autocorrelation as a new function of the KLT Final Instant (T). This is accomplished by adding one more row and one more column (i.e., bordering) to the autocorrelation matrix for each new positive eigenvalue, increasing the value of T. The idea is to use the BAM to exploit the dependence on the Final Instant on both sides of the relationship.

5. A BAM Example

Consider a pure tone (a sinusoid) buried in unitary white noise; write down the autocorrelations, assuming that the sinusoid and white noise are uncorrelated, and pull in the KLT of these two stationary processes. The expansion of both autocorrelations introduces the KLT eigenvalues for both processes and yields a new form of these KLT expansions from which unneeded terms can be deleted by executing two simple integrations from zero to T. The result, after the two integrations, is the relationship among the eigenvalues; it shows that by increasing "n" the two eigenvalues coincide, and that by increasing "T" the two eigenvalues coincide. But it also shows that the periodic part of the equation goes like the sine square of omega, i.e., with a period of twice omega, or most importantly twice the frequency of the signal of interest. In other words still, we have found the KLT eigenvalue of the pure signal as a function of "T". The frequency of the signal of interest is equal to half of the frequency of the inverse Fourier transform of the derivative with regard to T of the dominant KLT eigenvalue. This is the BAM-KLT method that allows for the isolation of frequencies of interest even for SNR very very low.

The Karhunen-Loeve Transform is not widely used today because of its mathematical complexity and processing costs; however, the development of BAM has offered a path of increased usage. Likewise, the advent of parallel computing techniques, coming in the near term, may revolutionize the algorithms for finding the eigenvalues and eigenvectors of symmetric matrices, enabling the KLT to become readily available to all scientists and engineers. 100 Year Starship should take on the task (perhaps in collaboration with IEEE) of encouraging more academic institutes to require courses in the KLT.

100 YEAR STARSHIP™

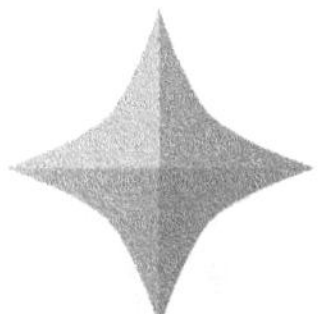

Chaired by Marianne Caponetto

Founder, MCW Group

and Dan Hanson, MILR, MCRP

Technology Innovation Group

Track Description

Technology progresses in small increments and by leaps and bounds. Often the biggest steps forward are through the invention and innovation required to meet grand challenges. Interstellar travel represents such a challenge that may spur new economies, combat climate change, address heretofore incurable diseases. This session asks "What are these innovations and how can we deploy these to enhance life here on Earth?"

Track Chair Biography

Marianne Caponnetto

MCW Group

Ms. Caponnetto's career has included the roles of global strategist, marketer and sales leader. As Founder of MCW Group, she acts as Strategic Advisor and Consultant with a focus on technology start-ups, Venture Capital and Fortune 500 organizations with significant transformational and growth objectives in both B2B and B2C businesses. She is also a Founding Partner in HyperBlue Lab, a technology IP consulting firm. From 2005-2008, as Chief Sales and Marketing Officer of DoubleClick, she was a key member of the management team that re-captured digital market leadership, resulting in the sale of the company to Google.

Track Chair Biography

Dan Hanson, MILR, MCRP

Technology Innovation Group

Dan Hanson has over 30 years experience addressing public policy, education, and economic development issues. His professional experience includes public private partnership establishment, investment portfolio management and risk management, capital raising through the bond and securitization markets, and public policy and program development for various government agencies.

Dan has a keen interest in leveraging art, science, and education infrastructure to promote economic development in regional economies. He helps build sustainable engines by forging partnerships among business, government, and nonprofit corporations. These partnerships often are a mix of organizations operating on a local, national, and international level. His efforts are realized through a combination of business and volunteer activities.

Dan is noted for his outstanding analytical and his commitment to promoting regional economic development and achieving equitable outcomes of government policy and program initiatives across communities. He has held volunteer board positions for local and state governments, nonprofit organizations, and foundations in Texas.

Dan is currently a principal with Technology Innovation Group, Inc. (TIG), which he co-founded in 2002. His company's mission is to connect innovation to societal needs. TIG pursues it mission through two primary service offerings: advising entrepreneurs, researchers, academics, technology professionals, and communities wanting to build technology-based economies; and serving as translational consultants with institutions and private companies to commercialize specific technologies, primarily those with public health or economic development benefits. TIG clients include local, regional, and national governments and economic development agencies, large corporations that manage portfolios of intellectual property, young companies that are developing products and services based on complex technologies, and universities and research institutions that desire to move discoveries from the laboratory to businesses.

Dan's academic qualifications include a Bachelor of Science and a Master of Industrial and Labor Relations from the University of the Oregon. He also attended the Kennedy School of Government at Harvard University and graduated with a Master of City and Regional Planning degree.

Molten Salt Reactors: A Case for Enhancing Life on Earth, and in Space

Kirk Austin Frenger

A Working Hypothesis, Inc, 814 Silvergate Drive, Houston, TX 77079

urdnotkirk@yahoo.com

Abstract

Today we face problems that could manifest themselves as real existential threats; energy is one of these. We seek newer, cheaper, cleaner and renewable sources of energy to wean ourselves from dependence on limited, carbon-based fossil fuels. Various spokesmen advocate reputed "green" technologies such as wind power, photovoltaics, and biofuels; unfortunately, these still are undeveloped and (like ethanol) energy inefficient. The author points out that a technological solution already exists, but its implementation has been hampered by formidable obstacles (government, industry and nuclear skeptics). The answer is Thorium power. Thorium is a much cleaner, safer, low radioactivity, plentiful nuclear fuel compared to the traditional Uranium and Plutonium. A variety of power plants can be constructed for this fuel: some such as the "molten salt reactor" (MSR) run at low pressures and high temperatures for failsafe operation. Some Thorium reactors breed more Uranium fuel and can permanently dispose of nuclear waste. This technology has numerous additional applications both on Earth and in space.

Keywords

thorium, power, nuclear, green energy, molten salt reactor

1. Introduction to the World Energy Problem

Today, there are over 7.2 billion people on Earth. Each person has needs: food, water, clean air, habitable land, sanitation, energy for warmth and transportation, education, employment and health care. Industrialized countries presently have the highest consumption of these resources, but developing countries may push demands much higher in the future. Any needed item may suddenly become a serious bottleneck, and competition for supplies could become problematic at any time (that is, threatening collapse of the regional economy with massive loss of life in their absence). Energy production is a case in point. In spite of calls for "green" (non-carbon based, renewable) fuels, there has been a recent worldwide increase in the use of oil and gas, continued use of coal and a retreat from nuclear power [1]. In 2011, world energy consumption was over 22 Terawatt-hours. We are literally burning up the Earth's combustible resources at an ever-increasing rate. This clearly is not sustainable.

2. The Failure of Green Technologies

The promise of non-carbon based "green" energy continues to hang-fire because of ongoing developmental problems. The most significant difficulty is that "green" proponents do not consider the environmental effects on all phases of the technology life cycle. This includes mining or collecting, purifying, fabricating, transportation and installation of various power conversion systems, their lifetime operation, then dismantling and recycling or neutralizing their toxic components. For example, Greenpeace International does not see any overall advantage of electric or hybrid vehicles at this time [2]. Wind power is hampered by a short equipment lifespan which leads to generator replacement before payback of the energy costs of their fabrication and installation occurs. Photovoltaics are limited by poor efficiency, expense of manufacture, and competition with agriculture for practical sunlit venues. Battery production relies too much on scarce elements and releases toxic heavy metals into the environment during manufacture and retirement. Sequestration of carbon dioxide from burned fossil fuels is impractical because of toxicity of the reactants, reduction of net energy produced when sequestration is employed, problems with suitable "safe" permanent storage sites, and huge costs [on December 4, 2014, Bloomberg television reported that carbon-capture applied to all power plants would cost $18 trillion]. Biofuels such as ethanol require more energy from petroleum in their production than they release upon combustion (ie: 3 barrels oil per 2 barrels ethanol), large amounts of fresh water to grow, and damage to equipment (such as automobile fuel systems). Industrial Hydrogen comes from the breakdown of limited reserves of petroleum or natural gas, not sustainably from electrolysis. Greenpeace does not consider traditional nuclear power generation to be either clean or renewable. The limited use of rebates by manufacturers or tax tricks by governments has not substantially altered the energy landscape. Something else is called-for now.

3. Nuclear Power the Thorium Way

Traditional nuclear reactors are expensive, dirty (because of the waste issue) and unsafe (one only needs look at Chernobyl, Three Mile Island and Fukushima). Governments in Germany and Japan are now moving away from pressurized water reactors. The United States, alone among developed nuclear nations, does not reprocess its spent fuel (which is 95% unused uranium). This results in higher fuel production costs and increased long-term radioactive waste storage problems.

Thorium is about four times as prevalent on Earth as Uranium. It consists almost completely as a single, low-radioactivity isotope (Th-232), a "fertile" but not "fissionable" material. Unlike Uranium, which needs additional U-235 for practical fuels, no isotopic enhancement is required. Th-232 is converted to fissionable U-233 by neutron capture and beta decay in a breeder reactor, with Protactinium-233 as a brief intermediate product. Although U-233 can be used mixed with U-235 in a conventional reactor, the best use of transmuted Th-232 seems to be in a molten salt reactor (MSR) where a fluoride salt is employed. MSRs run at or near atmospheric pressure, so explosions are not a problem. Their high temperature operation makes them more thermally efficient; cogeneration is a possibility. The fluoride salt does not react with air or water. The molten salt solution traps solid and gaseous fission byproducts. Fuel preprocessing into rods is not required. Its strong negative temperature coefficient makes the reactor self-regulating without control rods, and hot fuel can be drained from the core rapidly into a cooled reservoir in the event of a problem. Depending on reactor design details, they can be refuelled while operating (no shutdown required). MSRs use less fuel per megawatt of electricity produced. Thorium breeders work well, with an excellent neutron economy. Its wastes are dangerously radioactive for only about 300 years, compared to tens of thousands for conventional plants. Deliberate sabotage will not result in explosions or release of radioactive byproducts, which will be trapped in the cooled / congealed salt. MSR reactors are "walk-away safe" in that no power is required to cause shutdown; reactors are passively cooled and the molten salt can be drained by the melting of a fusible plug at the bottom of the reactor vessel when its active cooling fails.

Historically, the first MSR (the Aircraft Reactor Experiment) was designed for use in military aircraft. A 2.5 MW sodium-cooled core, it went critical November 3, 1954 and was revealed publicly in 1957 [3]. It operated successfully for 100 hours at 1600 °F. An advanced experimental civilian version of the MSR was built and went critical on June 1, 1965, operating successfully for 4.5 years. U-233, U-235 and Pu-239 were all used as test fuels at one time or another in this device. In 1974 the program was briefly restarted. Subsequent interest in MSRs in the United States waned in the face of a variety of obstacles (discussed below). There has been a recent resurgence of interest in these successful (if ancient, by nuclear standards) reactors. Kirk Sorensen, Flibe Energy, has been scouring the world looking for the elderly scientists and engineers familiar with these MSR experiments, to document this successful technology before their demise [4]. Another startup, Transatomic Power, has an MSR reactor on the drawing board which conforms to the old successful Oak Ridge National Laboratory designs, with detail improvements [5]. Thorium Power Canada and investigators in India, Russia and elsewhere are also researching Thorium. Groups such as these are racing to rediscover the success of the 1950's and 1960's before the passage of time destroys the remaining design and operational MSR information.

Thorium reactors can be made modular and scalable. Loren Kulesis, Laser Power Systems, has promoted a Thorium-powered Cadillac [6]. While cars may not be ideal applications for Thorium power, large trucks and locomotives are.

Cheap Thorium power would also make desalination of seawater economical for human use and for agriculture. As the need for fresh, unpolluted water supplies increase with rising populations, and as solar-driven cyclical climate change currently is reducing precipitation, desalination powered by Thorium MSRs placed near the coast could provide the water and avoid yet another existential threat to humanity. This application alone would justify the use of Thorium power [7].

Mankind should take Thorium reactors into space for powering large space vessels far beyond the ability of solar panels to be useful. Thorium retrieved from asteroids, comets and space debris picked up "along the way" on deep space missions will replenish fuel supplies. Thorium present on any other planets we colonize will be useful for terraforming and human settlement missions. Thorium-powered lasers will melt ice and rock, releasing oxygen gas. Water vapor may also be present in rock and can be laser-liberated.

4. Obstacles to Be Overcome

In spite of the success of the Aircraft Reactor Experiments of the 1950's, President Kennedy discontinued further MSR research after he took office [8]. The essential problem was that Thorium MSRs did not produce the Uranium / Plutonium isotopes needed for making nuclear weapons. Clear advantages for civilian use were ignored. Scaled-up variations of the pressurized water reactors used for Admiral Hyman Rickover's nuclear Navy became the model for all future atomic reactors, military and civilian [9]. In 1973 President Nixon fired physicist Alvin Weinberg (who held the patent for pressurized light water reactors) when the latter began promoting Thorium MSRs instead of the Administration's Liquid Metal Fast Breeder Reactor. Nixon cut funding for MSR research because (again) Thorium did not produce bomb-making materials [10]. Commercial reprocessing / recycling of plutonium from nuclear reactors was prohibited by President Gerald Ford in October 1976, and in April 1977 reprocessing of nuclear waste from reactors was blocked by President Jimmy Carter, fearful of nuclear proliferation [11]. In 1981 President Reagan stopped all further development on MSRs, rumored to be his reaction to the meltdown at Three Mile Island two years earlier. President Clinton supported the MSR, but only as a way to burn up excess Soviet-era Plutonium released by treaty (still a useful application). President Bush in 2001 did not fund further MSR research but revived the old Nixon Administration Liquid Metal Fast Breeder Reactor, but under a variety of different names. In 2006 Bush also terminated the MSR plan to burn up Soviet (now Russian) Plutonium. President Obama appears to support nuclear energy for the US, and a conference was held at Oak Ridge in October 2010 to explore MSRs, but nothing has come of this yet [12].

Not to go into fine detail, but established U.S. producers of conventional Atomic power plants, such as Babcock & Wilcox, General Electric and Westinghouse make it difficult for startups like Flibe Energy and Transatomic Power to take root. The Washington lobbying and financial support of political candidates by "big nuclear" is reflected in a scarcity of Federal research grants for alternative reactor designs. It appears that foreign companies, sometimes partnering with their governments (i.e.: China, India), will have the initiative in this area for the forseeable future.

5. Conclusion

Many "Green" energy solutions have been tried and found wanting: wind power, photovoltaics, biofuels and conventional pressurized nuclear reactors. None could live for long in the free market without the support of government in the form of subsidies, tax preferences and regulatory fiat. MSRs, a safer, cleaner nuclear alternative, was proven safe and effective 60 years ago. Plentiful Thorium could power Mankind's energy needs for 1000 years, by which time another source might be found (fusion?). Thorium should accompany man into deep space and onto future permanent colonies beyond Earth. The author suggests that this is the time for decisive action on MSR technology, before some natural or manmade disaster intervenes destructively to force our hand.

References

1. *World Energy Outlook 2012, Executive Summary*, International Energy Agency, pg.1. Available: http://www.iea.org/publications/freepublications/publication/English.pdf.

2. Greenpeace International, *Climate Change and Cars, Questions and Answers.* Available: http://www.greenpeace.org/international/en/campaigns/climate-change/cars/questions-answers.

3. Weinberg, A.M. "Preface: Molten-Salt Reactors", Oak Ridge National Laboratory, Oak Ridge, Tennessee, 30 August 1969.

4. Clark, D., "Thorium advocates launch pressure group", The Guardian Environment Blog, 9 September 2011. Available: http://www.guardian.co.uk/environment/blog/2011/sep/09/thorium-weinberg-foundation.

5. Strickland, E., "Start-up: Transatomic Power wants to build a better reactor; Its 'walk-away safe' nuclear reactor would run on spent fuel", IEEE Spectrum, June 2014. Available: http://spectrum.ieee.org/at-work/start-ups/start-up-transatomic-power-wants-to-build-a-better-reactor.

6. Jaynes, N., "This insane Cadillac concept remains a radioactive, laser-powered dream", *Digital Trends,* 21 January 2014, Available: http://www.digitaltrends.com/cars/cadillac-concept-powered-thorium-lasers-can-last-100-years.

7. I*ntroduction of Nuclear Desalination, A Guidebook.* Technical Report Series No.400, International Atomic Energy Agency, Vienna, 2000.

8. Energy from Thorium Foundation, "A Brief History of the Liquid-Fluoride Reactor", 22 April 2006. Available: http://energyfromthorium.com/2006/04/22/a-brief-history-of-the-liquid-fluoride-reactor.

9. Mahaffey, J., "Caught in the Rickover Trap", *Atomic Accidents: A History of Nuclear Meltdowns and Disasters,* Chap. 11, Pegasus Books, New York, 2014, pg.404-426.

10. Vey, G., "A Thorium Future?" *The View Zone,* 23 November 2014. Available: http://www.viewzone.com/thorium.html.

11. Nuclear Reprocessing, History. Wikipedia. Available: http://en.wikipedia.org/wiki/Nuclear_reprocessing.

12. Hoglund, B., "Why The Molten Salt Reactor (MSR) Was Not Developed By The USA", Moltensalt.org, 5 November 2010. Available: http://moltensalt.org/references/static/downloads/pdf/WhyMSRsAbandonedORNLWeinbergsFiringV3.pdf.

Removing Atmospheric Carbon Dioxide Using Terraforming Technology

Randall M. Chung

Initiative for Interstellar Studies

Laguna Niguel, California, USA

rmchung@gmail.com

Abstract

New genetic modification techniques can be used to modify organisms such as plants in ways that can be useful in building a closed ecology for star ships or for terraforming a planetary ecology at another solar system.

These same techniques can be used to help solve ecological problems on Earth. Global atmospheric carbon dioxide emissions have reached an estimated 35 billion tonnes in 2012, and the future level of CO2 may result in undesirable effects to Earth's ecosystem lasting for hundreds of years.

A new biological approach may be able to remove multi-gigatonnes of carbon dioxide from the atmosphere each year, at a cost much lower than other carbon dioxide removal approaches of similar size. Furthermore, this system could be implemented without the large capital outlays of other approaches.

In this approach, a pelagic seaweed would be genetically modified to add multiple new traits. Some of the genetic techniques that would be used include DNA sequencing to analyze existing plants, synthetic biology to generate desirable DNA sequences, and CRISPR/CAS to remove undesired sequences, silence existing genes, or to insert desired gene sequences. To provide as much environmental safety as possible, the lifetime of the plants can be limited to a set number of days. When the lifespan count is reached, the plant would die and sink, so that almost all of its carbon would be sequestered in the deep ocean for hundreds or thousands of years.

Keywords

Interstellar, genetic modification, carbon dioxide removal, pelagic seaweed

1. Terraforming a Planet

When we actually travel to another solar system some day, it may be desirable to terraform a planet to more closely resemble the Earth's environment. For instance, if we find a planet that lacks water, we might redirect water rich asteroids and comets to crash onto the planet, thus adding water. It might take a lot of time and energy to move enough asteroids and comets, but eventually it could be done.

For a planet that resembled primitive Earth, with plenty of liquid water and an atmosphere of carbon dioxide and methane but lacking oxygen, we might add cyanobacteria to the environment. Given enough sunlight and time, the cyanobacteria could convert most of the carbon dioxide and methane into oxygen, just as the Great Oxidation Event occurred on Earth 2.3 billion years ago [1].

For a planet that is too cold, greenhouse gases like methane or carbon dioxide might be added. The Earth went into a protracted ice age right after the Great Oxidation Event, because almost all of the methane and most of the carbon dioxide had been removed from the atmosphere by the cyanobacteria. The Earth warmed up and recovered from the ice age when atmospheric carbon dioxide levels increased after some of the cyanobacteria died off.

For a planet that is too hot, greenhouse gases could be removed, or space sunshades could be built. For a hot planet like Venus, the space sunshade might be the only approach that would work initially, until surface temperatures were reduced to the point where liquid water can exist and cyanobacteria could live.

There are few constraints in making terraforming changes to a planet that has no existing life, but there are enormous constraints in making changes to our own Earth, with the most important constraint being that our existing ecology should not be damaged by our terraforming efforts. The rest of this paper will concentrate on Earth's atmospheric carbon dioxide problem, and a way to mitigate the problem affordably, without damaging our environment.

2. Earth's Atmospheric Carbon Dioxide Level

The carbon dioxide level in Earth's atmosphere is estimated to be about 280 parts per million (ppm) in pre-industrial times, about three centuries ago. The first direct measurements of atmospheric carbon dioxide was started by Charles D. Keeling in 1958, when the carbon dioxide level was measured at 310 ppm with a growth rate of about .8 ppm/year. In 2014, the carbon dioxide level has reached 400 ppm with a growth rate of about 2.3 ppm/year [2], despite efforts in the last fifteen years to slow down the growth rate by increasing the use of renewable energy and the more efficient use of carbon and hydrocarbon based energy.

Someone born in 1944 would be 70 years old in 2014, and would have seen the atmospheric carbon dioxide level increase from 300 ppm to 400 ppm within their lifetime. If the atmospheric carbon dioxide growth rate does not slow down and stays at 2.3 ppm/year, the atmospheric carbon dioxide level will increase from 400 ppm to 500 ppm in 43 years.

The increase in atmospheric carbon dioxide level is definitely causing ocean acidification as the ocean absorbs some of the carbon dioxide [3], and the acidification is already affecting ocean organisms that form shells. It is difficult to predict the future effects on the ocean ecology, but it may have large scale effects. If negative effects do occur, it will be difficult to reverse the ocean's acidification. It may be possible to add chemicals to the ocean to raise the pH, but it will be extraordinarily expensive [4].

The increase in atmospheric carbon dioxide level is also increasing the Earth's average surface temperature, especially in the polar regions. There is much controversy over the precise amount of future warming and its timing, but many scientists believe that an atmospheric carbon dioxide level of 550 ppm could eventually cause at least a 2 degree Celsius rise in average global surface temperature, which could result in significant changes to the Earth's ecology as well as negative economic effects [5]. We may already be inadvertently terraforming our planet by burning so much carbon based fuel.

3. Biological Approach to Sequestering Carbon

In 2001, Metzger and Benford [6] proposed to sequester carbon by burying crop waste in the deep ocean. The low temperatures in the deep ocean would slow the decay of organic carbon back into inorganic carbon dioxide by bacteria and other organisms, a process called remineralization. Because of the slow rate of remineralization and the time it takes for currents to move the deep water up to the ocean's surface, carbon would be sequestered for hundreds or thousands of years [7]. However, it is expensive to collect and transport large amounts of crop waste from agricultural fields to a deep water ocean location, and there is also a limited amount of crop waste, which actually has many agricultural uses.

Instead, it may be possible to use terraforming techniques on our Earth to remove carbon dioxide, currently the dominant greenhouse gas, from our atmosphere and to do it in a way that does not disturb our ecology, to do it at relatively low cost, and be able to scale the process up to remove gigatons of carbon dioxide per year.

In this proposed approach, a pelagic seaweed (free floating macroalgae) would be genetically modified to grow quickly in wide areas where the ocean is deep and would be pre-programmed to die and sink, thus sequestering its carbon for hundreds or thousands of years. One particular ocean region, the Sargasso Sea, is named after the prolific floating seaweed Sargassum Natans and similar seaweed types. S. Natans can grow at the surface of deep ocean waters because it has no roots, unlike fast growing kelp, which typically grows close to shore. Kelp growth is restricted to waters less than 100 meters deep.

The S. Natans seaweed floats because it has many air bladders. To make it sink and sequester its carbon, the seaweed would be genetically modified to make the air bladders break off or split open on command. Many plants already have genes for the process called abscission, which the plants use to shed leaves in the fall, to drop dried flowers, or to drop ripe fruits. A similar process can occur in plants to order to open seams in flower buds or seed pods [8].

One approach to make the seaweed sink might be to add the abscission genes so that the air bladders all break off from the main plant when a triggering signal is generated. The air bladders would float to the ocean surface, and the bulk of the plant would sink to the ocean bottom. This approach might be preferable if the air bladders could be collected and re-used for some other purpose, or to recycle nutrients contained in the air bladders for use by other organisms.

A slightly different approach to seaweed sinking might be to have the air bladders split open to release the gas held in the bladder, so that the entirety of the plant would sink to the ocean bottom.

By controlling the seaweed buoyancy, the seaweed lifetime would also be limited, thus reducing the risk of uncontrolled growth.

4. Genetic Engineering Techniques

Adding this new functionality will require the use of genetic engineering techniques. Whole genome sequencing of various plants would allow the analysis of the DNA sequences in order to locate genes that can be copied from existing plants. Genome sequencing has dropped in price tremendously in the last 10 years. In 2004, the cost to sequence an 18 gigabase genome was about $20 million, and in 2014, the cost to sequence a 90 gigabase genome is about $5000, or 20,000 times less expensive overall [9]. The size of gene sequencing equipment has also shrunk tremendously. Both Craig Venter, head of Synthetic Genomics, and Jonathan Rothberg, founder of Ion Torrent, have said it is possible to send miniaturized DNA sequencers to Mars in the near future [10].

Synthetic biology would be used to generate entirely new gene sequences or to generate modifications of existing gene sequences. DNA sequences can be synthesized a base pair at a time from information in a computer file. The DNA sequence information might be an identical copy of some gene data sequenced from an organism, a slightly modified version of the gene, or even totally artificial genes to build logic circuits [11] [12]. For instance, it may be possible build logic circuits that could count day/night cycles, and could then trigger the air bladders to break off by abscission when the count reaches a certain value.

The synthetically generated DNA sequence can be inserted into the seaweed genome using multiple techniques. The most common techniques for inserting genes into plants use agrobacterium tumefaciens, which inserts DNA by infection, or use a "gene gun", which literally shoots DNA coated gold particles into plant tissues, or electroporation, where a pulse of electricity will force DNA into cells.

The newest technique uses CRISPR (Clustered Regularly Interspaced Short Palindromic Repeats) and Cas9 (CRISPR associated protein 9), an enzyme used to create a double stranded cleavage, to insert a genetic sequence into a very specific location of the genome determined by a "guide RNA" sequence [13] [14]. The CRISPR/Cas9 system can be used to delete genes, silence genes, or add genes.

The modified plant cells then can be grown into shoots, which are clones of the modified cells, using plant tissue culture methods. This would allow the large scale propagation of plants, all genetically identical, all containing the desired genetic modifications. The tissue culture method is commonly used to grow orchid plants, for instance.

5. Safeguards

Since the genetically modified plants must not damage the Earth's existing ecology, genetic safeguards will have to be added. Each safeguard may need to be implemented in multiple ways, so that if one genetic modification safeguard is lost, another safeguard path would be triggered.

1. Trigger the abscission gene to make the plants sink and die after a certain period of time, say after 365 day/night cycles. This would ensure that all of the modified plants would automatically expire after a certain time without any other action.
2. Cause cell suicide, trigger abscission, or inhibit growth if some genetic modifications are missing or damaged. Many types of mammalian cells can implement apoptosis, or programmed cell death, when DNA is damaged. It may be possible and would be desirable to add a similar function to the modified seaweed genome.

3. Make the plants sterile, to prevent genetic mixing of the modified plants and native plants. The native Sargassum Natans usually propagates by cloning when a branch breaks off, but it may also propagate with spores.
4. Generate compounds that bacteria digest very slowly to slow down decay and reduce the post-sinking decay oxygen consumption. Some plant compounds like lignin or cellulose decay much slower than other plant compounds.

6. Amount of Sequestered Carbon Dioxide

The Earth's surface area is 510 million km2. The oceans cover 70% of the Earth's surface, amounting to about 360 million km2. Half of the ocean might be warm enough (avoiding the polar regions) and deep enough (depth greater than 300 meters) for this terraforming approach, which would then amount to 180 million km2.

Native Sargassum natans carbon growth rate has been measured at 1 mg/m2/day [15], which is 200 times lower than phytoplankton production in the same area. It may be possible to enhance the Sargassum natans growth rate significantly without interfering with phytoplankton growth.

If Sargassum natans seaweed had a typical Relative Growth Rate (RGR) of 2% per day, it would grow from 100 grams to 100 kg in a year, a one thousand fold growth. The growth rate for a different variety of seaweed, Sargassum horneri, was measured at 2.25% per day to 4.95% per day [16]. If the RGR could be increased to 2.7% per day, a 100 g plant would grow to 1000 kg in a year, a ten thousand fold growth.

If Sargassum natans were modified to extend its growing range to half of the Earth's ocean, the amount of carbon taken up as plant growth with a 2% RGR would be 1.8 x1011 g/day or .22 gigatonne/year CO2 equivalent. With a 2.7% RGR, the amount of carbon taken up as plant growth would increase by a factor of 10, or 2.2 gigatonne/year CO2. If seaweed production could be increased to half that of phytoplankton by increasing the seaweed growth rate by a factor of 100, the growth would reach 22 gigatonne/year CO2 equivalent.

7. Enhancing Seaweed Growth Rate

It may be possible to make other genetic changes to enhance growth. Some plants such as legumes have nodules in their root structures that contain symbiotic bacteria called Rhizobia which fix nitrogen, so that the plant can grow with little or no additional nitrogen fertilizer.

To further enhance growth, it may be useful to attach slow release fertilizer to each plant when they are released into the ocean. Table 1 shows the mineral components of a typical plant [17].

Table 1: Mineral components of a typical plant

Component	% dry weight
Nitrogen	1.4%
Potassium	1.0%
Calcium	.5%
Magnesium	.2%
Phosphorus	.2%
Sulfur	.1%
Chlorine	.01%
Iron	.01%
Total	3.42%

If nitrogen does not have to be provided because the plant has nitrogen fixing capability, and potassium, calcium, magnesium, sulfur, and chlorine do not have to be provided because these compounds are common in seawater, phosphorus and iron would be only primary essential components that have be provided as a fertilizer, with the amounts shown in Figure 2. Additional minor components might need to be included as fertilizer, but the amounts would less than .01%.

Table 2: Essential mineral components for fertilizing modified seaweed

Compound	% dry weight
Phosphorus	.2%
Iron	.01%
Total	.21%

For a plant wet weight to dry weight ratio of 7:1, a 100 kg wet weight plant would require 30 g of phosphorus and iron if the plant did not obtain these nutrients from seawater. The assumption is that a slow release form of the fertilizer would then weigh 100 g.

8. Carbon Sequestering Costs

A million tonnes of plantlets at 100 g each, would amount to 10 billion plantlets. Distributed across the ocean, with an RGR of 2%/day, the plants would grow to 1 billion tonnes in one year (with no plant loss). The assumption is that the cost to culture and grow each plantlet, along with the amortized cost of an automated factory, is $.20 [18]. There is a further assumption that a 100 g fertilizer ring composed of 30% phosphorus and 1.5% iron would cost $.04 per 100 g.

Transporting the plants to the ocean destination might be done using a ship the size of a Very Large Crude Carrier (VLCC) supertanker. A VLCC with 200,000 deadweight ton capacity, might cost $120M, and might be chartered for $20M per year [19].

The amount of CO2 sequestered is calculated assuming 100% of the plants sink, a plant wet to dry weight of 7:1, and 42% carbon content of plant dry weight. One billion tons of wet seaweed would have .06 gigaton carbon content, or .22 gigaton of CO2 equivalent.

The estimated cost to culture, transport, and disperse the 10 billion plantlets and fertilizer rings for various growth rates are shown in Table 3.

A study by Dietz and Stern suggests a globally coordinated carbon tax level of at least $32/tonne CO2 in 2015, rising to at least $82/tonne CO2 in 2035, to keep global temperature rise to no more than 2 degrees C [21]. So, a country (or multi-national corporation) would be incentivized to research a solution that could have a substantial payoff. As an example, at a carbon tax level of $32/tonne CO2, the tax avoided would amount to $32 billion per year for each gigatonne of sequestered CO2, with an implementation cost of $4 to $5 billion per year.

Table 3: Cost breakdown for various growth rates

Seaweed Growth rate, normalized	Fertilizer weight, g each	Fertilizer cost, each	Transport costs	CO2-e GT	CO2-e cost / tonne	Total cost	CO2 Tax Value
Unmodified 1:1	100	$ 0.04	$ 20 M	.22 GT	$ 11.00	$ 2.4 B	$ 7.0 B
Modified 5:1	500	$ 0.20	$ 60 M	1.1 GT	$ 3.70	$ 4.1 B	$ 35 B
Modified 25:1	2500	$ 1.00	$ 260 M	5.5 GT	$ 2.25	$ 12.3 B	$ 176 B
Modified 100:1	10,000	$ 4.00	$ 1010 M	22 GT	$ 1.95	$ 43 B	$ 705 B

By comparison, the cost of power plant CO2 capture at the source is $43 to $80 per tonne CO2. This cost includes separating CO2 from exhaust gas, but does not include the additional cost of transportation and storage [20]. A cost comparison of other CO2 removal approaches [22] is shown in Figure 1.

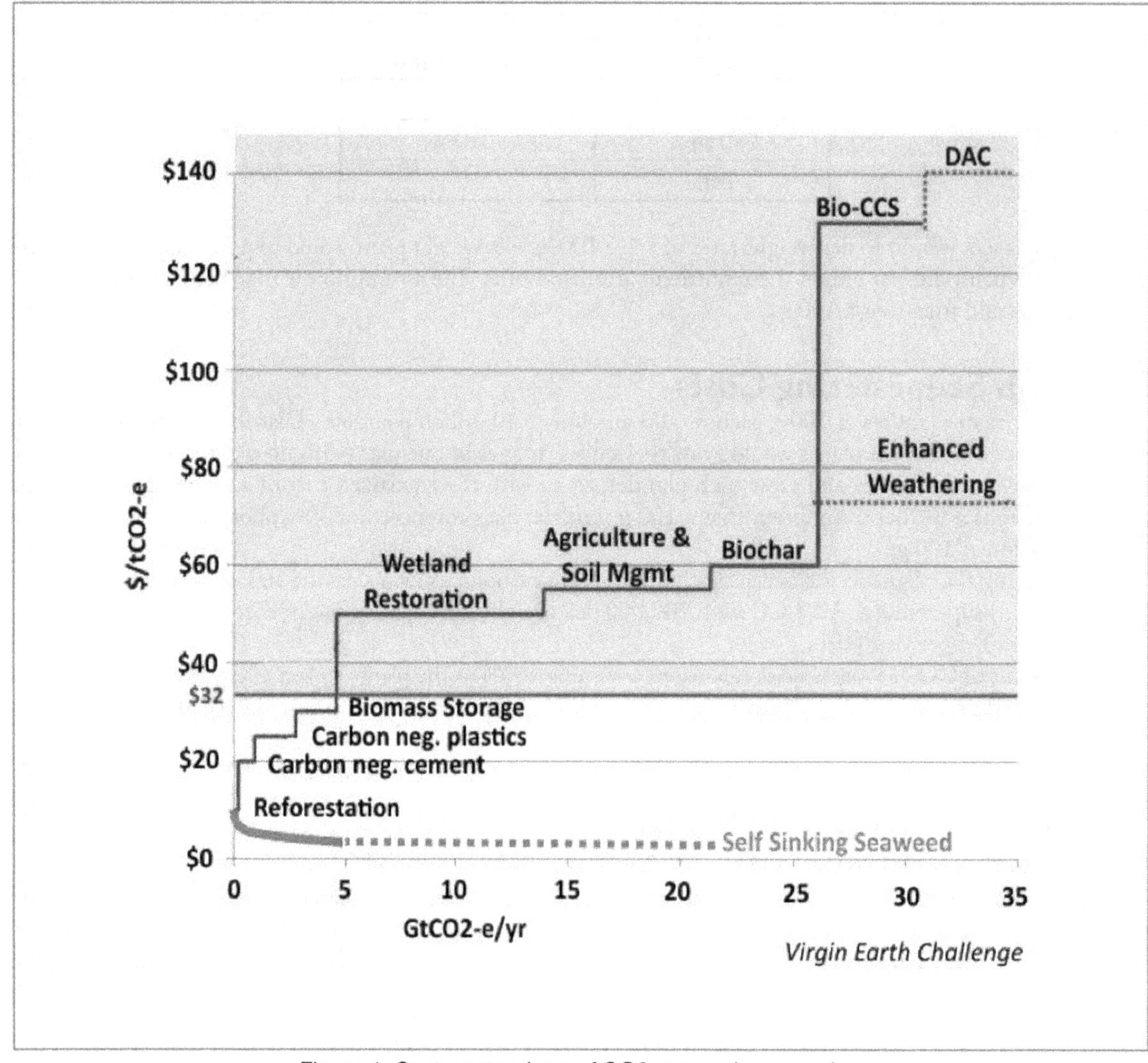

Figure 1: Cost comparison of CO2 removal approaches

9. Implementation Timeframe

The implementation might be broken into three major phases, each phase taking 10 to 15 years. In phase 1, many pelagic seaweeds would be collected and genetically sequenced. The target plants would have to be genetically modified, grown, and tested in a laboratory setting for safety and efficacy. The cycle of genetic modification and testing would be repeated multiple times to assure that all the desirable traits have been added. Near the end of the first phase, the target ocean region would be sampled and measured, to obtain an ecological baseline. The first phase would be complete when approval for deployment to the target ocean region by international regulatory bodies.

In the second phase, deployments might start in a closed lagoon, then move to multiple open ocean locations, building up to 1 million tonnes CO2 sequestered annually, with continual ocean monitoring and sampling.

In the third phase, deployments in the open ocean would build up to 1 billion tonnes CO2 or more, also with continual ocean monitoring and sampling.

From start to finish, the research phase and two deployment phases might take 30 years to 45 years. Seaweed based CO2 sequestration might continue for many decades, depending on the eventual emission level of CO2 emissions.

10. Conclusion

The needs and desires of our large and growing population have affected the ecology of the entire Earth. We have been inadvertently terraforming our own planet by burning enormous amounts of the remains of ancient plants. By using the latest genetic engineering techniques, this biological approach of growing and burying plants in the deep ocean should remove significant amounts of carbon dioxide from our atmosphere at a cost much lower than other approaches.

However, the CO2 sequestering amount using the genetically modified seaweed approach will still be much less than the 35 gigatonnes of CO2 emitted in 2012, which means it will also be necessary to cut world wide CO2 emissions substantially. When CO2 emissions are reduced enough, this self sinking seaweed method would be helpful in achieving net zero or even negative carbon emissions.

Finally, even if this approach is not needed, perhaps because Earth's temperature rise will not be as high as the worst case scenarios, or because CO2 emissions are brought down quickly, research in genetically modifying plants in general would be useful in the customization of many types of plants for the closed environment of a starship.

Acknowledgements

Thanks to Professor Gregory Benford, who gave a thought provoking talk on Geoengineering to the IEEE Orange County Computer Society in May 2011.

References

1. Holland, H. "The oxygenation of the atmosphere and oceans", Phil. Trans. R. Soc. B, vol. 361 no. 1470 903-915 (2006)

2. http://www.esrl.noaa.gov/gmd/ccgg/trends/#mlo_growth

3. Orr J., et al., "Anthropogenic ocean acidification over the twenty-first century and its impact on calcifying organisms", James, et al., *Nature* 437, 681-686, doi:10.1038/nature04095 (2005)

4. Köhler, P., Abrams, J., Völker, C., Hauck, J., and Wolf-Gladrow, D., "Geoengineering impact of open ocean dissolution of olivine on atmospheric CO2, surface ocean pH and marine biology", *Environ.* Res. Lett. 8 014009, doi:10.1088/1748-9326/8/1/014009 (2013)

5. Stern, N., "Stern review: The Economics of Climate Change", http://webarchive.nationalarchives.gov.uk/20130129110402/http://www.hm-treasury.gov.uk/d/Summary_of_Conclusions.pdf (2006)

6. Metzger, R., Benford, G., "Sequestering of Atmospheric Carbon through Permanent Disposal of Crop Residue", *Climatic Change*, Volume 49, Issue 1-2, pp 11-19 (2001)

7. Feely, et al., "Oxygen Utilization and Organic Carbon Remineralization in the Upper Water Column of the Pacific Ocean", *Journal of Oceanography*, vol. 60, pp. 45-52 (2004)

8. van Doorn, W., and Stead, A., "Abscission of flowers and floral parts", J. Exp. Bot. 48 (4): 821-837, doi: 10.1093/jxb/48.4.821 (1997)

9. "DNA Sequencing Costs", National Human Genome Research Institute, National Institutes of Health, http://www.genome.gov/sequencingcosts/

10. Regalado, A., "Genome Hunters Go After Martian DNA", *Technology Review*, (2012), http://www.technologyreview.com/news/429662/genome-hunters-go-after-martian-dna/

11. Bonnet, J. , Yin, P., Ortiz, M., Subsoontorn, P., Endy, D., "Amplifying Genetic Logic Gates", *Science*, vol 340 pp. 599-603 (2013)

12. Endy, D., "Transcriptors & Boolean Integrase Logic (BIL) gates, explained", https://www.youtube.com/watch?v=ahYZBeP_r5U

13. Jiang, W., Yang, B., Weeks, D., "Efficient CRISPR/Cas9-Mediated Gene Editing in Arabidopsis thaliana and Inheritance of Modified Genes in the T2 and T3 Generations", PLOS ONE, DOI: 10.1371/journal.pone.0099225 (2014)

14. Belhaj, K., Chaparro-Garcia, A., Kamoun, S., and Nekrasov,V., "Plant genome editing made easy: targeted mutagenesis in model and crop plants using the CRISPR/Cas system", *Plant Methods*, 9:39 doi:10.1186/1746-4811-9-39 (2013)

15. Carpentea, E., and Cox, J., "Production of pelagic Sargassum and a blue-green epiphyte in the western Sargasso Sea", Limnology and Oceanography 429 (1974)

16. Choi, H., Lee, K., Yoo, H., Kang, P., Kim, Y., Nam, K., "Physiological differences in the growth of Sargassum horneri between the germline and adult stages", *Proceedings of the 19th International Seaweed Symposium* (2007)

17. Edwards, D., "Concepts of essentiality and function of nutrients", (1971)

18. "Low cost options for tissue culture technology in developing countries", Proceedings of a Technical Meeting organized by the Joint FAO/IAEA Division of Nuclear Techniques in Food and Agriculture (2002), http://www-pub.iaea.org/mtcd/publications/pdf/te_1384_web.pdf

19. http://en.wikipedia.org/wiki/Oil_tanker

20. Finkenrath, M., "Cost and Performance of Carbon Dioxide Capture from Power Generation", working paper, International Energy Agency, http://www.iea.org/publications/freepublications/publication/cost-perf_ccs_powergen-1.pdf

21. Dietz, S. and Stern, S., "Endogenous growth, convexity of damages and climate risk: how Nordhaus' framework supports deep cuts in carbon emissions", Grantham Research Institute on Climate Change and the Environment, Working Paper No. 159, page 22, http://www.lse.ac.uk/GranthamInstitute/wp-content/uploads/2014/06/Working-Paper-180-Dietz-and-Stern-2014.pdf

22. Deich, N., "Costs and Supply of Greenhouse Gas Removal: Is GGR Affordable and Available at Scale?", Virgin Earth Challenge, http://www.virginearth.com/2014/07/costs-and-supply-of-cdr-approaches-is-cdr-affordable-and-available-at-scale/

Celebrity and Social Media in 100YSS Missions: Another Kind of "Dancing with the Stars"?

Jo Ann Oravec, MA, MS, MBA, PhD

University of Wisconsin at Whitewater

oravecj@uww.edu

Abstract

Space travel has often been reported, explained, promoted, and interpreted for fellow space voyagers as well as non-space travelers through the words and actions of celebrities. This article addresses how social media (such as Facebook, Twitter, and Youtube) and other channels of expression are becoming part of such celebrity initiatives in the context of space. Celebrities and their public relations teams have acquired an assortment of ways to stimulate interest in themselves as personalities as well as in their causes. Many individuals (especially young people) are influenced by celebrity expressions, whether as fans, detractors, or just consumers of online material. The article argues that the ways both celebrities and everyday individuals are using social media are often tightly coupled with notions of confessional discourse, which can raise particularly sensitive concerns in relation to long-term space missions such as 100yss. The article analyzes the benefits and hazards of celebrity-style interventions and initiatives in space-related expression and activism.

"Celebrity is the chastisement of merit and the punishment of talent."
—Emily Dickinson

1. Introduction

Celebrity status may not appear to have close couplings with the scientifically rigorous, highly detailed, and sometimes monotonous aspects of space travel and research. However, many of the psychological and interpersonal dimensions of space missions have strong celebrity components as mission participants relate to broader publics in order to educate, obtain funding, and maintain interest. Social media as well as more traditional modes of expression (such as books, print newspapers, and broadcast television) are being employed as ways to ensure that space travel is seen as an option for young people's futures and is also construed as a worthwhile expenditure of tax and corporate dollars. This article also argues that the ways both celebrities and everyday individuals are using social media are often tightly coupled with notions of confessional discourse, which can raise particularly sensitive concerns in relation to issues of personal privacy and authenticity. As addressed in the following sections, celebrity interventions in current events have often been criticized in terms of their appropriateness and usefulness. This article analyzes the benefits and hazards of involving celebrities in space-related expression and activism as well as utilizing the celebrity status that space travelers and tourists may obtain.

Consider the following segment of a recent news article about space shuttle astronauts:

> The head of NASA startled an Ottawa audience Thursday when he announced out of the blue: "Chris Hadfield is infamous." Charles Bolden, as NASA's administrator, is Hadfield's former boss. And over lunch with the Canadian Club he started to improvise. "My notes say he's famous. Chris is not famous! Chris is infamous," he said, drawing out the last word. "He has put pressure on everybody… Chris has revolutionized the way that people look at astronauts on orbit and he has brought space flight home to normal people down here. He has made them feel like: I can do this. I'm involved in this. And that's really, really, really special."
> —Spears, 2014, para. 2

In the perspective of this and other accounts, the success of a particular astronaut in attracting the attention of young people to space issues through an assortment of entertainment-style strategies has apparently translated into subtle pressures to alter the style of astronaut-Earth interaction. Astronaut Chris Hadfield's efforts to make space more comprehensible and available to individuals have been coupled with well-crafted musical and theatrical presentations that obtained for him a form of celebrity, both in social media platforms and in more traditional print and broadcast platforms. However, the expansion of these entertainment aspects of space travel could place demands on those individuals who are less well equipped to perform musically or provide other forms of amusement and distraction for the public.

Celebrity has profound influences on everyday life (as exemplified by the Hadfield story) as well as on larger political, cultural, and societal concerns. For instance, a number of devotees of Oprah Winfrey in the US have reportedly altered significant aspects of their lifestyles and appearance in the course of their fandom (Martin, 2013), some apparently changing their spiritual directions along with various consumer choices. Some fans have even placed tattoos of celebrities on their bodies, sometimes leading to intellectual property disputes (Beasley, 2011). Celebrities (especially in the entertainment and sport arenas) have often utilized space-related themes and issues in their commercial or publicity-related efforts; many of them have also integrated them into social activist campaigns in which they have considerable personal stakes. For example, a character portraying an astronaut has played a major part in the popular US comedy television show *The Big Bang Theory* for the past decade. A number of celebrities have also played considerable roles in 100YSS projects, including Levar Burton, an actor and entrepeneur, as a number of Star Trek and science fiction notables. Dr. Mae Jemison, leader of 100YSS, has herself gained considerable celebrity as the focus of a number of book and documentary treatments of astronauts (McGettigan, 2013). (A discussion of the definition of "celebrity" will follow in a later section.) Critical analysis of these celebrity initiatives can serve to illuminate how celebrities are influencing public discourse on space issues as well as expose the mechanisms through which they have societal influence. For example, Rojek (2001) contends that "We will not understand the peculiar hold that celebrities exert over us today unless we recognize that celebrity culture is irrevocably bound up with commodity culture" (p. 10). Today, space "commodities" can include tourist-style rides on low-earth-orbit space missions as well as books and movies about space, including the 2013 Oscar-winning *Gravity* (Cuarón, 2013). Allen and Mendick (2013) state that the need to comprehend the forms and extents of celebrity influences in the lives of individuals (especially young people) is "urgent" (p. 77) because of their potential impacts on identity development.

> Consider the potential influences upon young people of the following celebrity news relating to space travel: Lady Gaga will sing in outer space in 2015. *Us Weekly* reported the 27-year-old will blast off in a Virgin Galactic ship and let rip with a single track during the Zero G Colony high-tech musical festival in New Mexico. It makes sense. She already dresses like an astronaut. "She has to do a month of vocal training because of the atmosphere," said a source, who added that Gaga's glam squad will join her in the shuttle.
> —Hicks, 2013, para. 1

Thousands of social media expressions have been directly tied to the above announcement as people speculate as to what will happen and to the character of Lady Gaga herself. Other celebrities have also planned comparable space-based performances. The celebrity-related aspects of social media (including Facebook, Twitter, Youtube, and weblogs) can serve critical roles in space-related contexts. A widely-used characterization of social media is provided by Boyd and Ellison (2007): "web based services that allow individuals to construct a public or semi-public profile within a bounded system, articulate a list of users with whom they share a connection and traverse their list of connections and those made by others within the system" (p. 211). The examples in this article show how social media are incorporated into efforts to gain personal name recognition as well as promote particular social causes by celebrity figures and those who aspire to celebrity status. The examples also serve as exemplars of the influence celebrity social media utilization can have on everyday participants in social media, many of whom are just developing their personal styles of expression. Some non-celebrity social media participants may indeed eventually acquire forms of celebrity, and also obtain more influence in the arenas they care about, including space issues; however, most will not obtain such celebrity, possibly leading to their disappointment or cynicism (and possibly the abandoning of their efforts at expression). However, through the process of engaging in celebrity-related social media, individuals may gain awareness of some larger community and societal concerns, such as those linked with space travel and 100YSS missions.

Space has many attractive aspects in terms of imagery and textual expression, including those linked with weightlessness and spectacular planetary and lunar vistas. The prospect of increases in the numbers of individuals who produce exciting and readable material about space obtaining celebrity-style influence through social media holds some promise, as outlined in the sections to come, which could be of substantial assistance in directing societal attention and funding initiatives to space ventures.

2. Deconstructing Celebrity and Space in the Context of Social Media

Imagine a group of students who are interested in space. They are working on their Facebook pages, adding links to celebrity activities (such as those of the US performer Lady Gaga) and reflecting on their own and their friends' endeavors. The varieties of activities in which these students are engaged have been shown to have significant implications for their personal identity formation (Allen & Mendick, 2013; Greenhow & Robelia, 2009), providing feedback that can shape personality development. Will the online reflections of these students about their own experiences have the kinds of societal influence that celebrity-produced social media can have? The social media productions of these students may be as insightful as those of any celebrity, and the causes they support (including space missions) as important. Other issues involving the students' productions can also emerge, which will be addressed in later sections: what sorts of problems involving publicity and privacy may arise because of the students' online reflections? Will they be problems akin to those encountered by celebrities? It is clear that the students in our scenario will not have the elaborate public relations teams of many celebrity figures (such as Lady Gaga) to produce social media content and resolve problems dealing with publicity, such as the kinds described below:

> Popular discourse surrounding Lady Gaga, fueled by legacy media platforms like TV and print news, as well as a Gaga-promoted social networking site (littlemonsters.com), an extensive network of bloggers, academic/pop psychology websites (i.e., Gaga Stigmata), and various new media platforms of criticism, have created a rich fabric of explanations, interpretations, and predictions about the star's popular meanings.
> —Smit, 2014, p. 28

Although the Facebook page of a student may indeed have many of the surface qualities of those of a celebrity, a concentrated, professional team effort supports the online celebrity presence of a star such as Lady Gaga. Much of this effort is beneath the surface, however, and readers may not yet have sufficient critical literacy skills to reframe and interpret the situation. For example, many celebrities (including business leaders and political figures) have staff members who write their "personal" reflections in the form of Twitter tweets, blog posts, or other social media input; other staff members may choose the best photos or videos to include as well as modify these images using state-of-the-art software, and yet others may find ways to raise the search engine positions of the items involved (Oravec, 2013). Students who expect to have comparable influence through social media for the causes they care about as do celebrities, or who are concerned that their social media presence is not as professionally rendered, may not be aware of the extensive and expensive apparatus that supports celebrity online presence. Managing the expectations of individuals who are starting to develop their own social media presences may indeed involve the deconstruction of the societal mechanisms that currently support celebrity figures.

A growing number of entertainment celebrities utilize social media, providing some widely-disseminated models of how to proceed in integrating space insights into their online writings and images. For celebrities, space-related initiatives can emerge from specific personal or family experiences (such as plans to engage in space tourism) or be more diffusely related to social activism. Beck et al. (2014) analyzed an assortment of widely-distributed celebrity health and space narratives and found that the functions of education, inspiration, and activism were widely evident. Celebrities (especially those who gain fame through television, film, and sport) have had considerable impacts on major societal initiatives long before social media emerged in the past two decades. Some celebrities have been effective in influencing political campaigns in a number of Western countries (Marsh, Hart,

& Tindall, 2010); this possibly reflects broader societal trends involving disengaged voter bases that would be otherwise largely disconnected from electoral politics.

Celebrity events and broadcasts involving various social issues can have an international reach and convey significant themes (Cantor, 2001; Duenwald, 2002). For example, the efforts of celebrities such as the US actress Angelina Jolie and the Irish singer Bono have often served to stimulate discourse on various humanitarian issues, including poverty and human trafficking (Kapoor, 2013). Some critics of these efforts have given celebrity initiatives the label of "humanitarian optics," possibly providing more in terms of image than positive interventions in unfortunate situations (Weizman, 2012). However, the overall influence of celebrities has been immense: according to Brown and Frasier (2004), "The proliferation of entertainment media worldwide in recent decades has made celebrities powerful agents of social change" (p. 97), with television, radio and live theatre as vehicles as well as the Internet. Videos as well as text have been part of these expressions: the "It Gets Better" Internet campaign against bullying features a number of celebrities (along with some less famous individuals) recording messages of strength to individuals who face harassment (Rattan & Ambady, 2014).

As social media participants reflect on their philosophical and social action directions as well as their everyday lives, celebrity figures often play significant roles. Whether they are themselves celebrities, fans, or critics, social media participants often create hyperlinks to celebrity information and current events (or retweet or resend celebrity material), with some fan-produced social media platforms even focusing entirely on particular celebrity themes. For example, the Michael J. Fox fan page at https://www.facebook.com/pages/Michael-J-Fox-Fan-Club/185617404823484 discusses aspects of his Parkinson Disease-related social campaigns as well as various theatrical ventures, which include time travel and space themes rooted in his character in *Back to the Future* films (Moe, 2012). Many celebrities use social media to reach out to their fans, and thus create or support their fan networks, which can often be viewed as consumption networks (Ferris & Harris, 2011). Celebrity interaction with consumption networks can ensure that the commodities associated with the celebrities (music albums, movie tickets, or t-shirts, as well as particular causes and events) are participated in or purchased.

Since information about celebrities (either with a positive or negative spin) is ready-at-hand, its frequent integration into social media is not surprising. Hoffman and Tan (2013) state that the "influence of celebrity status is a deeply rooted process that can be harnessed for good or abused for harm" in communications (p. 7151). An example of the more problematic aspects of the celebrity-social media nexus is the case of Caleb Johnson, a finalist in *American Idol 13*. Mr. Johnson reportedly used the word "retarded" in reference to some of his fans, then made the situation more complex by posting the following message of apology on his Facebook page:

> For the record that juvenile comment I made in the interview was not directed towards my fans but to the wackos that send hundreds of hate messages a day to me ! You guys are amazing and I cannot thank you enough for your support. Sorry if it offended anybody it was the wrong choice of words . Also I greatly appreciate it when you guys give me song suggestions but it gets really overwhelming at the volume it comes in so please understand ! Rock on !:)

According to an assortment of critics, Mr. Johnson's social media apology was not adequate to deal with the situation (Amabile, 2014). Various remarks that are taken out-of-context or that emerge because of carelessness can be extraordinarily damaging. If delivered by individuals in long-term space missions, such remarks could produce reputational damage that could be especially difficult to repair. Use of social media to deal with crisis management involving celebrities can indeed be complex, and professionals in public relations are endeavoring to understand how such personal and organizational crises can be managed online (Colapinto & Benecchi, 2014).

3. What is "Celebrity" in the Context of Social Media?

Some historical and theoretical background on the notion of celebrity can be of assistance in equipping social media users to deal effectively with the kinds of issues described in this article. The concept of celebrity is evolving as the use of social media expands in society; research concerning celebrity events and influences in media has grown in the past decades, rooted in such disciplines as history, sociology and psychology (Boorstin, 1962; Blake, 2001; Oravec, 2006). A solid definition of "celebrity" is hard to come by since various figures gain popular attention and recognition in wide assortments of ways (Cantor, 2001). Although celebrity status can be fleeting for individuals, the matrix of celebrity figures of a society provides a readily-comprehensible stream of topics for everyday discourse (Rojek, 2012). Celebrities provide rich models of various combinations of behaviors and images, some being labeled as "deviant" in either positive or negative directions (Giles, 2000). Many space mission participants may feel that their lives are uninteresting and would want to remain relatively anonymous, whereas some space aficionados may want to learn even tiny details about them and consider them as heroes and celebrities.

Individuals are encouraged to explore the lives of celebrity figures by public relations professionals and celebrity handlers through following the successive revelation of layers of "secrets"; this is comparable to the process through which many individuals learn about each other over time (Dyer, 1987). In turn, celebrity gossip writers either online or in print formats endeavor to enhance the material they receive about certain celebrity figures (Tiger, 2013). The more information the public is given about the personal lives of celebrities the more it (apparently) wants, driven by professionals who make the successive stories ever more attractive. The "celebrity inversion" reflects the expanding numbers and kinds of individuals who have potential entry into the ranks of celebrity as well as the enlargement of the available kinds of technological and social platforms for celebrity-related discourse. Increasingly, celebrity status is not restricted to those of certain economic level or specific attainments (Oravec, 2006).

As described in this article, some forms of Internet-based celebrity can be established or reinforced through social media, which potentially widens the realm of celebrity to individuals pursuing their various interests (such as space travel and research). Themes related to space can be introduced into social media by linking to and commenting on Internet materials related to these issues, such as Twitter messages ("tweets"), Facebook or weblog postings, or online news articles. Unfortunately, along with the many positive models of such online behavior, creations of fictitious individuals have served to catalyze discourse on authenticity in social media communications. Space mission advocates or participants could be misquoted, parodied, or even fabricated. The example of "Joan," a woman with disabilities who was fabricated as an online character by a man on an Internet site, reportedly made many social media participants more cautious when encountering personal narratives, appeals, and advocacy (Ellcessor, 2012).

4. Space-related Online Revelations and Critiques

The notion of "celebrity" is itself often a topic of societal discourse especially as celebrities use their fame to promote political and social causes (such as space research) or some other desired end. The mechanisms of celebrity place a good share of societal attention and definition in the personage of a single individual or a small group-- although the construction of celebrity may often largely be a product of the efforts of public relations experts rather than the celebrity him or herself (Martin, 2013; Schickel, 1985). Some of the current notions of celebrity in Western nations were influenced by the artistic works of Andy Warhol (Mattick, 1998) and the writings of Gertrude Stein (Curnutt, 1999/2000; Goble, 2001). In a number of exhibitions, quotations, and publications through the 1960s and 70s, Warhol projected that celebrity often does not reside in particular accomplishments or heroic feats. Rather, Warhol construed celebrity as taking on a life of its own, being only loosely coupled with specific persons for various reasons related to media.

Gertrude Stein (1936) proposed that the inner/outer life dichotomy for US celebrities is rooted in a counterbalance to the public relations (PR) machine that supports celebrity. Through the past decades, individual celebrities have often employed a form of discourse that places their authentic "private" sexual and family-related lives in tension with the way these lives appear to the public, even when the celebrities involved are very much in synch with the PR apparatus that constructed those appearances. The public's interest in celebrities in part is linked to the successive discovery of their secret or hidden dimensions, as those secrets are constructed for the public's successive consumption. Confessional discourse is common in celebrity writing and broadcast presentation (Blake, 2001). Social media genres often fit in well with this confessional discourse, influencing participants to focus on private matters in their writings and image productions. Ellcessor (2012) writes that "Internet-based fame depends on the authenticity of a star's self-representation and on the notion of intimacy, experienced through the possibility of interaction rather than simple familiarity" (p. 51). Social media contributions that do not have an open and somewhat intimate approach can be uninteresting over time and even off-putting. However, being overly confessional can also release detailed information that can endanger individuals who are especially vulnerable to personal attacks, such as long-term space mission participants who have few ways to counter the negativity involved.

Social media have historical roots in the format of the daily diary, which is so powerful a tool for revealing aspects of self that it is often incorporated into therapy (Cushman, 1995; Oravec, 2002). Reflections on the writing process and its relation to self-- how various genres allow for the expression of personal identity and sexual emotions-- have a long history in the fields of psychology, philosophy, and literature (Barclay & Smith, 1993; Bruner, 1993; Fivush, 1991; Freeman, 1993; Oravec, 1996). In producing social media, individuals expose for analysis various aspects of their mental processing as well as material relating to detailed (and previously private) aspects of their social and personal lives. The chronological structure and simple design of many social media platforms allow participants to review their own personal evolution and those of others quickly.

Celebrity references about private and intimate aspects of life are common in social media; the retweeting of such celebrity Twitter contributions or sharing of links to other social media materials related to celebrities is fairly simple to accomplish. Relatively little explanation or embellishment is needed because of the background knowledge many people already have about the celebrities referenced. Because of the ubiquitous nature of sport, television, and film-based celebrity, many individuals with little acquaintance with historical or political events are often very much acquainted with narratives concerning the supposed sexual and other intimate activities of the currently-popular set of stars (Martin, 2013; Wilson, 2000). However, why do celebrities as well as non-celebrities choose to use social media to discuss minute aspects of their personal lives and social concerns on a daily basis? Motivations and strategies for obtaining online celebrity at the local or larger levels take a number of forms. Some people have reportedly performed horrific acts in order to obtain public attention and gain name recognition. Celebrity at the level of the individual has been manifested in a wide assortment of genres through the ages, including stage and screen as well as newspaper columns and travelogues (Dixon, 1999). Many of the newer genres have served to widen the pool of individuals who are selected for celebrity status, with "average" individuals being included who might otherwise have never had that option. For example, individuals can gain celebrity through the burgeoning international media trends of "reality TV" (as it is known in the US) or "factual TV" (Great Britain) such as *Big Brother* (Atkinson, 2006; Brunsdon, Johnson, Moseley, & Wheatley, 2001). The potential for long-term space missions such as 100YSS to become kinds of "*Big Brother* house" settings, displaying various social conflicts and intrigues among group members over time in a limited-mobility setting, should be carefully considered.

The impetus people have to read and participate in social media on a regular basis, thus sampling the personal lives and social concerns of others, is also a matter that has tight linkages to celebrity. Individuals often establish interaction patterns with celebrity figures that constitute a kind of imaginary or "pseudo-relationship" (often referred to as a "parasocial relationship"). In effect, they construct a one-sided relationship with the celebrity (Diekema, 1991), sometimes reinforced through organized fan networks. In a media-saturated society, innocuous one-sided relationships have been common for decades (Fraser & Brown, 2002; Leets, de Becker, & Giles, 1995). For example, many people have favorite talk show hosts or performers with whom they interact by calling a show or attending a performance. However, such behavior can also take more severe forms such as e-mail harassment, stalking, and other inappropriate conduct (Ferris, 2001). Sometimes these relationships are reinforced through celebrity websites that encourage fans to send messages to the celebrity involved, even though there is only a small likelihood that the message will be read by other than celebrity handlers. Space mission participants could well need to be aware of the potential for some non-mission individuals to form pseudo-relationships with them (or for the reverse to occur). The Internet has provided new vehicles for expressing and supporting these pseudo-relationships; social media participants are often voyeurs of aspects of others' private lives and can even obtain some aspects of involvement and interaction.

Given the volume and intensity of the online reflections often involved with social media, a major drawback in space-related contexts relates to the inappropriate release of information or opinion. Many celebrity participants in social media venues have staffs of individuals who produce and maintain their social media presences, and minor problems may be caught or rectified quickly. However, when producing streams of writing and images every day, items can be disseminated online without extensive consideration of their impacts for various audiences. Yet another drawback relates to the culture of critique that is associated with the emerging social media genres (as well as many other aspects of the Internet). Criticisms of the social media contributions of some individuals could trigger upsetting psychological incidents. The experience of celebrities in this regard provides clues as to what might occur on larger scales as social media proliferate: critiquing celebrity lives is entertaining for some individuals (as with the "worst dressed" list presented in some magazines and newspapers). Using comparable strategies to make comments about the personal lives and professional expressions of non-celebrities could be devastating, especially when the individuals involved are in long-term space missions such as 100YSS. Caustic criticisms are quite easy to produce via social media (Gardner & Davis, 2013; Oravec, 2006), which allows for the summary praising or "slamming" of particular Internet items. This may exacerbate already intense conflicts among individuals concerning space-related issues as well as other personal differences.

Through the analysis of celebrity social media strategies, individuals who advocate space travel and research can better understand the obstacles and hazards involved with these media such as the potentials for bullying and harassment, all of which can detract from the overall benefits of social media participation. They can also gain practical knowledge of how to restrict others' access to their social media sites through the use of privacy settings and other means (Gardner & Davis, 2013). By critically analyzing the social media activities of specific celebrities over time, non-celebrities as well as celebrities-to-be can gain a sense of how to generate and harness public attention to their causes and issues.

5. Some Conclusions and Reflections: Are Social Media the "New Paparazzi"?

Many celebrities are considering space travel as part of their efforts to gain public influence. Celebrity persuasion and impact can indeed be powerful and profound: Leo Braudy (1997) wrote that fame and celebrity have often reflected deeply-rooted societal perspectives about "the nature of the individual, the social world, and whatever exists beyond both" (p. 586). The months to come may elicit a great deal of celebrity attention to (and social media interaction about) space travel, as exhibited in the following discourse about musical pioneer Bob Geldof's planned space voyage:

> BEAM him up, Scotty! Businessman, father-of-four, Boomtown Rat and professional curmudgeon Bob Geldof recently announced, with uncharacteristic giddiness, that he will be the first Irishman in space. Confirming he has been given a seat on a commercial flight to the final frontier in 2015 with the company Space Expedition Corporation (SXC), he gushed: "Being the first Irishman inspace is not only a fantastic honour but pretty mind-blowing. The first rock astronaut --space rat! Elvis may have left the building but Bob Geldof will have left the planet, wild. Who would have thought it possible in my lifetime?"
> —Harrington, 2014, para. 1

As just exemplified, space travel issues indeed reflect critically-important themes that concern the survival of civilization and humanity as well as include matters of aesthetics and even playfulness. In the process of discussing space travel issues, everyday individuals are actively engaged in defining their identities, refining their personal styles, and presenting ideas and images to their families, colleagues, friends, and the public-at-large. Space travel discourses such as that associated with 100YSS missions also reflect individuals' dreams and visions about the future (McGettigan, 2013). Books such as Rose Blue's (2003) *Mae Jemison: Out of this World* can serve to inspire conversation about space among young people, as well as the social media venues discussed in this article. In engaging in this discourse, many of these individuals refer to the content produced by celebrities (such as Dr. Mae Jemison), contribute to celebrity fan and event websites, and possibly employ some celebrity-style approaches in their own writing and image-production efforts. Whether or not they are themselves celebrities, individuals who produce social media need to assess critically various aspects of celebrity social media, both to gain models of social media presence and to avoid various pitfalls such as privacy violations and bullying. The notion that social media could become the "new paparazzi" for many individuals can frame some of the discourse regarding the potential intrusiveness and the dangers of social media (McNamara, 2011). Whatever insights are gleaned, the fact should be considered that many celebrities participate in social media with the aid of public relations professionals and have hundreds of thousands of followers or fans. This situation is far different from that of individuals who are just starting off constructing Facebook pages or otherwise making social media contributions.

Individuals who follow certain celebrities through social media over time (such as Lady Gaga and Michael J. Fox) can get to know a great deal about the process of producing insights for public consumption and of selectively revealing aspects of self. Consumers of social media can benefit from a critical approach toward celebrity-related writings and images in social media contexts, whether or not they actively participate as fans or detractors. Readers of social media productions (whether celebrity- or non-celebrity produced) can gain valuable glimpses of individuals' day-to-day lives, information once "off limits" to all but a few family members or friends. With social media, individuals have a vehicle to build a kind of celebrity presence, but like celebrities can encounter serious social and interpersonal problems related to their increased levels of public exposure. As celebrities (or their publicists) write about their daily lives and their social causes critical comments often emerge, and whether or not these comments are intended to provoke they certainly can serve to create discord.

Since social media are still emerging as societal phenomena, it is still not clear what makes individuals especially attractive as reflected in their traditional and social media productions (such as those of the highly-popular astronaut Chris Hadfield). Photogenic individuals are reflected in complimentary ways by the camera, and young people often make dedicated efforts to look more attractive in the "selfies" they take (Gardner & Davis, 2013). In the effort to become more popular and widely-read (like the celebrities they read about online) individuals may alter their photographs, invent personal events that never occurred, and in other ways attempt to take on roles comparable to those of celebrity figures. A common theme in the analysis of social media is that of the effort it takes to maintain an interesting presence over extended periods of time and entertain an audience. Early enthusiasm wanes, creative juices dry up, and the quality of the material posted or tweeted can suffer. Individuals who are fearful of producing "boring" social media materials may indeed be tempted to embellish their activities. Incorporating details of particularly problematic situations into daily posts can make social media more interesting to some readers, but can potentially provide embarrassment to those being discussed or portrayed, even if efforts are made to ensure the anonymity of those involved. Social media can often foster attention on what is immediately available in terms of current events (on new and "hot" items, especially those involving celebrities) rather than encouraging a longer-term and more balanced perspective that is possibly more conducive to fostering the support of space travel and research over time.

Space issues are a growing part of celebrity expression and social media content. In turn, social media are part of the emerging media-literate culture that provides wide access to the means through which information and ideas are disseminated. The humility and mission-focus of many individuals in space programs through the years need also to be appreciated and admired, despite the current emphasis on more flamboyant and even glamorous celebrity style. A 2009 *Wall Street Journal* editorial entitled "Celebrity vs. The Right Stuff" reflects on the humble and direct style that characterized a great deal of the quality of astronaut and space mission support interaction: "much of what made the Apollo missions such a tribute to America was the character of the astronauts: their clipped exchanges between Houston and the spacemen; or Lovell, Anders and Borman reading from Genesis on Apollo 8; or the unflappable Flight Director Gene Kranz working the problems of Apollo 13 to triumph" (Stephens, 2009, p. A13). However, if engaged in critically and conscientiously, celebrity influence and social media participation can indeed provide some valuable venues for social activism in relation to space, particularly if some of the drawbacks related to online celebrity can be mitigated.

References

1. Allen, K., & Mendick, H. (2013). Young people's uses of celebrity: Class, gender and 'improper' celebrity. *Discourse: Studies in the Cultural Politics of Education*, 34(1), 77-93.

2. Amabile, M. (2014, May 5). American Idol contestant Caleb Johnson apologizes to fans. *Hollywood Reporter,* Retrieved July 1, 2014 from http://www.hollywoodreporter.com/idol-worship/american-idol-contestant-caleb-johnson-701469

3. Atkinson, R. (2006, July 10). *Big Brother's* 'freak show' has produced the first warts-and-all disabled person on TV - when will the soaps follow? *The Guardian*

4. Baker, M. L. D. (2013). Social software. *Law Librarianship in the Digital Age.* New York: Scarecrow Press, 215-231.

5. Barclay, C, & Smith, T. (1993). Autobiographical remembering and self-composing. *International Journal of Personal Construct Psychology,* 15, 1-25.

6. Beasley, M. (2011). Who owns your skin: Intellectual property law and norms among tattoo artists. *Southern California Law Review,* 85, 1137-1162.

7. Beck, C. S., Aubuchon, S. M., McKenna, T. P., Ruhl, S., & Simmons, N. (2014). Blurring personal health and public priorities: An analysis of celebrity health narratives in the public sphere. *Health communication,* 29(3), 244-256.

8. Blake, D. (2001). Public dreams: Berryman, celebrity, and the culture of confession. *American Literary History* 13(4), 716-36.

9. Blue, R. (2003). *Mae Jemison: Out of this World.* 21st Century Publishing.

10. Boorstin, D. (1962/1992). *The Image: A Guide to Pseudo Events in America.* New York: Vintage.

11. boyd, d., & Ellison, N.B. (2007). Social network sites: Definition, history and scholarship. *Journal of Computer-Mediated Communication.* 13(1), 210-230.

12. Braudy, L. (1997). *The frenzy of renown: Fame and its history.* New York: Vintage.

13. Brown, W. J., & Fraser, B. P. (2004). Celebrity identification in entertainment-education. In A. Singhal, MJ Cody, EM Rogers, and M. Sabido (eds.), *Entertainment-education and social change: History, research, and practice.* Mahwah, NJ: Lawrence Erlbaum Associates, 97-116.

14. Bruner, J. (1993). The autobiographical process. In R. Folkenflik (ed.), *The culture of autobiography: Constructions of self-representations* (pp. 38-56). Stanford, CA: Stanford University Press.

15. Brunsdon, C., Johnson, C., Moseley, R., & Wheatley, H. (2001). Factual entertainment on British television. *European Journal of Cultural Studies,* 4(1), 29-63.

16. Calvert, D. (2014). "A person with some sort of learning space": The aetiological narrative and public construction of Susan Boyle. *Space & Society,* 29(1), 101-114.

17. Cantor, P. (2001). *Gilligan unbound: Pop culture in the age of globalization.* New York: Rowman & Littlefield.

18. Cherney, J. L., Lindemann, K., & Hardin, M. (2013). Research in communication, space, and sport. *Communication & Sport,* 2(2), doi: 2167479513514847.

19. Colapinto, C., & Benecchi, E. (2014). The presentation of celebrity personas in everyday twittering: Managing online reputations throughout a communication crisis. *Media, Culture & Society,* 36(2), 219-233.

20. Cuarón, A. (director) (2013). *Gravity* (film), Warner Brothers.

21. Curnutt, K. (1999/2000). Inside and outside: Gertrude Stein on identity, celebrity, and authenticity. *Journal of Modern Literature,* 23(2), 291-308.

22. Cushman, P. (1995). *Constructing the self, contstructing America: A cultural history of psychotherapy.* Reading, MA: Addison-Wesley Publishing.

23. Diekema, D. (1991). Televangelism and the mediated charismatic relationship. *Social Science Journal,* 28(2), 143-163.

24. Dixon, W. (1999). *Disaster and memory: Celebrity culture and the rise of Hollywood Cinema,* New York: Columbia.

25. Dyer, R. (1987). *Heavenly bodies: Film stars and society.* New York: St. Martin's Press.

26. Ellcessor, E. (2012). Tweeting@ feliciaday: Online social media, convergence, and subcultural stardom. *Cinema Journal,* 51(2), 46-66.

27. Ellis, K. (2014). The Voice Australia: Space, social media and collective intelligence. *Continuum,* 28(4), 482-494. doi: 10.1080/10304312.2014.907874

28. Ferris, K. (2001). Through a glass, darkly: The dynamics of fan-celebrity encounters. *Symbolic Interaction.* 24, 25-47.

29. Ferris, K. O., & Harris, S. R. (2011). *Stargazing: Celebrity, fame, and social interaction.* New York: Routledge.

30. Fivush, R. (1991). The social construction of personal narratives. *Merrill-Palmer Quarterly,* 37, 59-82.

31. Fraser, B. & Brown, W. (2002). Media, celebrities, and social influence: Identification with Elvis Presley. *Mass Communication & Society,* 5(2), 183-206.

32. Freeman, M. (1993). *Rewriting the self: History, memory, narrative.* New York: Routledge.

33. Gardner, H., & Davis, K. (2013). *The app generation: How today's youth navigate identity, intimacy, and imagination in a digital world.* New Haven: Yale University Press.

34. Giles, D. (2000). *Illusions of immortality: A psychology of fame and celebrity.* New York: St. Martin's.

35. Goble, M. (2001). Cameo appearances; or, When Gertrude Stein checks into Grand Hotel. *MLQ: Modern Language Quarterly,* 62(2), 117-63.

36. Goggin, G., & Newell, C. J. (2004). Fame and space: Christopher Reeve, super crips, and infamous celebrity. *M/C: a journal of media and culture,* 7(5).

37. Greenhow, C., & Robelia, B. (2009). Informal learning and identity formation in online social networks. Learning, Media and Technology, 34(2), 119-140.

38. Haller, B., & Becker, A. B. (2014). Stepping backwards with space humor? The case of NY Gov. David Paterson's representation on "Saturday Night Live." *Space Studies Quarterly,* 34(1).

39. Harrington, P. (2013, Sep 21). The great celebrity space race. *Daily Mail.* Retrieved from http://search.proquest.com/docview/1434395226?accountid=14791

40. Hicks, T. (2013, Nov 06). Hicks: Lady gaga to perform in outer space. *Oakland Tribune.* Retrieved from http://search.proquest.com/docview/1448879656?accountid=14791

41. Hoffman, S. J., & Tan, C. (2013). Following celebrities' medical advice: Meta-narrative analysis. *BMJ: British Medical Journal,* 347, 7151.

42. Kapoor, I. (2013). *Celebrity humanitarianism: The ideology of global charity.* London: Routledge.

43. Leets, L., de Becker, G., & Giles, H. (1995). Fans: Exploring expressed motivations for contacting celebrities. *Journal of Language and Social Psychology,* 14(1-2), 102-123.

44. McNamara, K. (2011). The paparazzi industry and new media: The evolving production and consumption of celebrity news and gossip websites. *International Journal of Cultural Studies,* 10(2), 243–263. doi: 1367877910394567

45. McNeill, L. (2003). Teaching an old genre new tricks: The diary on the Internet. *Biography,* 26(1), 24-47.

46. Martin, L. (2013). Public policy, popular culture, celebrity, and the making of a spiritual-industrial marketplace in postwar America. *American Studies,* 52(3), 53-64.

47. Marsh, D., Hart, P., & Tindall, K. (2010). Celebrity politics: The politics of late modernity? *Political Studies Review,* 8(3), 322–340.

48. Mattick, P. (1998). The Andy Warhol of philosophy and the philosophy of Andy Warhol. *Critical-Inquiry,* 24(2), 965-87.

49. McGettigan, T. (2013, Spring). On the wings of a fantasy. *Contexts,* 12, 55-57. doi:http://dx.doi.org/10.1177/1536504213487700

50. Meeuf, R. (2014). The nonnormative celebrity body and the meritocracy of the star system: Constructing Peter Dinklage in entertainment journalism. *Journal of Communication Inquiry,* doi: 0196859914532947

51. Moe, P. (2012). Revealing rather than concealing space: The rhetoric of Parkinson's advocate Michael J. Fox. *Rhetoric Review,* 31, 443–460.

52. Oravec, J. (1996). *Virtual individuals, virtual groups: Human dimensions of groupware and computer network-*

ing. New York: Cambridge University Press.

53. Oravec, J. (2002). Bookmarking the world: Weblog applications in education. *Journal of Adolescent and Adult Literacy,* 45(7), 616-622.

54. Oravec, J. (2006). Weblogs and the professional as celebrity. *Professional Studies Review,* 2(2).

55. Oravec, J. (2013). Gaming Google: Some ethical issues involving online reputation management. *Journal of Business Ethics Education,* 10(1), 61-81.

56. Rattan, A., & Ambady, N. (2014). How "It Gets Better": Effectively communicating support to targets of prejudice. *Personality and Social Psychology Bulletin,* 40(7). 0146167213519480

57. Repo, J. & Yrjölä, R. (2011). The gender politics of celebrity humanitarianism in Africa, *International Feminist Journal of Politics,* 13(1), 44-62.

58. Rojek, C. (2001). *Celebrity.* London: Reaktion Books.

59. Rojek, C. (2012). *Fame attack: The inflation of celebrity and its consequences.* A&C Black.

60. Sastre, A. (2014). Towards a radical body positive: Reading the online "Body Positive Movement". *Feminist Media Studies,* (ahead-of-print), 1-15.

61. Schickel, R. (1985). *Intimate strangers: The culture of celebrity.* Garden City, NY: Doubleday & Company, Inc.

62. Smit, C. R. (2014). Body vandalism: Lady Gaga, space, and popular culture. *Forum: Popular Culture and Space,* 28.

63. Spears, T. (2014, Ocrober 3). "Chris Hadfield is infamous" among astronauts, NASA chief says: "We've got to learn how to sing." *National Post (Canada).* Retrieved on October 17, 2014 from http://news.nationalpost.com/2014/10/03/chris-hadfield-is-infamous-among-astronauts-nasa-chief-says-weve-got-to-learn-how-to-sing/

64. Stein, G. (1936). *The geographical history of America, or the relation of human nature to the human mind.* New York: Random House,

65. Stephens, B. (2009, July 21). Celebrity culture vs. the right stuff. *Wall Street Journal,* A-13.

66. Tiger, R. (2013). Celebrity gossip blogs and the interactive construction of addiction. *New Media & Society,* 1461444813504272.

67. Weizman, E. (2012). *The least of all possible evils: Humanitarian violence from Arendt to Gaza.* New York: Verso Books.

68. Wilson, C. (2000). *A massive swelling: Celebrity re-examined as a grotesque, crippling disease and other cultural revelations.* New York: Viking.

Minimal Requirements for a Fully Autonomous Health Care in a Closed System

Natalie Bordag, Ph.D.
Kurt Zatloukal, M.D.

Institute of Pathology, Medical University of Graz, Auenbruggerplatz 25, 8036 Graz, Austria

Abstract

Long-term human space flights pose several yet unsolved challenges such as adequate health care is of utmost importance for mission success. At the same time such flights provide a unique experimental environment for quantum leap innovations with the potential to revolutionize general health care. Space flights prohibit access to classical health care and therefore must have a fully autonomous health care system suitable for all known and yet unknown (e.g. resistant pathogens) disease conditions since health checks cannot exclude occurrence of diseases or health problems in the long-term. Since a starship cannot host all competencies of a classical health care system, new concepts are needed. An autonomous health care system starts with highly sensitive and specific diagnostics which flows into integrated data analysis yielding medical care instructions and outcome predictions coupled to drug assembly and personalized dosage up to robotic surgery. Although for most parts building blocks already exist or are on the immediate horizon, diagnostics remain the fundamental bottleneck and utilization of metabolomics is proposed. The metabolome is in itself the integration of the cascade of interactions that take place between environmental factors, individual (epi-)genetic makeup and his transcriptome, proteome and microbiome. Therefore metabolomics can detect essentially in real time the adaptive, multiparametric response of the organisms to pathophysiological stimuli even before clinical symptoms manifest and would be a perfect base for stand alone, autonomous health care systems.

Keywords

autonomous, health care, metabolomics, computational modelling, personalized, prevention

1. Introduction

For long-term space flights, health care cannot be provided on a remote basis. Therefore a fully autonomous health care system needs to be established. Such a health care system has to be suited to address all known and yet unknown disease conditions since health checks cannot exclude eventual occurrence of diseases or health problems in a long term perspective. Since it is not feasible to establish all competencies and infrastructure of a classical health care system, new concepts are needed. Furthermore, the shelf life of drugs is limited makes it necessary to produce drugs from stable precursors and requires capacities to produce new drugs for new diseases conditions (e.g. antibiotics or vaccines for newly emerging or drug resistant pathogens).

Building blocks for an autonomous health care system do already partially exist or are on the immediate horizon. Technologies such as specialized non- or minimally invasive diagnostic, computational modelling of disease conditions utilizing massive omics data and biosensor information, computer-designed drugs composed of individualized mixtures of functional components (drug precursors), stem cell-based regenerative medicine or robotic surgery will form the foundation for an autonomous health care system. Diagnostics are the starting point for any health care and are currently also the one fundamental bottleneck. Many diseases, ranging from metabolic syndrome over various cancer types, can be well treated in cases of sufficiently early detection. However, to cover the spectrum of current diagnostics a huge range of technologies and demanding infrastructure (e.g., for medical imaging) are needed and would for several major diseases still not achieve the necessary specificity and sensitivity. Therefore medical diagnostics in a closed system must integrate different relevant, easier obtainable and more sensitive parameters to inform on the health or disease status of individuals thereby giving guidance for personalized prevention or therapy. The approach should improve diagnostics up to a level where disease onset is safely detected before manifestation of clinical symptoms and irreversible organ damage occurs. Disease prevention and stabilizing management of unpreventable diseases is in a long-term enclosed environment such as space flight of critical importance for mission success.

2. Metabolomics a Key Pillar for Diagnostics

Metabolomics is a technology that provides the information on the cascade of interactions that take place between environmental factors, the individual genetic makeup and his microbiome which results in a highly informative metabolic state as defined by the sum and relative ratios of metabolites (small molecules <1.500 Da which are the products of metabolic reactions and define the functional state of an organ or whole body) (Figure 1). In this context, metabolomics has a preferential role with respect to other omic technologies because of its ability to detect essentially in real time the adaptive, multiparametric response of the organisms to pathophysiological stimuli or (epi-)genetic modifications.

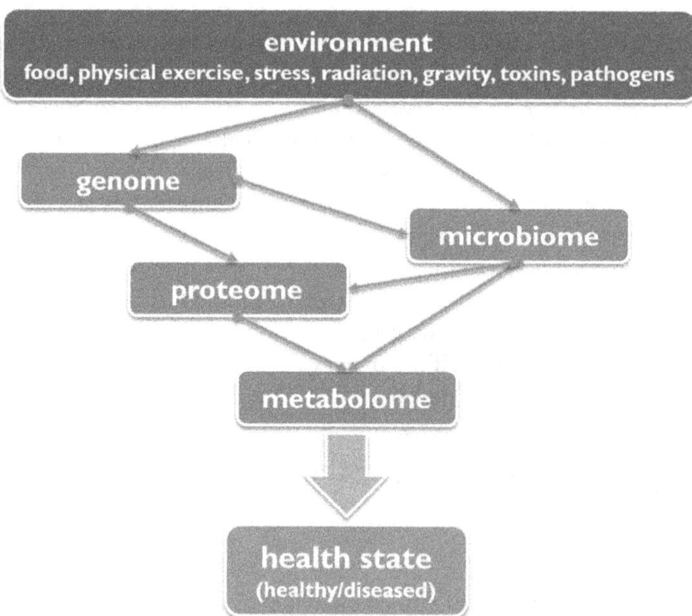

Figure 1: Schematic of the interaction of environment with genome, proteome, microbiome and metabolome which is the closest imprint of the individual's health state.

A specific advantage of metabolomics is that analyses can be performed on biological samples which can be collected by non invasive procedures (e.g. urine, saliva, exhale condensate) and further processed in automated work flows. Currently the stand alone and wide spread use of metabolomics for routine diagnostics is hindered by the demand for very high quality sampling often only possible in hospitals. Therefore the development of a sampling devise allowing for non invasive (e.g. urine, saliva, exhale condensate) self sampling by the patient or space traveler constitutes a major breakthrough towards personalized, stable, specific and most of all preventive health care. Another advantage of metabolomics is that the metabolic fingerprint is highly specific for a person and can only be interpreted reasonably in the context of the personal metabolic trajectory. This avoids several concerns typically discussed for genetic information and essentially eliminates the risk of misuse of data, e.g., by insurance companies or employers.

2.1 Metabolomics for Disease Prevention

The earliest available signs of any disease onset are alterations of the individual's metabolome, often resulting from compensatory mechanisms which start long before clinical symptoms and the disease manifest themselves. A specific and reliable detection of these early signs should allow for effective prevention even with mild counter measures like lifestyle changes or low dose treatments (Figure 2).

For metabolomics, two complementary analytical techniques exist, namely high-field proton nuclear magnetic resonance (1H-NMR) and liquid or gas chromatography coupled mass spectrometry (LC /GC MS). 1H-NMR spectra provide the overall metabolic profile of essentially all abundant (c>1 μM) metabolites in one sample, yielding a "metabolic fingerprint" which is highly specific for a person. LC-/GC-MS dissects and quantifies with high accuracy many, but not all, of the metabolites in a sample and provides the sum of those quantifications as metabolic profile in the form of a table with values per metabolite per sample. We propose exploiting the advantages and the added value from combining both technologies to establish the best possible protocol for monitoring the individual's metabolic trajectory changes.

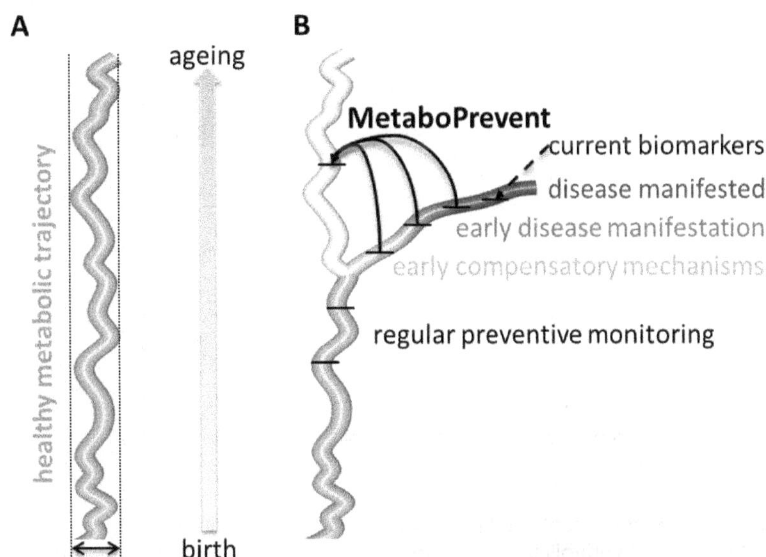

Figure 2: Schematic of an individual's relative metabolic profile variation during life time.

(A) Healthy metabolic profile trajectory (green) showing the intra-individual variability caused by factors, such as ageing, smaller health events (common cold), weight gain, change in dietary habits and others.
(B) Development of an individual's relative metabolic profile trajectory from healthy (green) to manifested disease (brown). Trajectory aberration start with early compensatory mechanisms (yellow) prior to manifestation of clinical symptoms, followed by early disease manifestation (orange) with first clinical symptom manifestation ending in a manifested disease with clear clinical symptoms (brown). Trajectory aberrations change the individual's metabolic profile stronger than intra-individual variability. Early detection allows for mild interventions and resumption of the healthy trajectory (black arrows). Different diseases may result in different metabolic profile trajectory aberrations (not shown).

2.2 Individualized Health Monitoring with Metabolite Patterns or Whole Spectra

Statistical robust detection of subtle metabolic changes from early compensatory mechanisms before disease onset based on single biomarkers is virtually impossible. However, the combination of several metabolites into complex biomarker patterns results in good statistically performance enabling early detection in the first place. The technical feasibility and relevance of biomarker patterns has been demonstrated with MetaMap® Tox, a rat plasma data base for early and systematic drug safety testing with currently more than 100 toxicological patterns deduced from more than 500 reference substances [2, 1]. The statistical power of metabolic patterns to detect early stages has been demonstrated for drug-induced toxicities and injuries[2]. A metabolic pattern contains 2-35 metabolites and improves statistical outcome in terms of sensitivity and specificity tremendously when compared to single biomarkers. This approach has been extended to use the whole information of metabolome NMR spectra to characterize the health status of individuals. NMR spectra have been shown to be highly specific for individuals and any deviation from this individual spectrum was indicative for a major lifestyle change or onset of disease[5, 4, 3].

2.3 Higher Specificity and Sensitivity with Personalized References

Currently the variability between individuals prohibits development of metabolomic biomarkers of sufficient sensitivity and specificity to detect early compensatory mechanisms (Figure 3).

The rationale for personalized approaches is the observation that intra-individual variability is at least two-fold lower than the variability between individuals[6]. Personalized assays would allow much higher sensitivity at equally high specificity. Studies have shown that individual metabolic fingerprints exist in the urine of healthy volunteers[7] and tend to remain sufficiently stable over time, and stability is not compromised by interfering factors like aging, lifestyle changes, smaller health events (common cold), weight gain, change in dietary habits and others[8, 9]. Early prediction of disease states based on referencing changes in metabolome to previous data from the same person (when he or she was still healthy) has also been shown for diabetes[3] or coeliac disease[5, 4].

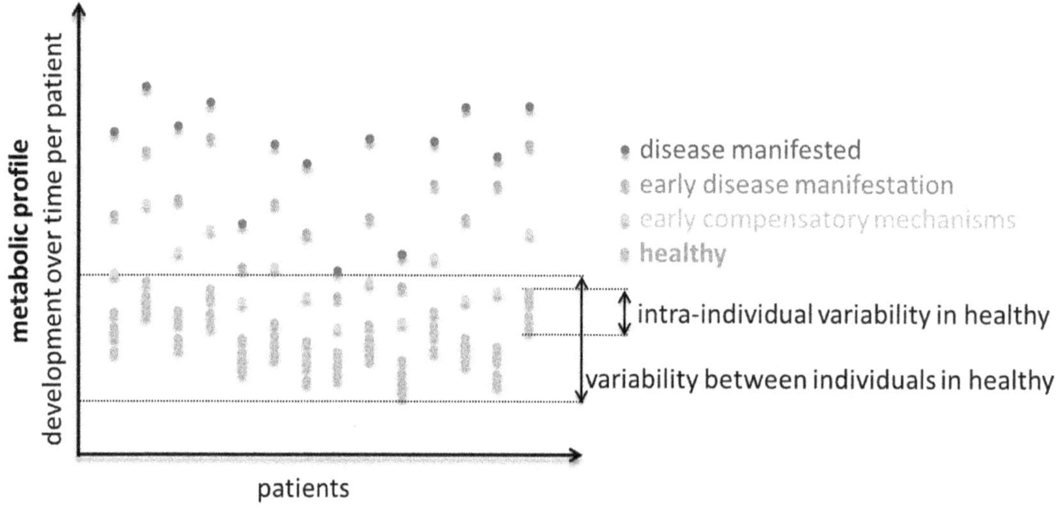

Figure 3: Schematic of development of metabolic profiles over time from healthy to disease manifested for various patients. The range of variability is larger for a healthy population than the intra-individual variability. Specific and sensitive detection of early disease manifestation is only possible with personalized cut offs requiring multiple measurements during lifetime. Personalized normalization and analysis allows specific and sensitive detection even of early compensatory mechanisms before disease manifestation.

2.4 The Virtual Human

The second step in an autonomous health care system after the establishment of metabolic diagnostics (e.g. via non invasive self sampling) is the automated measurement and analysis. The virtual human is the integrated computational model of all parameters impacting an individual's health status directly predicting disease/health status, and thereby giving individualized medical guidelines with treatment and therapy management instructions.

Currently, automated push-button metabolic measurement solutions and data analysis already exist, although not in the health care sector. They are able to deliver reliable targeted and untargeted screening results for complex biological samples like wine[10] or juice including automatic quantification of most decisive parameters, comparison to reference values and verification as well as prediction. A plethora of modeling techniques applied to metabolic data exist (e.g., elastic net, penalized logistic regression, random forest, support vector machines, knowledge discovery by accuracy maximization[11]) and move automated data modeling for the virtual human into reach. The metabolome is an ideal starting point for the virtual human as it comprises less complex data than other omics technology but is simultaneously in itself already an integration of the multiparametric response to pathophysiological changes. Addition of further data sources, such as proteomics, trascriptomics, metagenomics or environmental exposure data is envisioned by creating in a stepwise manner an increasingly complex and more reliable computational model of the human. Principal feasibility has already been demonstrated in integrated modelling of transcriptomics and metabolomics data in the context of metabolic liver disease[12].

3. Conclusions

Long-term human space flights pose yet unsolved health care challenges, and at the same time provide a unique experimental environment for quantum leap innovation of health care in general. An interdisciplinary think tank could work on the specification of key components of a fully autonomous health care system which can be established in a starship environment. Such a process could lead to a research agenda which has the potential to generate highly innovative concepts for addressing global health challenges related to ageing populations, remote locations, emerging and re-emerging infectious diseases and financial constrains in funding health care. Since innovation in medicine and health care is often hampered by complex regulatory processes, lack of flexibility in routine health care, traditional training and career models, the experimental environment of a starship could avoid several of these innovation roadblocks. Many building blocks exist or are on the immediate horizon but overcoming singular bottlenecks, optimizations of existing buildings blocks and most importantly integration of these buildings blocks into one whole autonomous system are needed.

Acknowledgements

This work was funded in part by the Christian Doppler Laboratory for Biospecimen Research and Biobanking Technologies. We sincerely thank Hans Lehrach, Berlin as well as Claudio Luchinat and Paola Turano, Florence for stimulating discussions.

References

1. Van Ravenzwaay, B., Cunha, G. C.-P., Leibold, E., Looser, R., Mellert, W., Prokoudine, A., … Wiemer, J. (2007) The use of metabolomics for the discovery of new biomarkers of effect. *Toxicology Letters,* 172(1-2), 21–28 doi:10.1016/j.toxlet.2007.05.021

2. Mattes, W. B., Kamp, H. G., Fabian, E., Herold, M., Krennrich, G., Looser, R., … Piccoli, S. P. (2013) Prediction of clinically relevant safety signals of nephrotoxicity through plasma metabolite profiling. *BioMed Research International,* 2013, 202497 doi:10.1155/2013/202497

3. Wang, T. J., Larson, M. G., Vasan, R. S., Cheng, S., Rhee, E. P., McCabe, E., … Gerszten, R. E. (2011) Metabolite profiles and the risk of developing diabetes. *Nature Medicine,* 17(4), 448–53 doi:10.1038/nm.2307

4. Bertini, I., Calabrò, A., De Carli, V., Luchinat, C., Nepi, S., Porfirio, B., … Tenori, L. (2009) The metabonomic signature of celiac disease. *Journal of Proteome Research,* 8(1), 170–7 doi:10.1021/pr800548z

5. Bernini, P., Bertini, I., Calabrò, A., la Marca, G., Lami, G., Luchinat, C., … Tenori, L. (2011) Are patients with potential celiac disease really potential? The answer of metabonomics. *Journal of Proteome Research,* 10(2), 714–21 doi:10.1021/pr100896s

6. Gruden, K., Hren, M., Herman, A., Blejec, A., Albrecht, T., Selbig, J., … Jeras, M. (2012) A "crossomics" study analysing variability of different components in peripheral blood of healthy caucasoid individuals. *PloS One,* 7(1), e28761 doi:10.1371/journal.pone.0028761

7. Assfalg, M., Bertini, I., Colangiuli, D., Luchinat, C., Schäfer, H., Schütz, B., & Spraul, M. (2008) Evidence of different metabolic phenotypes in humans. *Proceedings of the National Academy of Sciences of the United States of America,* 105(5), 1420–4 doi:10.1073/pnas.0705685105

8. Bernini, P., Bertini, I., Luchinat, C., Nepi, S., Saccenti, E., Schäfer, H., … Tenori, L. (2009) Individual human phenotypes in metabolic space and time. *Journal of Proteome Research,* 8(9), 4264–71 doi:10.1021/pr900344m

9. Heinzmann, S. S., Merrifield, C. A., Rezzi, S., Kochhar, S., Lindon, J. C., Holmes, E., & Nicholson, J. K. (2012) Stability and robustness of human metabolic phenotypes in response to sequential food challenges. *Journal of Proteome Research,* 11(2), 643–55 doi:10.1021/pr2005764

10. Godelmann, R., Fang, F., Humpfer, E., Schütz, B., Bansbach, M., Schäfer, H., & Spraul, M. (2013) Targeted and nontargeted wine analysis by (1)h NMR spectroscopy combined with multivariate statistical analysis. Differentiation of important parameters: grape variety, geographical origin, year of vintage. *Journal of Agricultural and Food Chemistry,* 61(23), 5610–9 doi:10.1021/jf400800d

11. Cacciatore, S., Luchinat, C., & Tenori, L. (2014) Knowledge discovery by accuracy maximization. *Proceedings of the National Academy of Sciences of the United States of America,* 111(14), 5117–22 doi:10.1073/pnas.1220873111

12. Pandey, V., Sultan, M., Kashofer, K., Ralser, M., Amstislavskiy, V., Starmann, J., … Wierling, C. (2014) Comparative Analysis and Modeling of the Severity of Steatohepatitis in DDC-Treated Mouse Strains. *PloS One,* 9(10), e111006 doi:10.1371/journal.pone.0111006

100 YEAR STARSHIP™

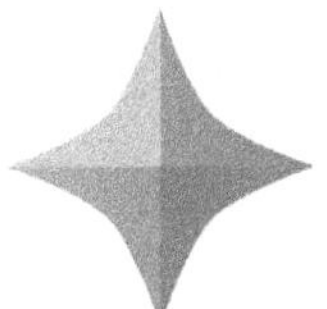

Chaired by Eric W. Davis, PhD and Hakeem Oluseyi, PhD

Track Description

How fast and how far can we travel? Fundamental breakthroughs in propulsion and energy are required for interstellar travel to be feasible. To overcome the formidable time-distance barrier for travel between stars, robust leaps in theory and engineering for energy production, control and storage must occur, as well as the advancement and demonstration of propulsion techniques.

Track Summary

Interstellar travel and exploration are not possible without starship propulsion and the energy required to power the propulsion system. How much elapsed time and how far through interstellar space can a starship go to deliver its mission payload is uniquely determined by the fastest type of propulsion system that achieves the shortest flight time over interstellar distances, and by how much energy can be generated to power the propulsion system. There are many other important engineering-physics parameters that have a strong impact on starship propulsion and energy. It is the purpose of the Propulsion and Energy Track to present the latest research and development on starship propulsion and energy concepts.

Session A of the track had one presentation that proposes the industrial production of antimatter using the Van Allen radiation belts as a source of copious amounts of naturally occurring antimatter. The second presentation was on the development of a new antimatter storage testbed to experimentally demonstrate the safe, long lifetime storage of antimatter. Antimatter is an exotic fuel comprised of elementary particles that are used in a matter-antimatter annihilation rocket engine, which is highly efficient and energetic enough to propel starships.

Session B of the track had one presentation that proposed using pulsed thermonuclear fusion micro-bombs to propel heavy-lift launch vehicles into orbit to deliver very large manned and related support payloads into space, which will in turn support manned or large robotic solar system planetary and interstellar missions. The second presentation proposed a new type of enhanced, very high specific impulse magnetoplasmadynamic thruster that can propel robotic interstellar flyby probes. The third presentation proposes a novel advanced ion thruster that uses a man-made version of the explosive energy released by the magnetic reconnection phenomenon that is observed in the solar atmosphere to highly energize and accelerate propellant ions to very high exhaust velocities, which will support interstellar flyby missions.

Session C of the track had a presentation on the fermion re-inflation of a Schwarzschild kugelblitz, which is proposed as a substantial source of energy to power a micro-black hole propulsion system for future interstellar flight. The second presentation was a review of the current status of the NASA advanced solar sail development program and its application to future interstellar sail flyby missions. The third presentation was a proposal to use macroscopic black holes to power a novel gravitational tandem spacecraft propulsion system for interstellar flight.

Most of the presentations given in this track were superb and of high professional quality. Many of them contributed valuable design studies and data that add to the ever-increasing knowledge base of interstellar flight science, as well as add to the foundations that will help guide mankind to the stars. However, only five of the presentations appear as full papers in this section of the proceedings.

Track Chairs Biography

Eric W. Davis, PhD

Senior Research Physicist, Institute for Advanced Studies

Eric Davis, PhD is a Senior Research Physicist at the Institute for Advanced Studies at Austin and is the co-editor/author of the first-ever academic research monograph on breakthrough propulsion physics for interstellar flight: Frontiers of Propulsion Science (AIAA Press). His research specializations include breakthrough propulsion physics, beamed energy and nuclear propulsion, general relativity theory, and quantum field theory. Since 1984, Dr. Davis has been a contractor/consultant to the USAF, Air Force Research Laboratory, Department of Defense, Department of Energy, and NASA. He has been featured in American and UK television and film productions, as well as in numerous news media articles, on interstellar flight.

Hakeem Oluseyi, PhD

Associate Professor, Florida Institute of Technology

Hakeem M. Oluseyi, PhD is an internationally recognized astrophysicist, science TV personality, and global science education activist. His research interests span the fields of astrophysics, cosmology, and technology development. He currently has 7 U.S. patents, 4 EU patents and over 60 scholarly publications in the areas of astrophysics, optics and detector technologies development; nanotechnology manufacturing; observational cosmology; and the history of astronomy. Dr. Oluseyi leads a group studying processes by which electromagnetic fields and plasmas interact in order to understand solar atmospheric heating and acceleration, which has resulted in a new in-space propulsion technology.

The Magnetic Reconnection Rocket: Advanced Ion Propulsion Inspired by Solar Particle Acceleration

David L. Chesny

Florida Institute of Technology, 150 W. University Blvd, Melbourne, FL 32901

OrangeWave Innovative Science, LLC, Moncks Corner SC, 29461

dchesny@my.fit.edu

Hakeem M. Oluseyi

Florida Institute of Technology, 150 W. University Blvd, Melbourne, FL 32901

Dave R. Valletta

RS&H, Merritt Island FL, 32953

N. Brice Orange

OrangeWave Innovative Science, LLC, Moncks Corner SC, 29461

Kareem Elashmawy

Florida Institute of Technology, 150 W. University Blvd, Melbourne, FL 32901

David L. Chesny, Hakeem M. Oluseyi, Dave R. Valletta, N. Brice Orange, and Kareem Elashmaw

Abstract

To achieve effective human travel within our solar system and beyond, we must move past our reliance on chemical propulsion. Chemical propulsion requires excessive fuel payloads, and its thrust magnitudes have reached a ceiling. Basic rocket science shows that higher spacecraft exhaust velocities directly contribute to increased thrust. Current ion propulsion technologies are effective for robotic spacecraft missions, provide advantages in lower fuel requirements, higher exhaust velocities, and higher efficiencies, but do not provide enough thrust to transport astronauts. Coulomb and Lorentz force-based ion thrusters still only induce exhaust velocities up to ~50 km s^{-1}. Thus, new mechanisms for accelerating ions to high speeds must be investigated. Solar flares and coronal mass ejections (CMEs) in the solar atmosphere are the most powerful explosions of plasma in the solar system, and are the result of a process known as magnetic reconnection. Particle speeds in the solar wind and CMEs range between 400-3,000 km s^{-1}, hinting at the potential of magnetic reconnection as a useful particle acceleration mechanism for in-space propulsion. In this feasibility study, it is shown that numerical results from the electromagnetic theory of reconnection about a magnetic neutral point can induce particle velocities of >3,000 km s^{-1}. The theory suggests that reconnection can be induced in the laboratory using existing technologies and result in a beam of high-speed plasma, contained within a volume of <1 m^3. This magnetic reconnection ion beam can supply thrusts of >3 N, bringing ion propulsion technology up to the thrust levels of chemical propulsion.

Keywords

propulsion, magnetic reconnection, plasma, interplanetary, interstellar

1. Introduction

In order to realistically expand the sphere of human influence to the outer solar system and beyond, spacecraft transit times must be reduced by an order of magnitude. Chemical, hypergolic, and compressed gas propulsion systems, as used during Apollo and the Space Shuttle programs, have reached their limit for the acceleration of spacecraft. Current electrostatic thrusters do not provide enough thrust to realistically transport astronaut payloads or astronauts. Systems close to their in-space testing phase like the Variable Specific Impulse Rocket (VASIMR) still only cut one-way Mars travel times to ~40 days with power requirements currently unavailable [1]. Thus, proposed human interplanetary missions have been forced to direct their selection criteria for astronauts away from scientific merit and physical ability to long-duration psychological endurance, forfeiting the former two along the way. For truly effective interplanetary missions, round trip travel time must be reduced from months or weeks to days. Electric ion thrusters currently provide thrust with exhaust velocities of ~30 km s^{-1} [2] while providing excellent specific impulses upwards of 3000 seconds. Thus, it is worth pursuing more powerful forms of ion propulsion that can provide enough thrust for human exploration. The key is to employ highly powerful and efficient mechanisms that generate extremely high exhaust velocities and that can be built using existing technology.

The most powerful explosions in our solar system are generated in solar flare phenomena. The extreme plasma heating and particle acceleration resulting from these flares is a consequence of the process of magnetic reconnection [3]. During magnetic reconnection, oppositely running magnetic fields in the presence of plasma experience non-ideal behavior which causes diffusion dynamics to change the connectivities of the field lines. A product of these dynamics is extreme particle acceleration resulting in plasma velocities proportional to the Alfvén speed [4]. Solar atmospheric phenomena contributing to coronal mass ejections, the solar wind, and coronal jet phenomena have been found to accelerate plasma at several hundreds of kilometers per second, at least an order of magnitude higher velocities than conventional ion rocket exhaust.

On Earth, harnessing this physical mechanism for practical use is only beginning Section 4.2, and solely for scientific study. However, there is one field that can benefit greatly from this type of plasma acceleration—spacecraft propulsion. In order to achieve interstellar travel, we must begin to expand our propulsion infrastructure to include the most powerful physical phenomena that we understand physically. The key to this concept comes directly from the simple equation for rocket thrust T,

$$T = u_{ex}\dot{m}_{ex} \tag{1}$$

where u_{ex} is the exhaust velocity with respect to the spacecraft and dm_{ex}/dt is the mass loss per unit time of the spacecraft, and is considered the mass loss rate of fuel. Thus, any order of magnitude increase in u_{ex} does the same for overall thrust on the spacecraft. This facet shows the importance of utilizing a mechanism that produces high exhaust velocities. The VASIMR ion propulsion system can produce exhaust velocities of ~50 km s^{-1} [5] and up to 5N of thrust [6]. Other advanced propulsion concepts have been proposed, but the construction of such conceptual systems are many years away from being feasible. The Alcubierre-White ``warp" drive [7] [8], capable of velocities up to 10*c*, is currently in the proof-of-concept phase, and may become the propulsion system for future generations. In addition, the ion propulsion concept of the quantum vacuum thruster (also from the White team) is in a similar conceptual phase.

In order to bridge the gap between current ion propulsion technologies and futuristic concepts, here it is proposed that an ion propulsion system with magnetic reconnection as the acceleration mechanism will be able to increase rocket thrust by at least one order of magnitude. The theory of magnetic reconnection (Sections 4.2 and 5) has advanced to the point where experimental setups have been built to directly observe the reconnection process. A logical next step will be to construct an experiment to observe the plasma trajectories resulting from reconnection at a magnetic neutral point (Section 5). It will be shown that a unique case of three-dimensional reconnection is ideal for the proposed ion propulsion system design.

This feasibility study will demonstrate that the technological readiness of the magnetic reconnection rocket (MRR) has advanced to the point that requires both extensive three-dimensional computational modeling and preliminary laboratory experimentation. The initial design described here can be refined and tested in order to be considered for future NASA missions.

2. Basics of Spacecraft Propulsion

The overall question is how to increase the propulsive force on a spacecraft to achieve higher transit velocities. The momentum equation gives rise to an expression for thrust (Equation 1) showing the importance of exhaust particle velocity. For the purpose of this study, to achieve a higher overall thrust the exhaust velocity must not only be orders of magnitude larger, but the mass loss per unit time must at least be comparable (same order of magnitude) to existing ion thrusters. Another important characteristic of ion rockets is their specific impulse. The specific impulse, I_{sp}, is a measure of the total impulse per unit weight of propellant provided for the spacecraft in Earth's gravitational field. A higher I_{sp} is desired as it is a measure of the efficiency of converting propellant kinetic energy to kinetic energy of the spacecraft. I_{sp} is defined as the exhaust velocity u_{ex} divided by the acceleration due to gravity at Earth's surface *g*,

$$I_{sp} = \frac{u_{ex}}{g_{\oplus}} \tag{2}$$

It can also be shown that the change in vehicle velocity Δu is dependent on the exhaust velocity and initial and final masses of the spacecraft after burnout M_o and M_b, respectively) as,

$$\Delta u = u_{ex} \ln \frac{M_o}{M_b} \tag{3}$$

Equation 3 shows the importance of utilizing powerful astrophysical processes to accelerate particles for the use of spacecraft propulsion. Any order of magnitude increase in u_{ex} will contribute significantly to the transit velocity of a vehicle.

For comparison, Table 1 gives some numerical specifications of current, in-use chemical and ion propulsion systems as a reference for comparison to the proposed system. It is these basic quantities that we will explore and solve for in order to deduce the usefulness of the proposed propulsion mechanism.

Table 1: Propulsion System Specifications

Propellant	uex (km s-1)	Input mass (g s-1)	Isp (s)	Thrust (N)	Max Velocity (km s-1)
Chemical	5–10	1	250–450	1	8–11
Ion	3–50	1.00E-003	2000–5000	1.00E-002	15
MRR	>3,600	1.00E-003	>3E5	>3	>7,400

3. Magnetic Reconnection

3.1 Two-Dimensional Reconnection

Non-uniform magnetic fields that come together undergo a topological change known as magnetic reconnection. Reconnection involving resistive plasmas converts magnetic energy into kinetic and thermal energy and induces particle acceleration. The first reconnection models were formulated independently by Sweet and Parker in 1956 and 1957, respectively, followed by the model of Petschek in 1964. The facet of these reconnection models of most interest here is the formation of a current sheet within the converging magnetic region (Figure 1). This X-point configuration is the most commonly studied reconnection scheme. In this region, incoming plasma is built up, forming what are known as magnetic islands [9]. Upon reaching a critical point, these plasma islands are liberated, expelling plasma in each horizontal direction at speeds proportional to the Alfvén velocity v_A (Figure 1),

$$v_A = \frac{B}{\sqrt{\mu_0 \rho}} \tag{4}$$

where B is the local magnetic field value, μ_o is the permeability of free space, and ρ is the mass density of the plasma.

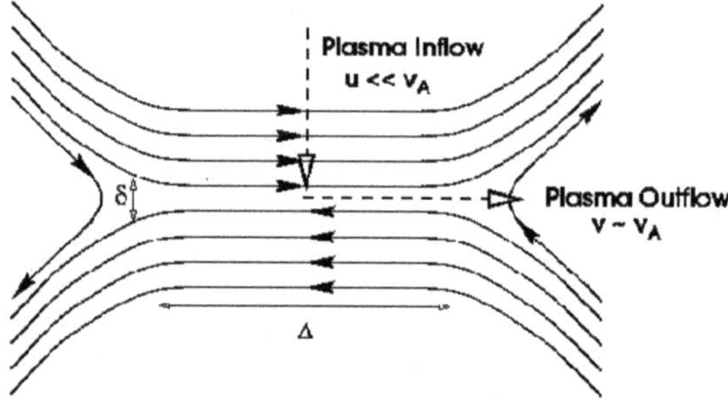

Figure 1: Topology of generalized 2D magnetic reconnection.

3.2 Three-Dimensional Reconnection

[10] gave a fully three-dimensional treatment for magnetic reconnection showing the resultant geometrical topologies. Using a three-dimensional stagnation point field, [10] considered solutions depending on three independent parameters, one exhibiting axisymmetry and one controlling stagnation point symmetry, without which curved field lines reconnect across a neutral point. They determine the field line equations close to the origin (reconnection site) as,

$$x^{\kappa}\left(y + \frac{a_1}{1+\kappa}x\right) = c_1 \tag{5}$$

$$x^{1-\kappa}\left(z + \frac{a_2}{2-\kappa}x\right) = c_2 \tag{6}$$

and solve for the separatrices by setting $c_1=c_2=0$. This results in the plane $x=0$ and the line $y=-a_1x/(1+\kappa)$, $z=-a_2x/(2-\kappa)$. These two results correspond to "fan" and "spine" reconnection, respectively. Numerous authors have employed the fan and spine configurations in explaining the physics of coronal jets (Section 4).

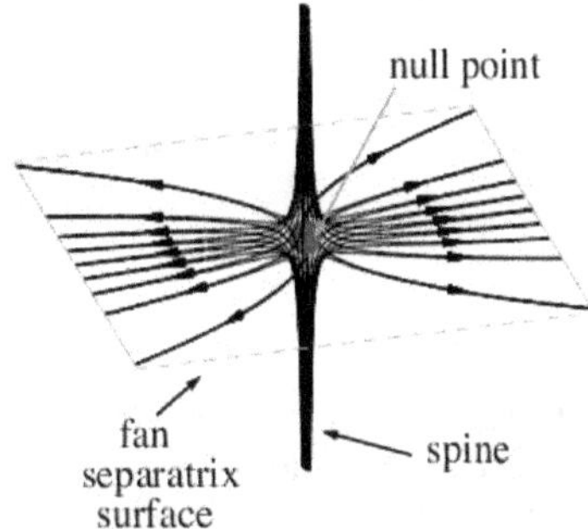

Figure 2: Geometry of three-dimensional torsional spine reconnection. Arrows show incoming field lines of the fan plane (negative neutral point: Section 5).

It is the geometry of torsional spine reconnection that is especially interesting (see Figure 2). Field lines coming in from the fan plane all reconnect at the neutral point forming narrow antiparallel cylindrical current structures. These compact, collimated collections of field lines are the basis for the design of the proposed propulsion mechanism (Section 9). It is our desire to produce a mechanical system that can induce the magnetic topology seen in Figure 2. Thus, the rest of this paper is dedicated to understanding the spatial dimensions, magnetic field components, and plasma instabilities that combined result in three-dimensional reconnection.

4. Observations of Magnetic Reconnection

4.1 Solar Atmospheric Fan and Spine Reconnection

As the heating source of the solar corona is an outstanding question in astrophysics, scientists have employed the dynamics of magnetic reconnection in both theoretical and computational studies of coronal heating scenarios. Due to the plasma conditions present in the corona, these magnetohydrodynamic (MHD) explanations have been shown to represent well the observed heating phenomena present in the solar atmosphere.

Recently, the basic topological signatures of reconnection about a magnetic neutral point (fan and spine) have been used in describing observations of solar coronal jets. The basic structure, as seen in Figure 3, is a canopy of field lines overlaying a bipolar magnetic field region. This canopy represents the fan plane, and after an instability forms a neutral point within the canopy, spine reconnection occurs and results in the well-documented observations of solar jet phenomena [11]. Numerous authors [12] [13] [14] [15] [16] have constructed three-dimensional simulations of the formation of 3D magnetic null points leading to fan and spine reconnection (Figure 4). These studies show null points are prolific in the solar atmosphere, many times resulting in non-potential magnetic field dynamics, and are prime regions for plasma heating and particle acceleration.

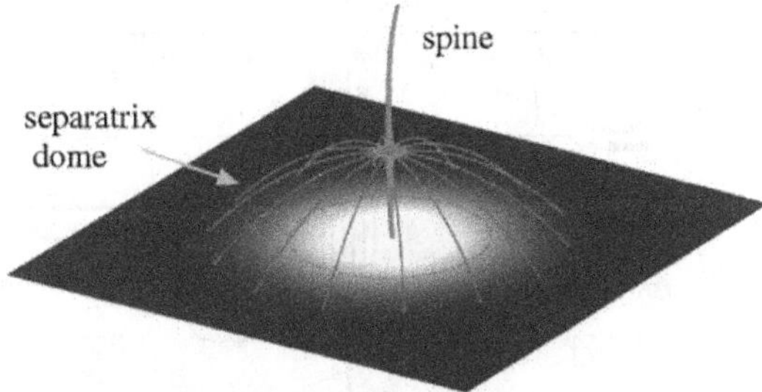

Figure 3: Three-dimensional representation of neutral point reconnection occurring above a solar bipolar region [11]. The fan plane is a canopy of field lines while the spine represents solar jet phenomena. Credit: Pontin et al. (2011).

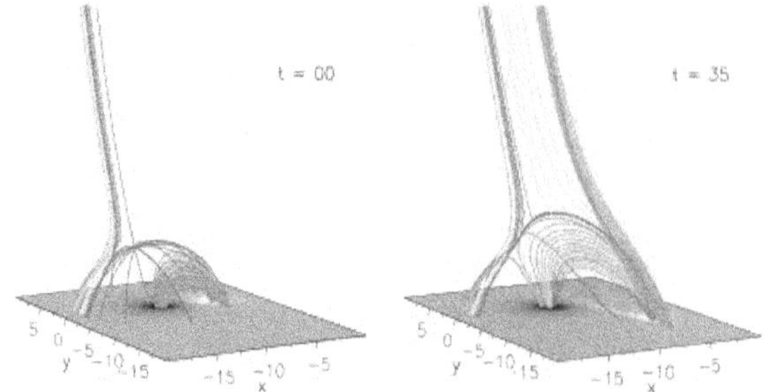

Figure 4: Null-point configurations at t=0 and t =35, respectively, from 3D simulation of a three-dimensional line-tied and initially current-free bipole, which is embedded in a non-uniform open potential field. Reconnection is induced via photospheric motions. The dark blue lines represent the fan and the spine separatrices passing through the null point. The colored field lines initially located in the outer connectivity domain are plotted from fixed footpoints and the group of colored field lines initially located below the fan surface are plotted from footpoints which are fixed in the advected flow [15]. Credit: Masson et al. (2012).

As it stands, the Sun is the best laboratory in which we can observe the powerful particle dynamics resulting from the physics of magnetic reconnection. Section 4.2 briefly describes efforts into inducing magnetic reconnection inside of toroidal cusps here on Earth, but the structure of null points has not been observed. This is mainly due to the fact that the experimental setups are not built to specifically produce individual null points, but only in hopes to observe particle acceleration of any kind resulting from reconnection. Thus, it is important to study the electromagnetic theory of magnetic neutral points which are known to pervade the solar corona. Understanding the concise mathematical form of magnetic neutral point reconnection leading to intense particle acceleration in a collimated form is central to the proposed ion propulsion mechanism.

4.2 Laboratory Reconnection

With the realization that numerous astrophysical processes involve magnetic reconnection, and that plasma confinement for fusion research is hindered by magnetic reconnection, major laboratory experiments have been constructed to study the plasma dynamics of the reconnection process.

The Princeton Plasma Physics Laboratory was one of the first to investigate the fundamental physics of magnetic reconnection (MRX experiment; Figure 5). With externally controlled boundary conditions, the device created an environment satisfying the criteria for magnetohydrodynamic (MHD) plasma. Initially it was found that two distinctive shapes of diffusion regions are possible during driven reconnection: "Y-shaped," as in 2D configurations such as Figure 1, on the size order of an ion gyroradius (~1 cm), and "O-shaped" of sizes 20–30 cm. However reconnection rates drop substantially due to toroidal field pressure and plasma incompressibility [17] [18] [19]. In addition, these types of reconnection were found to depend strongly on merging angles of source field lines.

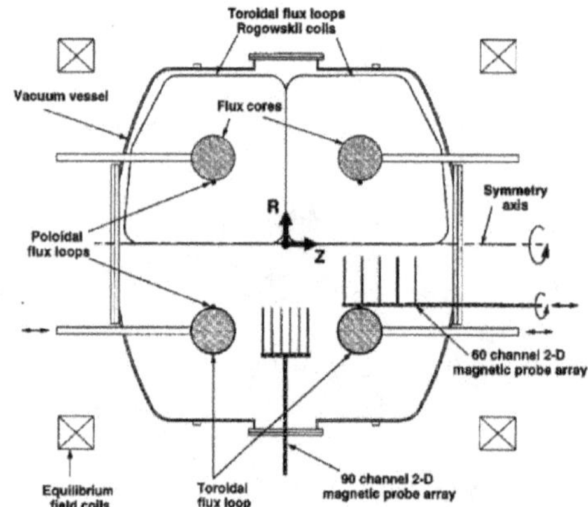

Figure 5: Cross-sectional view of the PPPL MRX spheromak. Credit: Yamada et al. (PPPL, 1997).

An experimental setup called the Versatile Toroidal Facility has been built at the MIT Plasma Physics Research Center [20] [21] [22]. This team has demonstrated that fast collisionless magnetic reconnection is routinely observed in their experiments, and that it occurs without the formation of macroscopic current sheets. It is only for large values of the toroidal magnetic field compared to the cusp field (cusp ~2% of toroidal) that a current channel is produced [23]. Reconnection is induced by sinusoidal current pulses in the center of the chamber (E~10 V m^{-1}). Thus, reconnection in this experiment is soley determined by imposing a given value of the induced toroidal electric field [23]. Their investigation of the dynamical plasma response to reconnection is that the size of the diffusion region is found to scale with the electron drift orbit width, and independent of both ion mass and plasma density. A cross-sectional sketch of the VTF is shown in Figure 6

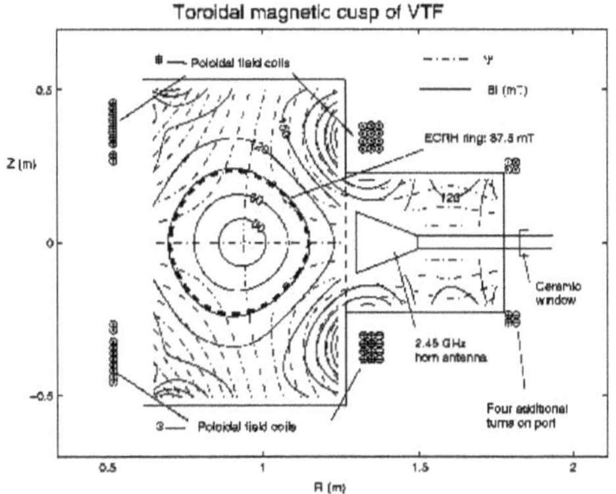

Figure 6: Poloidal cross section of VTF. The solid contour lines represent the poloidal magnetic field strength. The dashed contour lines correspond to constant levels of the poloidal magnetic flux which coincide with magnetic field lines. Credit: Edegal et al. (MIT, 2001).

These experiments demonstrate that macroscopic current sheets can be produced in a laboratory setting under a variety of boundary conditions. However, no experiment has been able to explicitly describe the particle acceleration or magnetic field topologies resulting from reconnection about any of the neutral points that formed during experimentation. Nevertheless, we know that mechanically producing a magnetic neutral point and reconnection is possible from these experiments. Studying the mathematical nature of neutral points may show us a way by which we can control reconnection and the resulting particle trajectories in a laboratory setting. In fact, the rest of the manuscript provides the framework with which we can set explicit boundary conditions that will allow us to control the site and topology of reconnection about an isolated magnetic neutral point.

5. The *M* Matrix

As seen in Section 3.2, magnetic reconnection can be induced about a magnetic neutral point. In order to mathematically describe a three-dimensional magnetic neutral point, we can assume the magnetic field approaches zero linearly. A first order Taylor expansion about the neutral point then shows the magnetic field forming the null can be expressed in matrix form,

$$\vec{B} \sim \vec{M} \cdot \vec{r} \tag{7}$$

where *M* is a 3x3 magnetization matrix and *r* is the position vector [24].

The trace of *M* must be zero as given by the constraint $\nabla \cdot \vec{B} = 0$. Thus, the sum of the eigenvalues, λ_1, λ_2, and λ_3, must also be zero and the associated eigenvectors are x_1, x_2, and x_3. Importantly, the matrix **M** determines all of the physical characteristics of the field including its structure, current, and associated Lorentz force [24].

[24] found that for three-dimensional magnetic neutral points, the number of free parameters determining the arrangement of field lines is four. The eigenvector x_3 with eigenvalue λ_3 defines the path of the spine, while the plane of the fan is defined by x_1, and x_2, and λ_2, and λ_3, respectively. The spine lies along the eigenvector of **M** that relates to the single eigenvalue whose sign is opposite to the real parts of the remaining eigenvalues, which define the fan plane. The field always has at least one real eigenvalue whose sign will always be opposite

to the real parts of the other two. Choosing a coordinate system so the spine eigenvector lies along the z-axis, [24] solved for the general matrix **M** as,

$$\vec{M} = \begin{vmatrix} 1 & \frac{1}{2}(q - j_{\parallel}) & 0 \\ \frac{1}{2}(q + j_{\parallel}) & p & 0 \\ 0 & j_{\perp} & -(p+1) \end{vmatrix} \tag{8}$$

where p and q are the necessary free parameters associated with the potential part of the field, and $j_{\parallel}$ and $j_{\perp}$ are the components of the current parallel and perpendicular to the spine, respectively. **M** has eigenvalues of the form,

$$\lambda_{1,2} = \frac{1}{2}(p+1) \pm \frac{1}{2}\sqrt{j_{thresh}^2 - j_{\parallel}^2} \tag{9}$$

$$\lambda_3 = -(p+1) \tag{10}$$

where a threshold current j_{thresh} is defined which depends solely on p and q as,

$$j_{thresh} = \sqrt{(p-1)^2 + q^2}. \tag{11}$$

The threshold current is an important factor of the magnitude of the current parallel to the spine.

It is important to note that the analysis of [24] investigated only the case $p \geq -1$ corresponding to positive neutral points (see below). It is a primary goal here to solve for a range of the free parameter $p < -1$, corresponding to negative neutral points.

An important result derived from Ampere's Law $\nabla \times \mathbf{B} = \mu_0 \mathbf{J}$ is that the source currents leading to the initial potential field of **M** are,

$$\vec{J} = \frac{1}{\mu_o}(j_{\perp}, 0, j_{\parallel}) \tag{12}$$

As we continue, we see that perpendicular source currents can lead to conditions ideal for the formation of a magnetic neutral point.

The direction of the field lines in the spine and fan are another constraint in this study. We want a configuration of fan field lines pointing into the null, and spine lines pointing away from the null. This is known as a negative neutral point. Conversely, spine lines pointing into the null and fan lines pointing out from the null are characteristics of positive neutral points. For a negative neutral point, the determinant of **M** must be less than zero, and the real parts of two eigenvalues must be negative, $\lambda_1, \lambda_2 < 0$. Here, we want to induce current parallel to the spine, so the number of free parameters is reduced to two [24].

At this point, we know a good amount of the constraints we want to impose on **M** in order to build a neutral point suitable for the proposed ion propulsion system. We wish to build a non-potential, negative neutral point. Thus, the det(**M**)>0, **M** must be asymmetric, $|\mathbf{J}|>0$, and $p<-1$. In a negative null, the real parts of two of the eigenvalues will be negative and the spine eigenvalue will be positive. The major axis of the fan is aligned along the vector associated with the fan eigenvalue with a larger real part. The threshold current value relative to $j_{\parallel}$ will determine whether **M** has real distinct, real repeated, or complex conjugate eigenvalues, which in turn determine the shape of the null [25]. It is necessary to apply these constraints to build a matrix **M** which will define the potential field leading to the neutral point by which we will induce torsional spine reconnection.

[25] states, "the parameters p and q determine the potential field of the null which is generated by the electric currents lying outside the considered domain. These currents are not unique, therefore, it is difficult to associate p and q with any real quantities apart from j_{thresh}." Therefore, it is conceivable to define an external current system which will produce the magnitude of **B** along the spine that will lead to an ion velocity of any

order of magnitude desired, further constraining the thrust in Equation 1.

6. Driving the Plasma Flow

The M matrix gives a magnetic field of the form,

$$\mathbf{B} = \left(\mathrm{x} + \frac{1}{2}(\mathrm{q} - \mathrm{j}_{\parallel})\mathrm{y},\ \frac{1}{2}(\mathrm{q} + \mathrm{j}_{\parallel})\mathrm{x} + \mathrm{py},\ \mathrm{j}_{\perp}\mathrm{y} - (\mathrm{p}+1)\mathrm{z} \right) \tag{13}$$

We wish to investigate how this initially potential magnetic field responds to small perturbations. Solutions of the linearized, low-β MHD equations give rise to expressions for the magnetic and velocity fields of neutral points from two types of perturbations [25]. Here they considered the magnetic field of Equation 13 with $q=j_{\perp}=0$. For the case of the initial perturbation inducing a current in the *z*-direction, *i.e.*, along the spine with only two free parameters p and $j_{\parallel}$ (Section 5), the current density has the solution,

$$\mathbf{J} = \mathrm{B}_0 \epsilon \mathrm{e}^{\omega \mathrm{t}}(0, 0, \mathrm{j}_{\parallel})/\mathrm{l} \tag{14}$$

where the growth rate ω is,

$$\omega = |p - 1|\, v_A / l \tag{15}$$

and l is the length scale of the parameter space, and v_A is the Alfvén velocity (Equation 4) for a hydromagnetic wave.

It was found by [26] that a circular magnetic field perturbation in the fan plane, centered around the spine, would induce current along the spine axis. The perturbation (Figure 7, left) would induce an ion flow down the fan lines into the null region and an intense current would propagate out from the null along the spine. Their numerical simulation showed such a response from the initially stationary ion field throughout the region (Figure 7, right). This is the type of perturbation that is necessary to induce in a mechanical system to drive such a collimated ion flow. It is important to consider the effects of plasma instabilities and turbulence in such studies, and the MHD equations used in the simulations of [26] do indeed include the necessary terms.

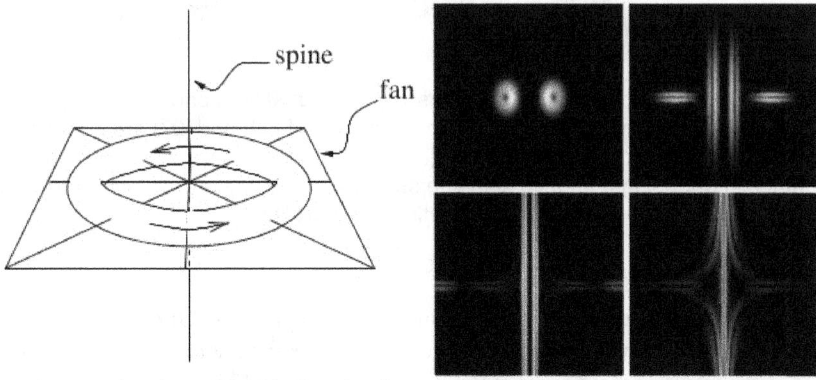

Figure 7: (left) Fan plane and spine axis with the circular magnetic field perturbation necessary to drive a current along the spine. (right) Numerical simulation of such a perturbation in an initially stationary ion field resulting in an intense current density along the spine (top to bottom, left to right). Credit: Pontin et al. (2007).

6.1 Reconnection and Release of Magnetic Energy

[27] produced simulations of 3D torsional spine reconnection where they imposed a magnetic field perturbation as a ring of magnetic flux centered on the null and lying in the fan plane, similar to [26]. The ingoing pulse gradually stretches out to form a cylindrical tube of intense current around the spine (Figure 8, left). [27] go on to state, "if it were feasible to drive the perturbation from the boundaries in a steady fashion, we would expect

this spine-localized current tube to persist in a quasi-steady state for a period controlled by the period of steady driving." This consequence is ideal for the proposed mechanical system.

The energy released (E) by the perturbation cannot be larger than the difference in the magnetic energy between the initial potential field configuration (B^f) and the final perturbed field (B^f). This is quantified by [28] in natural units as,

$$E \equiv \int (|\tilde{\mathbf{B}}^f|^2 - |\mathbf{B}^f|^2)d\mathbf{x} \tag{16}$$

Thus, the final energy released is dependent on the initial and final magnetic fields, and is thus also dependent on the change in magnetic flux induced by the presence of the current sheet,

$$\Delta E = -\frac{1}{c}\int_0^{\Delta\Psi_0} I(\Delta\Psi)d\Delta\Psi \tag{17}$$

These quantities can be informative in determining the plasma velocity in the spine and in calculating thrust, as well as determining the free magnetic energy that is available for the exhaust plasma. The induced field from the current sheet will reduce the strength of the reconnected field and will lower the resulting Alfvén velocity along the spine.

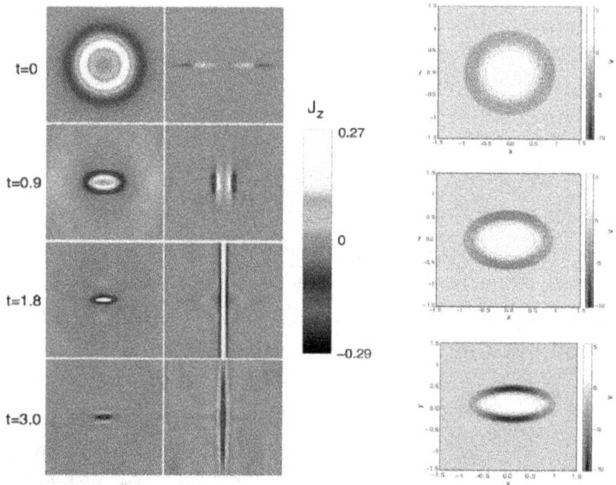

Figure 8: (left panel) Time sequence showing contours of j_z in the z=0 (left) and x=0 (right) planes for a torsional spine simulation with p=2. Note: the parameter p in the analyses of [24] and [27] are not necessarily equivalent. (right panel) Spine current cross-sections of [27]. More elliptical cross-sections correspond to higher asymmetries in opposing magnetic field line strength and a higher resulting current density. Credit: Pontin et al. (2011).

6.2 Boundary Conditions

[28] quantified current sheet formation along three-dimensional separators. One statement made by the authors is paramount to this proposed study. "The current sheet responsible for the reconnection must be posited as an initial condition or as the result of some outer magnetic configuration. This has led to the term *forced reconnection* since reconnection occurs in the local studies as an inevitable result of external boundary conditions...the current ribbon is a response to external changes imposed on the plasma," [28]. These comments allude to the driven formation of a null point in a mechanical system and that reconnection processes can provide the necessary dynamics for this type of an ion propulsion system. Reconnection does not occur "spontaneously," but as a consequence of local plasma perturbations that may or may not be controlled externally. However, our goal is to control the boundary conditions in such a way that the uncontrolled, or "spontaneous," perturbations are kept to a minimum.

7. Thrust

In order to find possible ion velocities resulting from reconnection about a negative neutral point, we investigate a range of values of the four free parameters comprising **M**. Solving Equations 13, 14, and 15 over a range of values for each parameter can give us a variety of types of neutral points to choose from. Negative neutral points are constrained to $p < -1$, so we solve over (-10.0, -1.1) in increments of 0.1. Without loss of generality, we can let $q=j_\perp=0$ [24] [25]. We vary $j_\|$ over a range of positive integers (1.0, 10.0) by increments of 0.1. The physically interesting solutions must satisfy three conditions: one, for negative nulls the real parts of the fan eigenvalues must be negative and the spine eigenvalue must be positive; two, the spine must be angled within 1° of normal to the fan plane to keep an ideal design geometry (Section 9); and three, the determinant of **M** must be greater than zero. In order to keep the angle within this 1° window, the calculation solves for the angle θ between the spine and fan plane [25] as,

$$\theta = \frac{\pi}{2} - \arccos\left(\frac{9(p+1)^2 - j_{thresh}^2 + j_\|^2}{|\mathbf{n}_{\text{fan}}|}\right) \tag{18}$$

where,

$$\mathbf{n}_{\text{fan}} = \left(2j_\perp(\sqrt{j_{\text{thresh}}^2 - (p-1)^2} + j_\|),\ -4j_\perp(p+2), 9(p+1)^2 - j_{\text{thresh}}^2 + j_\|^2\right) \tag{19}$$

Solving over these ranges and constraints, there exist 1,285 unique solutions out of 8,100 which produce negative neutral points within the necessary constraints. Equation 14 represents the resulting current density in the z-direction (along the spine). Converting this value to an ion velocity is done by simply dividing by the charge number density of the plasma and the charge of a singly ionized atom. We note that these are the properties of an ideal plasma, and are used as an approximation to determine the feasibility and usefulness of the proposed system. Using some specifications of known plasma systems, we use hydrogen gas with a charge number density of 10^{17} m^{-3} with the electric charge being $1.66\text{x}10^{-19}$ C in a magnetic field of 0.5 T. We consider Equation 14 at a time of 10^{-8} seconds. This time can be justified from Equation 15 for the growth rate. As we are considering a small scale system on the order of 1 m^{-3}, which will become clear in Section 9, the time it takes an Alfvén wave in the considered magnetic field strength to cross this length scale is in the order of 10^{-8} s.

Over our range of solutions, the resulting ion velocities are between 3,648–35,052 km s^{-1}. Thus, recalling current ion propulsion systems result in exhaust velocities of ~30 km s^{-1} (Section 1), our preliminary results provide evidence that increased spacecraft thrusts on the order of 10^2 can be achieved via magnetic reconnection. If we assume that a functioning magnetic reconnection rocket (MRR) can produce these ion velocities and has a similar mass loss rate dm_{ex}/dt as existing ion propulsion systems, we can achieve thrust values on the order of 3–30 N. Table 1 gives possible magnetic reconnection-based propulsion systems specifications. For the possible spacecraft speed, we use Equation 3 with the low-range exhaust velocity and assume the craft has the mass of the Voyager spacecraft plus a liquid hydrogen propellant tank the size of those used in the Space Shuttle external tank. The final velocity of a spacecraft using the MRR is 0.025c. Speeds of this magnitude will allow a spacecraft to reach the most nearby star systems in about 160 years, which is a vast improvement over the approximately 80,000 years it will take Voyager to reach those distances.

8. Modeling of Magnetic Neutral Points

In order to verify the theory of Section 5, we wish to visualize the potential field constructed from a given **M** matrix. Here, we employ the Python Enthought Tool Suite Mayavi application to produce a three-dimensional vector magnetic field which produces the desired negative neutral point. Mayavi is a GUI application that provides easy and interactive visualizations of 3D data inputs. Mayavi allows the user to easy visualize scalar and vector field data in two- and three-dimensions, and from any chosen angle.

The previous section gives the perturbed magnetic field with current induced along the spine from Equation 13. Using Python scripting, this is coded with each of the four parameters (p, q, $j_\perp$, $j_\|$) as free variables. Therefore, it is easy to switch between positive and negative neutral points, real and complex eigenvalues, skewed, critical, and spiral nulls, etc. Mayavi also allows for the construction of a custom parameter space, i.e. size of box and scaling of box. The authors [26] [27] carrying out neutral point reconnection simulations scale any of the *x*, *y*, or *z* axes in order to space out the plotted results for visual clarity. Such a technique will be employed here as well.

Using the parameters resulting in the lowest velocities from the 1,285 solutions, which are p=-3.8, q=0.0, $j_\perp$=0.0, and $j_\|$=4.7, we obtain the negative neutral point fields seen in Figure 9. These visualizations confirm the general form of a perturbed neutral point field exhibiting fan and spine structuring. Thus, as we proceed, we are confident the theory gives us the sufficient mathematical formulation needed to produce such a field in which to induce reconnection in the presence of plasma.

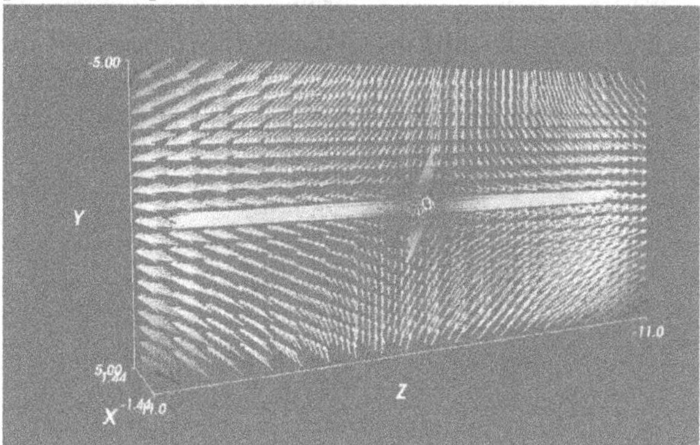

Figure 9: Vector magnetic field of perturbed negative neutral point in Mayavi with p=-3.8 and $j_\|$=4.7, corresponding to the lowest velocity investigated. The neutral point is located at (0, 0, 0). The streamlines show the path through the neutral point. The field is shown at [1, 3, 10] scale for visual clarity.

9. The Magnetic Reconnection Rocket

Translating the theory of magnetic neutral points into a mechanical system requires technology that can produce both the potential field and the necessary perturbation to that field. This section will present a preliminary design of a mechanical system that can induce the formation of a magnetic neutral point and lead to torsional spine reconnection.

9.1 Current Configuration

To provide the source currents that will form the necessary potential field, we simply utilize Equation 12. An interesting aspect of this result is that the source currents are perpendicular to one another. This is unambiguously similar to the classic "X-point" reconnection configuration in both theoretical two-dimensional studies and solar coronal reconnection schematics. We must note that we intentionally omit the displacement current source term in Ampere's Law, $\mu_o \epsilon_o \frac{\partial \vec{E}}{\partial t}$, due to the fact that we wish to isolate the production of the neutral point field to external currents alone. Were this propulsion system to be constructed and tested, this second term may play a role due to the presence of plasma in the neutral point region and the fact that the induced electric field of the reconnecting plasma directly contributes to the reconnection rate Ψ as,

$$\Psi = \int E_\| dz \tag{20}$$

where $E_\|$ is the value of the electric field along the spine, and *dz* is in the spine direction. This displacement current may be important due to the rapid nature of plasma production and evolution in the plasma production mechanism of a dense plasma focus apparatus, which is on the order of microseconds.

To produce the necessary perpendicular source currents, the existing technology of superconductors is ideal. Superconducting magnets are already in use in some propulsion systems, such as VASIMR, and produce magnetic fields of strengths up to 1 T. As our calculations in Table 1 were made with a source field value of 0.5 T, superconducting technology may certainly be utilized. We have produced COMSOL Multiphysics® simulations of the magnetic field produced by perpendicular current sources (Figure 10, left) offset by a few centimeters and show that the field in the center of the region where the neutral point will be induced is in a twisted shape (Figure 10, right). This is analogous to the twisted, helical fields in the solar atmosphere that have been shown to be prime sites for reconnection events.

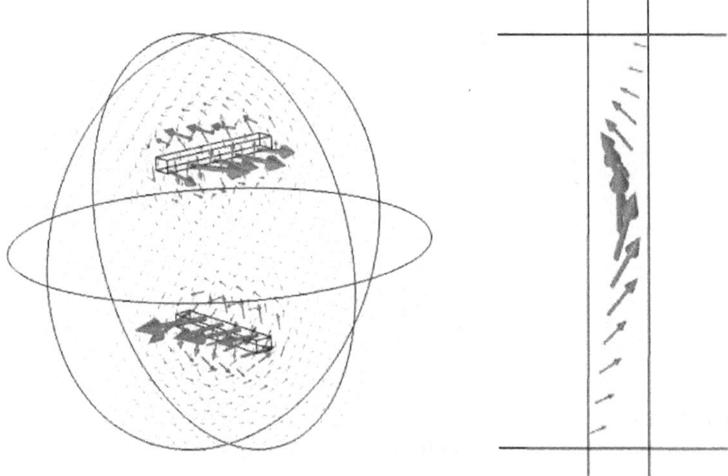

Figure 10: (left) COMSOL Multiphysics® simulation of the magnetic field produced by perpendicular source currents. Arrows depict vector direction and magnitude of the field at each point. (right) Zoomed in region along the spine axis of the source region. Field is strongest in the center and twisted along the spine axis, showing a prime site of reconnection.

9.2 Plasma Injection and Perturbation

We must identify a source of the necessary circular magnetic field perturbation to the initial potential field (Figure 10) that will result in current along the spine axis. It has been demonstrated with a dense plasma focus apparatus (DPF) that an azimuthal magnetic field can be formed under normal operation. DPFs are currently used for fusion research and remote sensing technology. They operate at very high power levels due to their requirement of initiating a fusion state. This existing technology can be modified to function at a lower initial power level with the specific geometry needed for the plasma flow from magnetic reconnection.

A DPF (Figure 11) consists of a central cylindrical electrode, typically a few centimeters in length and a centimeter or two in radius, surrounded by a sheath of outer cylindrical electrodes of the same geometry that are offset from the central electrode by two to three centimeters. The inner and outer electrodes are separated by an insulating material. Operation of a DPF occurs by charging a capacitor bank to a voltage of a few kilo-volts and discharging the potential difference across the inner and outer electrodes over a few microseconds. In the presence of an ambient gas, *e.g.*, H_2, a sufficient potential difference will ionize the gas between the electrodes and produce a plasma sheath (doughnut shape). The $\mathbf{J} \times \mathbf{B}$ force will then push the plasma sheath down the axial direction of the inner electrode, all the while the sheath produces an azimuthal magnetic field.

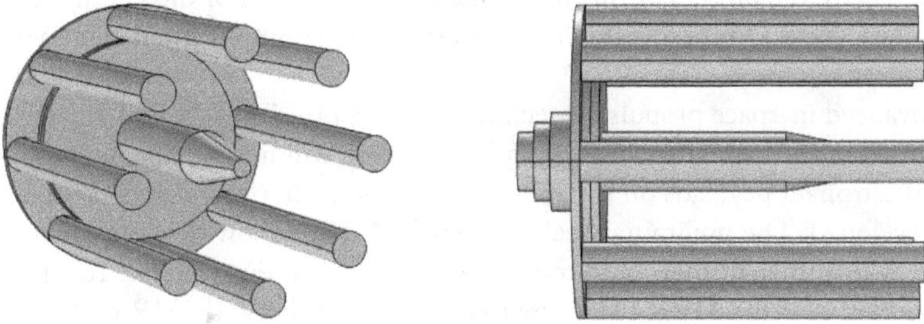

Figure 11: Geometry of a dense plasma focus made in COMSOL Multiphysics®. (left) From an angle, and (right) from the side. The inner and outer electrodes are 10 cm length.

9.3 Design

Combining these two technologies together is theorized to be able to produce the intense spine axis current shown in simulations by [26] and [27]. Figure 12 shows the preliminary design configuration of the MRR. The plasma sheath produced by the DPF will travel down the axis of the electrodes and enter the potential field region produced by the perpendicular source currents. The interaction of the two fields should induce reconnection between the two fields, and the resulting connectivities should point out of the region to the right. The plasma within the sheath should collapse into the newly formed neutral point and result in a beam of plasma with intense current density.

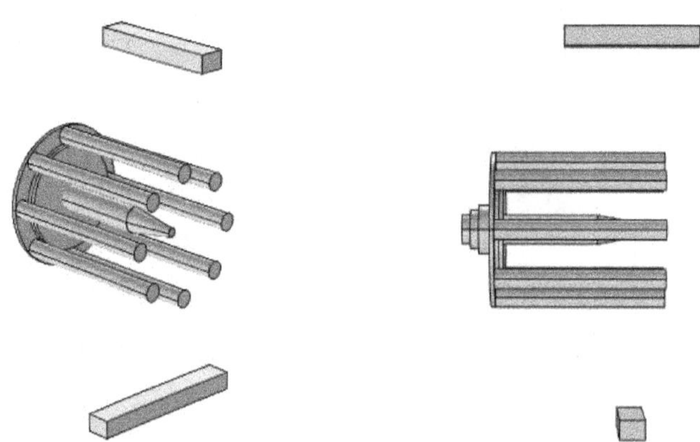

Figure 12: Preliminary design of the MRR. The DPF is located in the center, with the perpendicular source currents producing a potential field that is centered on the end of the DPF. (left) From and angle, and (right) from the side.

10. Conclusion

This feasibility study has investigated how fundamental magnetic reconnection processes can be utilized for a theoretical ion propulsion system. Particle acceleration in existing ion rockets is achieved with the Coulomb and Lorentz forces. While they produce exhaust velocities of ~30 km s^{-1} magnetic reconnection is the source of the most powerful explosions in the Solar System and can accelerate particles to hundreds, and even thousands, of km s^{-1} in solar atmospheric eruptions. Applying the theory of this scale-invariant reconnection process, it has been suggested that a mechanical system for inducing reconnection about a magnetic neutral point in an ideal laboratory setting can produce ion velocities two orders of magnitude higher than present propulsion capabilities. In addition, it is proposed that components of existing technology are sufficient for the production of the proposed system. Combined with the high theoretical velocity values arrived at here, it should be a high priority for this new type of ion propulsion system to be brought to a preliminary production phase in order to test the basic theory proposed here, and to address further engineering and plasma physics aspects, such as ion species populations and residual spacecraft charging.

Laboratory magnetic reconnection is only observed in tokamak-style experiments. In addition, no experiment exists that is able to examine the topological structure of magnetic neutral points. The preliminary design in Section 9 can be made to take up less than 1 m^3. If this setup is demonstrated to induce reconnection about a neutral point, it will be of interest to the solar and plasma physics communities to utilize this scaled-down setup for more in-depth study at comparatively lower cost.

While the theory suggests the usefulness of such an endeavor, it is imperative to provide further numerical modeling that can demonstrate the physics of this mechanical system. Much of the parameter space considered here is based on preliminary values. More investigation into the physical implications of varying these parameters is necessary.

Development of advanced in-space propulsion technologies is a priority of NASA operations. If the theory presented here is confirmed in the laboratory, this propulsion system will not only be powerful enough to transport astronauts and astronaut payloads on interplanetary missions, but could propel the highest velocity robotic spacecraft yet developed. The preliminary calculations in Table 1 show that the final speed of a Voyager-type spacecraft with sufficient fuel could travel at velocities exceeding 0.025*c*. It is left to the development phase to answer how the MRR compares to the thrust of the VASIMR rocket, which is in its testing phase. However, if the MRR can achieve the same mass loss rate as VASIMR, but with exhaust velocities two orders of magnitude higher, Mars round trip transit times can be reduced from ~100 days to only a few. The implications of this type of propulsion can only be speculated about. Ultimately, the final goal is still the same: decrease spacecraft transit times such that human interplanetary transport is achieved readily and our species becomes a citizen of the Cosmos.

Note: This technology is protected under the Florida Institute of Technology Intellectual Property Agreement; US patent pending, 61/939,120.

Bibliography

1. Ilin, A. V., Cassady, L. D., Glover, T. W., & Chang Diaz, F. C., 2011, in , 1-12

2. Charles, C. 2009, *Journal of Physics D Applied Physics*, 42, 163001

3. Cargill, P. 2013, Astronomy and Geophysics, 54, 030003

4. Parnell, C. E., Neukirch, T., Smith, J. M., & Priest, E. R. 1997, *Geophysical and Astrophysical Fluid Dynamics*, 84, 245

5. Squire, J., Olsen, C., Chang-Diaz, F., et al. 2012, in APS Meeting Abstracts, 3005

6. Bering, E. A., Longmier, B. W., Ballenger, M., et al. 2011, , 000, 26

7. Alcubierre, M. 1994, *Classical and Quantum Gravity*, 11, L73

8. White, H. G., & Davis, E. W. 2006, in *American Institute of Physics Conference Series, Vol. 813*, Space Technology and Applications International Forum - STAIF 2006, ed. M. S. El-Genk, 1382-1389

9. Saito, S., & Sakai, J. I. 2006, ApJ, 652, 793

10. Craig, I. J. D., Fabling, R. B., Henton, S. M., & Rickard, G. J., 1995, ApJ, 455, L197

11. Pontin, D. I., & Galsgaard, K. 2007, *Journal of Geophysical Research (Space Physics)*, 112, 3103

12. Torok, T., Aulanier, G., Schmieder, B., Reeves, K. K., & Golub, L. 2009, ApJ, 704, 485

13. Pariat, E., Antiochos, S. K., & DeVore, C. R. 2010, ApJ, 714, 1762

14. Liu, W., Berger, T. E., Title, A. M., Tarbell, T. D., & Low, B. C., 2011, ApJ, 728, 103

15. Masson, S., Aulanier, G., Pariat, E., & Klein, K.-L. 2012, Sol. Phys., 276, 199

16. Stanier, A., Browning, P., & Dalla, S. 2012, A&A, 542, A47

17. Yamada, M., Ji, H., Hsu, S., et al., 1997a, *Physical Review Letters*, 78, 3117

18. Yamada, M., Ji, H., Hsu, S., et al. 1997b, in *APS Meeting Abstracts*, 101

19. Yamada, M., Ji, H., Hsu, S., et al., 1997c, *Physics of Plasmas*, 4, 1936

20. Egedal, J., Fasoli, A., & Nazemi, J. 2001a, in *APS Meeting Abstracts*, 1006

21. Egedal, J., Fasoli, A., & Nazemi, J. 2001b, in *APS Meeting Abstracts*, 1135P

22. Egedal, J. 2004, in *APS April Meeting Abstracts*, B3003

23. Egedal, J., Fasoli, A., Tarkowski, D., & Scarabosio, A. 2001c, *Physics of Plasmas*, 8, 1935

24. Parnell, C. E., Smith, J. M., Neukirch, T., & Priest, E. R. 1996, *Physics of Plasmas*, 3, 759

25. Parnell, C. E., Neukirch, T., Smith, J. M., & Priest, E. R. 1997, *Geophysical and Astrophysical Fluid Dynamics*, 84, 245

26. Pontin, D. I., & Galsgaard, K. 2007, *Journal of Geophysical Research (Space Physics)*, 112, 3103

27. Pontin, D. I., Al-Hachami, A. K., & Galsgaard, K. 2011, A&A, 533, A78

28. Longcope, D. W., & Cowley, S. C. 1996, *Physics of Plasmas*, 3, 2885

Forecast of Breakthrough Trends in Technique and Technology of Interstellar Flights

Alexander Kazykin

Moskovskaya 228-41, Kaluga, Russian Federation, 248021

kazykin.kaluga@yandex.ru

Abstract

The experience of studying the problem of interstellar flights led the author to the conviction that the most powerful means of "fighting with space and time" is gravitation. Active usage of gravitation in spaceship's engines will mean qualitative leap in their development, opening the new era of space exploration - the era of interstellar flights. The author by means of theoretical modeling of mobile dynamic system with a compact concentrate of mass and a structural field coupling studied the features of a gravitational drive. The result of this study is the concept of a spaceship "Gravitational tandem" (GT). As a compact mass concentrate were considered low-mass black holes having masses of the order masses of asteroids ($10^{16} - 10^{20}$ kg). In the structure of GT such a black hole performs three functions: mass concentrate, a source of energy and a working body. Based on qualitative analysis and calculation modeling it was shown that the value of extreme acceleration of piloted systems type of GT is restricted only by tidal forces and may reach $10^4 - 10^5$ m/s^2 without causing excessive overloads in a spaceship.

The offered technology of interstellar flights allows to overcome the distance to the nearest star Proxima Centauri (4.3 light-years) in 9.5 hours; to the Andromeda Galaxy (2.2 mln light-years) - in 20.5 hours; to cross the Metagalaxy (13.7 bln light-years) in 28 hours.

Fundamental properties of gravity potentially confer spaceship GT with unique qualities, including inertialess principle of movement, non-rocket physical mechanism of acceleration, and the invariance of time in the Earth and spaceship systems of reference.

The combination of these three components into a single transportation technology leads to a drastic reduction in the duration of space flights and, as a consequence, to the unlimited expansion of humanity into the universe.

Keywords

acceleration, gravity, dynamic system, low-mass black hole

1. Introduction

Humanity is now taking its first steps into the universe as a space faring civilization. Thermo-chemical rocket engines are still used as the main "driving force" in space, which in the arsenal of space propulsion systems, are the most primitive. Therefore, the penetration into space at the present stage can be characterized as a "struggle with gravity". Increased penetration into space will require the creation of qualitatively new propulsion systems for which the priority is the "struggle with space and time". Paradoxically, gravity has the potential to become the powerful tool of "the struggle with space and time", which will provide humankind with the opportunity to breakthrough into deep space. The active use of gravity in spaceship propulsion would be a quantum leap in the development of space travel, opening a new era of space exploration—the era of interstellar flight.

2. Concept of Gravitational Tandem Starship

The idea of a Gravitational Tandem starship was born more than 30years ago in the ancient Russian city of Kaluga, the same place where in the early 20th century, the idea of the space rocket was first introduced by K.E. Tsiolkovsky, a brilliant self-taught scientist, recognized as the founder of theoretical astronautics.

2.1 Gravitational Tandem: the Idea of Active Use of Gravitation for Propulsion of Starships

Vast spaces that separate the stars are a major problem in the implementation of interstellar flight. Even light, which travels at nature's speed limit, requires many years to overcome the distance to the nearest stars. Many years of experience in the study of this issue has led the author to believe that the construction of a spacecraft with a gravitational drive will overcome the problems associated with interstellar distances.

Formally, the Gravitational Tandem (GT) is a mobile dynamic system with a compact concentrate of mass and a structural field coupling. In its simplest form, such a system consists of two independent modules: active and passive, between which there is no mechanical connection, except for the gravitational interaction. The active module is a compact concentrate of mass with a centrally symmetric gravitational field and is able to move in the range of linear accelerations of the order of 10^4 - 10^5 m/s^2. The physical mechanism of these accelerations can use either rocket or non-rocket mechanics (Fig. 1).

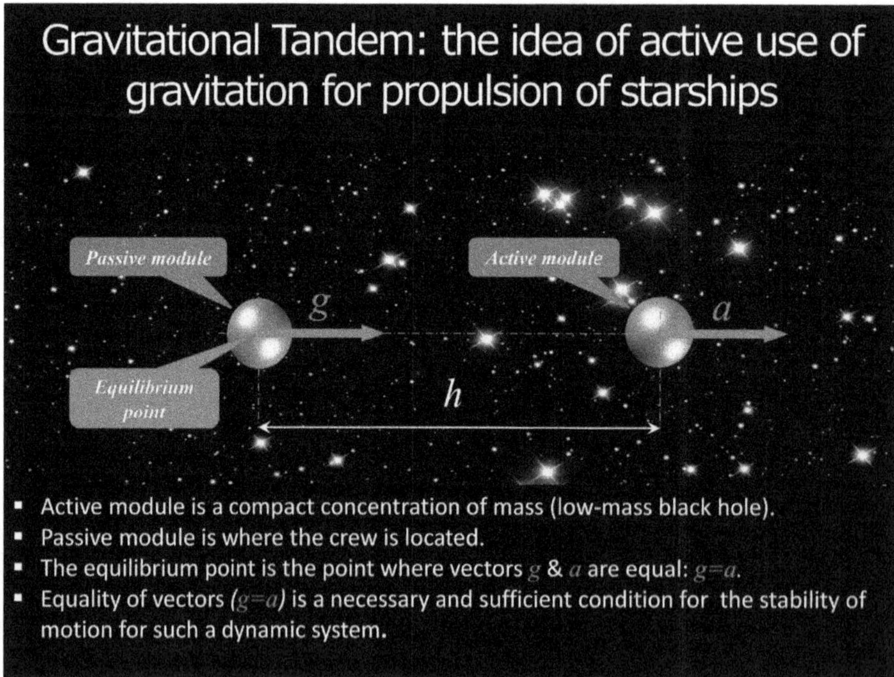

Figure 1: Principle of movement of GT Spaceship

At a distance h from the active module is a passive module, which holds the crew of the spaceship. The value of h is defined as follows: the gravitational acceleration g of the passive modules in the field of gravity and linear acceleration a of the active module are equal in magnitude and in the same direction. Strictly speaking, the center of mass of the passive module is in an equilibrium point of the system, where the equality of vectors:

$$g = a \quad (1)$$

Condition (1) defines the necessity and sufficiency of stability and structural integrity of GT. The same condition gives GT a unique quality - the ability to reach a huge acceleration without an overload in the crew compartment (in the passive module). Condition in the passive module, regardless of the magnitude of the linear acceleration of the system could be defined as close to weightlessness. This is due to the equality of inertial and gravitational forces. As part of a dynamic system, the passive module plays the role of a freely falling elevator: in the active phases of the movement, it is in a state of free fall in the gravitational field of the concentrate of mass.

2.2 Gravitational Tandem: Tidal Forces in Passive Module

Full mutual compensation of inertia and gravity takes place only in the equilibrium point of the system (in Fig. 1 it coincides with the point of application of the vector *g*). At all other points, there is a predominance of ether inertia or gravity, giving rise to so-called tidal forces. The field of tidal forces is a superposition of a centrally symmetric gravitational field of the concentrate of mass and a uniform field of inertial forces. An object placed in the equilibrium point undergoes longitudinal stretching and transverse compressing effect by tidal forces (Fig. 2).

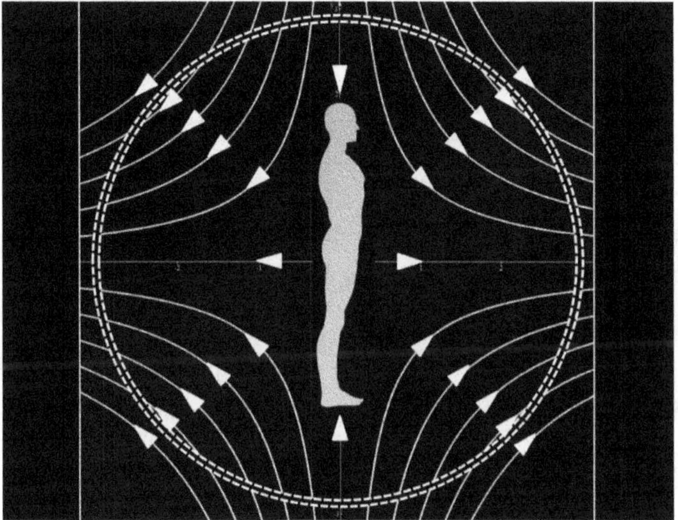

Figure 2: Structure of tidal forces in passive module

With increasing distance from the equilibrium point the gradient of gravity increases, so the magnitude of extreme linear accelerations of manned GT is limited only by the tidal forces. Increase acceleration g requires a reduction of the parameter h, but the approach to concentrate of mass leads to a rapid increase in the tidal forces that pose a serious danger to the astronauts and the structural integrity of the passive module (Fig. 3).

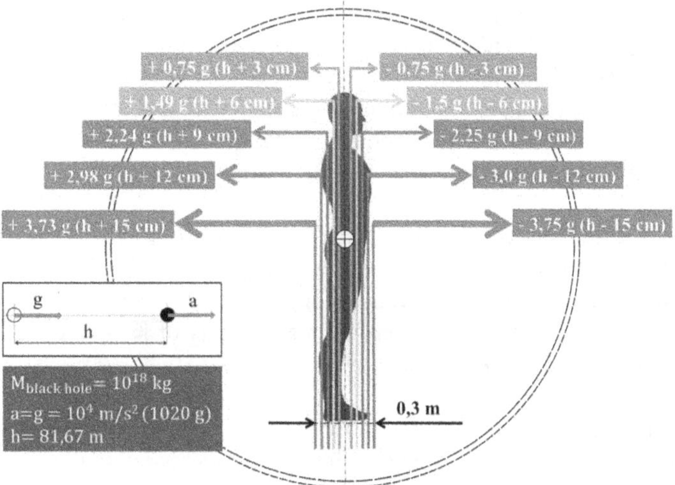

Figure 3: Magnitude of tidal forces in passive module

Table 1 shows the values of the masses of the active modules (black holes) and the corresponding maximum acceleration acceptable for a GT crew.

Table 1.

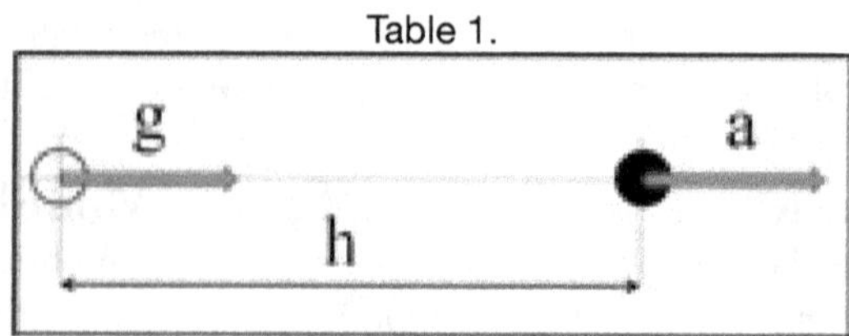

Mass of active module (black hole), kg	Acceleration (g = a), m/s2	Parameter h, m
1016	1500	21,1
1017	5000	36,5
1018	10000	81,6
1019	30000	149,1
1020	100000	258,2

Consider the following example. Let the mass of the active module be 10^{18} kg (1/260 of the mass of the asteroid Pallas). Then the linear acceleration of the system will be 10^4 m/s^2, which is more than a thousand times greater than the acceleration due to Earth's gravity. If the spacecraft is constantly moving with this acceleration, it will cross the orbit of the Moon in 270 seconds, the orbit of Mars in one hour, and the flight to Pluto, which is six billion miles from Earth, in a little more than nine hours. For comparison, light overcomes this distance in 5.5 hours. So gravity can be our great ally in spaceflight.

Even more so this applies to flights over interstellar distances. At a constant acceleration of 10^4 m/s^2 the distance to the nearest stars in the constellation of Centaurus (4.3 light years) takes about three days, and the diameter of our galaxy (100 thousand light years) takes 6.5 days. But this is not the limit. If the weight of the active module is increased to 10^{20} kg, acceleration of GT can be brought to the incredible value of 10^5 m/s^2. Then the length of uniformly accelerated flight to the Proxima Centauri will be reduced to 9.5 hours, and to the other end of the galactic disk would be down to 18 hours.

2.3 Gravitational Tandem: Use of Microscopic Black Holes for Propulsion Systems

As you can see, the active use of gravity opens up for astronautic a truly fantastic possibilities. But for this we need a concentrate of mass. It is play a key role in the propulsion system of the GT spacecraft. The hypothetical objects known as microscopiç black holes meet all the criteria of the concentrate of mass. Stephen Hawking postulated their existence in the 1970's. In contrast to the "normal" black holes formed in the gravitational collapse of stars, these objects are characterized by pronounced quantum properties and low weight (compared to the mass of a stars). However, they have a huge energetic potential, strong gravity, and are extremely compact [5]. In the structure of the GT spacecraft, a low-mass black hole has three main functions: a concentrate of mass, a working substance, and a source of energy.

Physical mechanisms of extracting energy from black holes have been studied for decades. For a rotating (Kerr) black hole, the Penrose mechanism was proposed. A Kerr black hole is characterized by two parameters - the mass and the angular momentum. If the black hole rotates with the maximum possible angular velocity, it is called extreme. The theory implies that 29% of the total mass-energy of Kerr black hole exists in the form of rotational energy [12, 14, 15]. Through the mechanism of the Penrose it can be completely removed and converted into useful work.

Physicists William Unruh and Robert Wald have offered a fundamentally different way of extracting energy from the black hole based on quantum phenomena occurring in strong gravitational fields. The polarization of the physical vacuum near the black hole leads to the phenomenon of quantum radiation. The power of this radiation is inversely proportional to the square of the mass. The greater the mass of the black hole, the more the huge gravitational force prevents spontaneous photons from flying into infinity. In other words, the quanta are behind the potential barrier. However, the intensity of the quantum evaporation of black holes can be controlled by external action [7, 9]. Quanta that are produced in large numbers from the vacuum near the event horizon, it is possible to push through a potential barrier in a special "box" with mirrored walls. If the box is lowered close to the black hole with an open lid, so that it will be filled with radiation, and then removed with the lid closed, we can "scoop out" energy from the black hole. This extractable power is not dependent on the mass of the black hole. Only the dimensions of the box determine it. In principle, it is limited to the Planck length (10^{-33} cm). In theory, the limit of power of such energetic machine reaches a truly monstrous magnitude - about 10^{52} watts [8, 9, 10]! In contrast to the natural course of the evaporation of a black hole, the "scooping" of radiation can be given an anisotropic character. If gamma – quanta is drawn not from the entire surface, but for example, only from one hemisphere, then it will lead to the phenomenon of the flow of negative energy impulse directed into a black hole. This flow force will act on the black hole and cause it to move. While lowering and lifting the box requires energy, the energy of the radiation contained in a raised box will be much more than these expenses. The extracted energy can be used for useful work, for example, to supply all spacecraft systems. Unspent energy would go back into the black hole, but from the opposite hemisphere. Repetition of such cycles provides a mechanism for linear acceleration of the black hole in the preferred direction. Changing the direction of the axis along which the box is lowering and lifted permits changing the direction of the force to control thrust vector. It is a closed traction system, which, with some reservations can be considered ecologically sterile as it emits almost nothing in the surrounding area (Fig. 4).

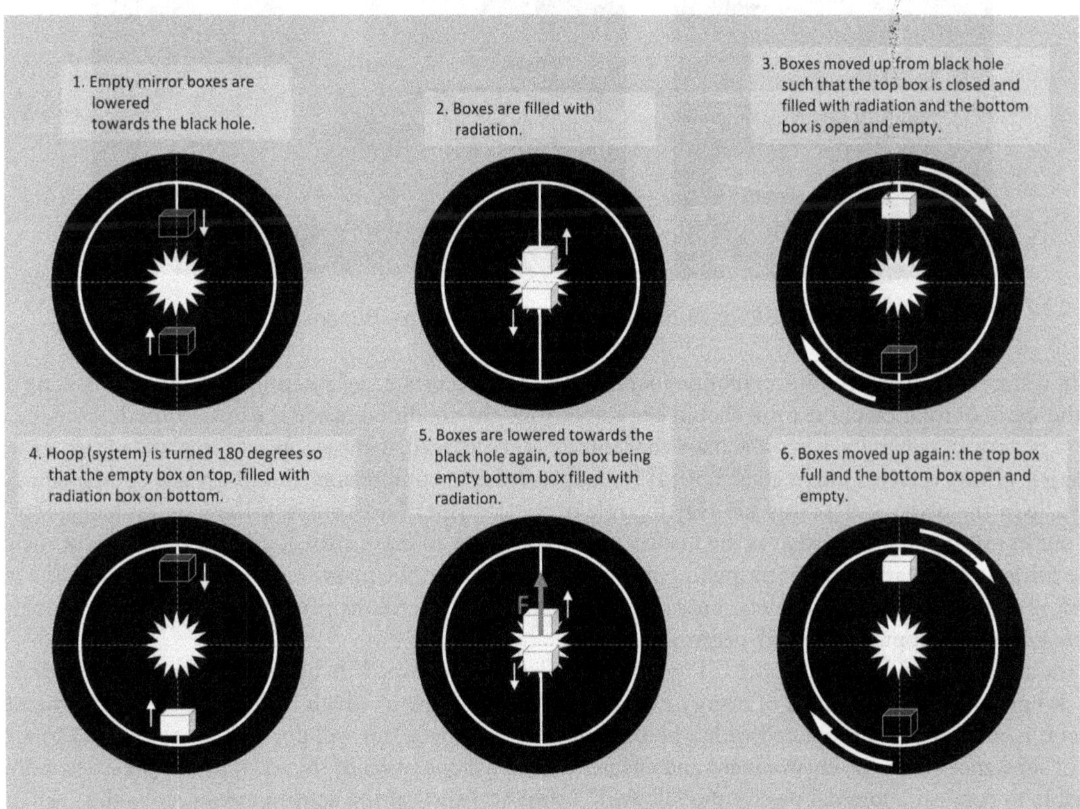

Figure 4: Extraction of energy from black hole by mechanism of W. Unruh and R. Wald

The huge energy expenses of a GT drive must be emphasized here: a 10^{20} kg black hole would need power on the order of 10^{33} watts for motion with constant acceleration of 10^5 m/s2 (for comparison, global power generation is limited to an order of 10^{13} watts). Although there are no fundamental restrictions here it is premature to discuss various technological nuances of a GT drive so, we only state that there is a theoretical framework in which to create it in the not too distant future.

2.4 Gravitational Tandem: Transformation of GT Structure

Scheme of spacecraft GT shown in Figure 1 is characterized by longitudinal axial instability and difficult transition from active to passive motion mode, and vice versa. The system will be dynamically stable and extremely easy to operate if the passive module is designed as a rotating hollow torus, as shown in Figure 5.

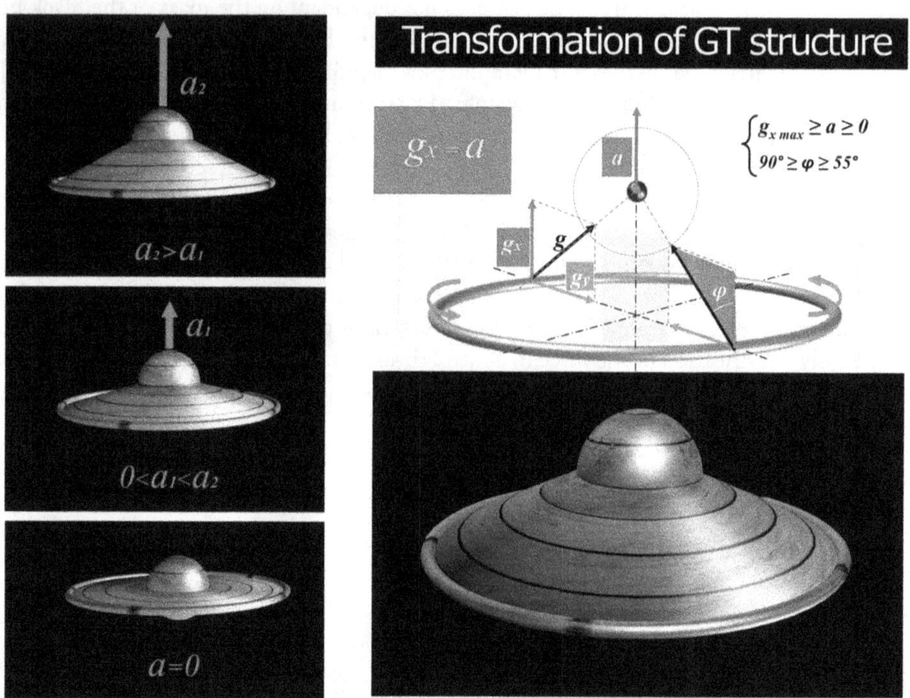

Figure 5: Evolution of Structural and Dynamic Scheme of GT

In rest mode, as well as uniform motion mode, the mass concentrate and rotating torus are on the same plane, and the speed of rotation of the torus should be orbital. Then the condition inside the passive module will be close to zero gravity. In dynamics, when the mass concentrate is accelerated, it comes out of the plane of the torus. The vector g is always directed into a black hole. If it is separated into components, the transverse component g_y will always lie in the plane of the torus and play the role of the centripetal acceleration. The longitudinal component gx is out in rest mode. But as soon as the mass concentrate begins to move with acceleration and out of the plane of the torus, it automatically appears, pulling passive module behind the mass concentrate. As a result, the whole system is set in motion with acceleration $g_x = a$. The value of g_x reaches its maximum when the angle between vectors gx and g is approximately 55 degrees.

The final shape of the spacecraft GT will be completed as follows. Firstly, the active module consisting of a black hole and the mechanism of energy extraction (the prototype of which is the "carousel" with the mirror boxes) it is necessary to surrounded with a hermetical spherical shell. This will prevent the absorption by a black hole of substances from the environment and ensure the normal operation of the traction complex. Secondly, it is advisable to connect a toroidal passive module with a spherical shell of the active module, since the principle of motion requires very precise positioning of the modules relative to each other. The mechanical connection between modules should not be rigid, but allow the ship to change configuration, as shown in Fig. 5. As a result of the evolution of structural and dynamic scheme and adaptation to the factors of space flight, the simple dynamical system with two modules is transformed into a rotating disk with a spherical surface in the center.

The Monoblock spaceship of variable geometry is adapted for flight in two environments - space and the dense layers of the atmosphere. However, here it is necessary to make an important caveat: for environmental security, only "light" vehicles with a weight not exceeding 10^{16} kg can dive into the dense layers of the atmosphere.

Figure 6: Gravitational Tandem Interstellar

3. Gravitational Tandem Interstellar

The duration of an interstellar flight is determined by the kinematic parameters - distance to the destination of travel and speed of the spacecraft. In addition, of considerable importance is the value of the acceleration in the active stages of flight: the greater the acceleration, the faster the spaceship will develop maximum speed. Using conventional rocket technology speed is limited to the speed of light, and the amount of acceleration is limited by the ability of the human body to carry long-acting overloads. Therefore, in the long interstellar flight, the proper acceleration of a rocket cannot significantly exceed the value of the acceleration of earth gravity - 9.8 m/s^2.

The situation has changed in the transition to the "inertialess" technology of movement. First, the use of a gravity drive permits the acceleration to be virtually any number. More importantly, the active use of gravity opens the possibility (in principle) to dramatically reduce the duration of spaceflight not only for the starship clock, but also for the Earth clock.

In special relativity (SR), the speed of light plays the role of a fundamental constant and is an insurmountable barrier [1]. In general relativity (GR), the speed of light is a function of the spatial coordinates. In the reference system where there is a gravitational field, the speed of light varies from point to point proportionally to the change of gravitational potential:

$$c = c0\ (1 + \Phi/c2) \qquad (2)$$

where Φ - gravitational potential,
c_0 - speed of light in the origin of coordinates [2].

Within the framework of general relativity, the speed of light as well as the speed of material bodies may take any numerical value [13]. The influence of the gravitational potential also applies to the speed of the clock: in areas with a large gravitational potential time goes faster, and it goes slower in an area with a small gravitational potential.

It is this fact that gives a full explanation of the "twin paradox" that if we take twin brothers, and one goes into interstellar travel at relativistic speeds, and the other remains on Earth, then when brother astronaut returns home, he would be much younger than his Earth brother. In the frame of special relativity, the "twin paradox" is not solved. The Earth is an inertial reference system and the spaceship is not because it is moving with acceleration or deceleration. However, the accelerated reference systems are not within the competence of the special relativity. Equality of inertial and accelerated reference systems is established in the general relativity. According to Einstein's equivalence principle, no one physical experiment can distinguish the state of rest in the gravitational field from the accelerated motion in free space, i.e. the field of forces of inertia and gravitation are absolutely identical [3]. Accelerated reference system can be considered as "hanging" in a homogeneous gravitational field, in which all surrounding bodies fall freely. Such a reference system is sometimes called "Rindler coordinate system" since Wolfgang Rindler introduced it. This is the reference system of the relativistic rocket during acceleration and deceleration. On defining stages of interstellar flight, the rocket with brother astronaut relatively to the Earth is located in areas with lower gravitational potential. Since the gravitational potential is directly proportional to the distance, the difference in age of the twin brothers will increase as the spacecraft travels farther into the depths of interstellar space.

Now consider the case when a vehicle of interstellar flight GT is used. The passive module housing the brother astronaut falls freely in the gravitational field of the active module. And his movement takes place throughout the entire flight in a homogeneous gravitational field. To an Earth-bound observer, the movement of spaceship GT looks like a freefall in the Rindler reference system. In the ship's reference system the Earth falls freely in an asymptotically homogeneous gravitational field. This implies a very important conclusion: the active use of gravity in motion technology sets the absolute equality between the Earth (fixed) and spaceship (accelerated) reference systems. As a consequence of this equality, the duration of the flight on spaceship is equal to the time passing on Earth.

Full-time flight under ideal conditions can define by the formula:

$$\tau = \frac{2c}{a} Arsh \frac{aS}{2c^2}, \qquad (3)$$

where a – acceleration,
c – speed of light,
S – distance.

Table 2 (a, b, c) shows the values of accelerations of the GT starship and the corresponding durations of interstellar flights according to (3).

Table 2 a

Object	Acceleration (a)	Distance, (S)	Full-time one-way flight (τ)
Proxima Centauri	105 m/c2	4,3 light-years	17,9 hours
Gliese 581 g	105 m/c2	20,3 light-years	20,42 hours
Vela Pulsar	105 m/c2	1000 light-years	26,94 hours
Diameter of the Galaxy	105 m/c2	105 light-years	34,62 hours

Table 2 b

Object	Acceleration (a)	Distance, (S)	Full-time one-way flight (τ)
Proxima Centauri	104 m/c2	4,3 light-years	5,86 days
Gliese 581 g	104 m/c2	20,3 light-years	6,91 days
Vela Pulsar	104 m/c2	1000 light-years	9,63 days
Diameter of the Galaxy	104 m/c2	105 light-years	12,8 days

Table 2 c

Object	Acceleration (a)	Distance, (S)	Full-time one-way flight (τ)
Proxima Centauri	0,38*104 m/c2	4,3 light-years	13,61 days
Gliese 581 g	0,38*104 m/c2	20,3 light-years	16,42 days
Vela Pulsar	0,38*104 m/c2	1000 light-years	23,57 days
Diameter of the Galaxy	0,38*104 m/c2	105 light-years	32 days

4. Conclusion

Fundamental properties of gravity potentially confer spaceship GT with unique qualities, including:

- the inertialless principle of movement,
- invariance of the rate of time in earthly and spaceship reference systems,

- gravitational mechanism of acceleration.

The combination of these three components into a single transportation technology leads to a drastic reduction in the duration of interstellar flights and, as a consequence, to unlimited expansion of the sphere of potential penetration of the humankind into the universe in future.

References

1. Einstein A., *Complete scientific works*, Moscow, Nauka, 1965, v. 1, pp. 7-35.

2. Einstein A., *Complete scientific works*, Moscow, Nauka, 1965, v. 1, pp. 165-174.

3. Einstein A., *Complete scientific works*, Moscow, Nauka, 1965, v. 1, pp. 616-624.

4. Hawking S.W., Mon. Not. RAS, 1971, v. 152, p. 152.

5. Hawking S.W., *Nature*, 1974, v. 248, p. 30.

6. Unruh W.G., Wald R.M., Phys. Rev. Ser. D, 1982, v. 25, p. 942.

7. Unruh W.G., Wald R.M., Phys. Rev. Ser. D, 1983a, v. 27, p. 2271.

8. Unruh W.G., Wald R.M., GRG, 1983b, v. 15, p. 195.

9. Новиков И.Д., Фролов В.П. Физика чёрных дыр - М.: Наука, 1986, стр. 262. (Novikov I.D., Frolov V.P., "Physics of the black holes", Moscow, Nauka, 1986, p. 262).

10. Новиков И.Д. Чёрная дыра как тепловая машина и квантовый источник энергии - Земля и Вселенная, 1986, т. 2, стр. 19. (Novikov I.D., "Black hole as a heat engine and the quantum energy source", Earth and the Universe, 1986, v. 2, p. 19).

11. Giacconi R., Ruffini R., "Physics and Astrophysics of Neutron Stars and Black Holes", North-Holland, Amsterdam, 1978.

12. Ruffini R., "About the gravitationally collapsed objects", Moscow, Mir, 1982, p. 397.

13. Born M., "Einstein's theory of relativity", Moscow, Mir, 1972, p. 347.

14. Kaufman W., "The cosmic frontiers of general relativity", Moscow, Mir, 1981, pp. 208-210.

15. Nicolson I., "Gravity, black holes and the universe", Moscow, Mir, 1983, pp. 117-121.

Simulation of the Enhanced Magnetoplasmadynamic Thruster for Interstellar Flight

C. A. Barry Stoute

Alumnus from Department of Earth and Space Science and Engineering, Lassonde School of Engineering, York University, Toronto, Ontario, M3J1P3, Canada

Abstract

Magnetoplasmadynamic (MPD) propulsion is a type of technology which the propellant is accelerated by electromagnetic forces. This technology has the potential of propelling spacecraft in deep-space for long periods of time. The fuel efficiency of MPDs is ten folds higher than conventional chemical rockets with the same propellant. The drawback is the thruster requires large power generators to have sustainable propulsion. Power generators such as fusion or antimatter reactors expel a significant amount of heat energy. The heat expelled from the generators can be recycled to increase the performance of the MPD. This maximizes the energy from the power and propulsion. The goal of this research is to boost the MPD performance by adding a converging-diverging variable hypersonic nozzle and transfer the excess heat from generators to the thruster. It is predicted that the performance increase is between 5 and 15 percent. In this work, the enhanced MPD thruster is simulated using computational fluid dynamics model which the MacCormack's method is implemented. One set of results show the inlets are varied from 0^0 to 15^0 without the heat exchanger. The second set of the results show the inlets are fixed at 5^0 with the heat exchanger from the power source. The quantities measured are the thrust and the specific impulse (the amount of force per mass flow rate at Earth's gravity). The results are the specific impulses varied between 964 seconds and 1060 seconds and the thrust varied between 1670 N and 2230 N using argon as the propellant. Adding the heat exchanger from the power plant of the spacecraft, the performance increases between 2% and 8%, assuming 100% efficiency of the heat exchanger.

Nomenclature

$e^{+,-}$	=	positron and electron ($\pm 1.602 \cdot 10^{-19}$ C)
ϵ_0	=	Permittivity of Free Space ($8.854 \cdot 10^{-12}$ F/m)
E	=	Electric Field (V/m)
E_i	=	Ionization Energy (eV)
F_{Thrust}	=	Overall Thrust (N)
Ψ	=	Power Density of Inelastic Collisions (W/m^3)

γ	=	Heat Capacity Ratio
g_o	=	acceleration of gravity at Earth's surface (9.806 m/s^2)
h	=	enthalpy (kJ/kg)
Δh_{kmol}	=	Change in enthalpy (kJ/kmol)
I_{sp}	=	Specific Impulse (s)
J	=	Current Density (A/m^2)
k_B	=	Boltzmann Constant ($1.38 \cdot 10^{-23}$ J/K)
$\dot{m}$	=	Mass Flow Rate (kg/s)
M	=	Atomic Mass (kg/kmol)
n_n/n_e	=	Neutral/Electron Density
$\dot{n}_e$	=	*Electron Production Rate*
v_e	=	Electron Neutrino
μ_0	=	Permeability of Free Space ($4\pi \cdot 10^{-7}$ H/m)
P_a	=	Atmospheric Pressure (Pa)
P_e	=	Exhaust Pressure (Pa)
P_t	=	Throat Pressure (Pa)
Q_{abs}	=	Power Absorption Density (W/m^3)
Q_{heat}	=	Heat Exchange Between Particles (W/m^3)
r	=	radius (m)
R	=	Universal Gas Constant (8314 J·kmol^{-1}·K^{-1})
T_e/T_h	=	Electron/Heavy Particles Temperature (K)
v_{ex}	=	Exhaust Velocity (m/s)

1. Introduction

In the early 21st century, propulsion technology shifted from chemical rockets to electric propulsion for deep space flight. Even with leaps of electric propulsion, such as ion and magnetoplasmadynamic (MPD) thrusters, the current technology is not sufficient for interstellar flight. The drawback of ion thrusters is the low thrust it provides [1,2,3]. It is excellent for unmanned spacecraft because the computers can be activated once the spacecraft has arrived at its destination. However, in manned flight, the lack of thrust makes interstellar flight time-consuming [3].

A MPD thruster produces and propels plasma with the use of a large direct current electric supply that runs continuously [3,4]. This method requires a large power source, in the order of megawatts, to have a sustainable propulsion at a relatively high thrust delivery (for an electric propulsion system). Alteriatively, the MPD thruster can run at quasi-steady state. Quasi-steady state is a state which the thruster operates in pulses within a 1 μs period, typically. This state uses fast-switching capacitors for the MPD propulsion [3].

One method of enhancing the MPD thruster is to add a hypersonic nozzle and recycle to excess heat from the power plants to the engine itself. This has been tested in a miniature electric thruster developed by the author of this paper, Stoute [5]. Figure 1 shows the operation of the miniature electric thruster, named hybrid electric thruster for its dual electrostatic and electromagnetic technologies.

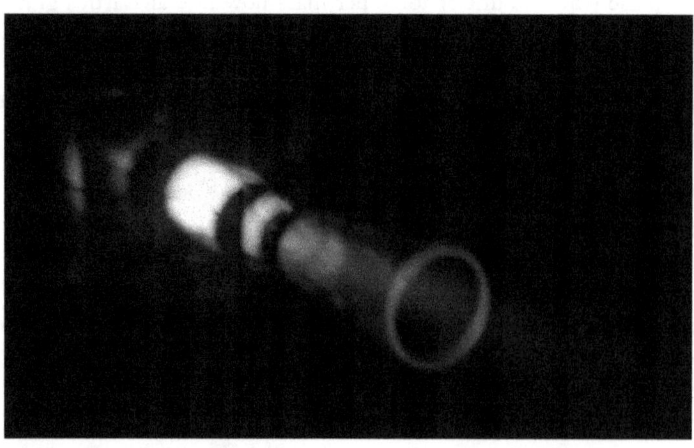

Figure 1: Hybrid Electric Thruster in Operation [5]

The technology has been tested using the radiating heat from the solenoid of the electronics. Reference [5] describes more about the technology.

The enhanced MPD thruster has thermionic generators placed at the throat which captures the heat from the nozzle and recycles the energy. The thermionic generators are not discussed further in this paper because of limits imposed in this publishing. The computer model of the enhanced MPD thruster is shown in figure 2.

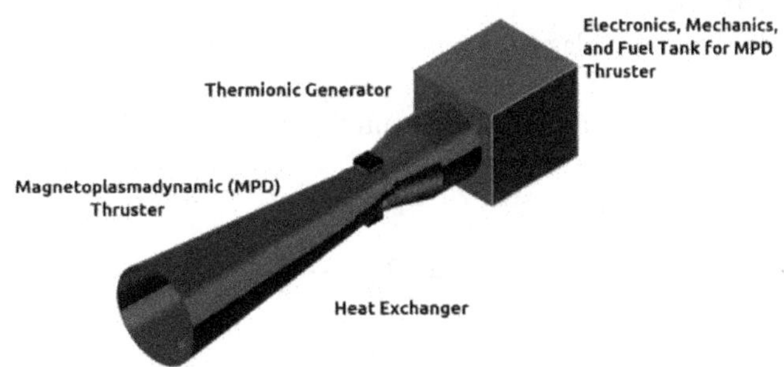

Figure 2: Computer Model of the Enhanced MPD Thruster

The paper discusses the performance of the EMPD thruster by its specific impulse and thrust. In additions, temperature distributions are also discussed. Boundary layers, shock waves, and other factors are not discussed in this paper. The paper first discusses the background theory: propulsion theory, plasma generation, and continuum mechanics [1,2,3,4]. The three sections of the background theory provide the reader an overview on how the simulation functions. Note that all the units are in SI. The methodology has the numeral analysis discussion, and an overview of the initial conditions used in the simulations. The results show selected computational fluid dynamics (CFD) results showing the fluid velocity and temperature. This section also discusses the results gathered from the CFD simulations and charted them. There is a discussion of the charted results. Finally, the conclusion of the simulations and what further work can be done, including experimental work.

2. Background Theory

2.1 Propulsion Theory

The propulsion theory discusses the nozzle theory. This entails the resulting pressure and area ratios from the heat capacity ratio and the Mach number of the propellant. These calculations yield the results for the exhaust velocity, thrust, and the specific impulse [4].

Following *Sutton & Biblarz*, The exhaust velocity determines the specific impulse of the rocket. The exhaust velocity v_{ex} is given by

$$v_{ex} = \sqrt{T \cdot \frac{R}{M_{molecular}} \cdot \frac{2\gamma}{\gamma+1} \cdot \left(1 - \left(\frac{P_e}{P_t}\right)^{\frac{(\gamma-1)}{\gamma}}\right)}, \tag{1}$$

where T, γ, P_e, P_t, R and $M_{molecular}$ are the gas temperature, heat capacity ratio, exhaust pressure, throat pressure, the universal gas constant, and the molecular mass, respectively. Given the exhaust velocity, the overall thrust, F_{thrust}, is [4]

$$F_{thrust} = \dot{m} \cdot v_{ex}, \tag{2}$$

where $\dot{m}$ is the mass flow rate.

Specific impulse describes the efficiency of a jet, rocket, or space engine. It is defined as the exhaust velocity of the propellant divided by the Earth's gravitational acceleration, g_0, at sea level; or, the amount of thrust divided by the accelerated mass flow rate of the propellant,

$$I_{sp} = \frac{v_{ex}}{g_0} = \frac{F_{thrust}}{\dot{m} g_0}. \tag{3}$$

Specific impulse indicates how efficiently the engine uses its fuel for propulsion. The higher the specific impulse, the more fuel efficient the engine is. A higher mass flow rate or exhaust velocity increases the thrust [4].

2.2 Plasma Generation

The first part of the simulation is the plasma generation. These equations are only time-dependent since the purpose is to find the temperature of the plasma. The plasma generation equations, which describe the temperatures and densities of the electrons, ions, and neutral particles [1,6,7],

$$\frac{d}{dt} n_e(t) = \dot{n}_e \ , \tag{4a}$$

$$\frac{d}{dt}\left(\frac{3}{2} n_e(t) k_B T_e(t)\right) = Q_{abs} - \Psi \ , \tag{4b}$$

$$\frac{d}{dt} n_n(t) = -\dot{n}_e \ , \tag{4c}$$

$$\frac{d}{dt}\left(\frac{3}{2} n_n(t) k_B T_h(t)\right) = Q_{heat} \ , \tag{4d}$$

where n_e, n_n, $\dot{n}_e$, k_B, T_e, T_h, Q_{abs}, Q_{heat}, and Ψ are the electron density, neutral density, electron production rate, Boltzmann constant, electron temperature, temperature of the heavy particles, absorption power density, the heat exchange between particles, and the power density of inelastic collisions, respectively. The temperature of the heavy particles is comprised of the temperatures of the ions and neutral particles. The absorption power density is the absorption power per unit volume [1,6,7]. Full disclosure of the equations is in reference [5].

The electron density and temperature are entered in the heat equation to solve for the overall temperature distribution of the plasma. The boundary conditions are the temperatures along the nozzle wall and cathode tip are kept constant, C_{temp}, at 500 K [8],

$$T(r)|_{r=anode wall} = T(r)|_{r=cathode wall} = C_{temp} \tag{5}$$

The heat equation is simplified to solve for the steady state of the plasma, and the equation varies along the radial direction such that,

$$\frac{1}{r}\frac{\partial}{\partial r}\left(r\kappa \cdot \frac{\partial T}{\partial r}\right) + Q_{abs} = 0 \tag{6}$$

where the value κ is the thermal conductivity of the plasma. After solving for the plasma temperature, the value is entered in the continuum mechanics equations [1,8].

2.3 Continuum Mechanics

This section discusses the continuum mechanics of the plasma, it is more commonly known as magnetoplasmadynamic (MPD) equations. In the MPD equations, the stress due to the walls of the nozzle, the force due to collisions, and any external forces are ignored for calculation purposes. The MPD equations, equation set 8, are used to simulate the plasma flow through the thruster [9].

$$\frac{\partial}{\partial t}\underbrace{\begin{bmatrix} \rho \\ \rho v \\ \mathcal{E} \\ \mathbf{B} \end{bmatrix}}_{\mathbf{V}} + \nabla \cdot \underbrace{\begin{bmatrix} \rho v \\ \rho v v \\ \mathcal{E} v - \overleftrightarrow{\mathcal{B}} \cdot v \\ v\mathbf{B} - \mathbf{B}v \end{bmatrix}}_{\mathbf{X}_{conv}} = \nabla \cdot \underbrace{\begin{bmatrix} 0 \\ \overleftrightarrow{\mathcal{B}} - \overleftrightarrow{\mathbf{p}} \\ \mathbf{q} \\ \overleftrightarrow{\mathcal{E}}_{res} \end{bmatrix}}_{\mathbf{X}_{diss}}. \tag{7}$$

Equation set 7 is derived from *K. Sankaran* in order to solve using the MacCormack's method (please see ref. [10]). The Lorentz force is in the form of the divergence of Maxwell stress tensor $\overleftrightarrow{\mathcal{B}}$ without the temporal derivative of directional energy flux density [9],

$$\overleftrightarrow{\mathcal{B}} = \epsilon_0 E_i E_j + \frac{1}{\mu_0} B_i B_j - \frac{1}{2}\left(\epsilon_0 E^2 + \frac{1}{\mu_0} B^2\right)\delta_{ij}. \tag{8}$$

The isotropic pressure tensor is represented by $\overleftrightarrow{\mathbf{p}}$. The divergence of the resistive diffusion tensor, $\overleftrightarrow{\boldsymbol{\mathcal{E}}}_{res}$ represents the resistive diffusion as an input to Faraday's law [9]. The full explanation of the continuum mechanics equations are in references [5,9].

The exhaust velocity from the MPD thruster reaches its peak velocity when the gas is fully ionized. A peak velocity is achieved when the kinetic energy is equal to the ionization energy of the particle. This phenomenon is called Alfvèn's critical velocity [3],

$$v_{critical} = \sqrt{2 \cdot \frac{E_i}{m_i}}. \tag{9}$$

3. Methodology

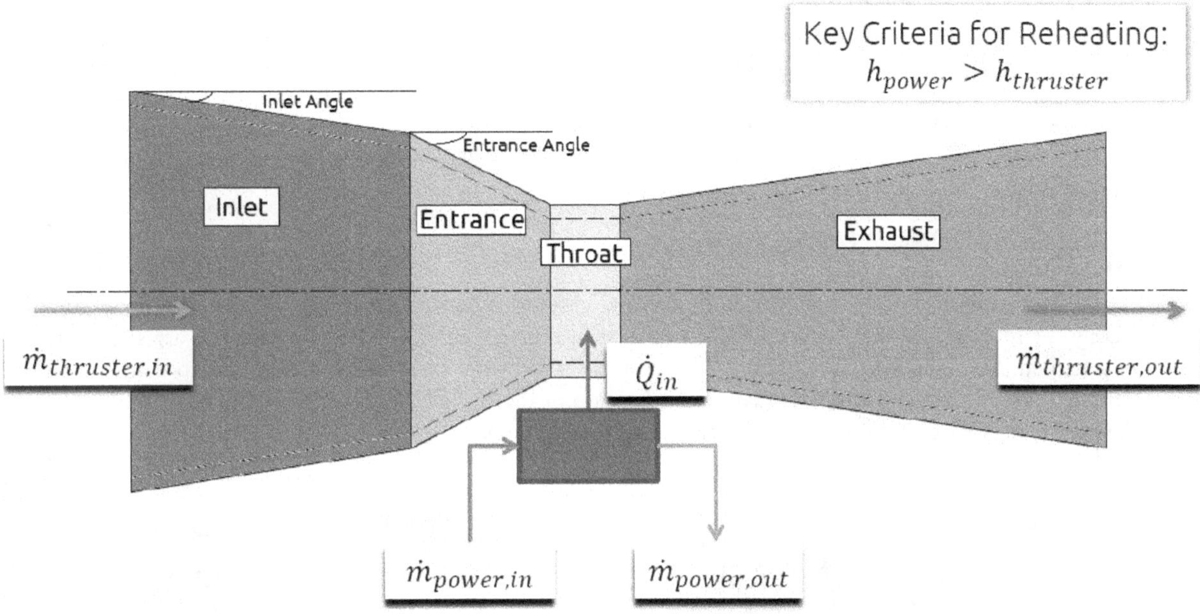

Figure 3: Diagram of the hypersonic nozzle of the Enhanced Magnetoplasmadynamic Thruster

Figure 3 illustrates the diagram of the hypersonic nozzle of the enhanced MPD. In the numerical model, the nozzle is divided into two sections. The first section consists of the inlet, entrance, and throat. This section models as the plasma propelled from the MPD and entered in the nozzle. The second section models the exhaust and its surrounding area. The results from the first section are transferred as the initial conditions for the second section. This ensures continuity of plasma flow model.

The heat exchanger is located at the throat, and the heat, $\dot{Q}_{in}$, flows from the exchanger to the nozzle. To ensure that the heat flows in the proper direction, the enthalpy of the exchanger, h_{power}, must be higher than the enthalpy of the thruster, $h_{thruster}$. Otherwise, the plasma cools down, and the process is rendered futile. The mass flow rates of both the power, $\dot{m}_{power}$, and the thruster, $\dot{m}_{thruster}$, are constant.

Table 1: Initial Conditions

Properties	Value
Propellant	Argon
Ambient Temperature	293 K
Ambient Pressure	100 Pa
Inlet Pressure	5000 Pa
Voltage	10 000 V
Current	9 200 A

Table 1 shows the initial conditions implemented in the simulations. The main propellant used in the simulation is argon. This choice is arbitrary; however, other propellants, such as nitrogen and helium, can be used. The ambient temperature in the simulation is chosen to be at room temperature, arbitrarily. It is known that the temperature in this solar system, near Earth, fluctuates between 230 K and 350 K. Ideally, the ambient pressure is 10^{-3} Pa but due to the finite precision of the numerical model, the program cannot go below 1 Pa. Therefore, the ambient pressure is set at 100 Pa to ensure accurate results, which are normalized. The inlet pressure is 5000 Pa, which corresponds to the ionization voltage of 10 000 V. The current is 9200 A simulating the discharge of a 330 μF capacitor rated at 10 000 V. The Alfven critical velocity and specific impulse using argon are 8.5 km/s or 870 seconds, respectively.

4. Results

The results sections consist of the flow velocity, temperature distribution, and the specific impulse results. Only the results from the 5^0 inlet – 9.25^0 entrance and the 15^0 inlet – 0^0 entrance models are shown in the flow velocity and temperature distribution. The purpose of showing only selected results is to compare the shocks entering the throat, and the temperature distribution of the throat. The specific impulses of the different inlet-entrance angles are charted, along with the heat exchange results.

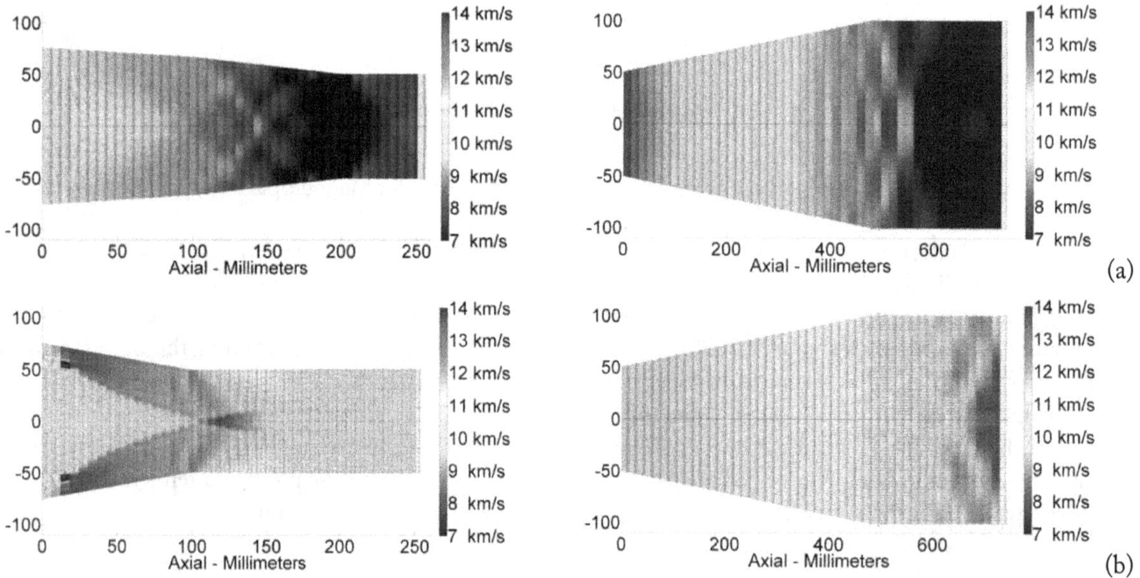

Figure 4: Velocity Models of the nozzle as (a) the inlet angle is at 50 and the entrance angle is at 9.250 and (b) the inlet angle is at 150 and the entrance angle is at 00

Figure 4 shows the velocity model of the nozzle. Part (a) where the inlet angle is at 5^0, there are significant shocks entering the nozzle. Further along the nozzle towards throat, the velocity slowed down. However, as the flow exits the exhaust portion, the velocities increase. Part (b) where the inlet angle is 15^0, the shocks are present, yet the flow velocity did not slow down as much as the 5^0 inlet. In addition, the exhaust velocity is significantly faster.

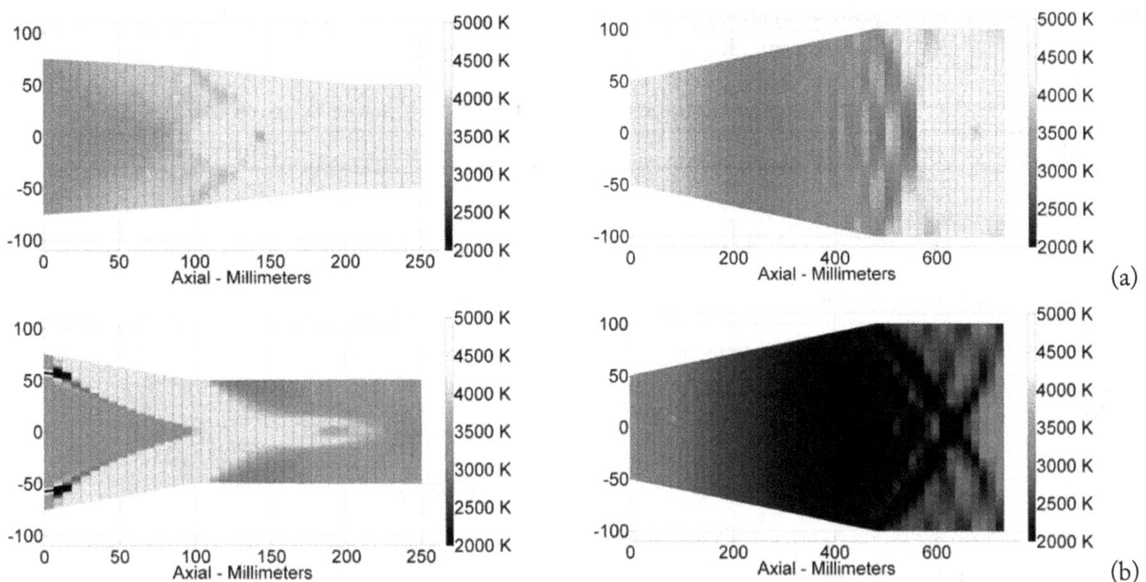

Figure 5: Temperature model of the nozzle as (a) the inlet angle is at 50 and the entrance angle is at 9.250 and (b) the inlet angle is at 150 and the entrance angle is at 00

Figure 5 shows the temperature distribution of the 5^0 and 15^0 inlet nozzles. The temperature of the 15^0 inlet at the throat is 3500 K. This is excellent to prevent any material degradation over a long period of time. In addition, the efficiency of the heat exchanger is higher than the 5^0 inlet because of the lower heat at the nozzle. In contrast, the temperature at the throat of 5^0 inlet is at least 5000 K. This does cause problems in maintenance of the nozzle over a long period of time. It is recommended to use carbon-fiber-reinforced polymer to withstand the heat inside. In addition, the heat exchange between the power plant and the nozzle would be less efficient, even a detriment because the plasma may cool down. However, at that temperature, other means of heat exchange can be achieved. The discussion and conclusion section of the paper examines the other means of the heat exchange.

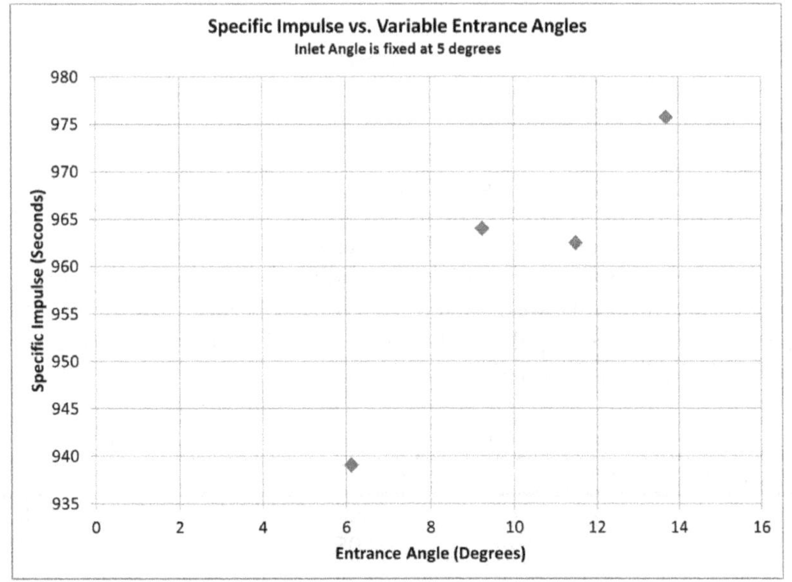

Figure 6: Specific Impulse vs. Entrance Angle where the Inlet Angle is set at 50

Figure 6 shows the specific impulse versus the entrance angle setting the inlet angle to 5^0. As the entrance angle increases, the specific impulse increases. However, between 9^0 and 12^0, the specific impulse drops slightly.

After 12^0, the specific impulse increases. The results signify that between 9^0 and 12^0, the shocks in the nozzle decrease the flow speed. The corresponding thrust is shown in table 2.

Table 2: Thrust vs. Entrance Angle where the Inlet Angle is set at 50

Entrance Angle (Degrees)	Thrust (N)	Thrust Increase (%)
6.13	3681	-6.50
9.25	3937	+0.00
11.5	4113	+4.47
13.7	4296	+9.11

Despite that the specific impulse stagnates between 9^0 and 12^0, the thrust increased by 4.5%. In addition, the thrust increased by another 4.61% as the entrance angle increases to 14^0. Therefore, the thrust increase is significant when increasing the entrance angle between 9^0 and 14^0.

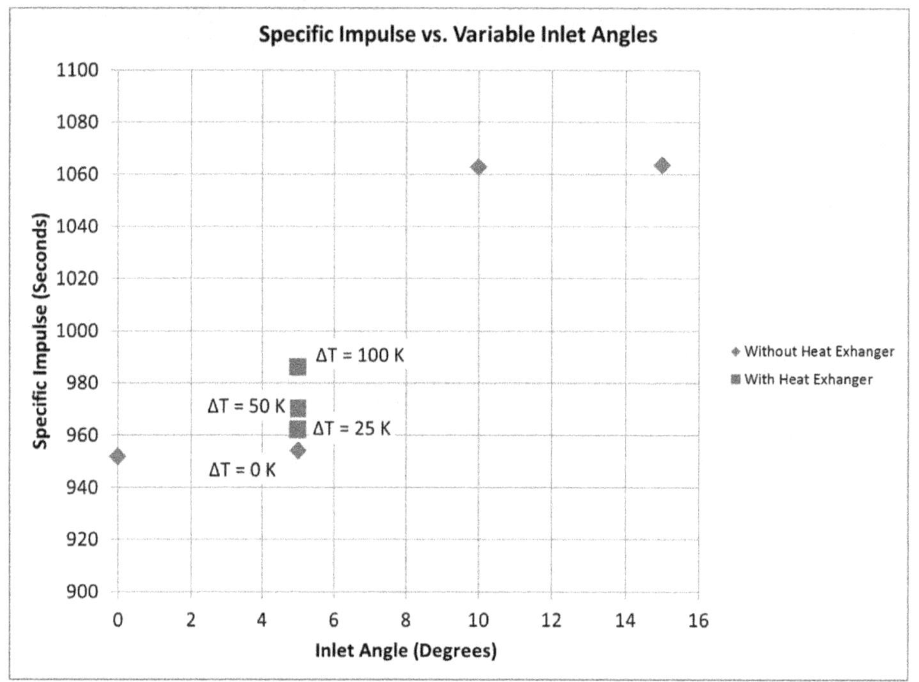

Figure 7: Specific Impulse vs. Inlet Angle

Figure 7 shows the specific impulse versus the inlet angle. The entrance angle in this case varies along with the inlet angle. However, showing the entrance angle is not significant in this case as it was discussed earlier. The specific impulse does not increase between 0^0 and 5^0; as well as between 10^0 and 15^0. Yet, between 5^0 and 10^0, there is a sharp increase in specific impulse. This signifies that the oblique shocks at 10^0 inlet angle do not stagnate the flow in the nozzle; therefore, a sharp increase in specific impulse.

Table 3: Performance Increase with the Simple Heat Exchanger

Change in Temperature (K)	Specific Impulse Increase (%)	Thrust Increase (%)
0	0.00	0.00
25	2.05	4.37
50	4.09	8.70
100	8.09	17.33

Table 3 focuses on the specific impulse and thrust increase when introducing a simple heat exchanger from the power plant. As seen in figure 7, the nozzle used in this simulation has inlet and entrance angles are 5^0 and 9.25^0 respectively. The specific impulse increases only by 8% when 100 K is added to the plasma flow. Note that the thrust increased by 17%. Therefore, despite that there is little change in the specific impulse, there is a significant increase in thrust.

5. Discussion and Conclusion

The simulations show that all of the resulting flow velocities are above the Alfven critical velocity of argon. It is positive because at supersonic speed going in the inlet and entrance areas, the flow velocity slows down and increases again as the exits the exhaust. Also, varying the angles show how the flow interacts in the nozzle. This is significant in preventing vibration and instability in the engine. The simple heat exchanger shows a trivial solution that increasing the heat increases the performance of the thruster. The heat exchanger in this case can come from the nuclear or antimatter power plant in the spacecraft. There are other ways to increase the performance of the thruster without a simple heat exchanger. These methods are near-future development; this means that these technologies are available for research and development.

One mean is to replace the simple heat exchanger with an oxidizer to produce a significant thrust [11]. Swapping the argon propellant with hydrazine (N_2H_4), combustion at the throat renders the enhanced MPD thruster into a high-thrust scramjet, similar to the X-43. The chemical breakdown of hydrazine in its gaseous state is

$$N_2H_4 + O_2 \rightarrow N_2 + 2H_2O \quad \Delta h_{kmol} = 0.577 \text{ kJ/kmol}. \tag{10}$$

where Δh_{kmol} is the enthalpy difference in kJ/kmol [12]. The estimated temperature, calculating by $Q_{in} = nC_p\Delta T$, the temperature increase per kilomole is 0.019 K/kmol. Plasma-assisted combustion is currently being research for the use of scramjets. When hydrazine is ionized and accelerated electromagnetically, the combustion efficiency increases since the propellant is already broken down into its individual species [11].

Another mean is to replace the simple heat changer with high-power lasers. Since there is significant heat lost from the nozzle, and the engine is design to collect the wasted heat via thermionic generators. The generators can power lasers and recycle the energy back to the nozzle, increasing the performance of the thruster. If a set of lasers deliver the total energy of 2 kJ, the estimated increase in temperature per kilomole is 0.069 K/kmol, over 3.5 higher than using combustion alone. This could initiate a proton-proton fusion reaction shown as the following [13]:

$$\begin{array}{ll} \{1\} & {}^1_1H + {}^1_1H \rightarrow {}^2_2He + \gamma \\ \{2\} & {}^2_2He \rightarrow {}^2_1H + e^+ + \nu_e \\ \{1\}+\{2\} & {}^1_1H + {}^1_1H \rightarrow {}^2_1H + e^+ + \nu_e + 0.42 \text{ MeV} \end{array} \tag{11}$$

The energy release after the proton-proton fusion {1} and the slow and rare beta positive decay {2} is 0.42 MeV. The drawback on using such process is it is difficult to sustain such process because the unstable Helium-2 decays back to hydrogen via proton emission [13].

A more sustainable fusion process is the deuterium-tritium cycle. Injecting deuterium-tritium pellets in the throat of the nozzle can be an excellent alternative to proton-proton fusion process. Using hydrogen or deuterium as the propellant would convert the enhanced MPD into, potentially, a deuterium-tritium fusion rocket. The deuterium-tritium fusion cycle is

$$\begin{array}{ll} \{1\} & {}^2_1H + {}^2_1H \rightarrow {}^3_2He + {}^1_0n + 3.27 \text{ MeV} \\ \{2\} & {}^2_1H + {}^2_1H \rightarrow {}^3_1H + {}^1_1H + 4.03 \text{ MeV} \\ \{3\} & {}^2_1H + {}^3_1H \rightarrow {}^4_2He + {}^1_0n + 17.6 \text{ MeV} \\ \{4\} & {}^2_1H + {}^3_2He \rightarrow {}^4_2He + {}^1_1H + 18.3 \text{ MeV} \end{array} \tag{12}$$

The total energy released per cycle is 43.2 MeV [14]. The process may be minor because even at 5000 K, the temperature is too low to produce a significant fusion rocket within a short period of time. Additional devices, such as electron accelerators or high-power lasers, may improve the chances of sustaining the fusion process. In this case, magnetic confinement is necessary to prevent the plasma from cooling down while propelling out of the nozzle [15,16].

In conclusion, the enhanced magnetoplasmadynamic thruster has potential to be developed today. The technology to make this is available now. The specific impulse is rather low for a plasma thruster. This is because of the selection of argon as the propellant. Using helium, hydrazine, or hydrogen would provide a significant increase in specific impulse. The thrust output in the electric magnetoplasmadynamic thruster is higher than the current ion and hall thrusters. Thus, the high thrust accelerates the spacecraft to the speed at which the ion and hall thrusters can continue the velocity increase. Overall, this technology could be available and fully functional before 2034.

References

1. Goebel, D. M.; Katz, I. *Fundamentals of Electric Propulsion*: Ion and Hall Thrusters. New Jersey: Wiley, 2008.

2. Jahn, R. G. *Physics of Electric Propulsion*. New York: McGraw-Hill, 1968.

3. Martinez-Sanchez, M. *Space Propulsion*. Boston: Massachusetts Institute of Technology, 2004.

4. Sutto, G. P.; Biblarz, O. *Rocket Propulsion Elements*. New Jersey: Wiley, 2010.

5. Stoute, C. A. B. *Hybrid Electric Propulsion using Gas Mixtures*. Toronto: York University, 2013.

6. Dinklage, A. et al. *Plasma Physics*. Berlin Heidelberg: Spring, 2008.

7. Howard, J. *Introduction to Plasma Physics*. Canberra: Australian National University, 2002.

8. Bose, T. K. *High Temperature Gas Dynamics*. Berlin Heidelberg: Springer, 2004.

9. Sankaran, K. et al. A flux-limited numerical method for solving the MHD equations to simulate propulsive plasma flows, v. 53, 2002.

10. Wendt, J. F. *Computational Fluid Dynamics*. Berlin Heidelberg: Springer, 2009.

11. Starikovskiy, A.; Aleksandrov, N. "Plasma-Assisted Ignition and Combustion." *InTech Aeronautics and Astronautics*, p. 331-368, 2013.

12. Gray, P.; Lee, J. C. "The Combustion of Gaseous Hydrazine." *Transactions of Faraday Society*, v. 50, p. 719-728, 1954.

13. Weiss, B.; Royden, L. *Physics and Chemistry of the Earth and Terrestrial Planets*. Boston: Massachusetts Institute of Technology, 2008.

14. Miyamoto, K. *Fundamentals of Plasma Physics and Controlled Fusion*. Tokyo: University of Tokyo, 2000.

15. Slough, J. et al. *Electromagnetically Driven Fusion Propulsion*. The 33rd International Electric Propulsion Conference. Washington, D.C.: [s.n.]. 2013. P. 1-14.

16. Miernik, J. et al. Fusion Propulsion Z-Pinch Engine Concept.

A Novel Method to Store Charged Antimatter Particles

Marc H. Weber, Kelvin G. Lynn,
Joshah Jennings, Kasey Lund, and Chandra Minnal

Center for Materials Research,

Dana Hall 102, Washington State University, Pullman, WA 99164-2711

m_weber@wsu.edu

Abstract
A new concept of storing positrons or other charged particles in an array of thousands of micro-tubes is explored. Each tube has 100 micrometer diameter and is 0.1 meter long. Metal walls of the tubes are designed to shield the electric repulsive forces between particles in neighboring tubes. This will lower the required confining forces of the trap and is expected to make storing macroscopic numbers of positrons for long times (years) possible in the future. Here a small demonstration system is tested with the goal of storing more than 10 billion positrons for up to 2 days.

Keywords
Energy, propulsion, positrons, long term storage, antimatter

1. Introduction
Space travel in general and interstellar travel in particular require large amounts of energy both for propulsion as well as internal consumption during flight and at the destination. At the same time this energy must be provided in high energy per mass formats to avoid the cost of lifting fuel out of the gravity wells of Earth and the Sun. Solar power is available free of weight cost beyond the collection and conversion hardware but fades quickly with distance from the sun. Nuclear fusion and fission energy as well as chemical energy require rapidly increasing amounts of energy to deliver to the location of use. Most of these fuels' mass is "dead weight". Antimatter fuel, on the other hand, can be annihilated with regular matter. 100% of its mass and that of the matter counterparts is converted into energy. Antiproton – proton annihilations result in energetic pions as well as photons.

Anti-electron – so called positron – and electrons always generate photons. Typically two 511 keV energy photons emerge in opposite direction. The emission of speed of light photons from positron-electron annihilations also offers the opportunity to accelerate a spaceship towards the speed of light. Considerations of this idea date back to the 1950-ties by E. Sänger[1]

Finding a method to store large macroscopic amounts of antimatter safely and for many years is paramount. The other two obstacles are to generate antimatter and to find efficient methods to convert annihilation power into desirable forms of energy for propulsion and electricity will likely be addressed once long term storage has been developed.

Here, a concept to store positrons for long terms is presented and the state of experiments towards that goal is discussed.

2. The Case for Antimatter Fuel

2.1 High Exhaust Velocity

Space travel to distant stars such as Canopus as discussed by Dr. Mae Jamison requires vast amounts of energy to lift a payload mass out of the gravity well of Earth and then the Sun and then to accelerate to near the speed of light. At the destination the spaceship would have to decelerate. Fuel would have to be found or carried along for the return trip. Current fuels used for space excursions into low earth orbit and the solar system are hopelessly inadequate. The rocket equation relates the achievable change in velocity (Δv) with the velocity of exhaust gasses from a rocket nozzle (vexhaust) and the ratio of the initial and final mass of the spaceship (M and m). Here the difference in mass is dominated by the consumption of fuel for the exhaust gas. The effect of gravity is ignored here.

$$\Delta v = v_{exhaust} \, ln(M/m)$$

Figure 1 shows the attainable change in velocity as a function of the fraction of mass available after the acceleration. The actually usable payload will be a fraction of that final mass. Horizontal lines show the Δv required to escape Earth gravity and the Solar system. The two exhaust velocities used in the figure are representative of chemical fuels such as the combustion of liquid hydrogen and oxygen and for much more energetic particles for which ion engines or more powerful devices are required. With chemical fuels 100 metric tons would be required to accelerate 1 kg of payload beyond the Solar system. With the Voyager spacecrafts 1 and 2 this was achieved after almost 40 years of travel time and several gravitational assists from planets. Much higher exhaust velocities are not possible with chemical reactions.

The exhaust velocity is limited when chemical reactions are used. Ion thrusters can deliver much higher speeds that could lead to more efficient uses of the available fuel in a spacecraft. However, currently they lack the thrust to be useful for larger systems.

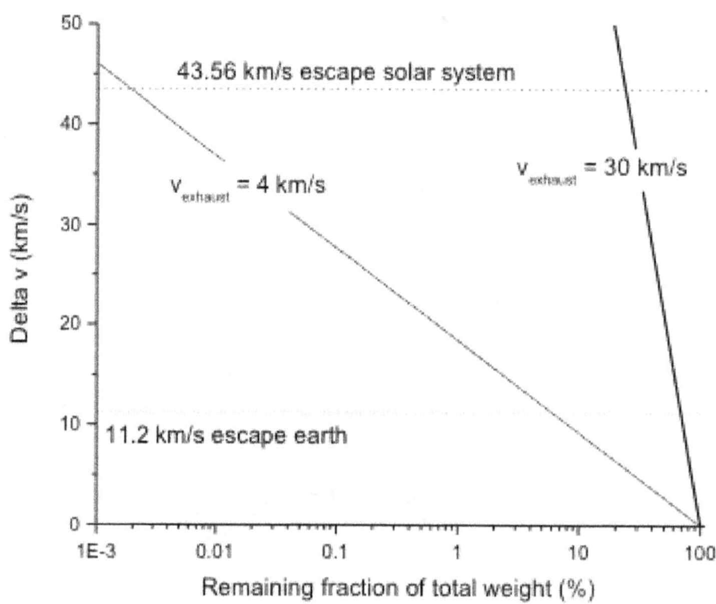

Figure 1: Δv as a function of the fraction of mass remaining for a given rocket nozzle exhaust velocity. Two example exhaust velocities are listed.

2.2 High Energy To Mass Ratios

The mass of the carried fuel is the dominant component of modern space craft at launch. Must is consumed to merely lift fuel a little bit higher and to slightly change Δv. The energy density per mass of the fuel must be increased. For travel significantly beyond the orbit of Earth solar power fades away. The Mars rover Curiosity and the Voyager probes are powered by radioisotope thermal generators (RTG). The heat from nuclear fission of a radioactive isotope is converted to electricity. A fraction of the mass of the parent isotope is converted to thermal energy and about 7% of that into electric power.[2] The energy per mass of such fission based generators is much larger than for chemical fuels as shown in Figure 2. Nuclear power contains a million times the energy per mass compared to chemical fuels. The highest energy per mass ratio by another 3 orders of magnitude is reached by antimatter-matter annihilations. In particular, positron annihilations with electrons generate 100% energetic photons. Nearly all annihilations result in two 0.5 million electronVolt (MeV, the energy an electron gains being accelerated across 1 million Volt potential step) photons emerging in opposite direction.

To illustrate the tremendous savings in mass the reader is referred to the large fuel tanks of the now retired Space Shuttle. In addition to the fuel contained in the solid rocket boosters about 100 metric tons of hydrogen are burnt to deliver the orbiter into low Earth orbit (LEO) at about half the Δv required to leave Earth behind. Burning hydrogen releases 142 MJ/kg (not counting the much heavier oxygen also required). The same energy is released by 84 mg (milligrams) of positrons. The total electric energy from the 3 RTGs installed in each Voyager craft of 113 kg mass is 1.9 MJ or 10.5 mg of positrons. [3] About half of this energy is generated during the first 87.7 year half-life of the ^{238}Pu isotope in the generators.

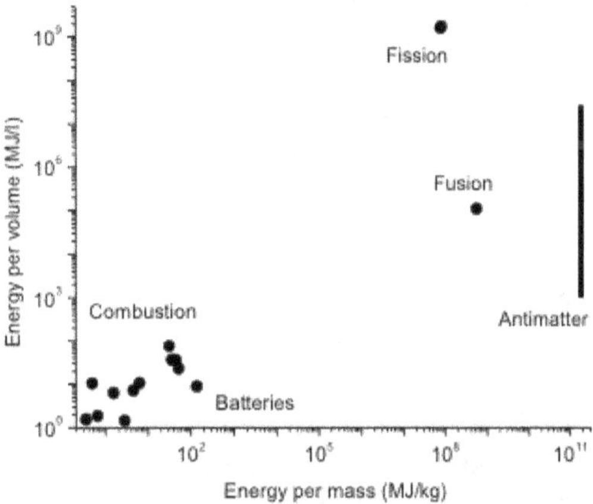

Figure 2: Comparison of ratios of energy per volume vs. energy per mass for different fuels. The bar for antimatter represents the authors' estimate of volume range for lack of an existing antimatter "tank". A number of chemical fuels and current batteries such as Li-ion make up the cluster of dots in the low energy corner.[3]

3. Storing Antimatter

3.1 The Standard Method

A milligram of antimatter stores a tremendous amount of energy. However, this mass equates to just about 10^{24} individual positrons. In contrast, current technology enables a small number of research laboratories to store ~10^9 positrons for hours.[4] All positrons carry the same charge and they tend to push each other apart. In Penning-Malmberg traps the particles are confined into a cylinder shape by a large axial magnetic field and electrostatic mirror potentials at the ends of the cylinder. The cylinder is maintained under vacuum such that the positrons cannot collide and annihilate with electrons of atoms and molecules. The magnetic field forces the positrons on a spiral motion around the field direction. This keeps them off the side walls of the cylinder. The mirror potentials at the ends of the cylinder reflect the positrons back towards the center of the cylinder as shown in Figure 3.

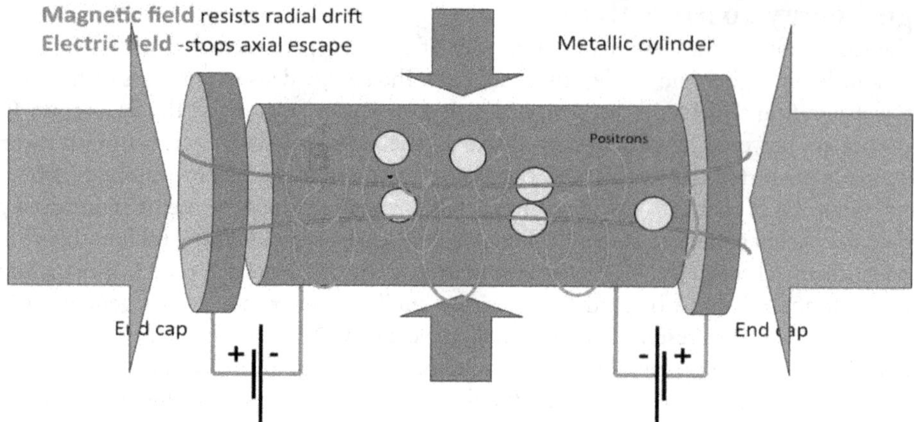

Figure 3: Schematic of a Penning-Malmberg trap. Positrons are confined in a cylinder under vacuum (blue colors). Magnetic field lines (red) force the stored positrons (green balls) onto circular motions and away from the cylinder sides (red arrows). The electric forces (green arrows) generated by the mirror potentials on each end of the trap (indicated by the battery symbols and connections) turn the positrons around at each end.

Squeezing more into such a single cylinder will generate ever more repulsive forces called space charge. This force must be overcome by the mirror potentials and the magnetic field. At very large densities of charge particles when the Brillouin Limit is reached the magnetic field can no longer hold the space charge together radially. The mirror potentials to reflect and hold the particles together axially exceeds practical limits at much lower densities. Figure 4 shows the required minimum mirror potentials for storing a certain amount of positrons in the cylinder. In this case the tubes are 100 micrometer (μm) in diameter and 0.1 meter long. The confining magnetic field is 7 Tesla. The dashed line refers to a single tube being filled with more and more positrons. The confined positrons form a plasma. Squeezing more and more together increases the plasma temperature and enhances the chance that small imperfections in the trap lead to outward motions and eventual losses by annihilation of the positrons.

3.2. A Novel Concept to Store Positrons

A novel concept proposed by Lynn and Weber and recently patented.[5] It aims at reversing the required increase in mirror potentials and the heating of the plasma. Instead of using a single tube the trap is divided into many parallel micro-tubes of equal length similar to a stack of soda straws. The walls of each tube are metalized. The metal forms an image potential. The metal shields the electric field from each particle. Positrons in any tube no longer "feel" the repulsive force from positrons in any other tube. The lower number of positrons in each tube decreases the temperature of the plasma and should lead to longer storing times.

In the new design, each micro-tube is to be filled with about 10 million positrons. The total stack is composed of 10^4 micro-tubes, each 100 micrometer in diameter and 0.1 meter long. This length to diameter ratio is about 100 times larger than what is currently used for positron traps. A total of 10^{11} positrons will be stored in a macro-cylinder comparable in size to a single soda can. Merely 10 Volts mirror potential are required. 100 times the number of positrons are stored compared to the current state-of-the-art trap in an overall volume that is thousands of times smaller.

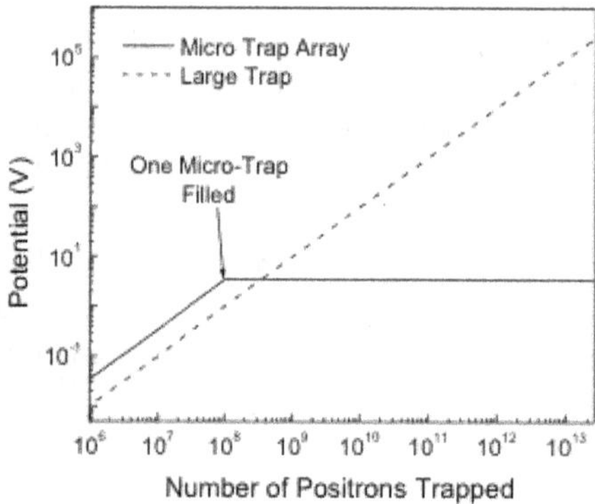

Figure 4: Minimum required mirror potential at each end of a cylindrical Penning-Malmberg positron trap. The dashed line refers to a single trap. The solid line refers to the novel concept discussed here.

4. Experiments

Such a micro-trap array is under construction and testing. Figure 5 shows a smaller version prior to installation in vacuum within a 7 Tesla magnet. A segment is composed of 10 5 cm diameter silicon wafers. The wafers are prepared by deep reactive ion etching (DRI) to hold a hexagonal pattern of 20 000 holes of 200 micrometer diameter each. The wafers are sputter coated with gold and then bonded with heat and pressure to form the stack. The pattern of the DRI holes must be aligned from wafer to wafer to form the micro-tubes. In Figure 4 four such stacks are mounted together to form. Precision ground sapphire rods passing through the larger holes (6 visible in the stack on the left of Figure 5) line up all the micro-tubes. More details of the design and fabrication are discussed in Reference [6].

An electron beam of 16 micrometer diameter was scanned across the micro-tube pattern of the quad stack with steering electromagnets. Electrons are loaded into a single micro-tube at a time for 1 millisecond. They are then stored for 1 millisecond and then released to fly towards the detector. This is repeated at a rate of 450 cycles per second as the electron beam is scanned with steering magnet current for horizontal steering and vertical steering. The detector sees either nothing when no electrons survive the trap time of 1 millisecond or more than one electron survived. Figure 6 show the detection rate coded in colors. Approximately 300 holes were tested. Some for large positive horizontal current did not work.

Figure 7 shows results where a single micro-tube was loaded with electrons. After a specified loading time the electrons were held in the trap for a trapping time and then the surviving electrons (if any) were released towards the detector. In this case the loading and holding times were equal. Initially as the loading time is raised, more electrons are stored in the micro-tube. Eventually around 10 milliseconds the decay due to losses in the trap take over. For long holding times the electron trap time is about 38 milliseconds. That is a far cry from the goal of years. However, experiments are at a very early stage. The alignment of the trap with the magnetic field, vacuum conditions, micro-tube surface imperfections, and vibrations all require fine tuning.

Figure 5: A segment of the trap (left) is assembled with three other segments into the test stack on the right. Each segment consists of 10 wafers of silicon coated with gold. It has about 20 000 holes which are aligned through all wafers and segments to for the micro-tubes. Separate isolation wafers with oxide surfaces allow for the mirror potentials.

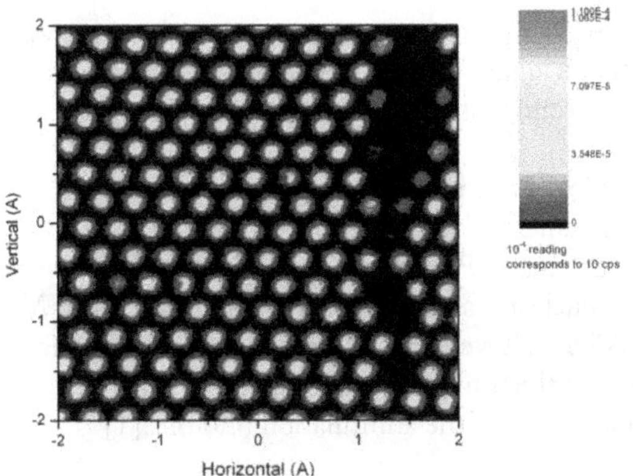

Figure 6: Trapping of electrons in the central region of the micro-tube array of a quad-stack. Black corresponds to no signal when the electron beam hits areas between tubes. Blue to red represent increasing signal rates from 1 to 10 pulses leaving the trap per second. The trap is cycled at a rate of 450 per second.

In a magnetic field the kinetic energy of charged particles stems from their motion along the field lines and motion perpendicular to them. The latter comprises the spiral motion mentioned earlier. The former component parallel to the field lines is easily measured. Preliminary data show that the incident electrons which are loaded with typically 5 eV kinetic energy appear to loose a large fraction of their kinetic energy within the first 10 microseconds of holding time in the trap. Alternatively the parallel component is converted to the perpendicular component. Further studies will resolve this.

Parallel to the experimental effort, a computer simulation effort is under way to predict ideal conditions for trapping positrons in a micro-trap array.[7] In the near future an in-house deuteron accelerator funded by a generous grant from the W.M. Keck Foundation combined with a positron production target will come on line. On the order of several million monoenergetic positrons per second are expected. Positrons are emitted during the beta-decay of ^{13}N which is produced by colliding 3 million eV deuterons with ^{12}C carbon atoms.[8]

5. Future

Once positrons can be filled into the micro-tubes and trapped, further experiments will cover filling multiple tubes and stacking more and more positrons into each of them. Eventually 10 stacks of silicon wafers will be installed for the full scale demonstration trap. The electric potential of the stacks can be raised and lowered to test pushing stored particles into the exit side of the tubes. This will free the entrance side half available to load more positrons. Subsequently the two bunches in each tube will be combined and the process can be repeated to further increase the number of positrons in each stack. Filling multiple micro-tubes in parallel will be tested. This is one of the key objectives of the project besides finding the maximum number of storable particles and extending the storage time towards days.

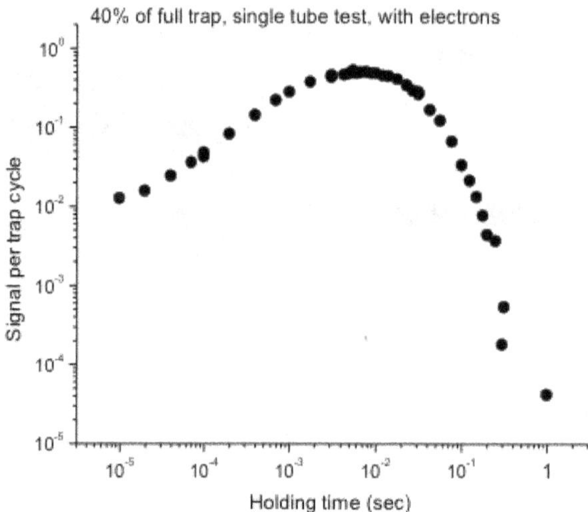

Figure 7: Signal from trapping electrons for equal times of loading a single micro-tube and holding them trapped. Up to 10 milliseconds longer loading times result in more signal. At holding times over 30 milliseconds the lifetime of the electrons in the tube results in a decay time on the order of 38 milliseconds.

Testing the storage capabilities of this micro-tube array is the objective of this project. Finding ideal methods of generating positrons as well as efficient means to convert annihilation power into Δv as well as internal spaceship power are the other two challenges. Positrons can be generated with powerful accelerator such as the ones at Jefferson Laboratories and at nuclear reactors like the research reactor in Munich Germany. Concepts of using energetic radiation in the van Allen belt were discussed by Sumontro Lal Sinha at this meeting.[10] Methods for using antimatter power range from the brute force use of gamma rays to heat a gas to more advanced methods of photon pressure propulsion of at least half the annihilation photons.[1]

6. Conclusions

A novel concept of storing positrons in many narrow long micro-tubes within a single large magnetic field is being evaluated experimentally as well as by computer modeling. A single large diameter tube with a length of about 10 times the diameter is being replaced with many micro-tubes of 100 micrometer diameter and much longer by a factor of 1000 tubes. This reduced the requirement for large mirror potentials of the trap and should lower the loss rate of particles. Significant obstacles of mechanical alignment, vacuum and homogeneity in the magnetic field need to be overcome. If successful this concept of storing antimatter may proof very attractive as a container for antimatter fuel for inter-planetary interstellar space travel.

The many contributions to the project by Bobby Riley, Christopher Baker, Alirezah Narimannezhad, Jeremy Eilers and Lloyd Pilant are greatly appreciated. They were vital in the design and construction of the trap, the beam lines and the development of the positron beam. This work is supported in part by the W.M. Keck foundation (for the accelerator to generate positrons) and the Army Research Laboratory under contract number W9113M-09-C-0075 (storing experiments) and the Office of Naval Research under award number N00014-10-1-0543 (numerical simulations).

References

1. Sänger, E., "Zur Mechanik der Photonen-Strahlantriebe", R. Oldenbourg, München (1956) p. 92.

2. Bennet, G. L., "Space Nuclear Power: Opening the Final Frontier", 4th International Energy Conversion Engineering Conference and Exhibit, 26-29 June 2006, San Diego (2006) http://www.fas.org/nuke/space/bennett0706.pdf.

3. Energy density per volume and per mass for selected fuels: http://en.wikipedia.org/wiki/Energy_density.

4. Jørgensen, L. V., et al, "New Source of Dense, Cryogenic Positron Plasmas", Phys. Rev. Lett. 95, 25002 (2005).

5. Lynn, K. G., Weber, M. H., "Positron Storage Micro-Trap Array", Unites States Patent No. 8,835,840, Issued September 16, 2014.

6. Verma, A., Jennings, J., Johnson, R. D., Weber, M. H., Lynn, K. G., "Fabrication of 3D charged particle trap using through-silicon vias etched by deep reactive ion etching". J. Vac. Sci. & Technol. B 31, 032001 (2013).

7. Narimannezhad, A., Baker, C. J., Weber, M. H., Jennings, J., and Lynn, K. B., "Simulation studies of the behavior of positrons in a microtrap with long aspect ratio", Eur. Phys. J. D 68, 351 (2014).

8. Weber, M. H., Riley, B. D., Baker, C. J., Lynn, K. G., "Positron beams from small accelerators and status of a novel positron storage project" in "16th International Conference on Positron Annihilation (ICPA-16)", Alam, A., Coleman, P., Dugdale, S., Roussenova, M. eds., J. Phys. Conf. Ser. 443, 012081 (2013).

9. Baker, C. J., Jennings, J., Verma, A., Xu, J., Weber, M. H., Lynn, K. G., "Progress toward the long time confinement of large positron numbers", Eur. Phys. J. D 66, 109 (2012).

10. Sinha, S. L., "Industrial Production of Antimatter in the Van Allen Belts", Sinhatech, USA, 100year Starship 2014 Public Symposium, Housten, 18-21 Septermber 2014.

Thermonuclear Operation Space Lift

Friedwardt Winterberg, Ph.D.

University of Nevada, Reno, Leifson Physics 204,

1664 N. Virginia Street, Reno, Nevada 89557-0220

Abstract

The "Project Orion" small fission bomb propulsion concept proposed the one-stage launching of large payloads into low earth orbit, but it was abandoned because of the radioactive fallout into the earth atmosphere. The idea is here revived by the replacement of the small fission bombs with pure deuterium-tritium fusion bombs, and the pusher plate of the Project Orion with a large magnetic mirror. The ignition of the thermonuclear fusion reaction is done by the transient formation of keV super-explosives under the high pressure of a convergent shock wave launched into liquid hydrogen propellant by a conventional high explosive.

Nomenclature

T = absolute temperature (K)
R, r = initial and final shockwave radius (cm)
p = pressure (dyn/cm^2)
E = energy of electron orbit (eV)
n = atomic number density (particles/cm^3)
d = inner atomic distance (cm)
Z = atomic number
τ_s, τ_c = time (s)
κ = opacity (cm^2/g)
ρ = density (g/cm^3)
λ = path length (cm)
ϕ = energy flux (erg/cm^{2}s)
τ = shear stress (dyn/cm^2)
$tan\ \rho$ = tangent of friction angle ρ
$sin\ \rho$ = sine of friction angle ρ

; = covariant derivative

Γ_{ik}^{l} = Christoffel symbols of the second kind (cm^{-1})

ds^2 = square of line element in spherical coordinates r, θ, ϕ (cm^2)

$\sqrt{g}$ = square root of determinant of metric tensor (cm^2)

t = thickness (cm)

$\sigma_{\|}$ = hoop stress (dyn/cm^2)

1. Introduction

A principal obstacle standing in the way of a large scale access to space is the small energy per unit mass which is released in chemical reactions, requiring multistage rockets. An early study by von Braun estimated that in an "Operation Space Lift" a few hundred rocket launches would be needed to bring into low earth orbit the parts to assemble two Mars space ships in space [1]. Von Braun had assumed that the trip to Mars would be done by chemical rockets, but it would still require a very large number of launches into low earth orbit if the trip to Mars would be done by some nuclear electric propulsion system instead.

Because of the million times larger energy in nuclear reactions, this game changing potential was realized early on. However, to fully utilize this potential requires going to much higher temperatures than the combustion temperatures of chemical reactions. It appeared that the only way to overcome this problem is a pulsed nuclear combustion, where the rocket motor is exposed only for a very short time, but where the waste-heat is released in the exhaust as in chemical rockets. This led directly to the idea to propel a spacecraft by a chain of small nuclear explosions [2]. Since up until now no thermonuclear fusion explosion has been ignited without a fission bomb trigger, such a bomb propulsion concept must involve a fission reaction as for fusion boosted fission bombs. However, as Dyson has pointed out, a fission reaction is subject to the tyranny of the critical mass, which means that small fission explosions with a small yield would be extravagant. For fusion explosions there is no tyranny of a critical mass, and in principle they could be made very small, but they are difficult to ignite. Attempts by Lawrence Livermore National Laboratory to ignite a small deuterium-tritium fusion explosion with a 2MJ laser have failed, but that does not mean that alternative non-fission ignition concepts would not work.

2. Ignition by a Convergent Shock Wave

For rocket propulsion the ignition of a thermonuclear micro-explosion by a convergent shock wave would be of great interest if the ignition temperature can be reached in the center of this wave. There the medium through which the convergent shock wave is propagating would become a propellant heated up to high temperatures by the thermonuclear micro-explosion in its center. Therefore, let us analyze this proposal. According to Guderley [3] the temperature T in a convergent shock wave propagating in hydrogen rises as

$$\frac{T}{T_0} = \left(\frac{R}{r}\right)^{0.9} \tag{1}$$

where R is its initial radius. To reach the ignition temperature $T \sim 10^8$ K at a radius of $r \sim 1$ cm would, at an initial temperature of $T_0 \sim 10^4$ K, supplied by a chemical high explosive, require that $R \sim 10$ m, which is unrealistically large.

The idea to ignite a thermonuclear micro-explosion by a convergent shock wave driven with high explosives was proposed by the author in a fusion workshop at the Max Planck Institute for Physics in Goettingen, October 23-24, 1956, organized by the fusion pioneer C.F. von Weizsäcker [4]. While an imploding spherical shell is subject to the Rayleigh-Taylor instability, a spherical convergent shock wave is stable. This has been demonstrated in the 15 Megaton 1952 "Mike" test, where a sphere of liquid deuterium was ignited by a plutonium (or uranium) bomb, with the X-rays from the exploding fission bomb launching a Guderley convergent shock wave into the deuterium. Apart from this demonstrated stability of the Guderley convergent shock wave solution, its stability has also been confirmed in an extensive analytical study by Häfele [5].

Guderley's convergent shock wave solution also predicts a rise in the pressure by

$$\frac{p}{p_0} = \left(\frac{R}{r}\right)^{0.9} \tag{2}$$

With high explosives producing pressures up to 1 Mb = 10^{12} dyn/cm^2 [6] and setting $R = 10^2$ cm, a pressure of 100 Mb = 10^{14} dyn/cm^2 would be reached at the radius $r \sim$ 1cm. Under these high pressures super-explosives can be formed on very short time scales, facilitating the ignition of a thermonuclear micro-explosion by a convergent shock wave [7, 8, 9].

3. Super-explosives

Under normal pressure the distance of separation between two atoms in condensed matter is typically of the order 10^{-8} cm, with the distance between molecules formed by the chemical binding of atoms of the same order of magnitude. As illustrated in a schematic way in Fig. 1, the electrons of the outer electron shells of two atoms undergoing a chemical binding form a "bridge" between the reacting atoms. The formation of the bridge is accompanied in a lowering of the electric potential well for the outer shell electrons of the two reacting atoms, with the electrons feeling the attractive force of both atomic nuclei. Because of the lowering of the potential well, the electrons undergo under the emission of eV photons a transition into lower energy molecular orbits.

Going still to higher pressures, a situation can arise as shown in Fig. 2, with the building of electron bridges between shells inside shells. There the explosive power would be even larger. Now consider the situation where the condensed state of many closely spaced atoms is put under high pressure, making the distance of separation between the atoms much smaller, whereby the electrons from the outer shells coalesce into one shell surrounding both nuclei, with electrons from inner shells forming a bridge. Because there the change in the potential energy is much larger, the change in the electron energy levels is also much larger, and can be of the order of keV. A very powerful explosive is formed, releasing its energy in a burst of keV X-rays. This powerful explosive is likely to be very unstable, but it can be produced by the sudden application of a high pressure in just the moment when it is needed. Because an intense burst of X-rays is needed for the ignition of a thermonuclear micro-explosion, the conjectured effect, if it exists, has the potential to reduce the cost of the ignition of thermonuclear micro-explosions by orders of magnitude.

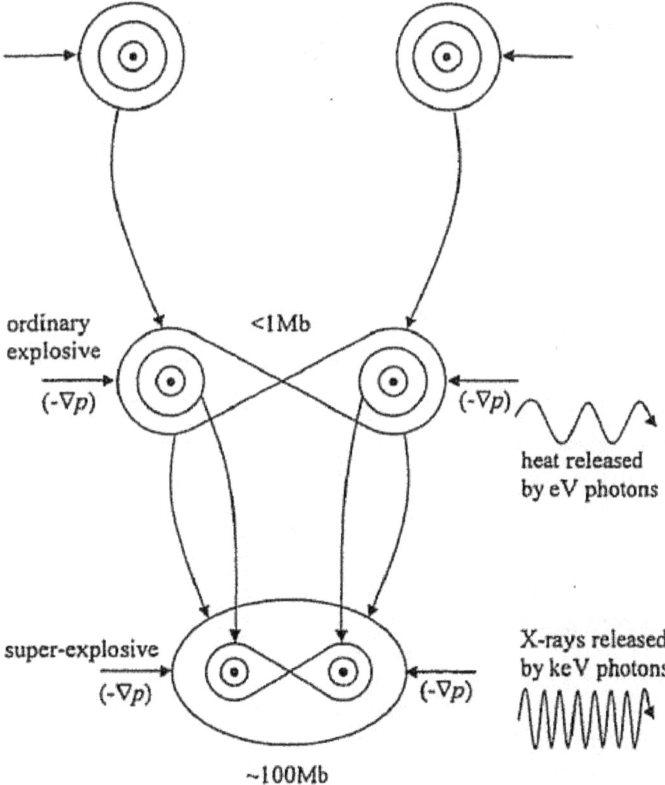

Figure 1. *keV* superexplosive formation

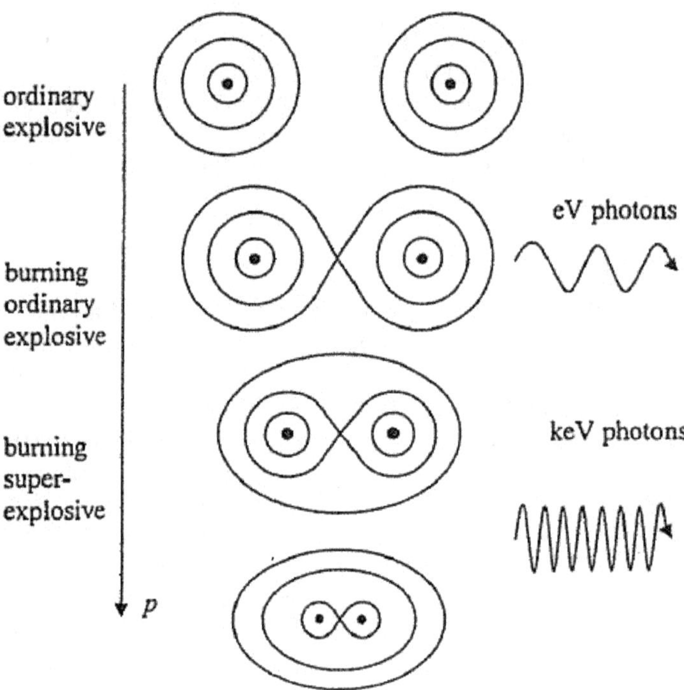

Figure 2. With increasing pressure electron-bridges are formed between shells melting into common shells.

The energy of an electron in the groundstate of a nucleus with the charge Ze is

$$E_1 = -13.6Z^2(\text{eV}) \tag{3}$$

With the inclusion of all the Z electrons surrounding the nucleus of charge Ze, the energy is

$$E_1^* \approx -13.6Z^{2.42}(\text{eV}) \tag{4}$$

with the outer electrons less strongly bound to the nucleus.

Now, assume that two nuclei are so strongly pushed together that they act like one nucleus with the charge $2Ze$, onto the $2Z$ electrons surrounding the $2Ze$ charge. In this case, the energy for the innermost electron is

$$E_2 = -13.6(2Z)^2(\text{eV}) \tag{5}$$

or if the outer electrons are taken into account,

$$E_2^* = -13.6(2Z)^{2.42}(\text{eV}) \tag{6}$$

For the difference one obtains

$$\delta E = E_1^* - E_2^* = -13.6Z^{2.42}\left(2^{2.42} - 1\right) \approx 58.5Z^{2.42}(\text{eV}) \tag{7}$$

Using the example $Z = 10$, which is a neon nucleus, one obtains $\delta E \approx 15$ keV. Of course, it would require a very high pressure to push two neon atoms that close to each other, but this example makes it plausible that smaller pressures exerted on heavier nuclei with many more electrons may result in a substantial lowering of the potential well for their electrons. For an equation of state of the form $p / p_0 = (n / n_0)^\gamma$, and a pressure of 100 Mb = 10^{14} dyn/cm^2, we may set $\gamma = 3$ and $p_0 = 10^{11}$ dyn/cm^2, where p_0 is the Fermi pressure of a solid at the atomic number density n_0, with n being the atomic number density at the elevated pressure $p > p_0$. With $d = n^{-1/3}$, where d is the lattice constant, one has

$$d / d_0 = \left(p_0 / p\right)^{1/9} \tag{8}$$

For $p = 10^{14}$ dyn/cm^2, $d / d_0 \sim 1/2$. Such a lowering of the inneratomic distance is sufficient for the formation of molecular states.

Calculations done by Muller, Rafelski and Greiner [10] show that for molecular states $_{35}$Br- $_{35}$Br, $_{53}$I- $_{79}$Au, and $_{92}$U- $_{92}$U, a twofold lowering fo the distance of separation leads to a lowering of the electron orbit energy eigenvalues by ~ 0.35, 1.4 keV, respectively. At a pressure of 100 Mb = 10^{14} dyn/cm^2 where $d/d_0 \cong 1/2$, the result of these calculations can be summarized by (δE in keV)

$$\log \delta E \cong 1.3 \times 10^{-2} Z - 1.4 \tag{9}$$

replacing Eq. 7, where Z is here the sum of the nuclear charge for both components of the molecule formed under the high pressure.

The effect the pressure has on the change in these quasi-molecular configurations is illustrated in Fig. 3, showing a $p - d$ (pressure-lattice distance) diagram. This diagram illustrates how the molecular state is reached during the compression along the adiabat a at the distance $d = d_c$ where the pressure attains the critical value $p = p_c$. In passing over this pressure the electrons fall into the potential well of the two-center molecule, releasing their potential energy as a burst of X-rays. Following its decompression, the molecule disintegrates along the lower adiabat b.

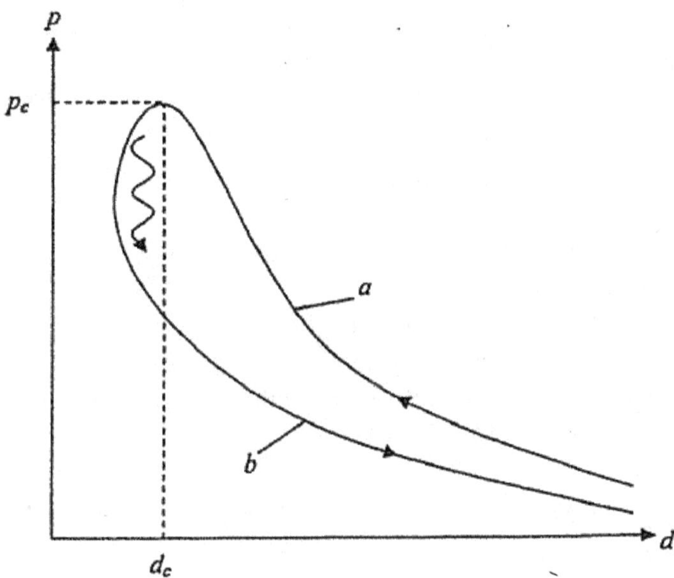

Figure 3. $p - d$, pressure-inneratomic distance diagram for the upper atomic and lower molecular adiabat.

The natural life time of an excited atomic (or molecular) state, emitting radiation of the frequency v is given by [11]

$$\tau_s \cong 3.95 \times 10^{22} / v^2 \text{(sec)} \tag{10}$$

For keV photons one finds that $v \cong 2.4 \times 10^{17}$ s^{-1}, and thus $\tau_s \cong 6.8 \times 10^{-14}$ s.

By comparison, the shortest time for the high pressure rising at the front of a shock wave propagating with the velocity v through a solid with the lattice constant d, is of the order

$$\tau_c \cong d / v \tag{11}$$

Assuming that $v \cong 10^6$ cm/s, a typical value for the shock velocity in condensed matter under high pressure, and that $d \cong 10^{-8}$ cm, one finds that $\tau_c \cong 10^{-14}$ s. In reality the life time for an excited state is much shorter than τ_s, and of the order of the collision time, which here is the order of τ_c.

The time for the electrons to form their excited state in the molecular shell is of the order $1/\omega_p \sim 10^{-16}$ s, where ω_p is the solid-state plasma frequency. The release of the X-rays in the shock front is likely to accelerate the shock velocity, exceeding the velocity profile of the Guderley solution for convergent shock waves.

A problem for the use of these contemplated super-explosives to ignite thermonuclear reactions is the absorption of the X-ray in dense matter. It is determined by the opacity [12]

$$\kappa = 7.23 \times 10^{24} \rho T^{-3.5} \sum_i \frac{w_i Z_i^2 g}{A_i t} \tag{12}$$

where w_i are the relative fractions of the elements of charge Z_i and atomic number A_i in the radiating plasma, with g the Gaunt and t the guillotine factor.

The path length of the X-ray is then given by

$$\lambda = (\kappa\rho)^{-1} \tag{13}$$

This clearly means that in material with a large Z value, the path length is much smaller than for hydrogen where $Z = 1$. This suggests placing the super-explosive in a matrix of particles, thin wires, or sheets embedded in solid hydrogen. If the thickness of the particles, thin wires, or sheets is smaller than the path length in it for the X-ray, the X-ray can heat up the hydrogen to high temperatures, if the thickness of the surrounding hydrogen is large enough for the X-ray to be absorbed in the hydrogen. The hydrogen is thereby transformed into a high temperature plasma, which can increase the strength of the shock wave generating the X-ray releasing pressure pulse.

If the change in pressure is large, whereby the pressure in the upper adiabat is large compared to the pressure in the lower adiabat, the X-ray energy flux is given by the photon diffusion equation

$$\phi = -\frac{\lambda c}{3}\frac{\partial}{\partial r} w \tag{14}$$

where w is the work done per unit volume to compress the material, and $w = p/(\gamma - 1)$. For $\gamma = 3$, one has $w = p / 2$, whereby (14) becomes

$$\phi = -\frac{\lambda c}{6}\frac{\partial}{\partial r} p \tag{14a}$$

Assuming that the pressure e-folds over the same length as the photon mean free path, one has

$$\phi \sim (c/6) p \tag{15}$$

For the example p = 100 Mb = 10^{14} dyn/cm^2 one finds that $\phi \sim 5 \times 10^{23}$ erg/cm^{2}s = 5×10^{16}W/cm^2, large enough to ignite a thermonuclear micro-explosion, and at a pressure of 100 Mb also large enough to satisfy for $r \leq 1$ cm the $\rho r > 1$ g/cm^2 condition for propagating burn.

If the conjectured super-explosive consists of just one element, as is the case for the $_{35}$Br- $_{35}$Br reaction, or the $_{92}$U- $_{92}$U reaction, no special preparation for the super-explosive is needed. However, as the example of Al-FeO thermite reaction shows, reactions with different atoms can release a much larger amount of energy compared to other chemical reactions. For the super-explosives this means as stated above that they have to be prepared as homogeneous mixtures of nano-particle powders, bringing the reacting atoms as close together as possible.

4. The Mini-fusion Bomb Configuration

As shown in Fig. 4, the deuterium-tritium (DT) fusion explosive positioned in the center is surrounded by a cm-size spherical shell made up of a super-explosive, which is surrounded by a meter-size sphere of liquid hydrogen. The surface of the hydrogen sphere is covered with many high explosive lenses, preferably of a high explosive made up of a boron compound. The DT fusion reaction releases 80% of the energy into neutrons. In absorbing the neutrons, the boron undergoes a nuclear reaction, each setting free a 2.31 MeV alpha particle, causing the boron shell to explode and further compress the deuterium inside the shell. Each explosive has an ignitor, and to produce a spherical convergent shock wave in the hydrogen the ignition must happen simultaneously, which can be done with just one laser beam split up into as many beamlets as there are ignitors.

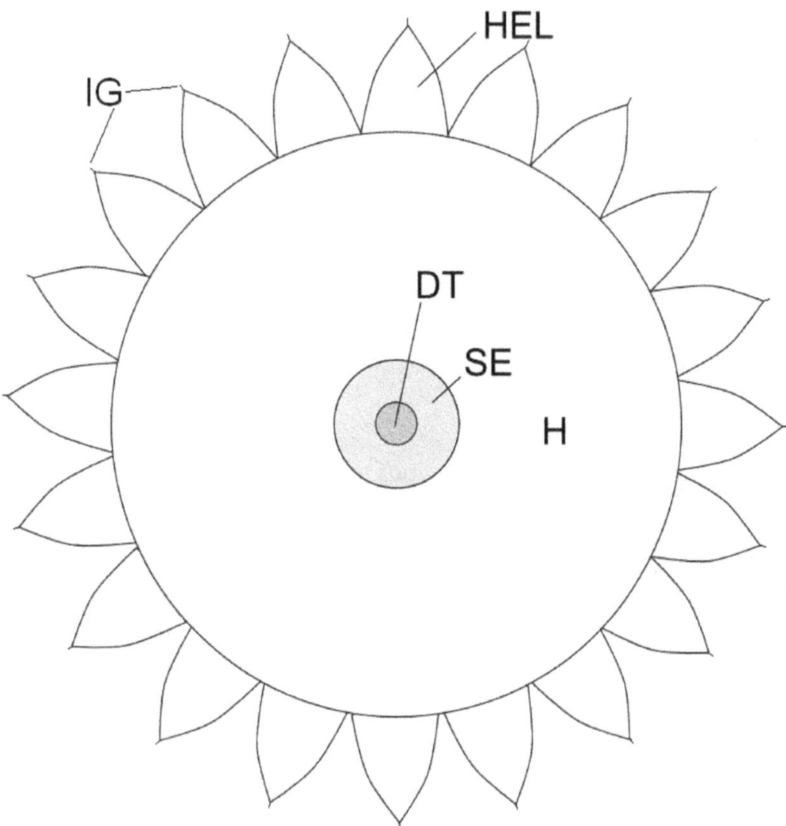

Figure 4. Pure fusion bomb assembly. HEL high explosive lens. IG ignitor. H liquid hydrogen. SE super-explosive. DT deuterium-tritium. (Not to scale)

5. The Propulsion Unit

The propulsion unit is very similar to the one in a previous publication [13], where the fusion bomb assembly is placed in the focus of a 10 meter-size large metallic reflector, positioned at the end of a magnetic mirror. The expanding fire ball by compressing the magnetic field will generate surface currents in the metallic reflector, producing a magnetized plasma layer protecting the reflector from the hot plasma. The meter-size hydrogen sphere of the mini-fusion bomb is transformed into a fireball with a temperature of ~ 10^5 K, or somewhat higher, with an expansion velocity of ~ 30 km/s (Fig. 5). Cooling the metallic reflector can be done with liquid hydrogen becoming part of the exhaust, as in chemical liquid fuel rocket technology. This is unlikely to amount to more than 10% of the liquid hydrogen heated by the neutrons of the fusion explosion.

A meter-size ball of liquid hydrogen heated to 10^5 K has a thermal energy of 10^{18} erg, equivalent to 25 tons of TNT. At this temperature the pressure is ~10^{11} dyn/cm^2. If the fireball expands from an initial radius of R_o ~1 m to R_1 ~10 m, the pressure goes down to 10^9 dyn/cm^2 ~10^3 atm, which is about 2 orders of magnitude smaller, and below the tensile strength of steel. At this pressure, the magnetic field at the surface of the steel is of the order 10^5 Gauss. The energy released by the eddy currents in the reflector can hardly be more than 10% of the energy released in the fusion explosion. The mass of a meter-size ball of liquid hydrogen is of the order 0.1 tons, such that 0.01 tons of liquid hydrogen would be available for the cooling of the reflector.

As with the Orion concept, the pressure of ~10^9 dyn/cm^2 acting on the metallic mirror is transmitted to the spacecraft by shock absorbers to produce thrust. If the thickness of the parabolic mirror is small in comparison to its radius, a large circumferential hoop stress is induced on the mirror, larger by a factor equal to the ratio of radius of the mirror to its thickness. This requires that the mirror be supported by external forces. These forces could be balanced by making the thickness of the mirror comparable to its radius, which for a mirror made from steel would make it very heavy.

We can adopt an idea by P. Schmidt and B. Pfau [14], who had shown that the wall thickness of cylindrical and spherical pressure vessels can be greatly reduced by surrounding them with a thick compact layer of a disperse medium composed of high tensile strength micro-particles. As shown in Fig. 5, to utilize this effect, the metallic mirror is placed inside a box tightly filled with a compactified disperse medium, such as SiC (carborundum) or AlO_3, with a tensile strength of 2×10^{11} dyn/cm^2, or carbon nanotubes, with a tensile strength of 10^{12} dyn/cm^2. Because of the friction between the particles of the disperse medium a shear stress is set up in the medium, reducing the radial stress in a spherical configuration.

Setting ρ as the friction angle between the particles of the disperse medium, one has for the maximum shear stress

$$\tau < \sigma_n \tan\rho \tag{16}$$

where σ_n is the normal component of the stress tensor. If plotted in a Mohr stress diagram as shown in Fig. 6, the maximum possible shear stress cannot exceed the line $\tau < \sigma_n \tan\rho$.

The maximum shear stress is

$$\tau_{\max} = \frac{\sigma_{\max} - \sigma_{\min}}{2} \tag{17}$$

which in the Mohr stress diagram is given by

$$\sin\rho = \frac{\sigma_{\max} - \sigma_{\min}}{\sigma_{\max} + \sigma_{\min}} \tag{18}$$

and one has

$$\frac{\sigma_{\min}}{\sigma_{\max}} = \tan^2(45° - \frac{\rho}{2}) \tag{19}$$

The friction angle ρ can be visualized by the slope of an imaginary "sandhill" made from particles of the disperse medium. For a "real" sandhill seen in nature, we estimate that $\rho = 45°$. Inserting this value into (19) one finds that $\sigma_{\min} / \sigma_{\max} \approx 0.1$:

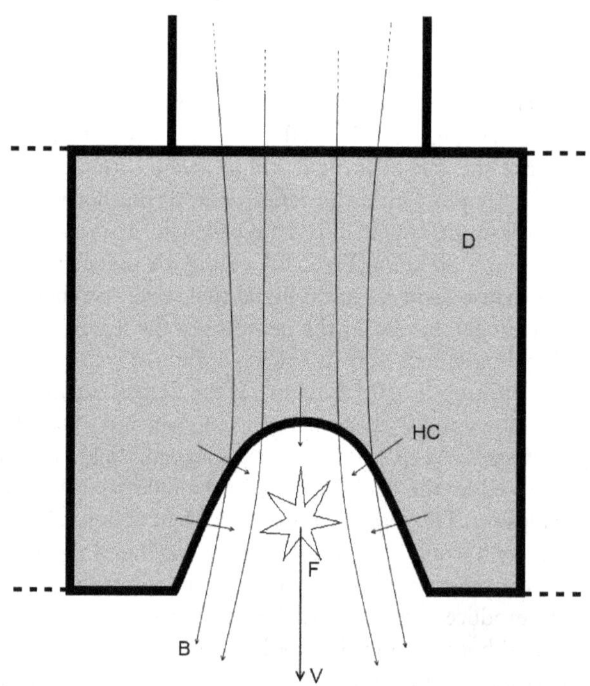

Figure 5. Propulsion unit: B magnetic field. F fireball. V exhaust jet, HC hydrogen coolant, D disperse medium

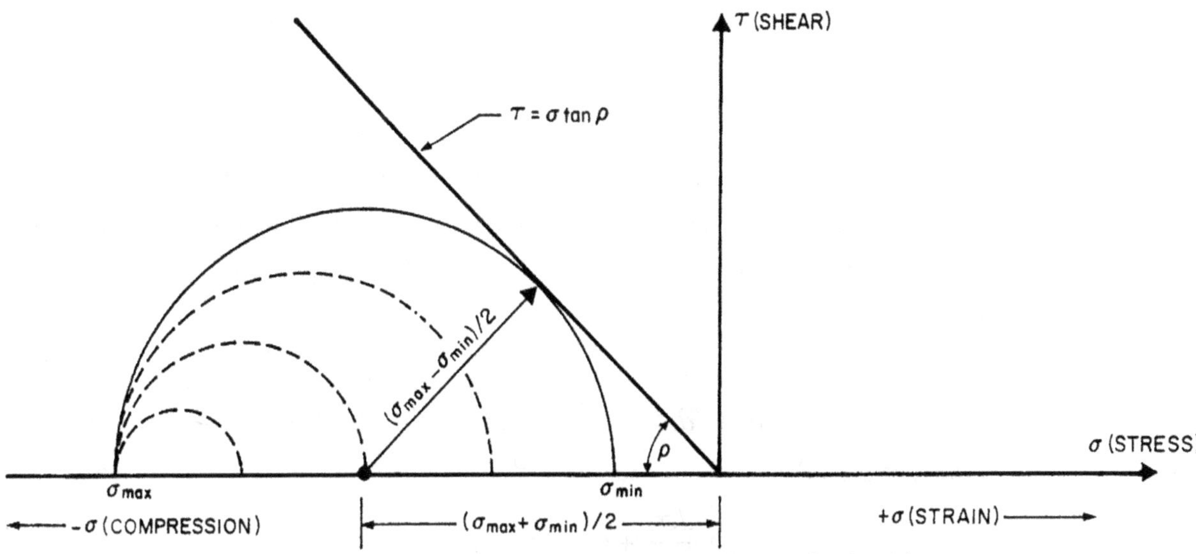

Figure 6. Mohr stress diagram for a disperse medium subject to frictional forces

The pressure distribution in the dispersive medium surrounding the metallic reflector is determined by the static equilibrium equation, which in Cartesian coordinates is given by

$$\frac{\partial \sigma_{ik}}{\partial x_k} = 0 \tag{20}$$

and in curvilinear coordinates by

$$\sigma^{k}_{i;k} = 0 \tag{21}$$

where the colon stands for the covariant derivative. For (21) one can also write

$$\frac{1}{\sqrt{g}}\frac{\partial}{\partial x^k}(\sqrt{g}\sigma^k_i) - \Gamma^l_{ik}\sigma^k_l = 0 \tag{22}$$

with the line element squared ($ds^2 = g_{ik}dx^i dx^k$) defining the metric tensor and $g = \det g_{ik}$. The Γ^l_{ik} are the Christoffel symbols of the 2nd kind. For simplicity we may approximate the metallic reflector by a spherical shell. Then, introducing in the dispersive medium spherical coordinates r, θ, ϕ, where the metric tensor is determined by the line element

$$ds^2 = dr^2 + r^2(d\theta + \sin^2\theta \times d\phi^2) \tag{23}$$

one has $\Gamma^1_{11} = 0$, $\Gamma^2_{12} = \Gamma^3_{13} = 1/r$ and $\sqrt{g} = r^2 \sin\theta$, and therefore from (22)

$$r\frac{d\sigma_r}{dr} + 2(\sigma_r - \sigma_\theta) = 0 \tag{24}$$

where σ_r and σ_θ are the components of the stress tensor in the radial and transverse direction.

Because $\sigma_r = \sigma_{\max}$ and $\sigma_{\min} \approx 0.1\sigma_r$ one has

$$r\frac{d\sigma_r}{dr} = -1.8\sigma_r \tag{25}$$

hence

$$\sigma_r = \sigma_r^{(0)} \left(\frac{r_0}{r_1} \right)^{1.8} \tag{26}$$

where r_0 is the radius of the reflector with $r_1 > r_0$. Assuming, for example, that $r_1 \approx 3r_0$, approximately shown in Fig. 5, one finds that $\sigma_r^{(1)} = 0.14\sigma_r^{(0)}$. For $\sigma_r^{(0)} = p_0 = 10^9 dyn/cm^2$, one has $\sigma_r^{(1)} = p_1 = 1.4 \times 10^8 dyn/cm^2 = 140 atm$. As in the Orion concept, this pressure is transmitted through shock absorbers to the spacecraft.

Because more than one propulsion unit is needed, a cluster of propulsion units are put together forming a disk as shown in Fig. 7. There, the pressure in the radial horizontal direction is computed in cylindrical coordinates r, ϕ. With $ds^2 = dr^2 + r^2 d\phi^2$, and $\partial / \partial\phi = 0$, one finds that $\Gamma_{11}^1 = 0$, $\Gamma_{21}^2 = 1/r$, and $\sqrt{g} = r$. From (22) one finds that here

$$r \frac{d\sigma_r}{dr} + \sigma_r - \sigma_\phi = 0 \tag{27}$$

or with $\sigma_\phi \approx 0.1\sigma_r$ that

$$r \frac{d\sigma_r}{dr} = -0.9\sigma_r \tag{28}$$

and hence

$$\sigma_r^{(2)} = \sigma_r^{(1)} \left(\frac{r_1}{r_2} \right)^{0.9} \tag{29}$$

where $r_2 > r_1$ is the radius of the disc.

For $r_2 \approx 3r_1$, as shown in Fig. 7, one has $\sigma_r^{(2)} \approx 0.37\sigma_r^{(1)}$. For $\sigma_r^{(1)} \approx 1.4 \times 10^8 dyn/cm^2$, one has $\sigma_r^{(2)} \approx 5.2 \times 10^7 dyn/cm^2 = p_2$.

Under these conditions the static equilibrium condition for a disc of radius r_2 and thickness t is given by

$$2 \times \pi r_2^{\,2} p_2 = 2\pi r_2 t \sigma_{\|} \tag{30}$$

where t is the thickness of the hoop put around the disc at its radius r_2, and $\sigma_{\|}$ the hoop stress. From (30) one obtains for the hoop stress

$$\sigma_{\|} = \frac{r_2 p_2}{t} \tag{31}$$

or

$$t = \frac{r_2 p_2}{\sigma_{\|}} \tag{32}$$

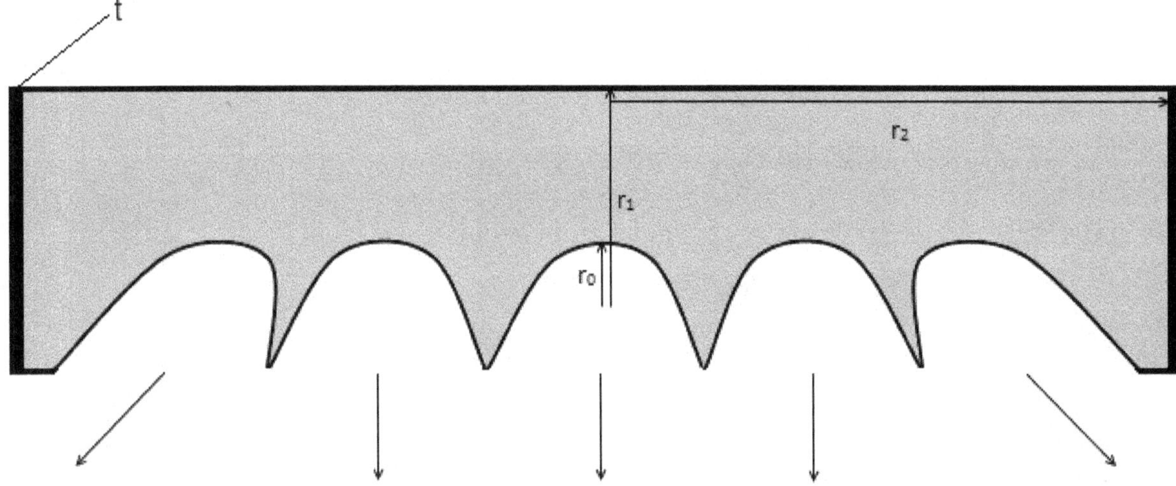

Figure 7. Radial cross section through assembly of propulsion units

Assuming that the material of the hoop has a tensile strength of about $\sigma_{\|} = 10^{10}\, dyn/cm^2$, then with a disc radius $r_2 \approx 10m = 10^4\, cm$, and for $p_2 = 5\times10^7\, dyn/cm^2$, one obtains $t \approx 50cm$. Therefore, a hoop with such a thickness would hold the disc together. The pressure in the vertical direction from the recoil of the fusion explosions is transmitted through shock absorbers to the spacecraft.

6. Thermonuclear "Operation Space Lift"

A thermonuclear space lift can follow the same line as it was suggested for Orion-type operation space lift, but without the radioactive fallout in the earth atmosphere. With a hydrogen plasma jet velocity of 30 km/s, it is possible to reach the orbital speed of 8 km/s in just one fusion rocket stage, instead of several hundred multi-stage chemical rockets, to assemble in space one Mars rocket, for example. At an exhaust velocity of 30 km/s = 3×10^6 cm, and the explosion of 1 ton hydrogen per second, the thrust would be $T = Vdm/dt = 3\times10^6$ cm/s $\times\, 10^6$ g/s = 3×10^{12} dyn = 3000 tons.

To stabilize the spacecraft against tilting, at least 3 propulsion units would be needed. With one propulsion unit expelling 0.1 tons of hydrogen at 30 km/s, this would require a cluster made up of 10 units, sufficiently large to stabilize the craft.

The launching of very large payloads in one piece into a low earth orbit has the distinct advantage that a large part of the work can be done on the earth, rather than in space. The proposed thermonuclear space lift would for this reason permit to launch the bulk of a large spacecraft directly into orbit.

Conclusion

The feasibility of the proposed concept would be of great significance for the future of space flight. It does not require a large number of chemical rockets to bring the parts of a Mars craft, for example, into low-earth orbit. And it would not require the Mars craft to be assembled there, which would need to be done by a large number of people in the weightless vacuum of space.

The fission-less ignition of a small thermonuclear explosion would be also of great interest if other controlled fusion concepts like laser or magnetic fusion turn out to be difficult to achieve or even unfeasible. There then, an economic incentive would support the research, regardless of its potential for spaceflight. Spaceflight with chemical rockets is too expensive. This is especially true for manned spaceflight. Thermonuclear space lift can fundamentally change the future of spaceflight.

References

1. W. v. Braun, *The Mars Project*, Urbana, University of Illinois Press, 1953.

2. Everett, C.J., Ulam S.M. *On a Method of Propulsion of Projectiles by Means of External Nuclear Explosions. Part I.* University of California, Los Alamos Scientific Laboratory, August 1955.

3. G. Guderley, Luftfahrtforschung **19**, 302 (1942).

4. Nachlass K. H. Höcker, Library University of Stuttgart, Konvolut 7.

5. W. Häfele, Z. Naturforsch. 10a, 1017 (1955).

6. R. Schall, *Physics of High Energy Density*, cd. P. Caldirola and H. Knoepfel, Academic Press 1971, p. 230ff.

7. F. Winterberg, Z. Naturforsch. **63a**, 35 (2008).

8. F. Winterberg, J. Fusion Energy DOI 10.1007/s 10894-008-9143-4 (2008).

9. Young K. Bae, Physics Letters A 372 (2008) 4865ff.

10. B. Müller, J. Rafelski, W. Greiner, Phys. Lett. 47B(1), 5 (1073).

11. R. W. Pohl, *Optik*, vol. 208 (Springer, 1940).

12. M. Schwarzschild, *Structure and Evolution of the Stars* (Princeton University Press, 1958), p.44.

13. F. Winterberg, Acta Astronautica. **89** (2013) 126-129.

14. Dr. Ing. Paul Schmidt and Ing. B. Pfau, unpublished report by the Technische Hochschule Stuttgart, 1964.

100 YEAR STARSHIP™

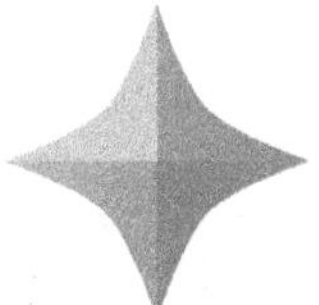

Chaired by Timothy Meehan, PhD

Saber Astronautics

Track Description

Great ideas arise through unique individual observations, from people of all ages and educational backgrounds. The Poster Sessions are an opportunity to present snapshots of these early concepts and experiments. Poster sessions are a great forum to communicate any commercial opportunities in space or here on earth and seek like-minded collaborators or investors. Presentation in the poster format allows in-depth discussion in a small group setting. Topics are open.

Track Chair Biography

Timothy Meehan, PhD

Saber Astronautics

Timothy Meehan, PhD has extensive expertise in biomolecular analytical science and diagnostics. He has applied nanotechnology approaches to bioengineering through a collaborative effort with the Australian Stem Cell Centre in order to achieve large scale human synthetic whole blood production. Tim is a seasoned small business leader with experience in genetic diagnostics, microbial parasite detection and commercial New Space startup ventures. At Saber Astronautics he is working with NASA performing reduced gravity hardware flight tests and developing the next generation of autonomous fault detection and recovery solutions for spacecraft.

The Way to Space

Marshall Barnes

SuperScience for High School Physics, 234 W. 18th Ave Columbus, OH 43210

superscience.hs@publicist.com

STDTS™ is the name of an electromagnetic technology, invented by Marshall Barnes, R&D Eng that was discovered through pure research in a laboratory in Grandview Heights, Ohio. in 2000. It produces an electromagnetic field that has a gravitic effect when in motion in the direction of the motion. This results in acceleration by a variable percentage based the field strength. Tests prove that the acceleration is due to the field's warping of space as it moves through space. This would connect it to the teleparallelism of Einstein's unfinished Unified Field Theory of gravity and electromagnetism as well as the 1967 prediction of Soviet scientist, N.A. Kozyrev, who stated that the way to space would be the manipulation of time through gravity, resulting in warp drive. Successful tests have brought STDTS™ rapidly to two breakthrough applications - warp drive without the energy requirements of General Relativity and wormhole generation without exotic matter and negative energy. This poster describes activities that are going on RIGHT NOW, to develop this technology towards these ends. These are 100% private sector, American developments without government funding or the need for it, nor the interest of involving foreign entities of any kind.

1. Testing STDTS™ for Warp Drive in Space 2016 - 2018

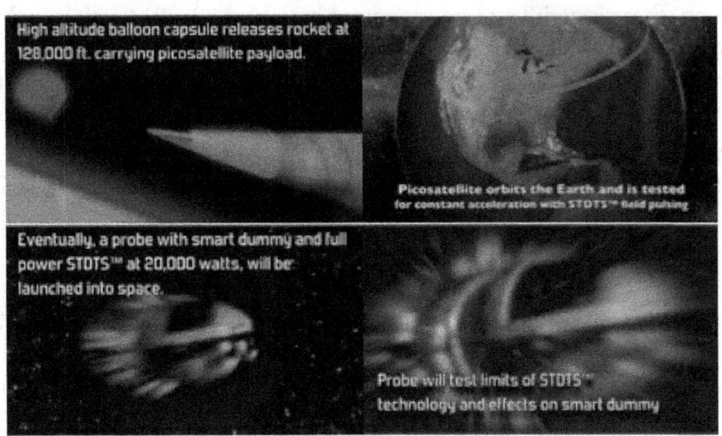

2. Torsion and Warped Spacetime through Verdrehung Fan™ Technology

One aspect of Einstein's incomplete theory of teleparallelism or distant parallelism (which was a unified field theory from the late 20s and early 30s) is the concept of torsion. Torsion is already cited as what gives a particle its angular momentum or spin, within the mathematical frame work of the torsion tensor. Torsion means, rotation, but in teleparallelism there is an application of torsion that extends beyond the behavior of particles and into space-time itself. In fact, the particles of matter itself are said to cause torsion in the local reference frames of space-time. If space-time is subjected to torsion to a strong enough degree, it will begin to compress, pleat, and then finally sheer. In 2007, Professor Jean Claude Ba of Columbus State Community College and I speculated what would happen if the STDTS™ field was rotated at high speed by being sent into a fan. This idea wasn't acted upon, however until five years later at the urging of a few members of the Mars Society and resulted in the creation of the Verdrehung Fan™

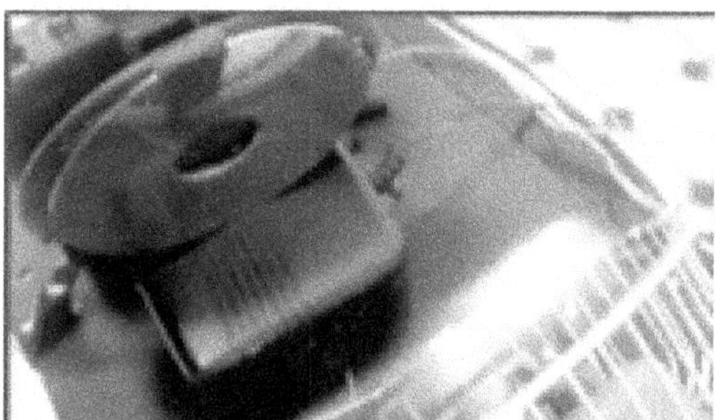

This is the Recoton TX RF transmitter in front of the Verdrehung Fan™ with a metal limiter in place.

The following is a series of photos from a session where this video transmitted image of Stephen Hawking is slowly disrupted and eventually swallowed by the Verdrehung Fan™. It takes 24 minutes before it begins to reach the stage where 80% signal loss happens and is quickly followed by 100%.

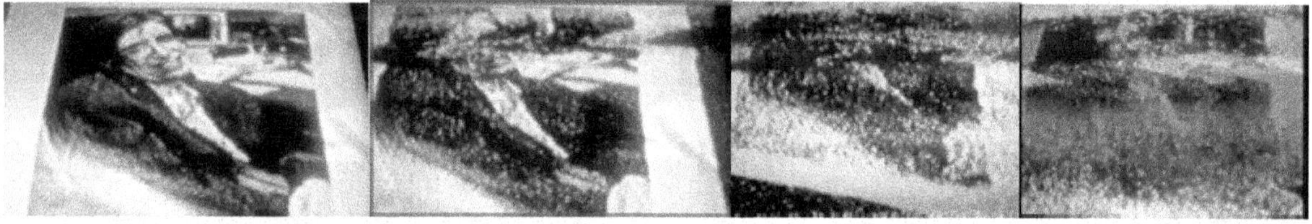

The fact this disruption in signal reception happens over time (roughly 30 minutes total) has astonished both physicists as well as TV/radio broadcast engineers. However, when applying the torsion model, the answer is clear - torsion in the center region of the fan is effecting the microwormholes in quantum foam, enlarging them so that the photons in the transmitted RF signal begin to disappear into them. This suggests that enlarging the fan and increasing the power of the field will result in more dramatic effects, such as the creation ultimately of a larger, single wormhole opening. These experiments are supported by a May 20, 2014 article in *New Scientist* magazine on the research of Luke Butcher of Cambridge University who posits that microwormholes could be held open via their own internal Casimer effect, long enough to send pulses of light through that might contain information. The article asserts that this could allow for sending signals to the past, making the Verdrehung Fan™ the 1st time machine in the world. A huge version in space suddenly makes the concept of the Tilpler Cylinder more viable if converted to this new configuration. Current power levels are only 195 watts. Imagine what a million would result in.

Acknowledgments

This project was sponsored in part by MUTUAL SPACE LIMITED, J. FELTRIC CONSULTING and XLN Systems.

The Adventures of Miss Q Spacecat

Erika Gasper

e@missqspacecat.com

Chanda Cummings

c@missqspacecat.com

Keywords
interstellar education, informal education, media, outreach

1. Introduction

How can we help shift public perception of interstellar travel? We believe that engaging a young audience with fun, creative media can help cultivate the inspiration needed to further space exploration. With this in mind we are developing a character-driven collection of resources that promote awareness and understanding of Earth's interstellar neighborhood.

Our heroine's story unfolds in a playful animated series *The Adventures of Miss Q Spacecat.* Related educational materials will leverage the excitement of Miss Q's adventures as motivation to further explore cutting edge and foundational content in space science. Topics emphasized across all Miss Q components include exoplanets, exobiology, space cartography, and methods of interstellar travel. Educational offerings will also include content related to the visionary arts and humanities.

The future depends on engaging young people. We hope that inspiring future generations will contribute to making interstellar travel a reality.

2. Animated Series: *The Adventures of Miss Q Spacecat*

Our Earth-based heroine, Miss Q, belongs to a species of spacecats. Millennia ago, an unknown event forced the spacecats to flee their home planet. Each episode takes Miss Q on a mission to a new exoplanet with the goal of discovering and reuniting lost spacecat colonies. This episode-based format creates a platform for developing recurring themes and imagery that will acclimate young viewers to our interstellar neighborhood. The goal of this playful and engaging animated series is to inspire imagination about the possibilities of exoplanets, exobiology, and interstellar travel.

The following are scenes from the pilot episode, which is currently in development.

Miss Q patiently awaits her mission

Miss Q naps in her sunny courtyard. By all appearances, she looks and lives like any other ordinary house cat. That is until her whiskers pick up the ever awaited signal . . .

Miss Q is summoned to Command Central

Cmdr Cat delivers the exciting news that the upgraded CAT-SCAN has detected a new signal that has definite spacecat stylings—a promising lead for the spacecat reunification program. Miss Q's mission is to visit the planet from which the signal originates, located in orbit around Gliese 251, which lies outside of the existing wormhole transit system serving Earth's interstellar neighborhood (Local Fluff Transit System, LFTS).

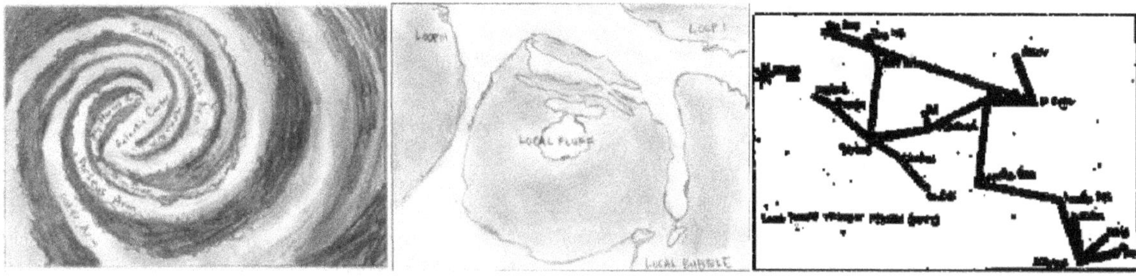

Cmdr Cat uses his 3-D map display to show Miss Q how to navigate to the planetary source of the signal by taking the LFTS to the end of the line at Luyten's star and then traveling the remaining distance using the warp drive on her kitty saucer.

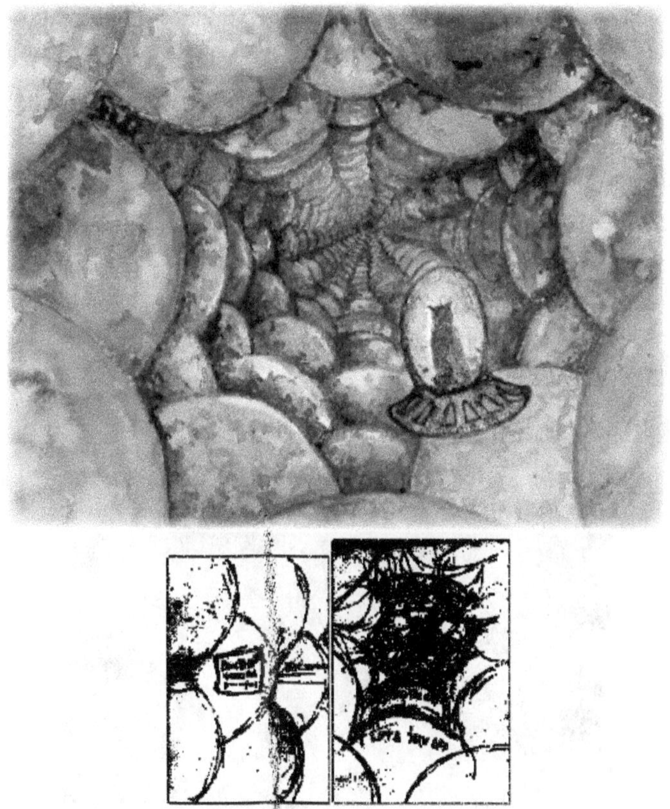

Through the wormhole

Miss Q sets off on her mission. She navigates the solar system in her kitty saucer, enters the local LFTS station past Saturn, and then settles into her commute to Luyten Station. At the end of the line she finds herself in an unfamiliar starscape. Using her Kitty Navigation System to find her bearings, she then dials her radio to the signal frequency. She locks in and heads towards its source.

'Insectoid City'

A dusty ochre planet comes into Miss Q's view. As she descends through the atmosphere she sees a city whose buildings resemble large, elaborate African termite mounds. Moving ever closer she discovers that the origin of the signal is a jazz radio broadcast. Miss Q enters a radio studio to find insectoid creatures playing this music. She introduces herself and states her mission. The insectoids react with surprise and explain that legends of spacecats are common in their culture. According to some accounts, insectoids learned their music from spacecats long ago. The band introduces Miss Q to a local professor who provides historical details that support the legends, including relics found during the construction of the insectoid city 500 years ago. The professor gives Miss Q an old coin with a spacecat image and some mysterious writing.

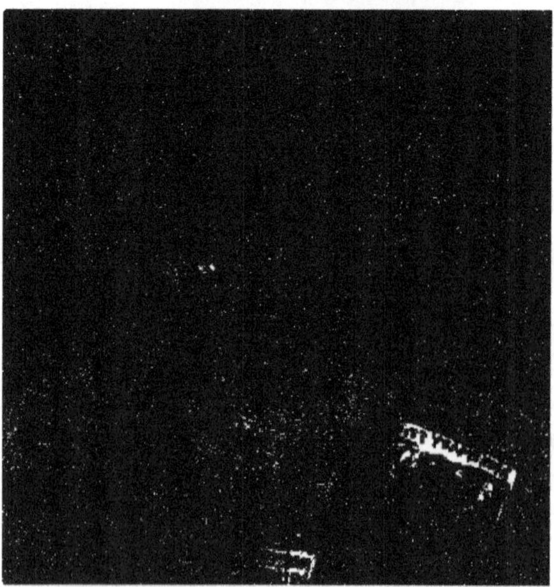

Next Adventure: 'Who is Bunny?'

As she departs the Insectoid planet, Miss Q fills in Cmdr Cat. He congratulates her on her successful mission to find clues about lost spacecat colonies. Cmdr Cat instructs Miss Q to stop by the interstellar market to restock exotic matter and other supplies and, while she is there, to show the mysterious coin to a merchant named Bunny.

2. Curriculum

Miss Q Space Explorations develop space science content with blended learning activities including modeling, recording and analyzing data, mathematical thinking, writing, and the arts. *Miss Q Space Explorations* include activities for multiple learning styles, including kinesthetic and visual modeling.

Miss Q Space Explorations are modular and flexible. They can be used in conjunction with a traditional textbook curriculum, as a stand-alone unit, or as a year-long theme. The format of *Miss Q Space Explorations* includes three main components:

- **Field observations**—observing positions and patterns in the movements of the Sun and Moon
- **Modeling activities**—fun kinesthetic and hands-on modeling activities with associated blended learning prompts that link conceptual development with visual, mathematical, and critical thinking
- **Spacecat Headquarters**—an ongoing classroom station in which students analyze their data, develop imaginative space themes in writing and art, and further explore Earth's interstellar neighborhood

Miss Q Space Explorations are currently available in K-2 and 3-6 grade band packages that address NGSS Space Systems standards. Additional curricula packages and components are in development.

2.1 Interactive Media

The following components are planned for future phases of development:

- *Earth's Interstellar Neighborhood Explore:* Earth's interstellar neighborhood with intuitive, simple 3-D models of stars within a 20-light year radius of sol. Model will include some familiar elements from the Adventures of Miss Q Spacecat, including featured stars and exoplanets and the Local Fluff Transit System wormhole network. Students will be encouraged to use the model to plan their own sci-fi adventure stories.
- *Sharing Space:* A moderated forum for sharing student work, including science fiction writing, art, multimedia, and research reports on exoplanets, exobiology, interstellar travel, and related topics.
- *Apps/Games:* Explore the interstellar neighborhood as a member of the Spacecat Federation in gaming adventures to search exoplanets for clues and trade for goods at the Local Fluff Interstellar Market.

 For more information or to contact us, please visit www.missqspacecat.com

Developing Pre-flight Culture: 100 Year Starship University-Level Student Organization Proposal

Katherine Herleman

PhD Candidate, Geological Sciences, Cornell University

kch227@cornell.edu

Abstract

100 Year Starship (100YSS) exists to make the capability of human travel beyond our solar system a reality with the next 100 years. In order to achieve this objective, 100YSS must engage in education and outreach activities which ensure that future generations of innovators will be inspired to continue research and development necessary for success. Effective organizational practices would ensure the development of a pre-flight 100YSS culture with strong intergenerational relationships which transmit continued project interest, encourage cooperative enterprise and, in accordance with its mission, push humanity "to accomplishments greater than any single individual". While additional sociological research on the development of pre-flight 100YSS culture needs to be undertaken in the long-term, 100YSS could increase research and development participation in the short-term by connecting future innovators in a naturally collaborative environment via the creation of a university-level student organization (ULSO). Additional benefits include addressing a growing concern over the lack of creative opportunities for post-secondary STEM students[1] and teaching students how to collaborate on transdisciplinary research teams.[2] At the institutional level, ULSO chapters could engage in educational and social activities such as hosting lecturers, screening movies, moderating discussions about the future of human spaceflight, and undertaking research to submit to the 100YSS annual public symposia. At the regional, national and international levels, 100YSS could bring events to its chapters by partnering with other organizations such as Google, IBM or the American Forensics Association, which sponsor pre-existing research competitions, coding challenges, ethics debates, etc. As ULSO participation grows, 100YSS could consider hosting exclusive regional, national and international events for its chapters. We propose a leadership and financial structure for this group based on successful experiences with other technical, project-based student organizations such as Engineers Without Borders (EWB) and Students for the Exploration and Development of Space (SEDS).

1. Mission & Objectives

100YSS ULSO will be a student-led organization that will enable aspiring researchers from any academic discipline to connect and create meaningful opportunities for transdisciplinary research and collaboration which will assist with the pre-flight endeavors of 100YSS.

Core Objectives:

1. Develop and sustain multigenerational interest in starship-related research, design, and construction
2. Provide opportunities for students to pursue careers in space-related endeavors
3. Engage broader university community to create institutional linkages which support 100YSS by strengthening research capacity and technology transfer

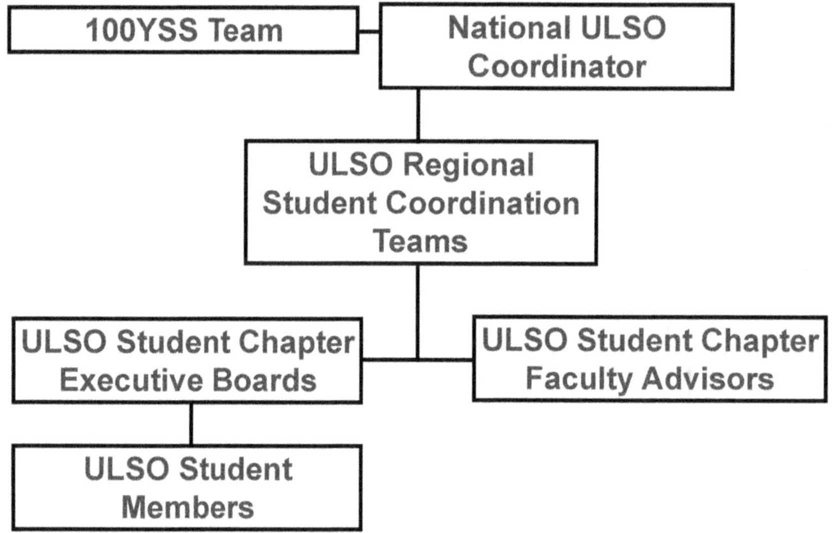

Figure 1. 100YSS ULSO Proposed Leadership Structure

2. Organizational Structure

- **100YSS Team**: Supports ULSO by identifying overarching research themes and projects in accordance with the master timeline
- **National ULSO Coordinator:** Facilitates communication and coordinates joint programming between 100YSS Team and ULSO regions
- **ULSO Regional Student Coordination Teams:** Create and maintain industry partnerships, plan programming, and provide organizational guidance to individual student chapters
- **ULSO Student Chapter Faculty Advisors:** Evaluate student project or programming proposals, assist in identifying and securing research funding, and provide guidance to student chapter executive boards
- **ULSO Student Chapter Executive Boards:** Set annual agendas, represent members, maintain student organization funding, and budget chapter-level programming
- **ULSO Student Chapter Members:** Participate in events, conduct research, build community and professional networks

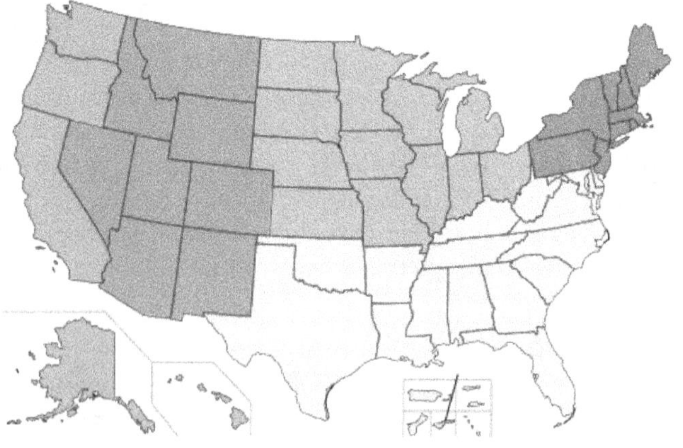

Figure 2. Potential ULSO Regional Structure Based on United States Administrative Divisions [3]

3. ULSO Responsibilities & Activities

Activities will be organized by organizational scale and scope

	Academic	Professional	Social
National	Manage ULSO research database	Manage mission and branding	Provide guidelines for social media
Regional	Host research conferences and poster competitions	Create and manage industry partnerships	Coordinate networking events Educate and engage public
Chapter	Form workgroups and conduct research	Host guest speakers Network with industry and government	Host inter-disciplinary mixers and sci-fi movie nights

4. Funding & Partnerships

ULSO chapters will seek social activities funding from their respective universities by registering as official student organizations. Therefore, primary ULSO support will remain independent of 100YSS contracts. Funding for academic and professional programming will be provided by cultivating relationships with industry and relevant government agencies. The National ULSO Coordinator and six ULSO Regional Student Coordination Teams would continue to evaluate strategies as the STEM research funding paradigm continues to change.

5. Research & Knowledge Sharing

The National Coordinator will establish a new model for collaborative research by creating an ULSO project database and discussion forum similar to NSF IGERT. ULSO Regional Student Coordination Teams will be responsible for assisting ULSO chapters with database management. Additional research needs to be undertaken on facilitating knowledge sharing in unique ways which promote innovation.

Acknowledgments

Thank you to Dr. Mason Peck, Department of Mechanical and Aerospace Engineering, Cornell University and Anitra Douglas-McCarthy, Director of Recruitment, Cornell University Graduate School, for their guidance and support.

References

1. Lee, Mee-Kyeong and Erdogan, Ibrahim. August 2007. The Effect of Science–Technology–Society Teaching on Students' Attitudes Toward Science and Certain Aspects of Creativity. International Journal of Science Education, 29(11), 1315-1327.

2. Morse, W., Nielsen-Pincus, M., Force, J., & Wulfhorst, J. D. 2007. Bridges and barriers to developing and conducting Interdisciplinary graduate-student team research. Ecology & Society, 12(2), 1-14.

3. Graphic by Wikimedia Commons user Scifiintel. Released under the GNU Free Documentation License.

Design & Layout

The wide variety of disciplines represented and information contained in the preceding papers provides unique design challenges. As in the previous year, the goal was to create a uniform presentation and easily accessible layout while still maintaining the individual author's needs and desires. The overall design is intended to be conducive to the experience of education and engagement. This year's improvements in design have focused on uniformity between the papers. Formulas, illustrations, and other non-text elements are more consistent (although not absolutely so). The challenge engaged and met is to provide a cross-discipline publication that highlights the variety of thought, research, and effort needed to achieve the audacious dream of interstellar travel.

Primary text is typeset in Adobe Caslon Pro. Headings are set in Helvetica Neue and League Gothic.

Design and layout by Jason D. Batt.

Editorial Board

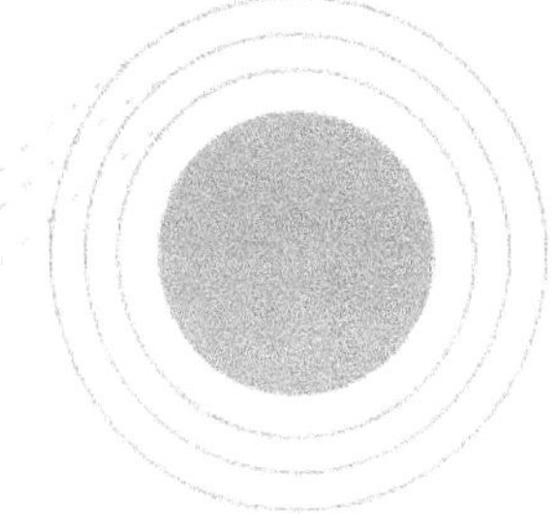

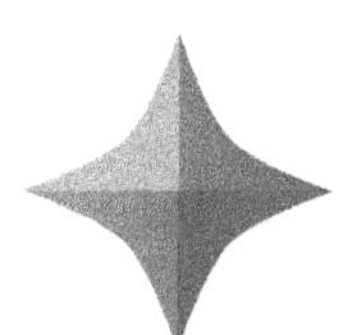